Nanogap Electrodes

Nanogap Electrodes

Edited by
Tao Li

WILEY-VCH

Editor

Dr. Tao Li
School of Chemistry and Chemical Engineering
Shanghai Jiao Tong University
Minhang District
200240 Shanghai
China

Library of Congress Card No.:
applied for

British Library Cataloguing-in-Publication Data
A catalogue record for this book is available from the British Library.

Bibliographic information published by the Deutsche Nationalbibliothek
The Deutsche Nationalbibliothek lists this publication in the Deutsche Nationalbibliografie; detailed bibliographic data are available on the Internet at <http://dnb.d-nb.de>.

Print ISBN: 978-3-527-33271-7
ePDF ISBN: 978-3-527-65959-3
ePub ISBN: 978-3-527-65958-6
oBook ISBN: 978-3-527-65956-2

Cover Design FORMGEBER Mannheim, Germany
Typesetting Straive, Chennai, India
Printing and binding CPI Group (UK) Ltd, Croydon, CR0 4YY

Printed on acid-free paper

C090866_300621

Contents

Preface

Molecular electronics describes a field that seeks to understand charge-transport properties of single molecules and their ensembles. Ultimately, this may pave the way for electronic components made of molecular building blocks. This visionary concept was first sketched out in 1974 by Arieh Aviram and Mark Ratner, who suggested that a single molecule could function as a rectifier in their theoretical work. If it comes true, computational architectures may be built atom by atom, or molecule by molecule, namely, in a "bottom-up" way. In the past 40 years, molecular electronics has generated more and more excitement supported by constantly growing theoretical and experimental achievements. On the one hand, the prospect of size reduction in electronics offers a potential alternative/addition to complementary metal–oxide–semiconductor (CMOS) technology using "top-down" methods, which is expected to face fundamental limitations on further scaling (e.g. direct source–drain electron tunneling) in the next decade or so. On the other hand, synthetic method promises a molecular-level control of properties. Well-designed organic molecules represent a class of inexpensive, flexible, and easy-processing materials compared with their inorganic counterparts. These advantages of organic materials have been well exploited in "bulk" thin-film devices, such as organic light-emitting diodes (OLEDs), organic field-effect transistors (OFETs), and organic photovoltaics (OPVs), some of which have already been successfully commercialized. On this basis, in spite of even more challenges, many efforts have been concentrated on pushing organic electronics into the nano/atomic realms. To date, individual molecules equipped with unique optoelectronic properties have shown their potential for molecular-scale electronics.

However, silicon-based technology is unlikely to be replaced anytime soon, and there are some major issues along the way to molecular electronics. Before we can actually set out to build computers out of molecules, understanding charge transport of single molecules or a small collection of molecules is of fundamental importance. This requires making reliable contacts to molecular components, which means a degree of control with atomic accuracy. Development of scanning probe microscopy (SPM) in the 1980s enabled some of the earliest attempts to image and measure individual molecules on surfaces. By using a scanning tunneling microscopy (STM) or conducting atomic force microscopy (cpAFM) tip, single molecules can be repeatedly interrogated, and in this way fundamental insights into how electron transport occurs within molecules and their contacts can be provided. SPM techniques are still among the most important testbeds today and continuously making contributions to our understanding of the principles in the field of molecular electronics. However, from an application perspective, integrated circuits of metal–molecule–metal junctions require robust electrode geometries that can be prepared over large areas. Given the size of a molecule, typically a few nanometers, making electrical contacts with a nanometer gap (dimension varies depending on the actual size of a target molecule) is the first crucial step for the realization of practical molecular electronic devices. To this end, a variety of platforms have been experimentally developed to wire molecules into circuits with atomically defined precision, which are generally referred to as "nanogap electrodes."

Fabrication of nanogap electrodes is a very challenging mission because it goes beyond the capability of traditional microfabrication technologies. Although the field of molecular electronics has been around for more than 40 years, it was not until late 1990s that researchers managed to measure single-molecule transport in a predefined nanogap. From then on, a collection of effective and creative methods for manufacturing nanogap electrodes with controlled spacing have been reported and some fundamental problems got overcome. Nowadays, researchers around the world are taking further steps toward more robust devices, deeper insight into molecule–electrode interfaces, multiterminal architectures, new materials for electrodes (e.g. carbon materials), and synthetic control on molecular functionalities. More significantly, both the experimental and theoretical advances have opened up a set of exciting possibilities of understanding and manipulating single-molecule properties beyond simple descriptions of electronic transport, such as quantum interference, thermoelectronics, mechanics, and spintronics.

This book focuses on the fundamental issues of nanogap electrodes and their applications in molecular electronics. It covers the topics of (i) introduction of molecular electronics and charge-transport mechanism in molecular junctions; (ii) techniques for the fabrication of nanogap electrodes and characteristics of various types of nanogap electrodes; (iii) characterization methods and analytical techniques for molecular transport junctions; (iv) experimental advances in electrical measurements of single molecules and self-assembled monolayers; (v) measurements in molecular junctions beyond simple electronic transport. The microfabrication methods for the preparation of nanogap electrodes and electrical measurements on the single-molecule level will be emphasized in this book. The state of the art of nanogap electrodes and their prospects (e.g. organic–inorganic hybrid structures, integration and application over large areas) will also be discussed.

We hope that this book will attract the attention of graduate students and young scientists who are interested in molecular electronics, as well as those more senior academic and industrial researchers related to nanoscience and technology. We also hope that this book will, in a more experimental perspective, improve people's understanding of this relatively new while immensely vivid field.

Finally, we would like to express our gratitude to Prof. Wenping Hu at Tianjin University for his full support on the book project, the contributors of each chapter for their great work, and editors Shirly Samuel, Daniela Bez, and Umar Saleem at John Wiley & Sons, Ltd for their patience and help through the editorial process.

20 January 2021

Tao Li
Shanghai

1

Nanogap Electrodes and Molecular Electronic Devices

Tao Li

School of Chemistry and Chemical Engineering and Key Laboratory of Thin Film and Microfabrication (Ministry of Education), Shanghai Jiao Tong University, 800 Dongchuan Road, Minhang District, Shanghai 200240, China

1.1 Introduction

The evolution of CMOS (complementary metal–oxide–semiconductor) technology has brought about unparalleled progress in semiconductor industry over the last 50 years. Today, the pace of exponential improvement on performance of digital electronics is still kept on. Integrated circuits (ICs) can be now made very compact, having up to several billion transistors and other electronic components in a fingernail-sized area. The proliferation of computers, communication, and other industrial and consumer electronics not only established a huge and ever-growing market of several hundred billion dollars per year [1], but also in an unanticipated way changed the world and our way of life. In August 2014, Intel formally unveiled its 14 nm manufacturing process, followed by Broadwell (Intel's codename for the 14 nm microarchitecture) based laptops available for sale in October. Subsequently, the 10 nm technique was released by Taiwan Semiconductor Manufacturing Company (TSMC) in 2016. And in April 2018, the company even has put the first generation of 7 nm processing technology into mass production. In this range, however, it will be more and more difficult to operate transistor structure that is utilizing MOS (metal–oxide–semiconductor) physics as the basic principle of operation. Moreover, manufacturing devices scaling to sub-10 nm dimensions goes beyond the capability of traditional lithography technologies. More advanced patterning techniques need to be developed, for example, the TSMC and Samsung are introducing extreme ultraviolet lithography (EUVL) to the subsequent products. Although it may very probably be too expensive and time-consuming for economics of chip production, the advanced technique can further decrease the power consumption and promote the operating frequency of the electronics.

How can we further increase the computation and memory capacity when the physical limits will be reached?

"There's plenty of room at the bottom" is a famous lecture given by American physicist Richard Feynman, who suggested that it should be possible to "arrange the

Nanogap Electrodes, First Edition. Edited by Tao Li.

atoms one by one the way we want them" [2]. This talk is often linked to the "birth of nanotechnology," as it inspired the conceptual beginnings of this field decades later. In view of his physical background, Feynman scored less strongly on the role of chemistry in nanoscale engineering, although it turned out that chemistry and molecular biology also play a crucial role on the engineering of small things, and Feynman himself conceded that chemists could, in principle, make anything from the bottom up. In the early 1970s, conductivity measurements through single layers of molecules were pioneered by Hans Kuhn and Dietmar Möbius [3]. More significantly, Arieh Aviram and Mark Ratner [4] in their theoretical work first suggested that a single molecule could mimic a semiconductor-like band structure by taking advantage of electron-rich and electron-poor moieties to achieve one-way conduction. In this sense, the possibilities for designing electronic functionality are proposed by bottom-up synthetic methods with atomically basic building blocks. It marks the origin of molecular electronics, which pointed out a possible way "beyond CMOS."

Understanding charge transport properties of single molecules or a small collection of molecules is of fundamental importance in the field of molecular electronics. Even though single-molecule devices are conceptually simple – they consist of functional molecular components and two or three electrodes – it presents significant experimental challenges [5, 6]. A number of different experimental strategies have been developed in an effort to wire molecules into circuits [7–21], among which "nanogap electrodes," namely a pair of electrodes with a nanoscale gap, are regarded as fundamental tools to study charge transport properties through single molecules and promisingly incorporate molecular components into electronic circuits. In this sense, the preparation of nanogap electrodes is as crucial as the design and synthesis of functional molecules for the construction of nanodevices.

1.2 Overview of Molecular Electronics

Molecular electronics is a relatively new while immensely vivid field, the major goal of which is to understand charge-transport properties of single molecules and their ensembles. Ultimately this will probably pave the way for electronic components made of molecular building blocks that can overcome the limits of conventional semiconductor technology [22, 23]. The research in molecular electronics is very much interdisciplinary (spans physics, chemistry, biology, computer technology, materials science, e.g.) and covers quite a broad range of topics. This chapter will begin with an overview of molecular electronics, including the origin and motivation of this field, materials and technologies involved, and fundamental challenges encountered, followed by the introduction of nanogap electrodes, which – as the main subject of this book – are the basic tools for studying molecular electronic properties and building blocks for manufacturing nanoelectronic devices. For a more comprehensive understanding of molecular electronics, several representative reviews [5, 24–27] and books [28–31] are recommended.

1.2.1 Why Molecular Electronics

It is widely accepted that we are now living in a period characterized by computer use and development, known as Computer Age (also known as Information Age or Digital Age). Capitalizing on computer-miniaturization advances has revolutionized almost every aspect of life, not only making it more efficient and convenient, but also changing the way people view the world. On the other hand, despite the great success, solid-state electronics faces a constant need for miniaturization in view of current and future technological developments, which is expected to approach fundamental limitations on further scaling based on MOS physics as the basic principle of operation. It is the basic motivation for developing beyond-CMOS devices. Molecular electronics is promisingly an option.

1.2.1.1 History of Computing

A computer is a general-purpose machine that can perform computations in a programmed way. The evolution of two separated techniques, automated calculation and programmability, is essential for the development of modern computers.

The Evolution of Calculator and Programmability in Embryo The history of computing can be traced back to the period of 2700–2300 BCE, when some data-processing tools in embryo came into being to meet the requirements of people's daily life (Figure 1.1a). Abacus, also called a counting board, was first devised by Sumerians and Egyptians. By using beans or stones moved in grooves in sand or on tablets of wood, stone, or metal according to some certain "programming rules" memorized by the user, simple operations such as counting, recording facts, addition, and subtraction can be made (Figure 1.1b). In the following thousands of years, various types of abacus were independently developed from ancient civilizations around the world and gradually matured as an important tool for merchants, traders, and clerks long before the adoption of the written modern numeral system. For example, the Chinese abacus, known as *suanpan*, usually has 13 rods and there are two beads on each rod in the upper deck and five beads each in the bottom for both decimal and hexadecimal computation. Skilled people can use *suanpan* to do addition, subtraction, multiplication, and division with very fast speed, handling numbers up to the tens of millions. It is still being in use by some old people nowadays.

Since 1600s, mechanical calculator came into being. Wilhelm Schickard, who was a universal scientist at the University of Tübingen, is known for his "calculating clock" evidenced by two letters from him to Johannes Kepler (well known for his eponymous laws of planetary motion) with a sketch drawing and building instructions of the calculating machine. Unfortunately, the design was considered to be incomplete and no Schickard's machine managed to survive to the present day. In 1642, Blaise Pascal designed and constructed a mechanical calculator capable of adding and subtracting numbers entered with dials to assist his farther, who had to do exhausting calculations as a tax collector. This machine, along with the ones modified for improvements by Pascal himself in the next following years, was generally recognized as the first digital

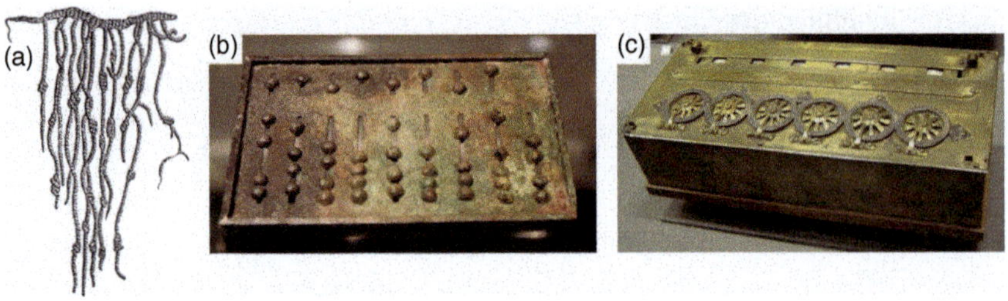

Figure 1.1 (a) A data-processing tool in embryo: keeping records by tying knots. Source: MKL online at Retro Bibliothek, work 149, 2009.Retrived from: https://commons.wikimedia .org/wiki/File:Meyers_b13_s0522.jpg. (b) The rudiment of abacus. (c) Pascal's calculator. Source: en.wikipedia.org/wiki/Pascal%27s_calculator.

calculating machines, called Pascal's calculator or the Pascaline (Figure 1.1c). Although these machines were too expensive to be commercialized, they provided the basic principles for subsequent computer engineering. Pascal's work was followed by Gottfried Leibniz, who devoted himself to building a machine that could execute all four arithmetical operations since 1670s. He invented the Leibniz wheel in 1673 and built a machine called the Stepped Reckoner based on this design in 1694, which was the first calculator that can perform multiplications and divisions. In addition, Leibniz documented the binary numeral system in his early life, which is at the foundation of virtually all digital computers.

More than 100 years later, the nineteenth century saw the first commercially successful mechanical calculator. By making use of the designing ideas from previous mechanical calculators such as Stepped Reckoner, Charles Xavier Thomas invented the Arithmometer and patented it in 1820. The manufacturing of Arithmometer started from 1851 and in the next 37 years, it was the only type of mechanical calculator in commercial production. More than 2500 Arithmometers had been sold all over the world by 1890. As the first mass-marketed calculating machine, Arithmometer marks the beginning of the mechanical calculator industry. Many clone makers appeared during the period of 1887–1915, and there had been about 20 independent companies manufacturing machines based on Arithmometer's design by 1920s.

While Thomas and some other engineers were putting their efforts to develop more and more powerful calculators, in 1837, Charles Babbage first conceptualized and designed a fully programmable calculator, his Analytical Engine. The machine can be programmed by punched cards (pieces of stiff paper that contained digital information represented by the presence or absence of holes in predefined positions, first introduced by Joseph Marie Jacquard in 1801 enabling his loom to automatically weave defined patterns) and different types of punch cards responsible for different operations (e.g. arithmetical, load, store operations) were used. The architecture consisting of both processing elements and memory units is very similar to a modern computer. Although Babbage was never able to complete construction of the machine himself due to inadequate funding and personality issues, he is credited with inventing the first general-purpose mechanical computer. The fusion of automated calculation with programmability started, which was one of the notable landmarks in the development of computers.

After Babbage's death, from late 1880s to early twentieth century, more and more interesting improvements came to the fore. Charles Babbage's son Henry Babbage built a simplified version of Analytical Engine in 1888 and gave a successful demonstration of how it worked in 1906. It to some extent made his father's design to become a reality, although not commercialized. Meanwhile, Herman Hollerith invented a mechanical tabulator based on punched cards and developed a tabulating machine to help process data for the 1890 United States Census. By using this machine, Hollerith's company accomplished the data-processing job, which was estimated to take around 13 years, in six weeks! Hollerith also introduced a machine-readable medium and the first key-punch machine (a punch operated by a keyboard). These inventions, together with the mechanical tabulator, were all among the foundations of the modern information processing industry. Hollerith's company merged with three other corporations to form the Computing Tabulating Recording Corporation (CTR) in 1911, which was renamed International Business Machines Corporation (IBM) in 1924. The tabulating equipment introduced by IBM in 1930s was capable of performing automatic sequences of advanced arithmetical operations in one second, enabling its clients (e.g. US Government) to process unprecedented amounts of data.

Timeline of Computer History On bases of the predecessors' achievements and promoted by the joint efforts of talents from almost all walks of life, the twentieth century witnessed the birth and explosive development of computer. It is generally accepted that there have been four generations of computers to date:

(i) First generation of computers (1946–1959) used vacuum tubes as the basic components for memory and circuitry, supporting machinc language only. They were huge and expensive. A lot of heat was produced by the tubes, giving rise to unreliable performances.
(ii) Second generation of computers (1960–1964) were transistor based. They were cheaper, consumed less power, more compact in size, more reliable, and faster than the first generation. Assembly language and high-level programming language were used.
(iii) Third generation of computers (1965–1970) were marked by the use of IC, which had many transistors, resistors, and capacitors along with the associated circuitry. As a result, computers became even smaller as more components were squeezed onto the chip. Multiprogramming operating system and high-level languages were used.
(iv) Fourth generation of computers (1970–) used Very Large Scale Integrated (VLSI) and Ultra Large Scale Integration (ULSI) technology. The computers became more powerful, compact, and reliable, giving rise to personal computer (PC) revolution. They were also marked by advancement in parallel processing, networks, artificial intelligence, user-friendly interfaces with multimedia features, and so on.

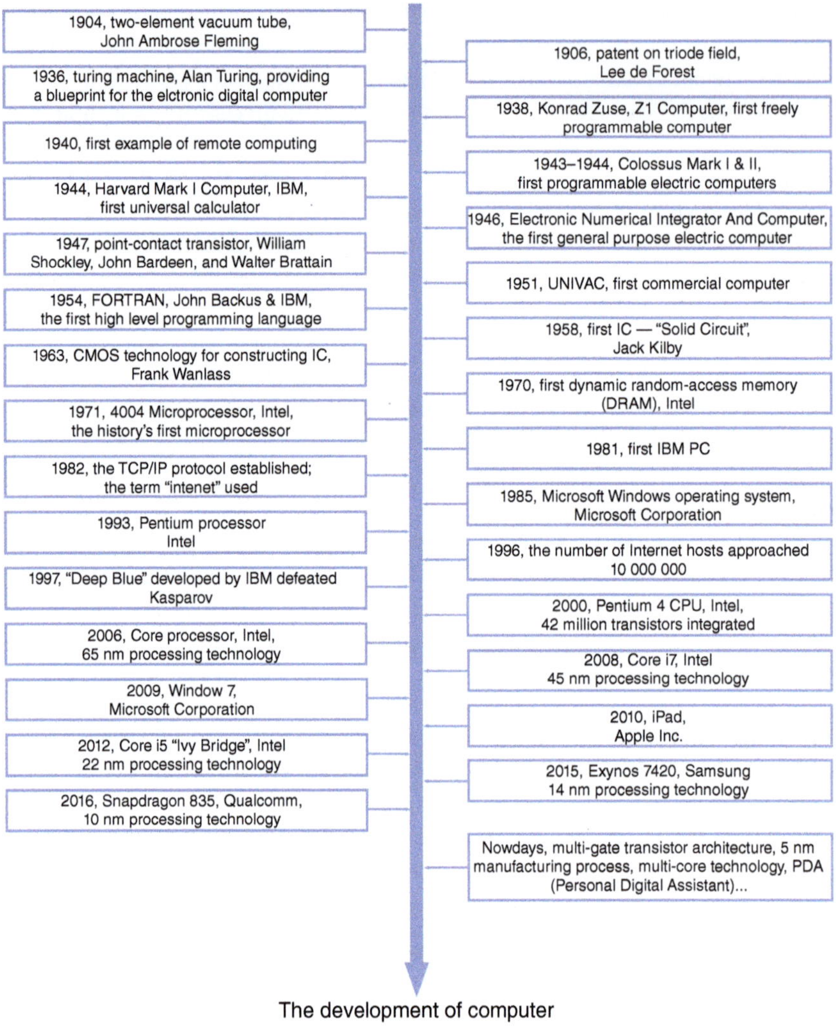

Figure 1.2 A general timeline of the important events in computer history.

A general timeline of the important events is provided in Figure 1.2.

1.2.1.2 Moore's Law

In 1965, only several years after the IC-based computer found its first customer, Gordon Moore, cofounder of Intel Corporation, noted that the number of transistors on a chip doubled approximately every two years. Therefore, he made a prediction that semiconductor technology will double its effectiveness every 18 months. This prediction was later named after his name, Moore's law [32]. It is not a physical or natural law; however, it has been held remarkably well for nearly 50 years (Figure 1.3) [33], in part because the law has been adapted as a driving force in the semiconductor industry to guide long-term planning for research and development.

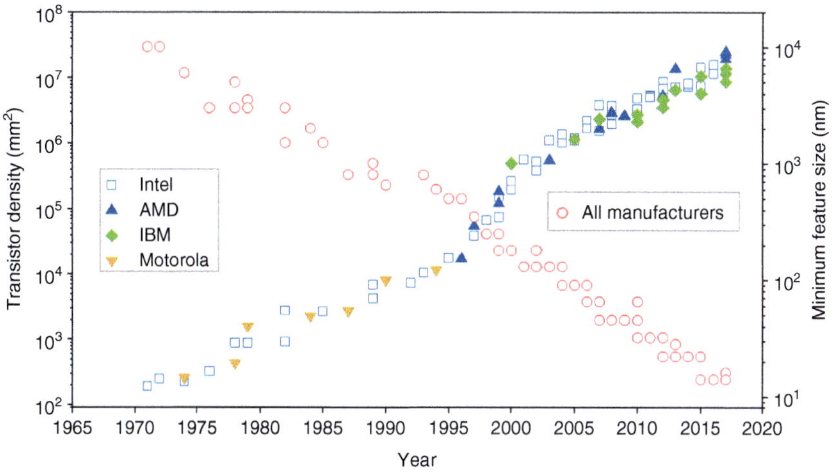

Figure 1.3 The density of transistors and the evolution of transistor gate length (minimum feature size) in microprocessors over time. Between 1971 and 2017, the gate length of metal–oxide–semiconductor field-effect transistors (MOSFETs) shrank from 10 000 to 14 nm (*y* axis, right), and the number of transistors per square millimeter increased from 192 to over 20 million (*y* axis, left). AMD, Advanced Micro Devices.

This exponential improvement has dramatically enhanced the impact of digital electronics in almost every aspect of the world.

On 2 May 2011, Intel announced its first 22 nm microprocessor, i.e. the average half-pitch (half the distance between identical features) of a memory cell shrunk to the 22 nm level. Over 1 billion transistors were integrated on one single processor die. Then around one year later, on 23 April 2012, consumer-level CPU (central processing unit) based 22 nm technology went on sale worldwide. And in 2014, Intel released its Broadwell chips, the first CPUs manufactured using the 14 nm processing technology. Based on international technology roadmap for semiconductors (ITRS) roadmap, Intel and its rivals are promising to go even smaller, packing even more transistors onto each chip, making them more powerful while draining less battery charge or powers. Moreover, TSMC announced that the 7 nm technology has been put into mass production in 2018, and its 5 nm manufacturing process has been carried out tentatively in 2019, in which up to 14 layers employed the EUVL technique. By following this trend, many physical dimensions are expected to be crossing the 10 nm threshold in the following decade. However, it is predicted that as dimensions scale to sub-5 nm, physical limits such as source-to-drain leakage currents stemming from thermal diffusion of electrons and gate leakage currents caused by quantum-mechanical tunneling will make it difficult to operate any transistor structure based on MOS physics [34]. On the other hand, new levels of miniaturization would be for sure more and more challenging due to fundamental geometrical limits, although some technical solutions such as 3D (three-dimensional) integration, multi-gate structures, high-κ gate dielectrics have been proposed/underway [35, 36]. Furthermore, continuous scaling calls for more sophisticated lithographic techniques (EUVL, multiple patterning, etc.) [37, 38], resulting in an exponential increase of the cost to manufacture chips (by a factor of 2 for every chip generation, sometimes known as Moore's second law). Therefore, the economics of chip production may also be a key issue that could bring Moore's law to an end.

1.2.1.3 Molecular Electronics: A Beyond-CMOS Option

It is widely acknowledged that Moore's law is approaching its threshold, as discussed earlier. On the one hand, many efforts have been devoted to the improvement of classical CMOS (e.g. ultrathin SOI (silicon-on-insulator) metal–oxide–semiconductor field-effect transistor (MOSFET) structure, high-κ gate dielectrics, strained silicon technology) [39], and expanding nonclassical CMOS technologies (e.g. multi-gate transistor, 3D integration, RF [radio frequency] interconnect) [36, 40]. On the other hand, people proposed many beyond-CMOS options, including spin device [41, 42], quantum computing [43], molecular device [14, 44, 45], nanotube transistor [46, 47], mechanical switch [48, 49], and so on. Although these emerging technologies are still very far from industrial maturity to replace CMOS architectures, they generally seek to use nanoscale components to build up more complex devices, namely in a bottom-up way, which is conceptually different from the well-established top-down approaches for the manufacture of products in silicon industry. These nanoscale building blocks, particularly molecules (dimensions could be as small as 1 nm), represent the ultimate limit that people can manipulate so far [50, 51]. If the new computing devices made up of these ultrasmall components (acting as channels and interconnects) do work, in principle, it can significantly extend Moore's law beyond the foreseen limit of integration.

The molecules are small enough to provide a spatial control that is currently impossible with top-down fabrication methods. We all know that. There is another thing we need to know, which is even more crucial: *will molecules compute*?

Conventional device for computing and information storage, put simply, is no more than an on–off switch. The switching between two (or more) clearly different states (normally characterized by electrical signal, i.e. high- and low-resistance states) defines binary "0" and "1" values (1 bit), the basic elements for information storage and complex logic operations. In 1974, Arieh Aviram and Mark Ratner [4] published their theoretical work entitled "Molecular Rectifiers" in the journal *Chemical Physics Letters*. They calculated charge transport through a single organic molecule consisting of donor and acceptor subunits separated by an insulating bridge (D–B–A system) and concluded that electrical conduction within this system would be favored from the donor subunit (electron rich) to the acceptor moiety (electron poor), while disfavored in the reverse direction. This work theoretically predicted the possibility of constructing a very simple electronic device, a rectifier/diode, based on single molecules, to potentially implement the basic on/off functionalities for computing (Figure 1.4) [52]. This is generally correlated to the origin of molecular electronics. More theoretical constructions came up afterward, including a detailed description of single-molecule field-effect transistor (FET) by Aviram [53] and the concept of single-molecule logic gates by Forrest Carter et al. in 1980s [54]. Concurrently, nature has been offering people much to learn. Taking human brain as an example, which is recognized as nature's computer,

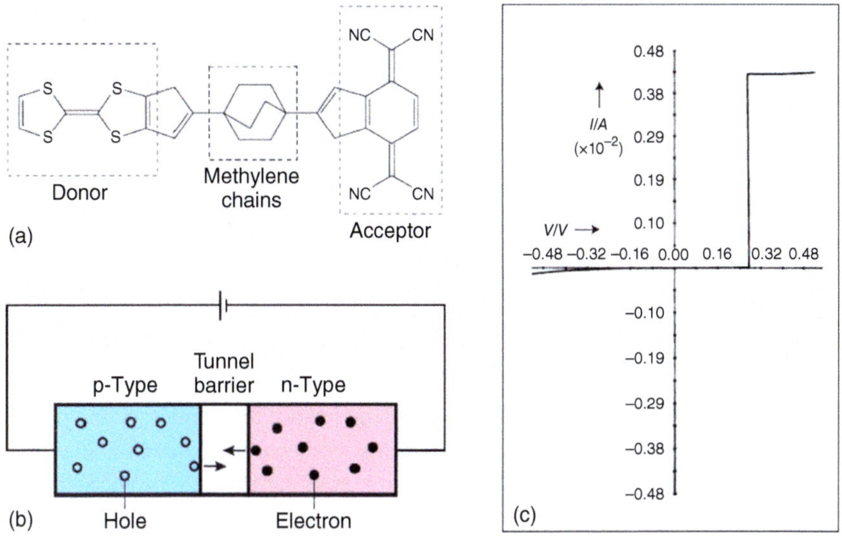

Figure 1.4 Aviram–Ratner single-molecule rectifier based on an acceptor tetracyanoquinodimethane (TCNQ) and a donor tetrathiofulvalene (TTF) separated by a triple methylene bridge. (a) Model molecule. (b) p–n-junction. (c) Calculated *I–V* characteristics. Source: (a, b) Kashimura and Goto [4]. (c) Aviram and Ratner [52]. Elsevier.

processing and storage of information (electrical and chemical signals) are executed by various biomolecular components (mainly nucleic acids and protein), thereby accomplishing complex physiologic processes. There are as many as 10^{11} neurons in this highly sophisticated system and each is connected to 10^3–10^4 others, giving a rough estimate of "bit count" up to 10^{15}. Although the working principle of a brain is different from solid-state microprocessors, the biological architectures inspired a wide range of ideas exploiting organic molecules in conjunction with electronics for the long-term perspective of molecular computers [55, 56].

In addition to the much smaller size of molecules for further scaling and promisingly higher efficiency compared to the electronic components currently used, as well as their potential capability for controllable charge transport and storage, several unique advantages are conductive to the incorporation of molecular components into microelectronic devices:

(i) Synthetic methods offer a high degree of freedom for tailoring molecular geometry and composition, realizing diverse optical and electronic functions;
(ii) Molecules have the ability to self-assembly and recognition, which is a key aspect of bottom-up approaches to nanotechnology. Well-organized structures for desired properties can be programmed with atomic precision by the specific interactions in between single molecules, which is previously unattainable with top-down methods;
(iii) Manufacturing with molecules promises to be low-cost and environmentally benign. Molecular devices are expected to incorporate with lighter and flexible electronics with promisingly lower power dissipation under operation.

In spite of theoretical foundation, manipulation of single molecules is no easy task. It was not until late 1990s that researchers managed to wire a limited number of molecules between contacts for direct measurement of their electronic characteristics [57], after the advances of experimental means in the field of nanotechnology. The great challenges imposed by device fabrication as well as data analysis and interpretation will be discussed in the following sections and chapters. Nevertheless, the appearance of concrete devices accelerated the progress of molecular electronics and substantially confirmed the theoretical predictions. Until today, a wide range of intrinsic transport properties from molecules including nonlinear characteristics have been reported. Embryonic forms of molecular resistors [58, 59], rectifiers [60, 61], switches [62, 63], and transistors [64, 65] were on this basis realized.

In the last 15 years or so, a rapid progress in experimental research on molecular electronic devices has further pushed forward molecular electronics as one potential option of beyond-CMOS technologies. However, electronic circuits made up of only molecular components are still very far from being realized. Molecular electronics has not yet prepared well itself for devices capable of optimum performance, parallel processing, defect tolerant, and low cost, all that competitive with similar CMOS solutions. There would be a long way to go with numerous challenges and opportunities lying ahead.

1.2.2 Molecular Materials for Organic Electronics

Molecular materials for organic electronics, in a broad sense is another branch of molecular electronics. Different (but related) to molecular-scale electronics, it mainly focuses on exploiting the "bulk" properties of molecules for electronic applications. As mentioned in the previous sections, organic materials possess intrinsic advantages compared to their inorganic counterparts, including light weight, flexibility, ease of functionalization, and capability to be processed in large areas at low cost. Since the discovery of conducting polyacetylene in 1977 (Nobel Prize in Chemistry in 2000), organic semiconductors began to attract a worldwide interest in electronic devices [66]. Over the last 30 years, much effort has been devoted to design, synthesis, and characterization of molecular systems with desired optoelectronic properties. On this basis, a variety of active devices based on organic thin films have been successfully fabricated, including organic light-emitting diodes (OLEDs) [67, 68], organic field-effect transistors (OFETs) [69, 70], organic photovoltaics (OPVs) cells [71], photodetectors [72], memory cells [73], chemical sensors [74], and actuators [75], as well as organic thermal electrics [76] and nonlinear optics [77]. In addition, concurrent advances in biotechnology have allowed for integration of biomaterials into the structural components of organic electronic devices [78]. For instance, semiconducting active layers can be fabricated from a variety of small-molecule dyes and pigments [79, 80]; hydrophilic biopolymers are perfectly suitable for gate dielectrics while a wide range of natural and synthetic polymers are appropriate for use as flexible bulk substrates [81–84]. Due to the limitation of this book, here we only very briefly introduce the state of the art of OLEDs, OFETs, and OPVs. For an extended discussion, we would like to refer the interested readers to some recent reviews and books [85–93].

1.2.2.1 OLEDs

OLEDs rely on organic materials (small molecules or polymers) that emit light in response to an electric current. The applications in displays and lighting are two dominant branches of OLED technology (Figure 1.5a). OLED display technology promises to deliver thin, flexible, power-efficient, and bright displays that will be competitive to mature LCD (liquid crystal display) technology. Substantial investments contribute to an increasing maturity in process technology, and today OLED technology has already been used in commercial applications as displays for computer monitors, television screens, and especially portable systems such as mobile phones. It is worth noting that, since Apple introduced OLED screens in its flagship smartphone iPhone X in the late 2017, almost all the cell phone companies, including Samsung, Huawei, and Apple, are launching new OLED produces in 2019. In addition to display, it is also possible to use the OLED technology to make white light panels for lighting. OLEDs can provide large area and hazard-free diffusing light sources and feature good color-rendering indexes and tunable color temperatures. They can be lightweight, flexible, transparent, energy-efficient, and recyclable [68, 94–96]. In 2013, Nippon Electronic Company (NEC) lighting Ltd. announced a new industry record for OLED efficiency as high as 156 lm/W at a color temperature of 2800–3000 K. In the first half of 2019, Jiangsu Yiguang Technology based in Liyang (China) upgraded its G2.5 OLED production line after taking the place of First-O-Lite in Nanjing (China), famous for its internal extraction strategy of OLED lighting panels. In the next three years, it will launch the world's only G6 OLED lighting production line, which further paves the way of OLED lighting technology to be a competitor in the general lighting markets.

1.2.2.2 OFETs

OFETs have received widespread interest since they were first reported by Tsumura et al. in 1986 using polythiophene as the semiconductor layer, because of their

(a) (b) (c)

Figure 1.5 (a) A curved OLED TV model employing 4-Color Pixel technology, produced by LG Electronics. Source: LG Electronics USA Inc. www.lg.com. (b) The organic electronic circuit based on OFETs. Source: http://en.wikipedia.org. (c) The integrated OPV array. Source: Fraunhofer ISE.

potential applications in low-cost, large-area, flexible, and biodegradable electronic products [97]. To date, thousands of organic semiconductors have been reported to exhibit FET characteristics and impressive improvements in device performance have been achieved [98–104]. For example, the field-effect mobility has increased from low values $<10^{-3}$ cm^2 V^{-1} s^{-1} to 1–10 cm^2 V^{-1} s^{-1} for thin-film transistors and to 15–40 cm^2 V^{-1} s^{-1} for single-crystal transistors over the last three decades. Remarkably, In 2014, Zhenan Bao and coworkers reported thin-film transistor hole mobility up to 43 cm^2 V^{-1} s^{-1}, which is a new record for all organic molecules [105]. Furthermore, Qian Miao and coworkers reported thin-film transistors electron mobility up to 27.8 cm^2 V^{-1} s^{-1}, which sets a new record for n-channel OFETs [106]. These values are now exceeding those of benchmark amorphous silicon devices (0.5–1 cm^2 V^{-1} s^{-1}), and there is little evidence that the progress in organic materials performance has reached fundamental limits [104]. On this basis, a wide range of applications can already be addressed, such as OFET-driven displays, mechanical sensors, and radiofrequency identification (RFID) tags (Figure 1.5b) [107–110]. Over the next few years, both materials' performance and processing methods (e.g. printing techniques) for OFET technology are expected to advance to the level that is compatible with full industrialization, allowing for more flexible electronic products, including the desired OFET-driven OLED displays [111–113].

1.2.2.3 OPVs

The development of photovoltaic (PV) cell that can transform inexhaustible solar energy into electricity represents one of the most promising solutions for providing clean and renewable energy (Figure 1.5c) [114]. Compared to the expensive silicon-based PV cells, solar cells based on organic materials (conductive polymers or small molecules) promise cost-effective PV applications, although OPVs generally suffer from relatively low efficiency and low stability. The first successful OPV device was reported as early as 1986 by Tang [115] with an impressive power conversion efficiency (PCE) of ~1% and high fill factor (FF) of 65%. Substantial progress has been made ever since in understanding the working mechanism of the OPV devices, molecular design, optimizing device architectures, and improving the device stability [116–118]. As a result, the maximum PCE of OPV cells has increased quickly to the level of over 15% [119, 120]. In January 2013, organic solar films developer *Heliatek* (Dresden, Germany) reported certified 12% PCE of a vacuum-processed small-molecule organic triple junction tandem solar cell with an active area of >1 cm^2. What's more, *Heliatek* achieved a record PCE of 13.2% for an OPV multi-junction cell in 2016. Remarkably, Zou and coworkers reported a high-performance single-junction organic solar cell with the PCE of 15.7% and FF of 74.8% in 2019. The work in this field is exciting, and undoubtedly there is still room for improving OPV technology to promote its industrial application. The challenge of its commercialization has always been the development of products with high efficiency, long operational lifetimes, high production yields, and low cost.

1.2.3 Molecules for Molecular-Scale Electronics

Molecular electronic devices (or junctions) are qualitatively similar to organic thin-film devices. The difference is molecular junctions only incorporate a limited number of molecules in electrical contact with two (or more) conductors (mostly metal). The molecular components within the junction could be either single molecules or molecular monolayers, whose feature size is generally no larger than a few nanometers. The overall performance of the molecular device is determined by the electrode/molecule interface and the electronic structure of the molecular kernel. For such tiny devices, the measured properties can also be largely influenced by variations of molecular geometries inside the junction and small changes of external environment. In order to operate these devices in a relatively controllable manner, chemical design and synthesis of molecular systems are of paramount importance.

Anchoring group and molecular backbone are two main parts of a molecular system [6, 121]. The anchoring group with an affinity for specific substrates (metals or other electrode materials) is responsible for binding the molecular backbone with electrodes. Both of its mechanical stability and electronic transparency (weak or strong coupling) are important considerations. A detailed discussion of anchoring groups can be found in the following chapters.

The molecular backbone is the key unit of the molecular system, which to a large extent defines the electronic functionality of the molecular device. Tailor-made molecules could function as resistors, diodes, or transistors depending on their electronic structures [122–124]. Linear alkanes, π-conjugated molecular wires, molecular switches, and biomolecules are major classes of molecules that have been widely studied [125–129]. Detailed descriptions of their specific electronic properties will be dispersed into the upcoming chapters that separately focus on charge transport mechanisms, testbeds, and analyzing methods. This section provides a broad overview of molecular wires.

In this book, the term "molecular wires" is used to describe rigid, linear, and highly conjugated molecules, although alkane chains as a benchmark molecular system are also "wire-shaped." Charge transport through molecular wires is more efficient because electrons can move freely in the delocalized orbitals over "long" distances [130, 131]. Typical molecular wires (Figure 1.6) are semiconducting oligomers with narrow HOMO–LUMO gaps (HOMO: highest occupied molecular orbital, LUMO: lowest unoccupied molecular orbital), such as oligo(*p*-phenylene ethynylene)s (OPEs) [14, 132], oligo(*p*-phenylene vinylene) [133], oligothiophenes [134], oligophenyleneimines (OPIs) [135], oligofluoreneimines (ONIs) [136], oligophenylenetriazoles (OPTs) [137], oligo(aryleneethynylene)s (OAEs) [138], oligoynes [139], and oligoanilines [140]. Measurements on these wires with a variety of testbeds revealed apparently smaller conductance decay coefficient (0.1–0.5 Å^{-1}) compared to that of nonconjugated alkanes (0.7–1.0 Å^{-1}) [141, 142]. The capability of delivering higher conductivities makes molecular wires potential candidates as building blocks in molecular electronic circuits.

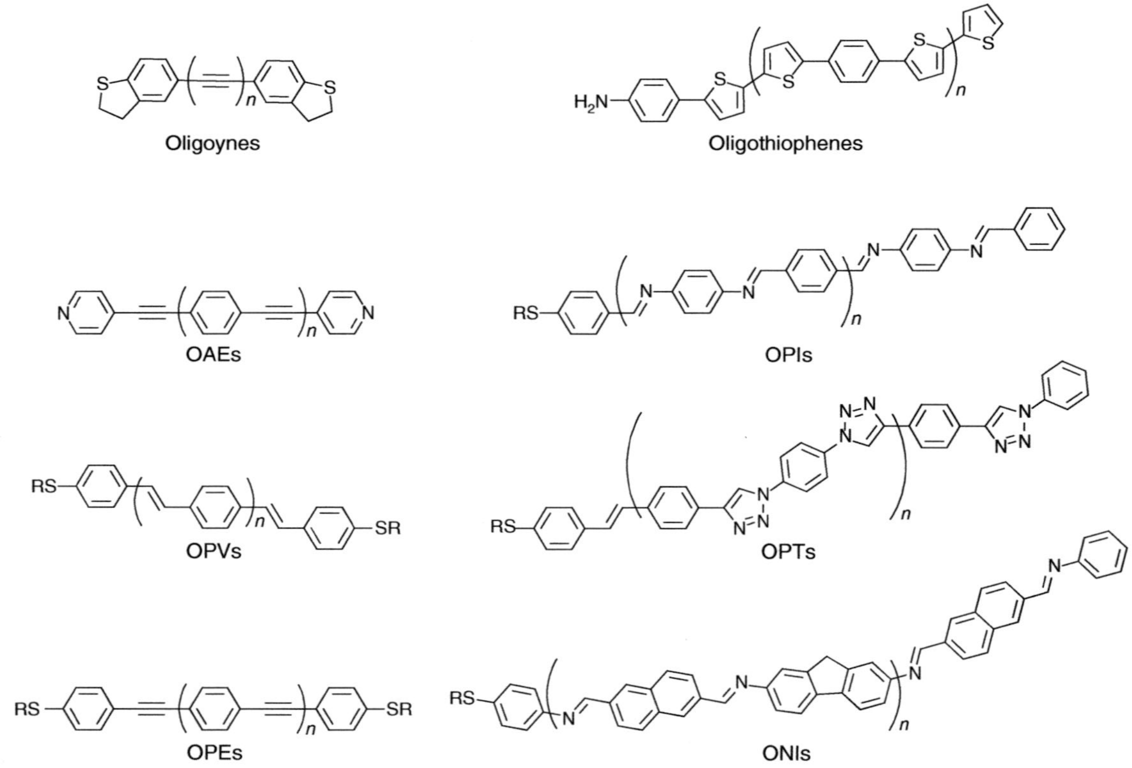

Figure 1.6 The structures of several typical molecular wires.

The length of the wires can be systematically changed by modulating the number of repeating units that constitute the molecular backbone. In general, within a given molecular series, the HOMO–LUMO gap of π-conjugated molecules decreases with an increase in conjugation length. It is thus possible to "manually" adjust the alignment of energy levels between contact electrodes (the Fermi level) and molecular components for ease of charge injection [14, 143]. More importantly, measurements on length-variable transport have been proved feasible to examine the charge transport mechanism for the molecular junction. This method was first proposed by Frisbie and coworkers, who synthesized a series of OPI wires (ranging in length from 1.5 to 7.3 nm) bonded to Au substrates via a thiolate anchoring group and measured their conductance using conducting probe atomic force microscope (cpAFM) [144]. The study indicated a clear transition of transport mechanisms from nonresonant tunneling for short molecules to thermally activated hopping for longer molecules (for the OPI system, the transition happens at around 4 nm), agreeing well with theoretical predictions. This phenomenon was observed in other molecular systems as well, independent of platforms and methods employed [136, 145].

In addition to oligomers, polymeric molecular wires with good conductivity and rigidity have also been reported [146]. A representative work was done by Hu and coworkers, who used poly(*p*-phenylene ethynylene)s (PPEs) (24 polymerization units, up to 18 nm in length) for the fabrication of molecular junctions [147]. It was interesting to see the long polymer has quantized electronic structures and charge hopping has been observed. Compared to the extensively studied short molecular wires, molecular-scale electronics calls for more attention on polymer wires with more extended conjugation length for potentially even larger charge transport capabilities.

Wires incorporating metal complexes are another group of interesting molecules [148–150]. Metal centers in the backbone can lead to multiple redox and spin states that can afford unusual current–voltage characteristics. Furthermore, gate control on molecular junctions is possible by directly modulating the specific optical, redox, and magnetic properties of the complex via external stimuli [151]. In this way, functional molecular junctions and logic gates can be potentially developed. At the present stage, there is still plenty of room for improvement of multifunctional molecular devices integrating organometallics as well as coordination compounds [152]. A recent review focusing on metal complexes in molecular junctions is recommended [153].

A molecular wire occurring in nature is DNA (deoxyribonucleic acid), which is a long polymer encoding the genetic instructions used by all living organisms and many viruses. Double-stranded DNA is a well-ordered, closely spaced (3.4 Å) arrangement of conjugated bases (adenine, thymine, guanine, and cytosine), suggesting its potential use as a bridge for charge transport [154]. Furthermore, the nature of DNA endows it with unique molecular self-assembly properties [155, 156]. It also exhibits good chemical and mechanical properties such as stability and rigidity. Today, advances in synthetic chemistry not only allow DNA to be synthesized in high yield with high purity, but also permit adding functionalities to it. What's more, the order of DNA bases can be sequenced to create arbitrary two- and three-dimensional architectures on the nanoscale. The challenge of implementing DNA as a molecular wire lies in reading its chemical and biological information based on electrical measurements [157]. Many groups have attempted to couple DNA with probing electrodes for conductance measurements, but the reported characteristics of DNA vary from insulator, semiconductor to conductor and superconductor. The reasons for these contradictory findings might be multiple. In addition to the DNA sequence and length, the measured conductance is also influenced by different experimental conditions (e.g. test environment and testbeds) [158, 159]. In recent years, researchers from different groups gradually reached a consensus that good coupling of electronic contacts to the DNA and preservation of the DNA π-stacking by maintaining physiological conditions are prerequisites to be met for observing high conductivity. What's more, DNA can also act as template for the seed-mediated growth of metal nanowires and nanogap architectures for the fabrication of nanodevices [160]. Collectively, understanding mechanisms of charge transport through DNA is fundamental for rational design of conductive DNA wires and programmable multistate DNA memory systems [161, 162]. For more complete discussions on biomolecules in the field of molecular electronics, several reviews and books are recommended [163–168].

1.3 Introduction to Nanogap Electrodes

At the current level of knowledge, a molecule is recognized as the smallest bit of substance that retains its chemical properties [51]. In the context of molecular electronics, synthetic molecules as small as 1 nm are expected to perform a variety of electronic tasks, aiming at the next limit of miniaturization of electronic devices when conventional silicon technology is foreseen to face fundamental limitations within a decade [50, 169]. As introduced in previous sections, molecular electronics has matured over the last 40 years since a paper from Arieh Aviram and Mark Ratner illustrated a theoretical molecular rectifier in 1974 [4]. However, in the first 20 years or so, research progresses in this field were all theoretical constructs [170] without concrete devices, because direct measurement of individual molecules is no easy task, awaiting the development of techniques for making molecular-scale contacts. The first experiment that successfully probed the conductance of a single molecule was reported by C. Joachim and J. K. Gimzewsky on a single C_{60} molecule in 1995 [171], followed by Mark Reed et al. on a limited number of self-assembled benzene molecules in 1997 [57]. Since then, the experimental branch of molecular electronics has developed rapidly, marked by much more reported testbeds for measuring the electronic characteristics of single molecules and their ensembles [57, 172, 173].

Nanogap electrodes are a big class of the testbeds featured by nanometer-sized spaces between paired (or more) contact electrodes, the architecture of which is well suited for molecular components to reside in [174]. Compared to the scanning probe microscopy techniques such as scanning tunneling microscopy (STM) [175] and cpAFM [176] that are powerful to address single molecules with precisely manipulated tips, the significance of nanogap electrodes lies in they facilitate the fabrication of solid-state devices for potentially practical applications besides fundamental research [17–19, 177]. Moreover, most nanogap electrodes present a planar configuration, making it easier to realize high-density integration of molecular devices and to take the underlying substrate as a gate contact to tune the electrical properties of the molecular components. Furthermore, nanogap electrodes are generally fabricated before the molecular components are subsequently inserted in, therefore they can be characterized with and without molecules in place, helping to distinguish intrinsic molecular properties from artifacts.

Despite the superiorities, the preparation of nanogap electrodes is a very challenging mission. Firstly, typical molecules are only several nanometers in size. This dimension is so small that it goes beyond the capability of traditional microfabrication technologies. Secondly, the size of different molecules largely varies from each other, thus for specific molecules, specific control on the nanogap is required, in case the molecules are posited in an undesired stretched/distorted state between the electrodes. Lastly but not least, from a practical point of view, nanogap electrodes should be reproducible and robust, preferably manufactured by reliable and cheap means. On this basis, mass production is a key issue that needs to be addressed for potential integration solutions.

Because of the major challenges, it was not until late 1990s that researchers managed to measure single-molecule transport in a predefined nanogap as introduced earlier [171]. The pioneering work has inspired a lot more experimental devices based on nanogap electrodes to come [57, 178, 179]. In the recent 15 years or so, a number of elaborate and creative methods for manufacturing nanogap electrodes have been reported [21, 180–182]. These methods cover a variety of techniques ranging from the state-of-the-art lithography techniques to chemically driven self-assembly techniques. Different methods are also frequently combined to use in order to obtain a desired dimension and configuration. For example, electron-beam lithography is often utilized to firstly prepare paired electrodes with spaces at 20–50 nm, and then other techniques, such as shadow mask evaporation, are implemented to further narrow the gap width to a 1–5 nm scale.

Today, nanogap electrodes have become fundamental tools for the construction of solid-state single-molecule devices and the research on charge transport mechanisms through the molecules of interest. Interesting functions performed by molecular components, including diodes [183, 184], switches [185, 186], and transistors [64, 65], have been successfully explored by using nanogap electrodes, and some of the theoretical predictions of the early researchers have been substantially confirmed. More significantly, both experimental and theoretical advances gradually open up a set of exciting possibilities of understanding and manipulating single-molecule properties beyond simple descriptions of electronic transport. Recent research on quantum interference [43, 187], spintronics [41, 188], electromechanics [48, 49], and thermoelectronics [189, 190] at the single molecule level could promisingly lead to fundamentally new concepts for device function beyond CMOS-like logics (Figure 1.7) [191]. On the other hand, although there have been plenty of publications on nanogap electrodes, development of an agreed-upon

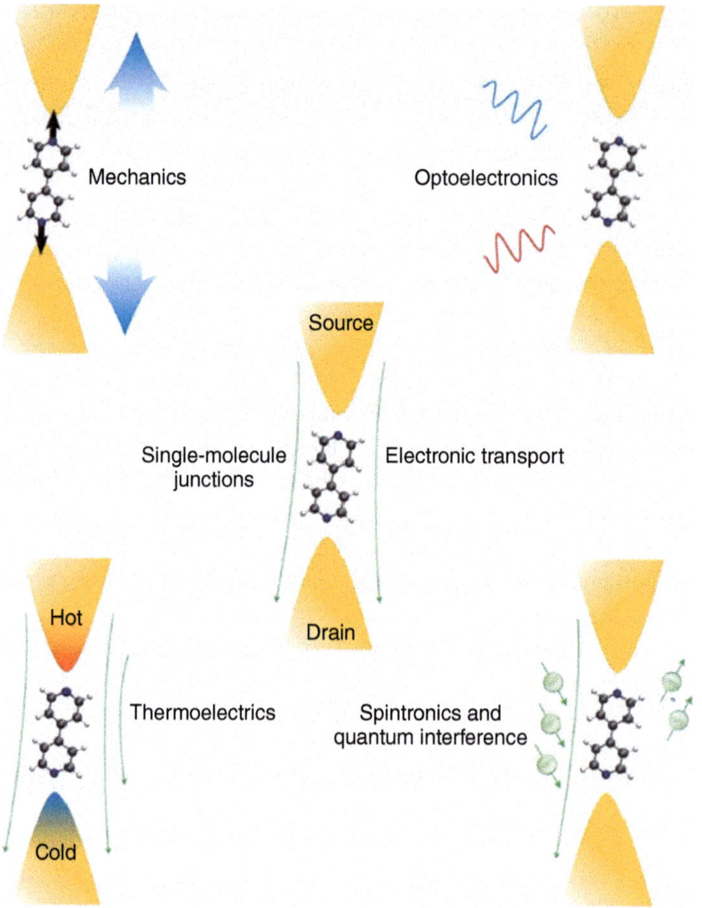

Figure 1.7 Probing multiple properties of single-molecule junctions. Beyond the electronic transport that led to their initial interest, recent studies are beginning to explore the rich physics of metal–molecule–metal junctions through measurements of mechanics, optical effects, and thermoelectric phenomena in addition to demonstrations of quantum mechanical spin- and interference-dependent transport concepts for which there are no analogues in conventional electronics. Source: Aradhya and Venkataraman [191]. 2013, Springer nature.

standard strategy for device fabrication and operation is still lacking. Only when such a standard junction can be reliably manufactured by different laboratories, are the researchers able to focus more on contact interface, molecular kernel, geometry, etc. and correlate their effect with electronic behavior. In the long run, mass production of robust molecular junctions for high-density device integration is a main issue awaiting to be addressed.

A thorough review of the techniques for the fabrication of nanogap electrodes and characteristics of various types of nanogap electrodes is a main focus of this book. Techniques for data collection and analysis, as well as physical phenomena observed in single-molecule devices, are involved. We will also include descriptions on contact methods for monolayered molecules (generally prepared by self-assembly process or Langmuir–Blodgett method) and in a broad sense view the vertical junctions based on monolayered molecules as a special type of nanogap devices.

1.4 Summary and Outlook

The visionary idea of molecular electronics has been sketched out for decades. If it comes true, the next generation of circuit elements made of molecular components is expected, preferably with lower power consumption, higher speed, and most importantly, an unprecedented level of integration. This chapter offers an overview of this cutting-edge interdisciplinary field, including its origin, motivation, and challenges, followed by a brief summary of molecular materials for electronics, which is an important branch of this field and closely related to the main topic of the book. On this basis, a general introduction to nanogap electrodes is made. The remaining chapters are set up around this theme.

References

1 KPMG (2019). Global Semiconductor Industry Outlook: Semiconductors: As the backbone of the connected world, the industry's future is bright.
2 Feynman, R.P. (1992). *J. Microelectromech. Syst.* 1: 60.
3 Kuhn, H. and Möbius, D. (1971). *Angew. Chem. Int. Ed. Engl.* 10: 620.
4 Aviram, A. and Ratner, M.A. (1974). *Chem. Phys. Lett.* 29: 277.
5 Cui, A., Dong, H., and Hu, W. (2015). *Small* 11: 6115.
6 Su, T.A., Neupane, M., Steigerwald, M.L. et al. (2016). *Nat. Rev. Mater.* 1: 16002.
7 Andres, R.P., Bein, T., Dorogi, M. et al. (1996). *Science* 272: 1323.
8 Tian, W., Hong, S., Reifenberger, R. et al. (1997). *Phys. Rev. Lett.* 79: 2530.
9 Dhirani, A., Lin, P.H., Guyot-Sionnest, P. et al. (1997). *J. Chem. Phys.* 106: 5249.
10 Venkataraman, L., Klare, J.E., Nuckolls, C. et al. (2006). *Nature* 442: 904.
11 Wu, S.W., Ogawa, N., and Ho, W. (2006). *Science* 312: 1362.
12 He, J., Fu, Q., Lindsay, S. et al. (2006). *J. Am. Chem. Soc.* 128: 14828.
13 Wei, Z., Wang, X., Borges, A. et al. (2014). *Langmuir* 30: 14868.
14 Wei, Z., Li, T., Jennum, K. et al. (2012). *Langmuir* 28: 4016.
15 Kushmerick, J.G., Naciri, J., Yang, J.C., and Shashidhar, R. (2003). *Nano Lett.* 3: 897.
16 Fracasso, D., Valkenier, H., Hummelen, J.C. et al. (2011). *J. Am. Chem. Soc.* 133: 9556.
17 Bourgoin, J.P., Palacin, S., Esteve, D. et al. (1999). *Phys. Rev. B* 59: 12505.
18 Liang, W., Shores, M.P., Bockrath, M. et al. (2002). *Nature* 417: 725.
19 Qing, Q., Chen, F., Li, P. et al. (2005). *Angew. Chem. Int. Ed.* 44: 7771.
20 Nagase, T., Kubota, T., and Mashiko, S. (2003). *Thin Solid Films* 438–439: 374.
21 Jain, T., Westerlund, F., Johnson, E. et al. (2009). *ACS Nano* 3: 828.
22 Lörtscher, E. (2013). *Nat. Nanotechnol.* 8: 381.

23 Haedler, A.T., Kreger, K., Issac, A. et al. (2015). *Nature* 523: 196.
24 Xiang, D., Wang, X., Jia, C. et al. (2016). *Chem. Rev.* 116: 4318.
25 Astruc, D., Boisselier, E., and Ornelas, C. (2010). *Chem. Rev.* 110: 1857.
26 McCreery, R.L. (2004). *Chem. Mater.* 16: 4477.
27 Perrin, M.L., Burzurí, E., and van der Zant, H.S.J. (2015). *Chem. Soc. Rev.* 44: 902.
28 Delhaes, P. (2017). *Materials and Thermodynamics: Living and Economic Systems*. ISTE Ltd.
29 Alkire, R.C., Bartlett, P.N., and Lipkowski, J. (2016). *Electrochemistry of Carbon Electrodes*. Weinheim: Wiley-VCH.
30 Wolf, E.L. and Medikonda, M. (2012). *Understanding the Nanotechnology Revolution*. Weinheim: Wiley-VCH.
31 Schwoerer, M. and Wolf, H.C. (2006). *Organic Molecular Solids*. Weinheim: Wiley-VCH.
32 Moore, G.E. (1965). *Electronics* 38: 114.
33 Ferain, I., Colinge, C.A., and Colinge, J. (2011). *Nature* 479: 310.
34 Jacob, A.P., Xie, R., Sung, M.G. et al. (2017). *Int. J. High Speed Electron. Syst.* 26: 1740001.
35 Park, S., Song, I., Lee, H., and Yoo, K. (2018). *IEEE Trans. Electron Dev.* 65: 2917.
36 Dixit, A., Kottantharayil, A., Collaert, N. et al. (2005). *IEEE Trans. Electron Dev.* 52: 1132.
37 Blanco Carballo, V.M., Bekaert, J., Mao, M. et al. (2017). *Extreme Ultraviolet (EUV) Lithography VIII*. Single exposure EUV patterning of BEOL metal layers on the IMEC iN7 platform, Volume 10143, 1014318.
38 Hamed Fatehy, A., Kotb, R., Lafferty, N. et al. (2018). *Extreme Ultraviolet (EUV) Lithography IX*. Exploring EUV and SAQP pattering schemes at 5 nm technology node, Volume 10583, 105831X.
39 Colinge, J. (2004). *Solid State Electron.* 48: 897.
40 Hisamoto, D., Lee, W.-C., Kedzierski, J. et al. (2000). *IEEE Trans. Electron Dev.* 47: 2320.
41 Zyazin, A.S., van den Berg, J.W.G., Osorio, E.A. et al. (2010). *Nano Lett.* 10: 3307.
42 Osorio, E.A., Moth-Poulsen, K., van der Zant, H.S.J. et al. (2010). *Nano Lett.* 10: 105.
43 Solomon, G.C., Andrews, D.Q., Van Duyne, R.P., and Ratner, M.A. (2009). *ChemPhysChem* 10: 257.
44 Li, T., Hauptmann, J.R., Wei, Z. et al. (2012). *Adv. Mater.* 24: 1333.
45 Li, T., Jevric, M., Hauptmann, J.R. et al. (2013). *Adv. Mater.* 25: 4164.
46 Qi, P., Javey, A., Rolandi, M. et al. (2004). *J. Am. Chem. Soc.* 126: 11774.
47 Tsukagoshi, K., Yagi, I., and Aoyagi, Y. (2004). *Appl. Phys. Lett.* 85: 1021.
48 Quek, S.Y., Kamenetska, M., Steigerwald, M.L. et al. (2009). *Nat. Nanotechnol.* 4: 230.
49 Bruot, C., Hihath, J., and Tao, N. (2011). *Nat. Nanotechnol.* 7: 35.
50 Ratner, M. (2013). *Nat. Nanotechnol.* 8: 378.

51 Tour, J.M., Reinerth, W.A., Jones, L. et al. (2006). *Ann. N.Y. Acad. Sci.* 852: 197.

52 Kashimura, Y. and Goto, T. (2009). *NTT Tech. Rev.* 7: 1–6.

53 Aviram, A. (1988). *J. Am. Chem. Soc.* 110: 5687.

54 Carter, F., Siatkowski, R., and Wohltjen, H. (1988). *Molecular Electronic Devices*. Amsterdam etc.: North-Holland.

55 Kahan, M., Gil, B., Adar, R., and Shapiro, E. (2008). *Physica D* 237: 1165.

56 Schneider, T.D. (1994). *Nanotechnology* 5: 1.

57 Reed, M.A., Zhou, C., Muller, C.J. et al. (1997). *Science* 278: 252.

58 Seo, S., Min, M., Lee, J. et al. (2011). *Angew. Chem. Int. Ed.* 51: 108.

59 Stan, M.R., Franzon, P.D., Goldstein, S.C. et al. (2003). *Proc. IEEE* 9: 1940.

60 Kornyshev, A.A., Kuznetsov, A.M., and Ulstrup, J. (2006). *Proc. Natl. Acad. Sci. U.S.A.* 103: 6799.

61 Metzger, R.M., Chen, B., Höpfner, U. et al. (1997). *J. Am. Chem. Soc.* 119: 10455.

62 Collier, C.P., Mattersteig, G., Wong, E.W. et al. (2000). *Science* 289: 1172.

63 Blum, A.S., Kushmerick, J.G., Long, D.P. et al. (2005). *Nat. Mater.* 4: 167.

64 Park, H., Park, J., Lim, A.K.L. et al. (2000). *Nature* 407: 57.

65 Kubatkin, S., Danilov, A., Hjort, M. et al. (2003). *Nature* 425: 698.

66 Shirakawa, H., Louis, E.J., MacDiarmid, A.G. et al. (1977). *J. Chem. Soc., Chem. Commun.* (16): 578.

67 Dodabalapur, A. (1997). *Solid State Commun.* 102: 259.

68 Reineke, S., Lindner, F., Schwartz, G. et al. (2009). *Nature* 459: 234.

69 Horowitz, G. (1999). *Adv. Mater.* 10: 365.

70 Meijer, E.J., de Leeuw, D.M., Setayesh, S. et al. (2003). *Nat. Mater.* 2: 678.

71 Li, Z., He, G., Wan, X. et al. (2011). *Adv. Energy Mater.* 2: 74.

72 Peumans, P., Yakimov, A., and Forrest, S.R. (2003). *J. Appl. Phys.* 93: 3693.

73 Masopust, D., Vezys, V., Marzo, A.L., and Lefrançois, L. (2001). *Science* 291: 2413.

74 McQuade, D.T., Pullen, A.E., and Swager, T.M. (2000). *Chem. Rev.* 100: 2537.

75 Baughman, R.H., Cui, C., Zakhidov, A.A. et al. (1999). *Science* 284: 1340.

76 Chu, C.W., Ouyang, J., Tseng, J.H., and Yang, Y. (2005). *Adv. Mater.* 17: 1440.

77 Wolff, J.J. and Wortmann, R. (1999). *Advances in Physical Organic Chemistry*, vol. 32, 121. Academic Press.

78 Bettinger, C.J. and Bao, Z. (2010). *Polym. Int.* 59: 563.

79 Hug, H., Bader, M., Mair, P., and Glatzel, T. (2014). *Appl. Energy* 115: 216.

80 Głowacki, E.D., Irimia-Vladu, M., Kaltenbrunner, M. et al. (2012). *Adv. Mater.* 25: 1563.

81 Singh, T.B., Meghdadi, F., Günes, S. et al. (2005). *Adv. Mater.* 17: 2315.

82 Li, Y., Neoh, K.G., and Kang, E.T. (2005). *J. Biomed. Mater. Res. A* 73A: 171.

83 Yumusak, C., Singh, T.B., Sariciftci, N.S., and Grote, J.G. (2009). *Appl. Phys. Lett.* 95: 263304.

84 Yuan, H. and Ma, Z. (2006). *Appl. Phys. Lett.* 89: 212105.

85 Lipomi, D.J. and Bao, Z. (2017). *MRS Bull.* 42: 93.

86 Root, S.E., Savagatrup, S., Printz, A.D. et al. (2017). *Chem. Rev.* 117: 6467.

87 García De Arquer, F.P., Armin, A., Meredith, P., and Sargent, E.H. (2017). *Nat. Rev. Mater.* 2: 16100.
88 Pavase, T.R., Lin, H., Shaikh, Q. et al. (2018). *Sens. Actuators, B* 273: 1113.
89 Di Maria, F., Lodola, F., Zucchetti, E. et al. (2018). *Chem. Soc. Rev.* 47: 4757.
90 Chen, D. and Pei, Q. (2017). *Chem. Rev.* 117: 11239.
91 Huang, H., Yang, L., Facchetti, A., and Marks, T.J. (2017). *Chem. Rev.* 117: 10291.
92 Roth, S. and Carroll, D. (2015). *One-Dimensional Metals: Conjugated Polymers, Organic Crystals, Carbon Nanotubes and Graphene.* Weinheim: Wiley-VCH.
93 Osaka, I. and Takimiya, K. (2017). *Adv. Mater.* 29: 1605218.
94 Uoyama, H., Goushi, K., Shizu, K. et al. (2012). *Nature* 492: 234.
95 Kulkarni, A.P., Tonzola, C.J., Babel, A., and Jenekhe, S.A. (2004). *Chem. Mater.* 16: 4556.
96 Wong, M.Y. and Zysman-Colman, E. (2017). *Adv. Mater.* 29: 1605444.
97 Tsumura, A., Koezuka, H., and Ando, T. (1986). *Appl. Phys. Lett.* 49: 1210.
98 Mei, J., Diao, Y., Appleton, A.L. et al. (2013). *J. Am. Chem. Soc.* 135: 6724.
99 Zhou, K., Dong, H., Zhang, H., and Hu, W. (2014). *Phys. Chem. Chem. Phys.* 16: 22448.
100 Shan, B. and Miao, Q. (2017). *Tetrahedron Lett.* 58: 1903.
101 Zhou, Z., Zhang, Z., Wu, Q. et al. (2018). *ACS Appl. Mater. Interfaces* 10: 35395.
102 Lamport, Z.A., Haneef, H.F., Anand, S. et al. (2018). *J. Appl. Phys.* 124: 71101.
103 Luo, C., Kyaw, A.K.K., Perez, L.A. et al. (2014). *Nano Lett.* 14: 2764.
104 Sirringhaus, H. (2014). *Adv. Mater.* 26: 1319.
105 Yuan, Y., Giri, G., Ayzner, A.L. et al. (2014). *Nat. Commun.* 5: 3005.
106 Chu, M., Fan, J., Yang, S. et al. (2018). *Adv. Mater.* 30: 1803467.
107 Paterson, A.F., Singh, S., Fallon, K.J. et al. (2018). *Adv. Mater.* 30: 1801079.
108 Loi, A., Basiricò, L., Cosseddu, P. et al. (2013). *IEEE Sens. J.* 13: 4764.
109 Krishnamoorthy, K., Facchetti, A.F., Hu, W., and Bao, Z. (2015). *Phys. Chem. Chem. Phys.* 17: 26509.
110 Gay, N. and Fischer, W.J. (2017). OFET-based analog circuits for microsystems and RFID-sensor transponders. *Presented at Polytronic 2007 - 6th International Conference on Polymers and Adhesives in Microelectronics and Photonics.*
111 Baeg, K., Caironi, M., and Noh, Y. (2013). *Adv. Mater.* 25: 4210.
112 Sirringhaus, H. (2009). *Adv. Mater.* 21: 3859.
113 Boddaert, X., Bensaid, B., Benaben, P. et al. (2010). *Microelectron. Reliab.* 50: 1884.
114 Barr, M.C., Rowehl, J.A., Lunt, R.R. et al. (2011). *Adv. Mater.* 23: 3500.
115 Tang, C.W. (1986). *Appl. Phys. Lett.* 48: 183.
116 Gao, K., Zhu, Z., Xu, B. et al. (2017). *Adv. Mater.* 29: 1703980.
117 Grant, T.M., Gorisse, T., Dautel, O. et al. (2017). *J. Mater. Chem. A* 5: 1581.
118 Li, X., Wang, H., Schneider, J.A. et al. (2017). *J. Mater. Chem. C* 5: 2781.
119 Yuan, J., Zhang, Y., Zhou, L. et al. (2019). *Joule* 3: 1140.
120 Meng, L., Zhang, Y., Wan, X. et al. (2018). *Science* 361: 1094.
121 Wang, L., Wang, L., Zhang, L., and Xiang, D. (2017). *Top. Curr. Chem.* 375: 61.
122 Dell, E.J., Capozzi, B., Xia, J. et al. (2015). *Nat. Chem.* 7: 209.

123 Ye, L., Zhang, S., Huo, L. et al. (2014). *Acc. Chem. Res.* 47: 1595.
124 Nakano, M., Osaka, I., Hashizume, D., and Takimiya, K. (2015). *Chem. Mater.* 27: 6418.
125 Su, T.A., Li, H., Steigerwald, M.L. et al. (2015). *Nat. Chem.* 7: 215.
126 Smith, C.E., Odoh, S.O., Ghosh, S. et al. (2015). *J. Am. Chem. Soc.* 137: 15732.
127 Sukegawa, J., Schubert, C., Zhu, X. et al. (2014). *Nat. Chem.* 6: 899.
128 Sakamoto, R., Wu, K., Matsuoka, R. et al. (2015). *Chem. Soc. Rev.* 44: 7698.
129 Guo, C., Wang, K., Zerah-Harush, E. et al. (2016). *Nat. Chem.* 8: 484.
130 Murphy, A.R. and Fréchet, J.M.J. (2007). *Chem. Rev.* 107: 1066.
131 Wang, C., Dong, H., Hu, W. et al. (2012). *Chem. Rev.* 112: 2208.
132 Xiao, X., Nagahara, L.A., Rawlett, A.M., and Tao, N. (2005). *J. Am. Chem. Soc.* 127: 9235.
133 Moth-Poulsen, K. and Bjørnholm, T. (2009). *Nat. Nanotechnol.* 4: 551.
134 Martin, P., Della Rocca, M.L., Anthore, A. et al. (2012). *J. Am. Chem. Soc.* 134: 154.
135 Sangeeth, C.S.S., Demissie, A.T., Yuan, L. et al. (2016). *J. Am. Chem. Soc.* 138: 7305.
136 Choi, S.H., Risko, C., Delgado, M.C.R. et al. (2010). *J. Am. Chem. Soc.* 132: 4358.
137 Luo, L. and Frisbie, C.D. (2010). *J. Am. Chem. Soc.* 132: 8854.
138 Zhao, X., Huang, C., Gulcur, M. et al. (2013). *Chem. Mater.* 25: 4340.
139 Moreno-García, P., Gulcur, M., Manrique, D.Z. et al. (2013). *J. Am. Chem. Soc.* 135: 12228.
140 Cai, L., Cabassi, M.A., Yoon, H. et al. (2005). *Nano Lett.* 5: 2365.
141 Lee, T., Reed, M.A., and Wang, W. (2003). *Phys. Rev. B* 68: 35416.
142 Wang, G., Kim, T., Jang, Y.H., and Lee, T. (2008). *J. Phys. Chem. C* 112: 13010.
143 Chen, X., Jeon, Y., Jang, J. et al. (2008). *J. Am. Chem. Soc.* 130: 8166.
144 Ho Choi, S., Kim, B., and Frisbie, C.D. (2008). *Science* 320: 1482.
145 Luo, L., Choi, S.H., and Frisbie, C.D. (2011). *Chem. Mater.* 23: 631.
146 Wang, Z., Dong, H., Li, T. et al. (2015). *Nat. Commun.* 6: 7478.
147 Jiang, J., Nakashima, H., Luo, Y. et al. (2006). *Phys. Rev. Lett.* 96: 27801.
148 Liu, K., Wang, X., and Wang, F. (2008). *ACS Nano* 2: 2315.
149 Seo, K., Konchenko, A.V., Lee, J. et al. (2008). *J. Am. Chem. Soc.* 130: 2553.
150 Luo, L., Benameur, A., Brignou, P. et al. (2011). *J. Phys. Chem. C* 115: 19955.
151 Bao, Z., Lovinger, A.J., and Dodabalapur, A. (1996). *Appl. Phys. Lett.* 69: 3066.
152 Higgins, S.J. and Nichols, R.J. (2018). *Polyhedron* 140: 25.
153 Milan, D.C., Vezzoli, A., Planje, I.J., and Low, P.J. (2018). *Dalton Trans.* 47: 14125.
154 Porath, D., Lapidot, N., and Gomez-Herrero, J. (2015). *Introducing Molecular Electronics*, 411. Berlin, Heidelberg: Springer-Verlag.
155 Cohen, H., Nogues, C., Naaman, R., and Porath, D. (2005). *Proc. Natl. Acad. Sci. U.S.A.* 102: 11589.
156 Göhler, B., Hamelbeck, V., Markus, T.Z. et al. (2011). *Science* 331: 894.
157 Liu, S., Clever, G.H., Takezawa, Y. et al. (2011). *Angew. Chem. Int. Ed.* 50: 8886.
158 Xu, B., Zhang, P., Li, X., and Tao, N. (2004). *Nano Lett.* 4: 1105.

159 Wohlgamuth, C.H., McWilliams, M.A., and Slinker, J.D. (2013). *Anal. Chem.* 85: 8634.

160 Kim, K., Kim, J., Huang, X. et al. (2008). *Small* 4: 1959.

161 Li, Y., Xiang, L., Palma, J.L. et al. (2016). *Nat. Commun.* 7: 11294.

162 Teschome, B., Facsko, S., Schönherr, T. et al. (2016). *Langmuir* 32: 10159.

163 Genereux, J.C. and Barton, J.K. (2010). *Chem. Rev.* 110: 1642.

164 Zwang, T.J., Tse, E.C.M., and Barton, J.K. (2018). *ACS Chem. Biol.* 13: 1799.

165 Arya, S.K., Solanki, P.R., Datta, M., and Malhotra, B.D. (2009). *Biosens. Bioelectron.* 24: 2810.

166 Artés, J.M., López-Martínez, M., Díez-Pérez, I. et al. (2014). *Electrochim. Acta* 140: 83.

167 Greulich, K.O. (2005). *ChemPhysChem* 6: 2458.

168 Willner, I. and Katz, E. (2005). *Bioelectronics: From Theory to Applications.* Weinheim: Wiley-VCH.

169 Carroll, R.L. and Gorman, C.B. (2002). *Angew. Chem. Int. Ed.* 41: 4378.

170 Aviram, A. (1989). *Angew. Chem.* 101: 536.

171 Gimzewski, J.K., Schlittler, R.R., Chavy, C., and Joachim, C. (1995). *Phys. Rev. Lett.* 74: 2102.

172 Park, J., Pasupathy, A.N., Goldsmith, J.I. et al. (2002). *Nature* 417: 722.

173 Bonifas, A.P. and McCreery, R.L. (2010). *Nat. Nanotechnol.* 5: 612.

174 Li, T., Hu, W., and Zhu, D. (2010). *Adv. Mater.* 22: 286.

175 Tao, N.J. (2006). *Nat. Nanotechnol.* 1: 173.

176 Smith, C.E., Xie, Z., Bâldea, I., and Frisbie, C.D. (2018). *Nanoscale* 10: 964.

177 Saifullah, M.S.M., Ondarcuhu, T., Koltsov, D.K. et al. (2002). *Nanotechnology* 13: 659.

178 Chen, W., Ahmed, H., and Nakazoto, K. (1995). *Appl. Phys. Lett.* 66: 3383.

179 Morpurgo, A.F., Marcus, C.M., and Robinson, D.B. (1999). *Appl. Phys. Lett.* 74: 2084.

180 Chen, Y., Hong, M., and Huang, G.S. (2012). *Nat. Nanotechnol.* 7: 197.

181 Strobel, S., Harrer, S., Penso Blanco, G. et al. (2009). *Small* 5: 579.

182 Jiang, L., Dong, H., Meng, Q. et al. (2011). *Adv. Mater.* 24: 694.

183 Iwane, M., Fujii, S., and Kiguchi, M. (2017). *Sensors* 17: 956.

184 Elbing, M., Ochs, R., Koentopp, M. et al. (2005). *Proc. Natl. Acad. Sci. U.S.A.* 102: 8815.

185 Wu, Q., Hou, S., Sadeghi, H., and Lambert, C.J. (2018). *Nanoscale* 10: 6524.

186 Whalley, A.C., Steigerwald, M.L., Guo, X., and Nuckolls, C. (2007). *J. Am. Chem. Soc.* 129: 12590.

187 Andrews, D.Q., Solomon, G.C., Goldsmith, R.H. et al. (2008). *J. Phys. Chem. C* 112: 16991.

188 Galbiati, M., Barraud, C., Tatay, S. et al. (2012). *Adv. Mater.* 24: 6429.

189 Mahan, G.D. and Sofo, J.O. (1996). *Proc. Natl. Acad. Sci. U.S.A.* 93: 7436.

190 Tan, A., Sadat, S., and Reddy, P. (2010). *Appl. Phys. Lett.* 96: 13110.

191 Aradhya, S.V. and Venkataraman, L. (2013). *Nat. Nanotechnol.* 8: 399.

2

Electron Transport in Single Molecular Devices

Lei-Lei Nian, Liang Ma, and Jing-Tao Lü

School of Physics and Wuhan National High Magnetic Field Center, Huazhong University of Science and Technology, Wuhan 430074, People's Republic of China

2.1 Introduction

The physical and chemical properties of single molecular junctions sensitively depend on how the single molecule connects to the nanogap electrodes. However, it is quite difficult to visualize this geometrical information directly by using available experimental technique. Electron transport measurement is an important indirect tool to characterize a single molecular junction. From the current–voltage or differential conductance–voltage characteristics, it is possible to deduce the geometrical information of the single molecular junction. It is also important to understand its electrical transport properties for its use as electronic devices. Thus, electron transport across single molecular junctions has received intense research attention since the birth of molecular electronics. In many cases, the interpretation of the experimental measurement is challenging. The development of theoretical methods is thus important to understand the experimental results and to propose novel device concepts using single molecular junction.

In this chapter, we give an introduction to typical theoretical approaches that can be used for these purposes. We focus on the study of electrical transport properties using nonequilibrium Green's function (NEGF) method, although we also introduce the density matrix (DM) or master equation approach. We start from noninteracting electron transport. Afterward, we extend the model to consider the effect of electron-vibration (e-vib), electron–electron (e–e) interaction on the transport properties of single molecular junctions. Based on the method we introduced, we give a brief account of the thermoelectric transport. Before closing, we introduce briefly the combination of the transport theory with density function theory-based electronic structure calculation. This has proven to be an efficient tool to simulate realistic molecular junctions and compare with experimental measurement.

Nanogap Electrodes, First Edition. Edited by Tao Li.

2.2 General Methods

2.2.1 Transport Mechanisms

In this section, we introduce general methods that can be used to study transport properties of nano-junctions. We adopt the traditional two-probe setup, where we divide the whole system into two metal nanogap electrodes (Left – L and Right – R) and one central device region (D). Depending on the relevant time/energy scales, the transport process can be divided into two typical regimes. When the device–electrode coupling is strong, the electron dwell time on the central device is short, such that the phase information of the electron wavefunction is preserved during transport. The wave nature of the electrons is important in this regime, which should be considered in the theoretical modeling. In the opposite limit where the device–electrode coupling is weak, the electron may stay longer time in the center before it hops into the other electrode. If the dwell time in the center is longer than the dephasing time, the phase information of the electron is lost during the transport process. Thus, a semi-classical treatment of the electron transport is enough. In the following, we will give a brief introduction to the theoretical approaches normally used to study transport in nano-junctions.

2.2.2 Nonequilibrium Green's Function Method

The power of NEGF lies in the ability to consider many-body effects on the transport. But we start from the discussion of the simplest case that does not include any many-body interaction and the electrons tunnel across the junction coherently. Thus, the phase information of the conduction electrons is preserved. In this case, electron transport can be considered as wave scattering process, where the energy-dependent transmission coefficient becomes the key quantity that determines the junction transport property. This method is pioneered by Landauer and Büttiker [1, 2], now termed the Landauer–Büttiker approach. The NEGF method can be used to calculate the transmission coefficient (Figure 2.1).

To make further discussions, we consider a simple model to describe the transport process. In terms of tight binding representation, the total model Hamiltonian is given by

$$H = H_L + H_R + H_D + H_T, \tag{2.1}$$

$$H_\alpha = \sum_k \varepsilon_{k\alpha} c^\dagger_{k\alpha} c_{k\alpha}, \quad \alpha = L, R, \tag{2.2}$$

$$H_D = \sum_n \varepsilon_n d^\dagger_n d_n, \tag{2.3}$$

$$H_T = \sum_{k\alpha,n} (V_{k\alpha,n} c^\dagger_{k\alpha} d_n + h.c.). \tag{2.4}$$

Here, $c^\dagger$ (c) and $d^\dagger$ (d) are the creation (annihilation) operators for electrons in the two electrodes (L, R) and device subspace (D), respectively. The parameters k and n

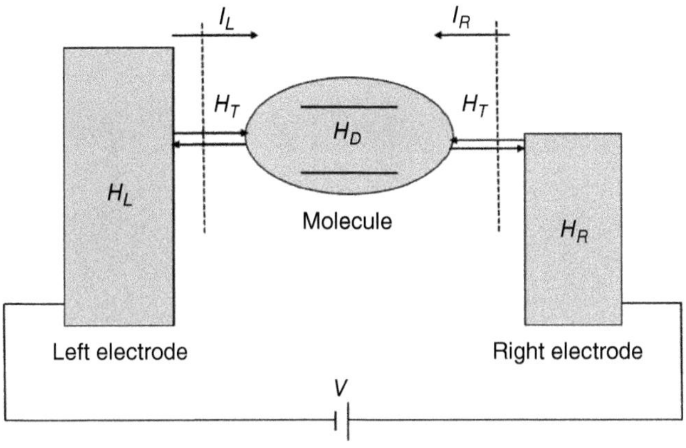

Figure 2.1 Schematic of model for describing electron tunneling through a center junction sandwiched by two metallic nanogap electrodes.

are collective indices representing the spatial quantum numbers of the electronic states in the electrodes and the device region, respectively. $V_{k\alpha,n}$ is the coupling matrix element between left (right) electrode and device region. Here we ignore the direct coupling between two electrodes. We consider spin-independent transport and ignore the spin degree of freedom.

In order to express the electrical current across the junction in terms of the Green's functions of the central device region, we should first give the definitions of different Green's functions, including retarded, advanced, greater, and lesser Green's functions

$$G^r(1,2) = -\frac{i}{\hbar}\Theta(t_1 - t_2)\langle[c_H(1), c_H^\dagger(2)]_+\rangle, \tag{2.5}$$

$$G^a(1,2) = \frac{i}{\hbar}\Theta(t_2 - t_1)\langle[c_H(1), c_H^\dagger(2)]_+\rangle, \tag{2.6}$$

$$G^>(1,2) = -\frac{i}{\hbar}\langle c_H(1)c_H^\dagger(2)\rangle, \tag{2.7}$$

$$G^<(1,2) = \frac{i}{\hbar}\langle c_H^\dagger(2)c_H(1)\rangle. \tag{2.8}$$

They are all defined in time domain. For steady state transport, one can perform Fourier transform to convert them to energy domain. Here, $c_H^\dagger$ (c_H) is the electron creation (annihilation) operators in the Heisenberg picture. The operators in Heisenberg and Schrödinger pictures are related by $A_H(t) = e^{\frac{i}{\hbar}Ht}A_S e^{-\frac{i}{\hbar}Ht}$. The contracted indices 1 and 2 could include time, spatial, and other indices. $\Theta(x)$ is the Heaviside step function. The brackets mean quantum-mechanical average over the full system Hamiltonian.

To calculate the electrical current through the molecule, we follow the approach of Meir and Wingreen [3]. We consider the change rate of electron number in one of the electrodes. Taking the left electrode as an example, the number of electrons in electrode L is

$$N_L = \sum_k c_{kL}^\dagger c_{kL}. \tag{2.9}$$

Its equation of motion is

$$\begin{aligned}\dot{N}_L &= \frac{i}{\hbar}[H, N_L] \\ &= \frac{i}{\hbar}[H_T, N_L] \\ &= \sum_{kn}[V_{kL,n}G^<_{n,kL}(t,t_+) - V_{n,kL}G^<_{kL,n}(t,t_+)].\end{aligned} \tag{2.10}$$

Hereafter, we will omit the subscript H in the Heisenberg operator. We use the fact that N_L commutes with H_L, H_R, and H_D and define the mixed lesser Green's functions

$$G^<_{n,kL} = \frac{i}{\hbar}\langle c_{kL}^\dagger(t_+)d_n(t)\rangle, \tag{2.11}$$

$$G^<_{kL,n} = \frac{i}{\hbar}\langle d_n^\dagger(t_+)c_{kL}(t)\rangle. \tag{2.12}$$

Here, t_+ means time slightly later than t. The electrical current from the left lead into the device region can be expressed as the decrease of electron number in the left lead times the charge of electron (e)

$$\begin{aligned}I_L &= -e\dot{N}_L(t) \\ &= e\sum_{kn}\left[V_{n,kL}G^<_{kL,n}(t,t_+) - V_{kL,n}G^<_{n,kL}(t,t_+)\right] \\ &= \frac{e}{h}\int d\varepsilon \sum_{kn}\left[V_{n,kL}G^<_{kL,n}(\varepsilon) - V_{kL,n}G^<_{n,kL}(\varepsilon)\right] \\ &= \frac{e}{h}\int d\varepsilon \mathrm{Tr}[G_D^>(\varepsilon)\Sigma_L^<(\varepsilon) - G_D^<(\varepsilon)\Sigma_L^>(\varepsilon)].\end{aligned} \tag{2.13}$$

At this point, we have expressed the electrical current in terms of the Green's function of the device region and the electrode self-energies, due to coupling between the device region and the electrodes. The self-energy can be defined as $\sum_\alpha = \sum_{kn} V_{n,k\alpha} g_{k\alpha,k\alpha} V_{k\alpha,n}$. And, $g_{k\alpha,k\alpha}$ is the surface Green's function of the lead.

Equation (2.13) is the Meir–Wingreen formula for the electrical current. We need to multiply the factor 2 to take into account the spin degeneracy. This formula is quite general and applies even when there is many-body interaction in the device region. For noninteracting case, it can be arranged to another form [4]

$$\begin{aligned}I_L &= \frac{e}{h}\int d\varepsilon\, \mathrm{Tr}[G_{0D}^r(\varepsilon)\Gamma_L(\varepsilon)G_{0D}^a(\varepsilon)\Gamma_R(\varepsilon)][f_L(\varepsilon) - f_R(\varepsilon)] \\ &\equiv \frac{e}{h}\int d\varepsilon\, T(\varepsilon)[f_L(\varepsilon) - f_R(\varepsilon)],\end{aligned} \tag{2.14}$$

with $\Gamma_\alpha = i(\Sigma_\alpha^r - \Sigma_\alpha^a)$ and

$$T(\varepsilon) = \mathrm{Tr}[G_{0D}^r(\varepsilon)\Gamma_L(\varepsilon)G_{0D}^a(\varepsilon)\Gamma_R(\varepsilon)]. \tag{2.15}$$

This is the Landauer formula that gives the expression of the transmission coefficient $T(\varepsilon)$ in terms of the Green's function, and

$$f_{L,R}(\varepsilon) = \frac{1}{e^{\frac{\varepsilon-\mu_{L,R}}{k_B T_{L,R}}} + 1},$$

is the Fermi–Dirac distribution of the electrode. The subscript 0 means single electron Green's functions without any many-body interactions. The retarded and advanced Green's function of the device region can be written as

$$G^r_{0D} = G^{a\dagger}_{0D} = [(\varepsilon + i\delta)I - H_D - \Sigma^r_L - \Sigma^r_R]^{-1}. \tag{2.16}$$

2.2.3 Master Equation Method

Apart from the NEGF method, DM approach [5–7], especially the resulted rate equation, is another useful tool to investigate electrical transport. It is suitable for dealing with the case of weak coupling between device and electrodes, i.e. the device–electrode coupling can be regarded as a perturbation. Starting from the full DM, the electrodes degrees of freedom are traced out. We are left with the reduced density matrix (RDM) of the device. Its equation of motion is described by the master equation, which has been widely used to study the quantum transport in quantum dot, single molecule and other low-dimensional systems. Below we sketch briefly the derivation of the master equation.

In the Schrödinger picture, the time evolution of the full DM can be expressed as

$$\frac{\partial \sigma_s(t)}{\partial t} = \frac{1}{i\hbar}[H, \sigma_s(t)], \tag{2.17}$$

where the Hamiltonian H describes the full system consisting of device, electrodes, and their coupling. Because the number of degrees of freedom in the electrodes is infinite, the RDM is normally introduced

$$\overline{\sigma}_s(t) = Tr_{\text{elec}}\{\sigma_s(t)\} = \sum_{l\in\text{elec}} \langle l|\sigma_s(t)|l\rangle. \tag{2.18}$$

The tracing operation for electrodes can reduce the dimension of the problem to the dimension of the device.

Next, we need to transfer the full DM from Schrödinger to the interaction picture

$$\sigma_I(t) = e^{H_0 t(i/\hbar)}\sigma_s(t)e^{-H_0 t(i/\hbar)}. \tag{2.19}$$

Combining Eqs. (2.17) and (2.19), we have

$$\frac{\partial \sigma_I(t)}{\partial t} = \frac{1}{i\hbar}[H^I_T(t), \sigma_I(t)], \tag{2.20}$$

where H^I_T describes the time-dependent coupling of device and electrodes given by

$$H^I_T(t) = e^{H_0 t(i/\hbar)}H^I_T e^{-H_0 t(i/\hbar)}. \tag{2.21}$$

We can also get the RDM in the interaction picture from Eq. (2.18)

$$\overline{\sigma}_I(t) = Tr_{\text{elec}}\{\sigma_I(t)\}. \tag{2.22}$$

Equation (2.20) can be rewritten as an integral form

$$\sigma_I(t) = \sigma_I(0) + \frac{1}{i\hbar}\int_0^t [H_T^I(t_1), \sigma_I(t_1)]dt_1. \tag{2.23}$$

By iterating the above equation to the second order in H_T, we can get

$$\frac{\partial \sigma_I(t)}{\partial t} = \frac{1}{i\hbar}[H_T^I(t), \sigma_I(0)] - \frac{1}{\hbar^2}\int_0^t [H_T^I(t), [H_T^I(t_1), \sigma_I(t_1)]]dt_1, \tag{2.24}$$

where H_T up to the second-order perturbation, i.e. the so-called the Born approximation (BA). In the case of weak coupling between electrodes and device, the full DM can be simplified to the following form

$$\sigma_I(t) \approx \overline{\sigma}_I(t) \otimes \sigma_{\text{elec}}. \tag{2.25}$$

Tracing out the electrodes degrees of freedom at both sides of Eq. (2.24), we get

$$\frac{\partial \overline{\sigma}_I(t)}{\partial t} = -\frac{1}{\hbar^2}\int_0^t Tr_{\text{elec}}[H_T^I(t), [H_T^I(t_1), \sigma_I(t_1)]]dt_1. \tag{2.26}$$

If we assume a product initial state $\sigma_I(0) = \overline{\sigma}_I(0) \otimes \sigma_{elec}$, we have

$$\frac{\partial \overline{\sigma}_I(t)}{\partial t} = -\frac{1}{\hbar^2}\int_0^t \text{Tr}_{\text{elec}}[H_T^I(t), [H_T^I(t_1), \overline{\sigma}_I(t_1) \otimes \sigma_{\text{elec}}]]dt_1. \tag{2.27}$$

Equation (2.27) is the non-Markovian master equation. If the memory time of the electrodes (environment) is short, one can get the Markovian master equation by replacing the $\overline{\sigma}_I(t_1)$ with $\overline{\sigma}_I(t)$

$$\frac{\partial \overline{\sigma}_I(t)}{\partial t} = -\frac{1}{\hbar^2}\int_0^t \text{Tr}_{\text{elec}}[H_T^I(t), [H_T^I(t_1), \overline{\sigma}_I(t) \otimes \sigma_{\text{elec}}]]dt_1. \tag{2.28}$$

From the above equation, we can evaluate the diagonal elements of the RDM. This results in a rate equation for the population of each electronic state in the following form

$$\frac{dP_n^q}{dt} = \sum_{q',n'} (W_{nn'}^{qq'} P_{n'}^{q'} - W_{n'n}^{q'q} P_n^q). \tag{2.29}$$

Here, $W_{nn'}^{qq'}$ is the transition rate from state (q', n') to state (q, n). Here, n' and n are electron numbers and q' and q represent other state indices, i.e. molecular orbital index, vibrational state index, and so on. In the simplest case, the transition rate can be calculated from the Fermi golden rule. At steady state, the left side of the equation is zero, we get a set of linear equations, from which we can solve for P_n. The transport properties, for example the electrical current, can be obtained from the occupation probabilities P_n^q as

$$I_L = e\sum_{n,q,q'} (W_{n,n+1;L}^{q,q'} - W_{n,n-1;L}^{q,q'})P_n^q, \tag{2.30}$$

where $W_{n,n+1;L}^{q,q'}$ represents the transition rate from the left electrode to the molecule which results in the change of molecular state from (q, n) to $(q', n+1)$.

2.3 Single Electron Transport Through Single Molecular Junction

2.3.1 Coherent Transport

The coherent electronic transport can be well described by the Landauer–Büttiker formalism. The transmission coefficient can be calculated from the NEGF method as discussed in Section 2.1.

A simplified Anderson–Newns model is normally used to describe the single molecule conductance [8, 9]. In this model, the center part includes only one electronic state at ε_0. It represents the lowest unoccupied molecular orbital (LUMO) or highest occupied molecular orbital (HOMO) that participates the transport dominantly. The tunnel coupling between this electronic state with the two metal electrodes is characterized by two level broadening parameters γ_L and γ_R. The transmission coefficient can be written as

$$T(\varepsilon) = \frac{\gamma_L \gamma_R}{(\varepsilon - \varepsilon_0)^2 + (\gamma_L + \gamma_R)^2/4}. \tag{2.31}$$

The electrical current can then be obtained from the Landauer (2.14). This result has been widely used to fit the experimentally measured current or conductance of single molecular junctions to extract the LUMO or HOMO level position and its coupling to the electrodes.

If there are more than one paths that electrons can take during transport, quantum interference effect could be important. One can still use the Landauer formula Eq. (2.14) to consider this effect, but the transport has to be modeled with multiple paths. The transmission coefficient is then obtained from Eq. (2.15). Simple tight-binding model and first-principles density functional theory (DFT) calculations have been used to study this effect [10–15]. According to classical Kirchhoff's circuit law, thc net conductance of two parallel components in an electronic circuit is the sum of the individual conductance. However, quantum interference effects in molecular junctions can enhance (constructive interference effects) or diminish (destructive interference effects) the molecular conductance.

Long conjugated molecules are good candidates for observing a clear quantum interference effects in molecular junctions. H. Vazquez et al. [13] synthesized a series of molecular systems consisting of single or two backbone connected to common sulfur atoms at each end. In both cases, the terminal sulfur groups bind to the gold electrodes. The conductance of the molecular junctions was measured using the scanning tunneling microscope (STM)-based break junction technique. The experimental data and transport calculations based on DFT show that the conductance of a double-backbone molecular junction can be more than twice (~2.8) that of a single-backbone junction, providing clear evidence for constructive interference (Figure 2.2).

Guédon et al. provide direct evidence for destructive quantum interference in two-terminal molecular junctions at room temperature [14]. To investigate the influence of quantum interference on molecular conductance, they synthesized a few rigid π-conjugated molecular wires (shown in Figure 2.3) and created

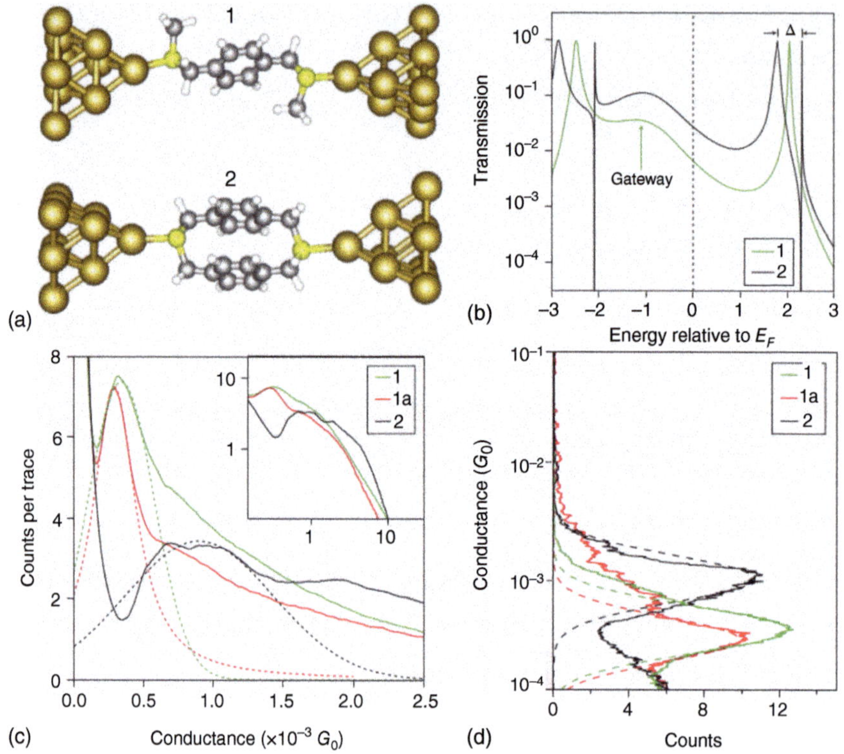

Figure 2.2 Probing quantum constructive interference effects of single molecular junction. (a) Chemical structure of exemplary molecules with single-(1) and double-(2) backbone. (b) Transmission spectra shows the low bias conductance for the double-backbone species is more than double that for single-backbone species. (c) Conductance histograms (on a linear scale) for each molecule: 1, 1a (is a two-backbone molecule with one side conjugated and the other saturated C_4H_8.) and 2. Dashed lines indicate Lorentzian fits to the pecks. Inset: same data on a log scale. Giving a conductance ratio of $G(2)/G(1) \approx 3$. (d) Conductance profile determined from two-dimensional histograms for 1, 1a, and 2 over a width of 0.1 nm centered at about 0.5 nm displacement. Dashed lines indicate Gaussian fits to the peaks. Giving a conductance ratio of $G(2)/G(1) = 2.8$. Source: Vazquez et al. [13]. 2012. Springer nature.

self-assembled monolayers (SAMs) of each molecule type on thin gold layers. Next, a conducting atomic force microscopy (AFM) was brought into close contact with a SAM to measure charge transport. The experimental data combining with theoretical calculation show that the large difference in conductance is instead indicative of destructive interference.

2.3.2 Hopping Transport

When the electron dwell time on the molecule is longer than the dephasing time, electrons lose their phase information during transport. This happens when the coupling to electrodes is weak, or the molecule is longer than the electron coherent length. In this case, the electron transport process can be described by classical

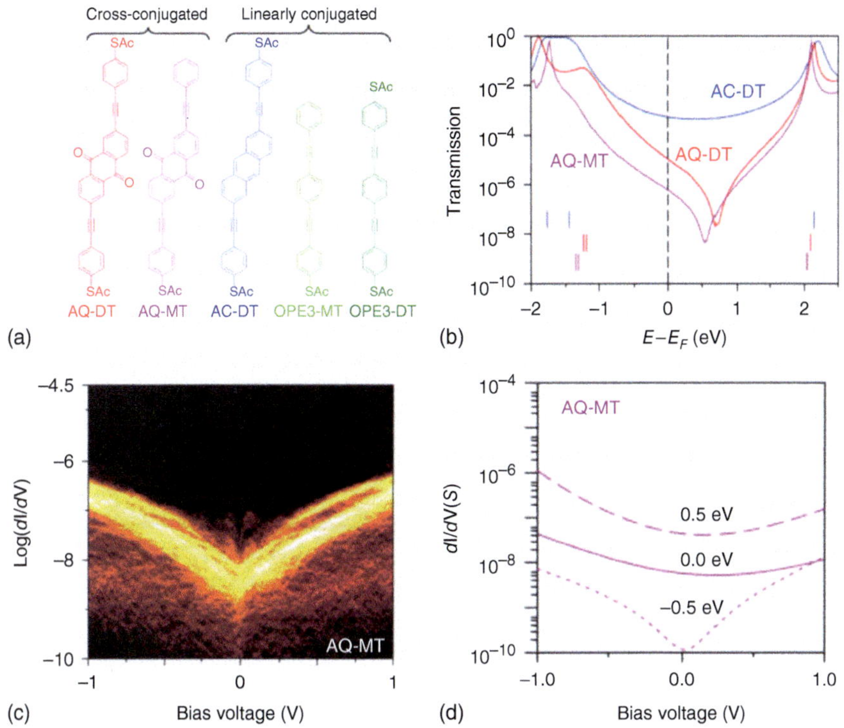

Figure 2.3 Observation of quantum destructive interference effects of single molecular junction. (a) Chemical structure of exemplary molecules (AQ-DT, AQ-MT, AC-DT, OPE3-MT, and OPE3-DT). (b) Transmission function calculated with DTF+Σ. (c) Logarithmically binned two-dimensional histograms of dI/dV vs. bias voltage V for AQ-MT. (d) dI/dV for AQ-MT computed from the transmission function in (b). Source: Guédon et al. [14]. 2012, Springer nature.

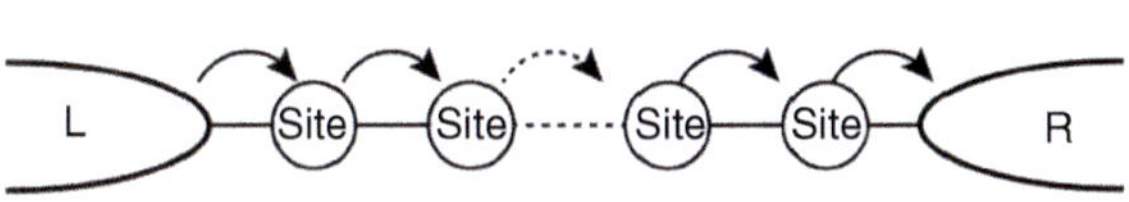

Figure 2.4 Hopping transport mechanism in a single molecular junction consisting of a single molecule and two electrodes (L and R).

hopping from site to site, as illustrated in Figure 2.4. Note that the definition of the hopping sites determined by intrinsic electric structures of molecular units is very crucial when one uses this model to describe the practical molecular systems because the coherent transport and hopping transport may coexist. In general, the electron/hole will lose its coherence in the hopping transport regime due to its interaction with the other degrees of freedom in the molecule or environment. However, the coherence may be maintained locally over several hopping sites. Semi-classical rate equation in the form of Eq. (2.30) can be used to describe the transport process in this case.

Due to the long dwell time in the molecule, the electrons have enough time to interact with other degrees of freedom. Thus, one normally needs to consider different types of interactions in this transport regime. For example, the e-vib interaction may renormalize the hopping parameter between different sites. This effect can be easily incorporated in the transition rates appeared in the rate equations.

One immediate question one could ask is how we distinguish coherent transport and hopping transport experimentally? In fact, the dependence of electric conductance (G) on molecular length (L) and temperature (T) is quite different in the two regimes. The electric conductance usually decreases exponentially with L. It can be expressed as $G = G_0 \exp(-\beta L)$, where G_0 is a system dependent constant and β is a decay factor [16]. This is the typical behavior of tunneling transport and the decay factor depends on the barrier height, i.e. the position of the HOMO or LUMO level, the coupling to electrodes and other factors. In hopping transport regime, the dependence of G on L exhibits a quasi-Ohmic behavior, i.e. G is inverse proportional to L. Thus, it is possible to observe a crossover from tunneling to hopping regime by looking at the length dependence of the conductance. This behavior has been observed by several groups [17–22]. They attributed this phenomenon to the transition from coherent to hopping mechanism. For example, the rate of charge transfer through a short double strand DNA (dsDNA) molecule with the sequence $(G:C)(A:T)_M(G:C)_3$was measured [22]. It is found that the length dependence has a crossover at $M = 3$, i.e. the charge transfer displayed a strong distance dependence when $M < 3$,while a weak distance dependence could be observed when $M > 3$ (see Figure 2.5).

Temperature dependence of the conductance in the coherent tunneling regime and the hopping regime are also different. For hopping transport, increasing

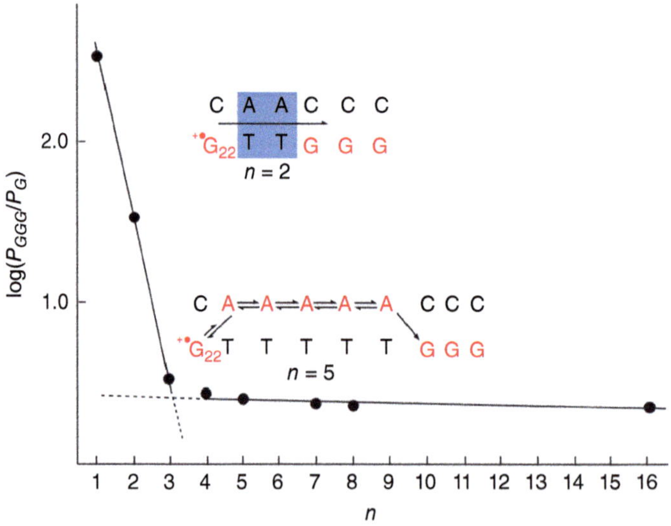

Figure 2.5 Observation of the transition from coherent to hopping mechanism in a dsDNA molecule by measuring the efficiency of the charge transfer against the number n of A:T base pairs. Source: Giese et al. [22]. 2001, Springer Nature.

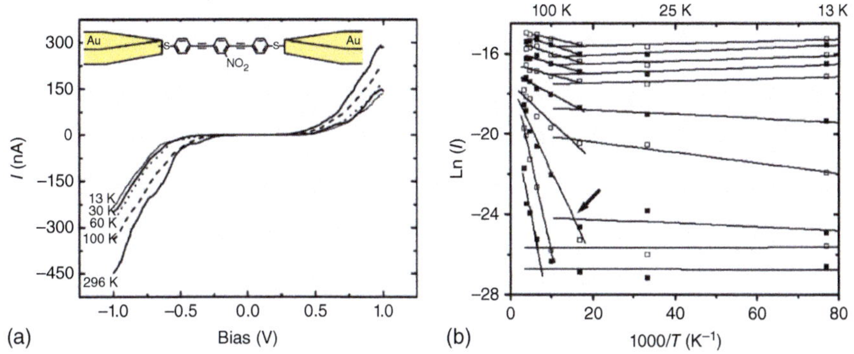

Figure 2.6 (a) The current (*I*) as a function of bias voltages with different values temperature (*T*) in which the inset is the molecular junction under consideration. (b) The Arrhenius plots of Ln(*I*) vs. inverse *T* for different values bias voltages. Source: Selzer et al. [23]. 2004, American Chemical Society.

temperature may lead to higher hopping probability between sites, while coherent tunneling transport has a much weaker temperature dependence. The thermally activated transport in a single molecular junction consisting of a 1-nitro-2,5-di(phenylethynyl-4′-mercapto)benzene molecules and two Au nanogap electrodes was investigated by the Selzer et al. [23]. It is found that the current changes significantly with the increase of the temperature (see Figure 2.6). Figure 2.6 displays an Arrhenius plot for a typical junction. One can find a characteristic transition from temperature-independent behavior at low temperature region where the conduction is dominated by coherent superexchange tunneling to the incoherent temperature-dependent hopping behavior in the region of high temperature. By increasing the bias voltage, the temperature-dependent crossover can be shifted to low temperature region. This is due to the fact that the activation energy is decreased by increasing the bias, and the current is proportional to $\exp(-\Delta E/k_B T)$ (ΔE is the system-dependent activation energy) in the hopping transport regime. Another possible reason is effective temperature of the molecule can be enhanced by increasing the bias. Both of two reasons can lead to the emergence of the transition.

The change of conductance with length is not always linear or exponentially decaying. A parity oscillation in the conductance of Au, Pt, and Ir single atom chain with respect to the number of atoms forming the atomic chain has been observed, as shown in Figure 2.7. This is consistent with odd–even oscillation of the conductance with the number of atoms predicted theoretically [24, 25].

2.4 Effect of Many-Body Interactions

2.4.1 Electron-Vibration Interaction

So far, we have not considered the interaction of electrons with other degrees of freedom. Actually, when the electrons traverse the molecular junction, they may

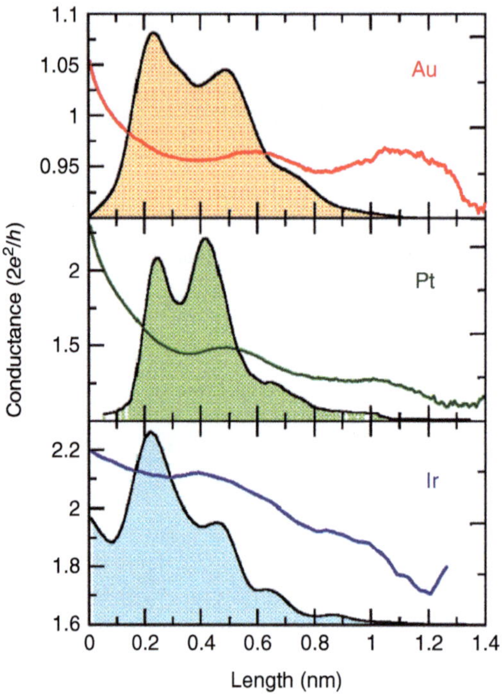

Figure 2.7 Averaged plateaus of conductance for chains of atoms of the three metals: Au, Pt, and Ir. Source: Smit et al. [24]. 2003, American Physcial Society.

interact with molecular vibrations or other electrons. In general, the manifestation of e-vib interaction in the transport characteristics is described by three kinds of energies, including the molecule–electrode coupling Γ_α ($\alpha = L, R$), the electron-vibration coupling m_{ep}, and the thermal energy $k_B T$ (T is the temperature).

In the case of $\Gamma_\alpha \ll m_{ep}$ (weak coupling between molecule and electrode) and $k_B T \ll \Gamma_\alpha$ (low temperature), the electrons will stay in the molecule for long enough time to interact with the vibration. Their interaction may lead to the formation of polaron. The polaron density of states shows several peaks. Corresponding steps in the current–voltage (I–V) curve can be observed. Each step corresponds to the participation of one vibronic state in the transport.

For $\Gamma_\alpha \gg m_{ep}$ (the strong coupling between molecule and electrodes) and $k_B T \ll \Gamma_\alpha$ (low temperature), the dependence of the conduction and the current on e-vib coupling is weak. However, the signal of e-vib coupling shows up in the inelastic electron tunneling spectroscopy (IETS), defined as $(d^2I/dV^2)/(dI/dV)$. For relatively low voltage $eV < \hbar\omega$ (ω is a typical molecular vibration frequency), the current is elastic. When $eV > \hbar\omega$, electrons may tunnel inelastically by emitting one vibrational energy. This opens an additional inelastic transport channel. These processes are depicted schematically in Figure 2.8. In the second-order differential conductance, the inelastic transport gives rise to a clear vibrational feature. In the following, we introduce theoretical methods to deal with these two different parameter regimes.

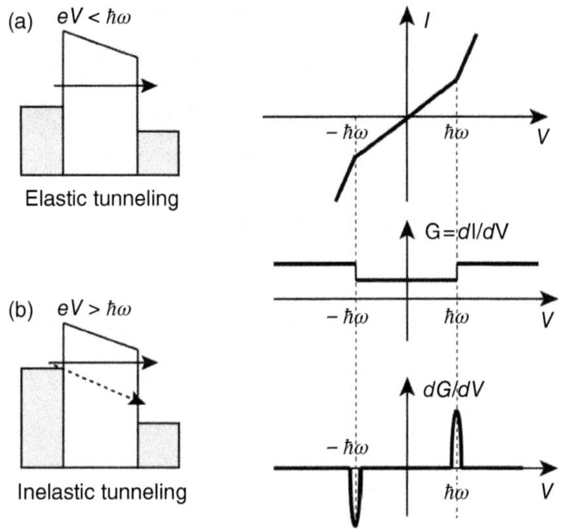

Figure 2.8 Schematic illustrations of the IETS. The solid arrows in (a) and (b) represent the elastic tunneling ($eV < \hbar\omega$), and the dotted arrows stand for the inelastic tunneling ($eV > \hbar\omega$). The right plane display the corresponding current, conductance (dI/dV), and the IETS (dG/dV).

2.4.1.1 Weak Coupling Regime

In the weak coupling regime, the e-vib interaction can be treated perturbatively. The NEGF approach within the self-consistent Born approximation (SCBA) or the lowest order expansion (LOE) has been widely used to study the IETS spectra in molecular junctions [26–37].

We can add the following Hamiltonian to Eq. (2.1)

$$H_{ep} = \sum_q \omega_q a_q^\dagger a_q + \sum_{q,ij} M_{ij}^q (a_q^\dagger + a_q) c_i^\dagger c_j, \tag{2.32}$$

where the first and second terms describe the vibrations and e-vib interaction, respectively. M_{ij}^q is the interaction matrix element.

To consider the e-vib interaction, we need to define vibration Green's functions

$$D_q^r(1,2) = -\frac{i}{\hbar}\Theta(t_1 - t_2)\langle [Q_q(1), Q_q(2)]_- \rangle, \tag{2.33}$$

$$D_q^a(1,2) = \frac{i}{\hbar}\Theta(t_2 - t_1)\langle [Q_q(1), Q_q(2)]_- \rangle, \tag{2.34}$$

$$D_q^>(1,2) = -\frac{i}{\hbar}\langle Q_q(1) Q_q(2) \rangle, \tag{2.35}$$

$$D_q^<(1,2) = -\frac{i}{\hbar}\langle Q_q(2) Q_q(1) \rangle, \tag{2.36}$$

where $Q_q = a_q + a_q^\dagger$ is the displacement operator. The retarded and lesser Green's functions satisfy the Dyson equations and can be expressed in the energy space

$$D_q^r = [(\omega + i\eta)^2 - \omega_q^2 - \Pi_q^r]^{-1}, \tag{2.37}$$

$$D_q^< = D_q^r \Pi_q^< D_q^a. \tag{2.38}$$

Here ω_q is the angular frequency of the vibrational mode q, Π_q is the corresponding self-energy due to e-vib interaction. Correspondingly, the e-vib interaction will also introduce the electron self-energy. Standard Feynman diagram technique can be used to calculate these self-energies.

Considering the lowest order Feynman diagrams, we have Hartree- and Fock-like terms for the electrons and a polarization term for the vibrations. The self-energies for electrons can be expressed as

$$\Sigma_{mn}^{F,<}(\varepsilon) = iM_{mi}^{k} \int G_{0,ij}^{<}(\varepsilon - \omega) D_{0,kl}^{<} M_{jn}^{l} \frac{d\omega}{2\pi}, \tag{2.39}$$

$$\Sigma_{mn}^{F,r}(\varepsilon) = iM_{mi}^{k} \int [G_{0,ij}^{r}(\varepsilon - \omega) D_{0,kl}^{<}(\omega) + G_{0,ij}^{<}(\varepsilon - \omega) D_{0,kl}^{r}(\omega) + G_{0,ij}^{r}(\varepsilon - \omega) D_{0,kl}^{r}(\omega)] M_{jn}^{l} \frac{d\omega}{2\pi}, \tag{2.40}$$

$$\Sigma_{mn}^{H,r} = -iM_{mn}^{i} \int D_{0,ij}^{r}(\omega = 0) M_{kl}^{j} \int G_{0,kl}^{<}(\varepsilon) \frac{d\omega}{2\pi}, \tag{2.41}$$

where labels F and H represent the Fock- and Hartree-term. The less Hartree self-energy is zero. The self-energies for vibrations take the form

$$\Pi_{mn}^{<}(\omega) = -iM_{lk}^{m} \int G_{0,ki}^{<}(\varepsilon) G_{0,jl}^{>}(\varepsilon - \omega) M_{ij}^{n} \frac{d\varepsilon}{2\pi}, \tag{2.42}$$

$$\Pi_{mn}^{r}(\omega) = -iM_{lk}^{m} \int [G_{0,ki}^{r}(\varepsilon) G_{0,jl}^{<}(\varepsilon - \omega) + G_{0,ki}^{<}(\varepsilon) G_{0,jl}^{a}(\varepsilon - \omega)] M_{ij}^{n} \frac{d\varepsilon}{2\pi}. \tag{2.43}$$

In the above equations, the Green's function with subscript 0 means whose without including e-vib interaction. Based on Dyson equation, the full Green's functions for electrons and vibrations can be written as

$$G^{r}(\varepsilon) = G_{0}^{r}(\varepsilon) + G_{0}^{r}(\varepsilon)[\Sigma^{H,r}(\varepsilon) + \Sigma^{F,r}(\varepsilon)] G^{r}(\varepsilon), \tag{2.44}$$

$$G^{<}(\varepsilon) = G^{r}(\varepsilon)[\Sigma^{F,<}(\varepsilon) + \Sigma_{L}^{<}(\varepsilon) + \Sigma_{R}^{<}(\varepsilon)] G^{a}(\varepsilon), \tag{2.45}$$

$$D^{r}(\omega) = D_{0}^{r}(\omega) + D_{0}^{r}(\omega) \Pi^{r}(\omega) D^{r}(\omega), \tag{2.46}$$

$$D^{<}(\omega) = D^{r}(\omega)[\Pi^{<}(\omega) + \Pi_{L}^{<}(\omega) + \Pi_{R}^{<}(\omega)] D^{a}(\omega). \tag{2.47}$$

This is the so-called Born approximation (BA). The SCBA can be obtained by replacing all the bare Green's functions G_0 and D_0 in Eqs. (2.39)–(2.43) with the full G and D (the D_0 in the expression of $\Sigma^{H,r}$ does not need to be replaced).

One important application of the above formula is to study the IETS of molecular junctions, especially the molecular vibration spectra. Historically, Jaklevic and Lambe first measured the IETS spectra of molecules embedded in the metal–oxide–metal tunnel junctions in the 1966s [38]. Afterward, W. Ho's group applied this technique in the characterization of single molecule on metal surface using STM [39]. Later, it has proven to be an extremely useful tool to study single molecular junctions. Unlike the STM measurement, the atomic details of the single molecule between two metal electrodes are unknown in molecular junction made from break-junction technique. The vibrational spectrum thus measured can serve as a fingerprint of the molecule. By comparing with numerical calculations, it is possible to deduce the possible molecular conformation of the junction from electrical measurement.

In fact, the underlying physics of the IETS is very simple. An inelastic channel opens when the applied voltage exceeds the vibrational energy. This opening of additional channel can be seen in the second-order differential conductance (d^2I/d^2V). Normally, a sharp peak appears at the threshold bias, although dips or other complicated line shapes are also possible. Based on the above theory, practical numerical schemes have been developed to study IETS of single molecular junctions [34, 40]. The second derivative of the current near the vibrational threshold can be expressed as

$$\partial_V^2 I(V) = \sum_\lambda \gamma_\lambda \partial_V^2 I^{\mathrm{sym}}(V, \hbar\omega_\lambda, T) + \kappa_\lambda \partial_V^2 I^{\mathrm{asym}}(V, \hbar\omega_\lambda, T), \tag{2.48}$$

with two universal functions (Figure 2.9)

$$I^{\mathrm{sym}} = \frac{e}{2\pi\hbar}\sum_{\sigma=\pm}\sigma(\hbar\omega_\lambda + \sigma\mathrm{eV})\left(\coth\frac{\hbar\omega_\lambda}{2k_BT} - \coth\frac{\hbar\omega_{\lambda+\sigma eV}}{2k_BT}\right), \tag{2.49}$$

$$I^{\mathrm{asym}} = \frac{e}{\hbar}\int_{-\infty}^{+\infty}\frac{d\varepsilon}{2\pi}[f(\varepsilon) - f(\varepsilon - \mathrm{eV})] \times \mathscr{H}_{\varepsilon'}\{f(\varepsilon' + \hbar\omega_\lambda) - f(\varepsilon' - \hbar\omega_\lambda)\}(\varepsilon), \tag{2.50}$$

$f(\varepsilon)$ is the Fermi–Dirac distribution. The symbol $\mathscr{H}$ stands for Hilbert transform, defined as $\mathscr{H}_{\varepsilon'}\{f(\varepsilon')\}(\varepsilon) = \frac{1}{\pi}\mathscr{P}\int\frac{f(\varepsilon')}{\varepsilon-\varepsilon'}d\varepsilon'$. The system-dependent information is contained in the two coefficients γ_λ and κ_λ. They can be expressed from the Green's functions [36]

$$\gamma_\lambda = \mathrm{Tr}[M^\lambda \widetilde{A}_L(\mu_L) M^\lambda A_R(\mu_R)] + \mathrm{Im}\, B_\lambda, \tag{2.51}$$

$$\kappa_\lambda = 2\,\mathrm{Re}\, B_\lambda, \tag{2.52}$$

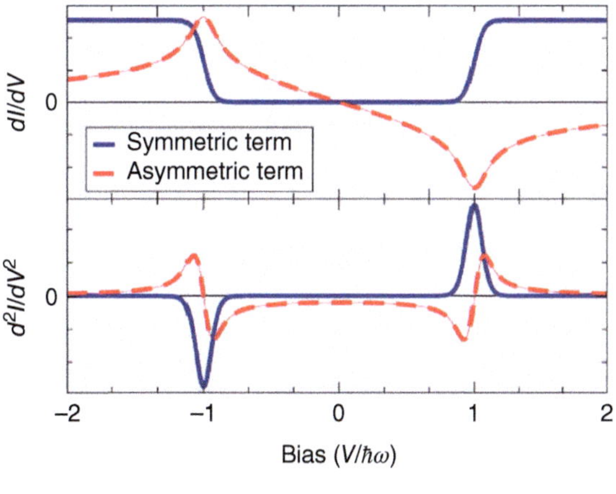

Figure 2.9 Universal functions giving the phonon contribution to the current. The dI/dV and d^2I/dV^2 spectroscopy are shown for one phonon mode with the bias in units of the phonon energy at a temperature $k_BT = 0.025\hbar\omega$. Source: Paulsson et al. [40]. 2005, American Physical Society.

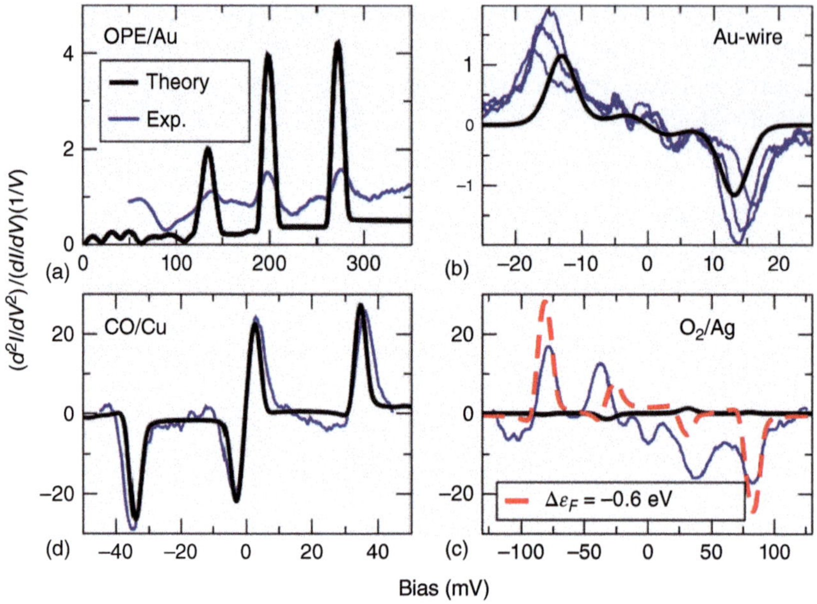

Figure 2.10 Calculated (black lines) and experimental (blue lines) IETS. (a) Oligophenylethylene (OPE) molecule bonding to Au(111) electrode. (b) Au chain connected to Au(100) leads. (c) O_2 molecule on Ag(110). (d) CO molecule on Cu(111). In case (c) the Fermi energy has been shifted manually to match the experiment (dashed line). For the STM configurations (c) and (d), the calculated IETS is compared with a rescaled d^2I/dV^2. Source: Paulsson et al. [35]. 2008, American Physical society.

with

$$\begin{aligned} B_\lambda &= \mathrm{Tr}[M^\lambda A_R(\mu_L)\Gamma_L(\mu_L)G^r(\mu_L)M^\lambda A_R(\mu_R) \\ &\quad - M^\lambda G^a(\mu_R)\Gamma_L(\mu_R)A_R(\mu_R)M^\lambda A_L(\mu_L)], \\ A_\alpha &= G^r\Gamma_\alpha G^a, \ \widetilde{A}_\alpha = G^a\Gamma_\alpha G^r. \end{aligned} \tag{2.53}$$

We have set eV $= \mu_L - \mu_R$. These results are derived base d on an expansion over M to the second order. These results have been used to interpret recent experimental results on near resonant IETS [37, 41]. For off resonant case, all the electronic properties can be calculated at the equilibrium Fermi level to reduce the numerical calculations. In Figure 2.10, theoretical results are compared to experiments for four cases [35]. Good agreement between the experimental and numerical results is obtained.

2.4.1.2 Strong-Coupling Regime

In this regime, the strong e-vib coupling has much larger effect on the transport property. Nonperturbative methods are needed to describe the interaction. In general, a polaron transformation, i.e. the Lang–Firsov (LF) transformation [42], is performed first on the Hamiltonian. It is defined as

$$\widetilde{H} = e^S H e^{-S}, \tag{2.54}$$

with

$$S = m(a_\lambda^\dagger - a_\lambda)d^\dagger d. \tag{2.55}$$

We have included only one vibrational mode and one electronic state here. If there are more than one, we need to sum over all the electronic states and vibrational modes. We note that, this kind of transformation only works for Holstein-type on-site e-vib interaction, which means $M_{ij}^\lambda = 0$ if $i \neq j$ and $M_{ii}^\lambda = m$. The effect of the transformation is as follows:

$$\widetilde{d} = de^{m(a_q - a_q^\dagger)}, \tag{2.56}$$

$$\widetilde{d}^\dagger = d^\dagger e^{-m(a_\lambda - a_\lambda^\dagger)}, \tag{2.57}$$

$$\widetilde{a}_\lambda = a_\lambda - md^\dagger d, \tag{2.58}$$

$$\widetilde{a}_\lambda^\dagger = a_\lambda^\dagger - md^\dagger d. \tag{2.59}$$

The electrode operators are not affected by the transformation. After this transformation, the explicit e-vib coupling term $m(a_\lambda^\dagger + a_\lambda)d^\dagger d$ disappears. Instead, the tunneling coupling between the electrodes and the molecule becomes vibrational-state dependent, due to the first and second equation, i.e. $d^\dagger c_j \to d^\dagger c_j\, e^{-m(a_\lambda - a_\lambda^\dagger)}$. This means that the electron tunneling between the molecule and the electrode involves vibrational-assisted process, through absorption or emission of vibrational quanta. This needs to be taken into account when calculating the transition matrix element. Normally, one performs a direct decoupling of the electronic and vibronic degrees of freedom, and results in the well-known the Franck–Condon (FC) matrix elements between different vibrational states [43–45]. Similarly, in the NEGF method, the electron Green's function transforms into new forms given by

$$\langle d(t)d^\dagger(t')\rangle \to \langle d(t)e^{m(a_\lambda(t) - a_\lambda^\dagger(t))}d^\dagger(t')e^{-m(a_\lambda(t') - a_\lambda^\dagger(t'))}\rangle. \tag{2.60}$$

It is calculated as

$$\langle d(t)e^{m(a_\lambda(t) - a_\lambda^\dagger(t))}d^\dagger(t')e^{-m(a_\lambda(t') - a_\lambda^\dagger(t'))}\rangle \approx \langle d(t)d^\dagger(t')\rangle\langle e^{m(a_\lambda(t) - a_\lambda^\dagger(t))}e^{-m(a_\lambda(t') - a_\lambda^\dagger(t'))}\rangle. \tag{2.61}$$

Several interesting effects have been predicted or observed in this regime. Among them is the FC blockade [43, 46–48]. The FC effect is a result of strong e-vib coupling. The presence of an extra electron on the molecular orbital leads to a finite displacement along the vibrational mode that couples to this orbital (Figure 2.11). This vibrational shift changes the overlap integral between different vibrational states, due to the spatial separation between them. As a result, the FC matrix elements between low-lying vibrational states are suppressed. This could lead to a significant current suppression at low bias voltages in transport through a single molecular junction, termed FC blockade. A large Fano factor in the current noise is also expected in this regime [43]. Since its prediction in 2005 (Figure 2.12), several experimental results have confirmed this physics in single molecular junctions [49, 50].

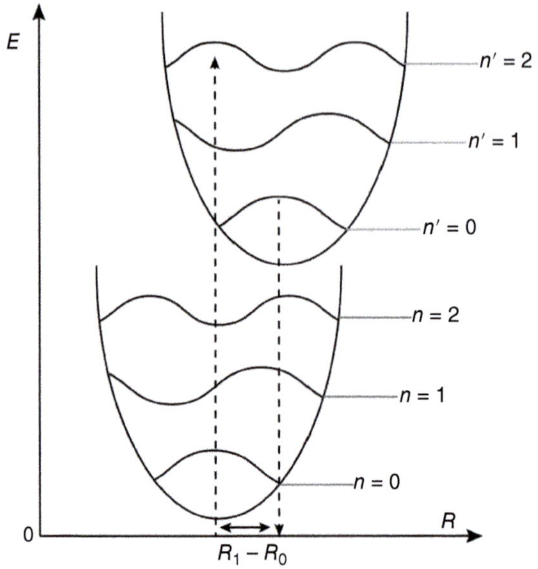

Figure 2.11 Schematic the FC principle. The change of electronic occupation induces a shift of the vibrational mode. This changes the wavefunction overlap between different vibrational states.

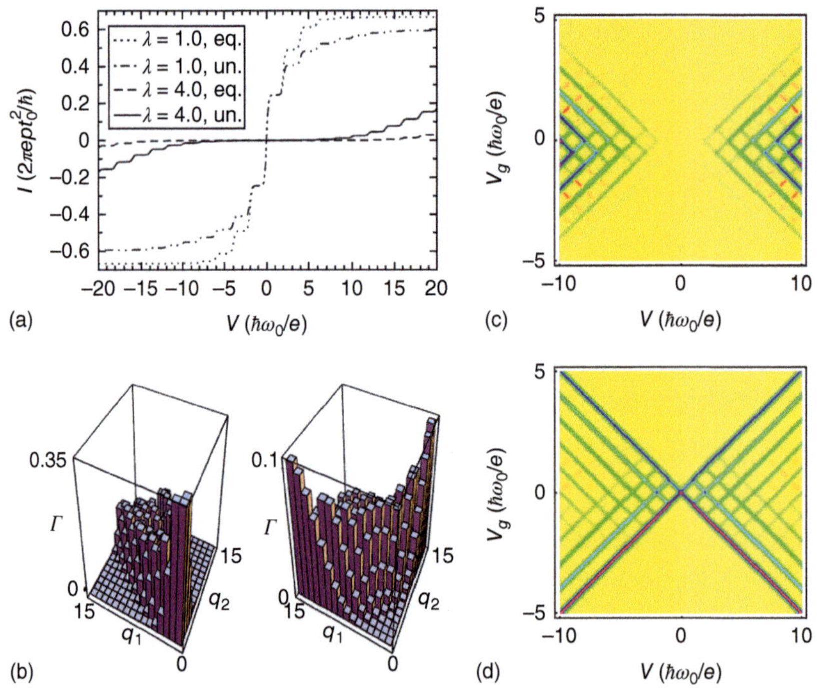

Figure 2.12 (a) The current vs. voltage for weak and strong e-vib coupling ($\lambda = 1$ and $\lambda = 4$) calculated with equilibrated and unequilibrated vibrations, respectively. (b) The Franck–Condon rates of vibration transitions from q_1 to q_2 are plotted at $\lambda = 1$ (left) and $\lambda = 4$ (right). (c) The dI/dV vs. the gate voltage V_g and voltage V calculated for unequilibrated vibrations at $\lambda = 4$. (d) Similar to (c), but for $\lambda = 1$. Source: Koch and von Oppen [43]. 2005, American Physical Society.

2.4.2 Electron–Electron Interaction

2.4.2.1 Coulomb Blockade

The Coulomb interaction between electrons can have significant effect on the electron transport across a single molecular junction. This is true especially when the coupling to the source and drain electrodes is weak. In this case, the molecule has well-defined charging state. When the molecular orbital is already occupied by one electron, there will be an energy cost for the second electron to hop in. The energy is supplied by the applied voltage bias or a third gate electrode. Otherwise the transport is blocked. This is termed Coulomb blockade. Typical feature of the Coulomb blockade transport is shown in Figure 2.13. When the electronic state has already been occupied by one electron, it takes an addition energy U_a for the second electron to enter. This addition energy can be provided by either the source–drain or the gate electrode. The phenomenon of Coulomb blockade has been observed in many small systems in which an "island" is nested within the tunnel junction. The addition energy for typical island can be expressed as

$$U_a = \frac{e^2}{2C}, \tag{2.62}$$

where e is the electron charge, C is the capacitance of the structure, and U_a is the charging energy. The threshold voltage can be written as

$$V_t = \frac{U_a}{e} = \frac{e}{2C}. \tag{2.63}$$

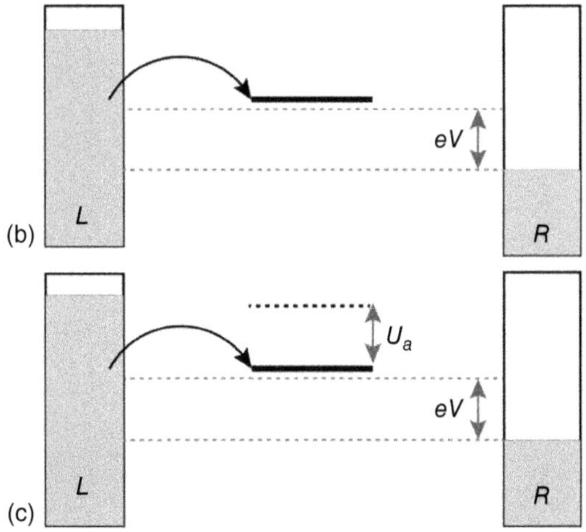

Figure 2.13 A molecular junction for indicating the behavior of Coulomb blockade. (a) No voltage bias case. (b) Electron can transfer from one of the leads to molecule in presence of an external bias. (c) The addition energy U_a is needed for a second electron to hop into the molecule when the molecular level is already occupied by one electron.

For single molecule junction, the capacitance is small. As a result, the charging energy is much larger than semiconductor quantum dot.

From a theoretical point of view, to consider Coulomb blockade, it is necessary to include the e–e interaction in the Hamiltonian

$$H_U = \sum_{i,j} U_{ij} d_i^\dagger d_i d_j^\dagger d_j, \tag{2.64}$$

where U_{ij} represents e–e interaction, i and j label the electronic states. Combining Eq. (2.64) with the Hamiltonian of the electron, one can study the transport properties of a single junction with consideration of the e–e interaction based on the method introduced in Section 2.2. Since the molecule–electrode coupling is weak, one can start from the isolated molecule and treat the molecule–electrode coupling as a perturbation. For a single molecular orbital at energy ε_0, we have the following simple Hamiltonian for the molecule

$$H_D = \sum_{\sigma=\uparrow,\downarrow} \varepsilon_{0\sigma} d_\sigma^\dagger d_\sigma + U d_\uparrow^\dagger d_\uparrow d_\downarrow^\dagger d_\downarrow. \tag{2.65}$$

We have to include the spin index here. To use the Meir–Wingreen formula, we need to calculate the Green's function of the molecule. Simple equation of motion calculation shows that for the retarded Green's function

$$i\hbar \frac{\partial}{\partial t} g_\sigma^r(t-t') = \varepsilon_{0\sigma} g_\sigma^r(t-t') + U g_\sigma^{r(2)}(t-t') + \delta(t-t'), \tag{2.66}$$

$$i\hbar \frac{\partial}{\partial t} g_\sigma^{r(2)}(t-t') = (\varepsilon_{0\sigma} + U) g_\sigma^{r(2)}(t-t') + n_{\bar{\sigma}}\, \delta(t-t'). \tag{2.67}$$

We have defined one higher order Green's functions

$$g_\sigma^{r(2)}(t-t') = -\frac{i}{\hbar} \Theta(t-t') \langle d_\sigma(t) d_{\bar{\sigma}}^\dagger(t) d_{\bar{\sigma}}(t) d_\sigma^\dagger(t') \rangle. \tag{2.68}$$

Solving for g_σ^r, we have in the frequency space,

$$g_\sigma^r(\varepsilon) = \frac{n_{\bar{\sigma}}}{\varepsilon + i0^+ - \varepsilon_{0\sigma} - U} + \frac{1-n_{\bar{\sigma}}}{\varepsilon + i0^+ - \varepsilon_{0\sigma}}. \tag{2.69}$$

Here, $\bar{\sigma}$ means electron spin different from σ. To include coupling to the electrodes, we introduce the electron self-energy. The final retarded Green's function is obtained from the Dyson equation

$$G_\sigma^r(\varepsilon) = \frac{1}{(g_\sigma^r(\varepsilon))^{-1} - \Sigma_L^r(\varepsilon) - \Sigma_R^r(\varepsilon)}. \tag{2.70}$$

The electron occupation is expressed in terms of the lesser Green's function

$$n_{\bar{\sigma}} = \langle d_{\bar{\sigma}}^\dagger d_{\bar{\sigma}} \rangle = -\frac{i}{2\pi} \int_{-\infty}^{+\infty} d\varepsilon\, G_{\bar{\sigma}}^<(\varepsilon), \tag{2.71}$$

which is given by

$$G_\sigma^<(\varepsilon) = G_\sigma^r(\varepsilon) \Sigma_\sigma^<(\varepsilon) G_\sigma^a(\varepsilon). \tag{2.72}$$

These equations have to be solved self-consistently, since G_σ^r depends on $n_{\bar{\sigma}}$, while $n_{\bar{\sigma}}$ depends on $G_{\bar{\sigma}}^<$. The electrical current can then be obtained from the Green's

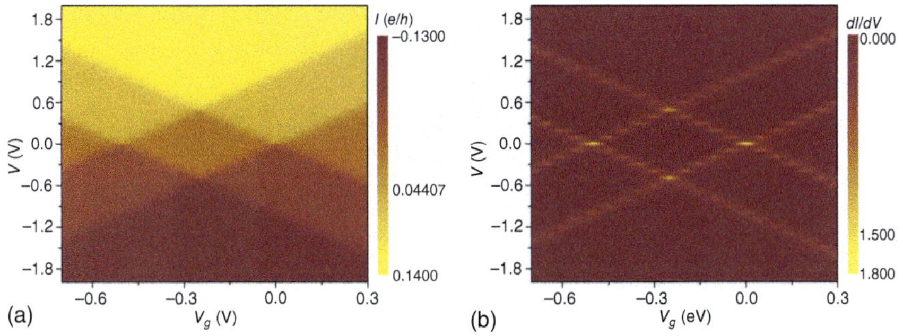

Figure 2.14 (a) Contour plot of the electrical current (I) as a function of the source–drain voltage (V) and gate voltage V_g. (b) The corresponding curve of differential conductance dI/dV, showing the Coulomb diamond, which is a typical feature of the Coulomb blockade transport. The calculated parameters are taken as: $k_BT = 2.6$ meV, $U = 0.5$ eV, $\Gamma_L = \Gamma_R = 0.02$ eV, $\varepsilon_{0\sigma} = 0$.

function using Eq. (2.13). The dependence of the Green's function and thus the electrical current on electron occupation is obvious from the above analysis.

Based on Eqs. (2.65)–(2.72), we calculate the current (I) and the differential conductance (dI/dV) vs. voltage (V) and gate voltage (V_g), as shown in Figure 2.14. It is found that the current is significantly suppressed around the zero bias voltage region, which determined by the gate voltage (see Figure 2.14a). The corresponding dI/dV curve is displayed in Figure 2.14b, the red areas indicate the regions of no current, while the yellow areas correspond to the current step, which is the signature of Coulomb blockade, as explained previously. The bright lines form the so-called Coulomb diamond.

Coulomb blockade effects have been observed in single molecular junctions experimentally [51–54]. One typical example is shown in Figure 2.15, where an individual C_{60} molecule is connected to gold electrodes. The current–voltage curves were measured at $T = 1.5$ K for different gate voltages. As can be seen, the current was strongly suppressed in the zero bias voltage region and showed a sudden-turned on of the current at higher voltage regions. This is attribute to the Coulomb blockade, i.e. an addition energy is required to drag an electron that transfer between the electrodes and C_{60}.

2.4.2.2 Kondo Effect

Electron–electron interaction may also lead to strong correlations between electrons and affect the transport property of single molecular junction. The Kondo effect is one of the most important examples. Kondo effect manifests in normal metals as a low temperature resistance peak. This low-temperature anomaly in resistance is caused by exchange interaction of spins from the localized magnetic impurity with the itinerant conduction electrons. At low temperature, the spin magnetic moment is screened by itinerant electrons through the exchange interaction. The impurity serves as a scattering center and changes the resistance of the metal. This is now termed Kondo effect, since Kondo gave the first hint to explain the

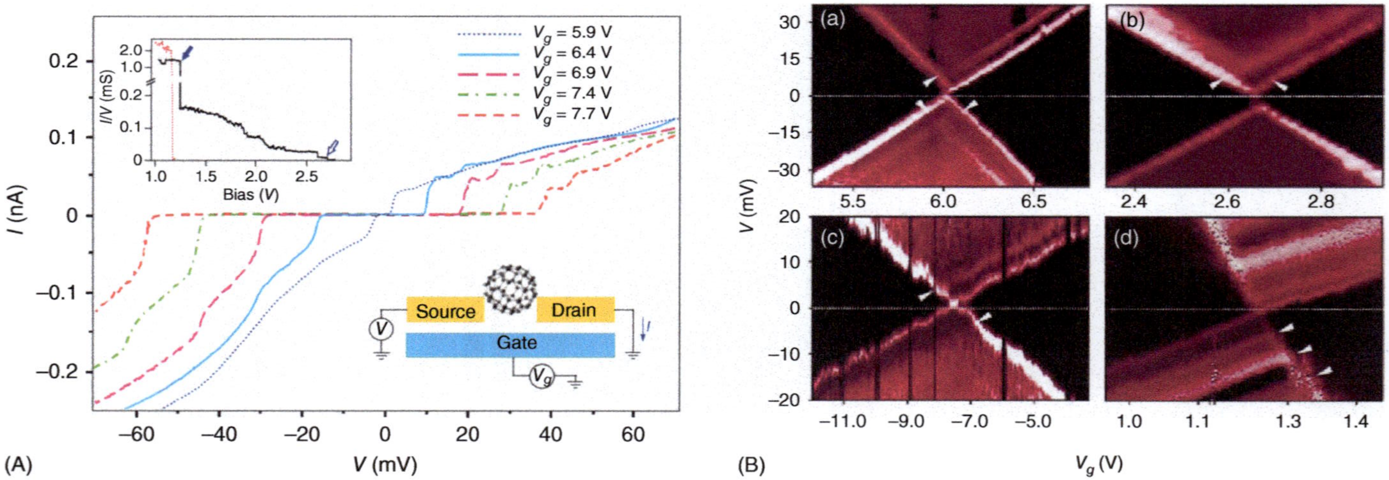

Figure 2.15 (A) The current–voltage curves for different values gate voltages obtained from a single-C_{60} transistor. Upper inset, a large voltage is applied to the source and drain electrodes. Lower inset, an idealized diagram for this system. (B) Two-dimensional contour plot of the differential conductance showing the Coulomb blockade for several C_{60} junctions. The dark area represents low conductance, while the bright lines show high conductance. Source: Park et al. [51]. 2000, Springer nature.

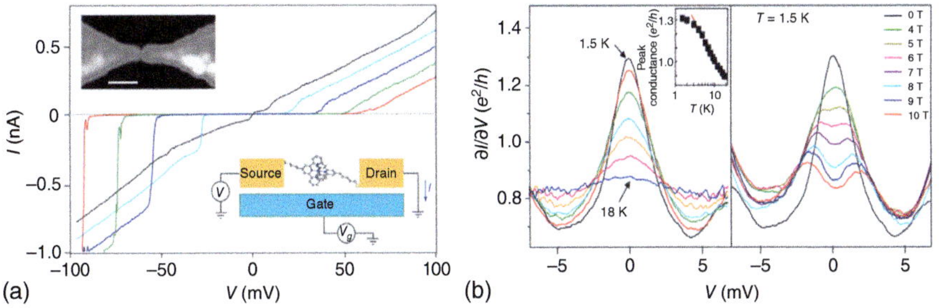

Figure 2.16 (a) The current–voltage ($I-V$) curves for different values of gate voltages, the insert is the experimental setup. (b) The temperature and magnetic field dependence of differential conductance ($\partial I/\partial V$). Source: Park et al. [52]. 2002, Springer nature.

phenomenon by using perturbation theory [55]. The complete understanding of the problem relies on the development of renormalization group theory of Wilson [56].

With the continuous development of nanotechnology, Kondo effect has also been revealed in single molecular junctions [52, 57–59] and quantum dots [60, 61] experimentally. In these systems, a net magnetic moment from unpaired electron in molecule or quantum dot can be regarded as a magnetic impurity. The exchange interaction between the impurity spin and the conducting electrons gives rise to a virtual state at the Fermi level. It manifests as a conductance maximum at zero bias, termed Kondo resonance.

In experiment, Coulomb blockade and Kondo effect often show up in pairs. For example, Park et al. reported the observation of both the Coulomb blockade and Kondo effect in a single-atom transistor coupled to a gate and two metallic nanogap electrodes [52]. As can be seen from Figure 2.16, a threshold voltage displays in the current–voltage (I–V) curve. This is interpreted as Coulomb blockade effect. When there is unpaired electron in the molecule, a zero-bias peak is also observed in this junction, which is attributed to the Kondo effect originated from the Co ion. This is further proved by checking the temperature and magnetic field dependence of the peak. A similar Kondo resonance in a single divanadium molecule transistor also has been observed by Liang et al. [59].

On the other hand, spin degrees of freedom will play a key role in determining the transport behaviors of molecular junctions when ferromagnetic electrodes are introduced [52, 62–65]. For example, Pasupathy et al. measured the transport properties of a C_{60} molecule coupled to two ferromagnetic nickel electrodes. When the magnetic moments of the electrodes are parallel (P), the suppression of Kondo anomaly in conductance can be observed, while the phenomenon does not exist for antiparallel (AP) configuration of the electrodes [62]. This is attribute to the presence of the effective exchange field induced by the coupling between the molecule and the ferromagnetic electrodes, which can lead to a nonzero spin splitting of Kondo peak in the conductance in the P configuration, whereas the splitting does not exist for AP configuration because the coupling of the molecule and the electrodes is independent of the spin orientation in this case.

As mentioned earlier, the master equation can well deal with the electron transport through the molecular junctions when the molecule is weakly coupled to the electrodes. In this case, the e–e interaction is larger than molecule–electrodes coupling. However, simple standard rate equation ignoring the electronic correlation between molecule and leads cannot account for the Kondo effect. One needs to go back to the original DM formalism to include correlation effect between electrons in the electrode and the molecule. Equation of motion method within the NEGF formalism can also be used for this purpose. Higher order Green's functions between the molecular and the electrode need to be introduced through the equation of motion. Different truncation schemes have been introduced in the literature [66–69]. Please refer to the original literature for the details.

2.5 Thermoelectric Transport

The measurement of current–voltage (*I–V*) characteristics is a powerful tool to understand the transport properties of molecular junctions. However, still, some important information of the molecule cannot be obtained directly from electrical transport measurement. For example, distinguishing whether electronic transport is dominated by HOMO or LUMO of the molecule is a challenging task due to the uncertainties in the microscopic details of the contacts, also due to the different electronic structure of the molecule and metal electrodes. It has been shown that the sign of thermoelectric Seebeck coefficient (*S*) can be used to resolve this issue [70]. The Seebeck coefficient is defined as

$$S = -\frac{\Delta V}{\Delta T}. \tag{2.73}$$

Here, ΔT is the temperature difference between the two electrodes and ΔV is the generated voltage difference due to the temperature bias. $S > 0$ indicates that the Fermi energy is nearby the molecular HOMO level, while $S < 0$ indicates the Fermi energy is nearby the LUMO level. In addition, studying the thermoelectric transport in molecular junctions is a key step to develop thermoelectric devices based on molecules. Reddy et al. made the first measurement on the Seebeck coefficient of single molecular junction coupled to two gold electrodes [71]. By comparing the experimental measurement and the theoretical calculation on the Seebeck coefficient, they confirmed that the sign of the Seebeck coefficient can be used to clarify the type of the charge carrier, as shown in Figure 2.17. Apart from the Seebeck effect, other thermoelectric effects, including the Peltier effect, have also been studied in molecular junctions [72].

The energy conversion efficiency of thermoelectric materials is normally characterized by a dimensionless figure of merit $ZT = GS^2T/\kappa$, where G represents the electrical conductance, S is the Seebeck coefficient, T is the temperature, and κ is the thermal conductance including the phonon and electron contributions. In general, a large ZT can be achieved by reducing thermal conductivity while keeping Seebeck coefficient and electrical conductance intact. However, the Wiedemann–Franz law states that the three key factors that determine the ZT are not independent,

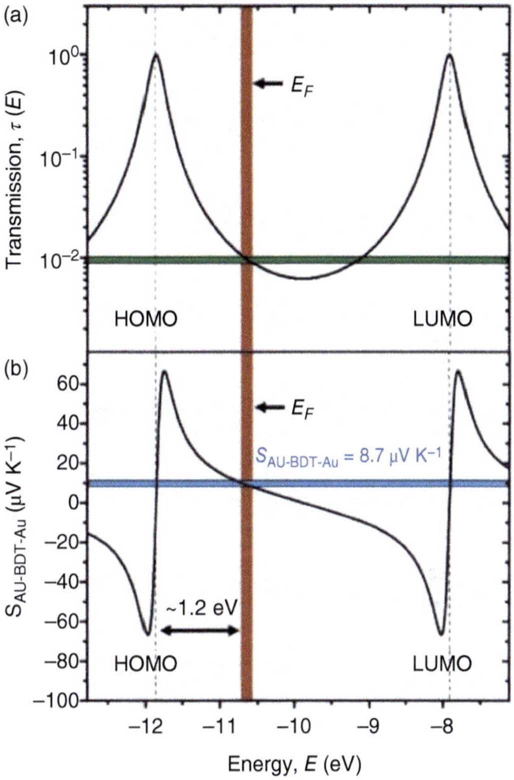

Figure 2.17 (a) Transmission function for a Au–benzenedithiol–Au junction. (b) The predicted Seebeck coefficient as a function of the relative position of the Fermi level E_F. The blue band is the measured value of Seebeck coefficient with 8.7 ± 2.1 mV K^{-1}. Source: Reddy et al. [71]. 2007, AAAS.

e.g. large electrical conductance is often accompanied by a large electron thermal conductance. Initially, the thermoelectric effect was studied in bulk materials, from metals to semiconductors. Nanostructured materials are proposed later to enhance the thermoelectric figure of merit by engineering the electronic density of states [73–75]. Of the various candidates, the single molecular junction has become one of promising materials for future high-performance thermoelectric materials. Additional details on this topic can be found in Section 7.3.2.

In the following, we will provide a theoretical framework to describe the thermoelectric transport in molecular junctions in the coherent transport regime. Currently, most of the experimental results are explained within this framework.

Considering a molecular junction (L–D–R) as mentioned above, our starting point is the celebrated Landauer formula

$$I = \frac{e}{h}\int_{-\infty}^{+\infty} d\varepsilon\ T(\varepsilon)[f_L(\varepsilon) - f_R(\varepsilon)], \tag{2.74}$$

$$I_E = \frac{1}{h}\int_{-\infty}^{+\infty} d\varepsilon\ \varepsilon T(\varepsilon)[f_L(\varepsilon) - f_R(\varepsilon)], \tag{2.75}$$

where I and I_E are the charge and energy currents, respectively. $f_{L(R)}$ is the Fermi–Dirac distribution function in the left (right) electrode.

We start from the fundamental thermodynamic quantity

$$dE = T\,dS - P\,dV + \mu\,dN, \tag{2.76}$$

with S, V, N the extensive properties, entropy, volume, electron number; and T, P, μ the intensive properties, temperature, pressure and chemical potential. Then we have

$$I_E = TI_S + \mu\,I, \tag{2.77}$$

if we assume the volume (V) of our system does not change. The heat current is

$$I_h = TI_S = I_E - \mu I = \frac{e}{h}\int_{-\infty}^{+\infty} d\varepsilon\,(\varepsilon-\mu)T(\varepsilon)[f_L(\varepsilon) - f_R(\varepsilon)]. \tag{2.78}$$

Assuming small electrical and temperature bias, we can do a linear expansion of the Fermi distribution function

$$f(\mu, T) \approx f(\mu_0, T_0) + \frac{\partial f}{\partial T}\delta T + \frac{\partial f}{\partial \mu}\delta\mu = f(\mu_0, T_0) - \frac{\varepsilon-\mu}{T}\frac{\partial f}{\partial \varepsilon}\delta T - \frac{\partial f}{\partial \varepsilon}\delta\mu. \tag{2.79}$$

The electrical current takes the form

$$\delta I = \frac{e}{h}\int_{-\infty}^{+\infty} d\varepsilon\;T(\varepsilon)\left[(\varepsilon-\mu)\frac{\delta T}{T} + \delta\mu\right]\left(-\frac{\partial f}{\partial \varepsilon}\right). \tag{2.80}$$

In the same way, the heat current is

$$\delta I_h = \frac{1}{h}\int_{-\infty}^{+\infty} d\varepsilon\,(\varepsilon-\mu)T(\varepsilon)\left[(\varepsilon-\mu)\frac{\delta T}{T} + \delta\mu\right]\left(-\frac{\partial f}{\partial \varepsilon}\right). \tag{2.81}$$

Writing them in matrix form, we have

$$\begin{pmatrix}\delta I/e \\ \delta I_h\end{pmatrix} = \begin{pmatrix}L_0 & L_1 \\ L_1 & L_2\end{pmatrix}\begin{pmatrix}\delta\mu \\ \delta T/T\end{pmatrix}. \tag{2.82}$$

We have defined the transport coefficients

$$L_n = \frac{1}{h}\int_{-\infty}^{+\infty} d\varepsilon\,(\varepsilon-\mu)^n\,T(\varepsilon)\left(-\frac{\partial f}{\partial \varepsilon}\right). \tag{2.83}$$

In principle, if we know the transmission spectrum, from Eq. (2.82) we can calculate all the linear thermoelectric transport coefficients within the noninteracting electron approximation. Note that the derivative of Fermi function is an even function about the chemical potential, with width on the order of $k_B T$ (room temperature corresponds to around 25 meV). So we only need to know the transmission spectrum in the small energy range near μ. The electrical conductance G, electron thermal conductance κ_e, Seebeck coefficient S, Peltier coefficient P, and thermoelectric figure of merit ZT are

$$G \equiv \frac{\partial I}{\partial V} = e^2 L_0, \tag{2.84}$$

$$\kappa_e \equiv \left(\frac{\partial I_h}{\partial T}\right)_{I=0} = \frac{1}{T}\left(-\frac{L_1^2}{L_0} + L_2\right), \tag{2.85}$$

$$S \equiv -\frac{\partial V}{\partial T} = \frac{L_1}{eL_0T}, \tag{2.86}$$

$$P = ST, \tag{2.87}$$

$$ZT = \frac{GS^2T}{\kappa}. \tag{2.88}$$

Note that the electron thermal conductance is defined as the derivative of heat current over temperature difference when there is no electrical current. The Seebeck coefficient can be further simplified at low temperature, known as Mott formula

$$S = \frac{\pi^2}{3e}k_B^2T\left(\frac{\partial \ln \tau(\varepsilon)}{\partial \varepsilon}\right)_\mu. \tag{2.89}$$

We can learn from this analysis that (i) the sign and magnitude of the Seebeck coefficient is determined by the slope of the logarithmic of the transmission around Fermi level. In order to get a large S, we need the transmission changes sharply near μ. (ii) One only needs the transmission coefficient to calculate the thermoelectric properties of the molecular junction.

2.6 First-Principles Simulations of Transport in Molecular Devices

Many experimental technologies have been developed to study the transport property of single molecules. However, the experiments are often hard to interpretative and technically demanding. Generally, it is difficult to characterize and control the realistic contact details between the molecules and the electrodes with atomic accuracy. Theoretical modeling of current–voltage characteristics of realistic molecular devices is needed to complement the experimental observations and plays an important role in progressing the field.

It is challenging to develop general methods for quantitative simulations of realistic devices taking into account the atomistic details. The main difficulties lie in the following aspects: (i) molecules are contacted to two (or even more) semi-infinite electrodes forming realistic molecular junction. The unit cell of the simulation normally includes hundreds of atoms. It is a huge burden from the computational viewpoint. (ii) The typical device geometry is neither isolated nor periodic. We must deal with an open system where the molecule can exchange particles with the electrodes. (iii) When a voltage bias is applied across the molecular junction to make a current flow, the system is driven to a nonequilibrium states. We should adopt nonequilibrium statistical mechanics to calculate the properties of this system. Some methods have been developed to deal with these problems. The most popular one is the NEGF formalism combined with DFT, which provides a practical scheme and has been implemented in several packages [76, 77]. These packages have been successfully used to interpret the experimental measurement.

2.7 Conclusions

In this chapter, we have introduced the basic theoretical tools to understand the transport properties of single molecular junction. The effect of e–e, e-vib interaction may have important effect on the transport characteristics. They can be used as a spectroscopy tool to characterize the molecular junction. We have focused on the principles of the theory and omitted the technical details. These details can be filled in from the original literature. It is hoped that the materials presented here are accessible to experimentalists and can be useful in the analysis of the experimental results.

References

1 Landauer, R. (1957). Spatial variation of currents and fields due to localized scatterers in metallic conduction. *IBM J. Res. Dev.* 1 (3): 223–231.

2 Büttiker, M. (1986). Four-terminal phase-coherent conductance. *Phys. Rev. Lett.* 57 (14): 1761–1764.

3 Meir, Y. and Wingreen, N.S. (1992). Landauer formula for the current through an interacting electron region. *Phys. Rev. Lett.* 68 (16): 2512–2515.

4 Harbola, U. and Mukamel, S. (2006). Nonequilibrium superoperator GW equations. *J. Chem. Phys.* 124 (4): 044106.

5 Schoeller, H. and Schön, G. (1994). Mesoscopic quantum transport: resonant tunneling in the presence of a strong Coulomb interaction. *Phys. Rev. B* 50 (24): 18436–18452.

6 König, J., Schoeller, H., and Schön, G. (1997). Cotunneling at resonance for the single-electron transistor. *Phys. Rev. Lett.* 78 (23): 4482–4485.

7 Splettstoesser, J., Governale, M., König, J., and Fazio, R. (2006). Adiabatic pumping through a quantum dot with coulomb interactions: a perturbation expansion in the tunnel coupling. *Phys. Rev. B* 74 (8): 085305.

8 Anderson, P.W. (1961). Localized magnetic states in metals. *Phys. Rev.* 124 (1): 41–53.

9 Newns, D.M. (1969). Self-consistent model of hydrogen chemisorption. *Phys. Rev.* 178 (3): 1123–1135.

10 Ke, S.-H., Yang, W., and Baranger, H.U. (2008). Quantum-interference-controlled molecular electronics. *Nano Lett.* 8 (10): 3257–3261.

11 Markussen, T., Stadler, R., and Thygesen, K.S. (2010). The relation between structure and quantum interference in single molecule junctions. *Nano Lett.* 10 (10): 4260–4265.

12 Sangtarash, S., Sadeghi, H., and Lambert, C.J. (2018). Connectivity-driven bi-thermoelectricity in heteroatom-substituted molecular junctions. *Phys. Chem. Chem. Phys.* 20 (14): 9630–9637.

13 Vazquez, H., Skouta, R., Schneebeli, S. et al. (2012). Probing the conductance superposition law in single-molecule circuits with parallel paths. *Nat. Nanotechnol.* 7 (10): 663–667.

14 Guédon, C.M., Valkenier, H., Markussen, T. et al. (2012). Observation of quantum interference in molecular charge transport. *Nat. Nanotechnol.* 7 (5): 305–309.

15 Cardamone, D.M., Stafford, C.A., and Mazumdar, S. (2006). Controlling quantum transport through a single molecule. *Nano Lett.* 6 (11): 2422–2426.

16 May, V. (2004). *Charge and Energy Transfer Dynamics in Molecular Systems.* Berlin: Wiley-VCH.

17 Luo, L. and Frisbie, C.D. (2010). Length-dependent conductance of conjugated molecular wires synthesized by stepwise "Click" chemistry. *J. Am. Chem. Soc.* 132 (26): 8854–8855.

18 Ho Choi, S., Kim, B., and Frisbie, C.D. (2008). Electrical resistance of long conjugated molecular wires. *Science* 320 (5882): 1482–1486.

19 Choi, S.H., Risko, C., Delgado, M.C.R. et al. (2010). Transition from tunneling to hopping transport in long, conjugated oligo-imine wires connected to metals. *J. Am. Chem. Soc.* 132 (12): 4358–4368.

20 Yan, H., Bergren, A.J., McCreery, R. et al. (2013). Activationless charge transport across 4.5 to 22 nm in molecular electronic junctions. *Proc. Natl. Acad. Sci. U.S.A.* 110 (14): 5326–5330.

21 Asai, Y. and Fukuyama, H. (2005). Theory of length-dependent conductance in one-dimensional chains. *Phys. Rev. B* 72 (8): 085431.

22 Giese, B., Amaudrut, J., Köhler, A.-K. et al. (2001). Direct observation of hole transfer through DNA by hopping between adenine bases and by tunnelling. *Nature* 412 (6844): 318–320.

23 Selzer, Y., Cabassi, M.A., Mayer, T.S., and Allara, D.L. (2004). Thermally activated conduction in molecular junctions. *J. Am. Chem. Soc.* 126 (13): 4052–4053.

24 Smit, R.H.M., Untiedt, C., Rubio-Bollinger, G. et al. (2003). Observation of a parity oscillation in the conductance of atomic wires. *Phys. Rev. Lett.* 91 (7): 076805.

25 Kim, T.-S. and Hershfield, S. (2002). Even-odd parity effects in conductance and shot noise of metal--atomic-wire--metal (superconducting) junctions. *Phys. Rev. B* 65 (21): 214526.

26 Jauho, A.-P., Wingreen, N.S., and Meir, Y. (1994). Time-dependent transport in interacting and noninteracting resonant-tunneling systems. *Phys. Rev. B* 50 (8): 5528–5544.

27 Datta, S. (1995). *Electronic Transport in Mesoscopic Systems.* Cambridge University.

28 Haug, H. (1996). *Quantum Kinetics and Optics of Semiconductors.* Springer Verlag.

29 Ryndyk, D.A., Hartung, M., and Cuniberti, G. (2006). Nonequilibrium molecular vibrons: an approach based on the nonequilibrium Green function technique and the self-consistent Born approximation. *Phys. Rev. B* 73 (4): 045420.

30 Galperin, M., Ratner, M.A., and Nitzan, A. (2007). Molecular transport junctions: vibrational effects. *J. Phys. Condens. Matter* 19 (10): 103201.

31 Lü, J.T. and Wang, J.-S. (2007). Coupled electron and phonon transport in one-dimensional atomic junctions. *Phys. Rev. B* 76 (16): 165418.

32 Wang, J.-S., Wang, J., and Lü, J.T. (2008). Quantum thermal transport in nanostructures. *Eur. Phys. J. B* 62 (4): 381–404.

33 Frederiksen, T., Brandbyge, M., Lorente, N., and Jauho, A.-P. (2004). Inelastic scattering and local heating in atomic gold wires. *Phys. Rev. Lett.* 93 (25): 256601.

34 Frederiksen, T., Paulsson, M., Brandbyge, M., and Jauho, A.-P. (2007). Inelastic transport theory from first principles: methodology and application to nanoscale devices. *Phys. Rev. B* 75 (20): 205413.

35 Paulsson, M., Frederiksen, T., Ueba, H. et al. (2008). Unified description of inelastic propensity rules for electron transport through nanoscale junctions. *Phys. Rev. Lett.* 100 (22): 226604.

36 Lü, J.-T., Christensen, R.B., Foti, G. et al. (2014). Efficient calculation of inelastic vibration signals in electron transport: beyond the wide-band approximation. *Phys. Rev. B* 89 (8): 081405.

37 Guo, J., Lü, J.-T., Feng, Y. et al. (2016). Nuclear quantum effects of hydrogen bonds probed by tip-enhanced inelastic electron tunneling. *Science* 352 (6283): 321–325.

38 Jaklevic, R.C. and Lambe, J. (1966). Molecular vibration spectra by electron tunneling. *Phys. Rev. Lett.* 17 (22): 1139–1140.

39 Stipe, B.C. (1998). Single-molecule vibrational spectroscopy and microscopy. *Science* 280 (5370): 1732–1735.

40 Paulsson, M., Frederiksen, T., and Brandbyge, M. (2005). Modeling inelastic phonon scattering in atomic- and molecular-wire junctions. *Phys. Rev. B* 72 (20): 201101.

41 Song, H., Kim, Y., Jang, Y.H. et al. (2009). Observation of molecular orbital gating. *Nature* 462 (7276): 1039–1043.

42 Mahan, G.D. (2000). *Many Particle Physics*. New York: Springer Science + Business Media, LLC.

43 Koch, J. and von Oppen, F. (2005). Franck–Condon blockade and giant Fano factors in transport through single molecules. *Phys. Rev. Lett.* 94 (20): 206804.

44 Koch, J., von Oppen, F., and Andreev, A.V. (2006). Theory of the Franck–Condon blockade regime. *Phys. Rev. B* 74 (20): 205438.

45 Koch, J., Semmelhack, M., von Oppen, F., and Nitzan, A. (2006). Current-induced nonequilibrium vibrations in single-molecule devices. *Phys. Rev. B* 73 (15): 155306.

46 May, V. and Kühn, O. (2008). Optical field control of charge transmission through a molecular wire. II. Photoinduced removal of the Franck–Condon blockade. *Phys. Rev. B* 77 (11): 115440.

47 Lüffe, M.C., Koch, J., and von Oppen, F. (2008). Theory of vibrational absorption sidebands in the Coulomb-blockade regime of single-molecule transistors. *Phys. Rev. B* 77 (12): 125306.

48 Ryndyk, D.A., D'Amico, P., Cuniberti, G., and Richter, K. (2008). Charge-memory polaron effect in molecular junctions. *Phys. Rev. B* 78 (8): 085409.

49 Leturcq, R., Stampfer, C., Inderbitzin, K. et al. (2009). Franck–Condon blockade in suspended carbon nanotube quantum dots. *Nat. Phys.* 5 (5): 327–331.

50 Burzurí, E., Yamamoto, Y., Warnock, M. et al. (2014). Franck–Condon blockade in a single-molecule transistor. *Nano Lett.* 14 (6): 3191–3196.

51 Park, H., Park, J., Lim, A.K.L. et al. (2000). Nanomechanical oscillations in a single-C_{60} transistor. *Nature* 407 (6800): 57–60.

52 Park, J., Pasupathy, A.N., Goldsmith, J.I. et al. (2002). Coulomb blockade and the Kondo effect in single-atom transistors. *Nature* 417 (6890): 722–725.

53 Dorogi, M., Gomez, J., Osifchin, R. et al. (1995). Room-temperature Coulomb blockade from a self-assembled molecular nanostructure. *Phys. Rev. B* 52 (12): 9071–9077.

54 Porath, D., Bezryadin, A., de Vries, S., and Dekker, C. (2000). Direct measurement of electrical transport through DNA molecules. *Nature* 403 (6770): 635–638.

55 Kondo, J. (1964). Resistance minimum in dilute magnetic alloys. *Prog. Theor. Phys.* 32 (1): 37–49.

56 Wilson, K.G. (1975). The renormalization group: critical phenomena and the Kondo problem. *Rev. Mod. Phys.* 47 (4): 773–840.

57 Yu, L.H. and Natelson, D. (2004). The Kondo effect in C_{60} single-molecule transistors. *Nano Lett.* 4 (1): 79–83.

58 LeRoy, B.J., Lemay, S.G., Kong, J., and Dekker, C. (2004). Electrical generation and absorption of phonons in carbon nanotubes. *Nature* 432 (7015): 371–374.

59 Liang, W., Shores, M.P., Bockrath, M. et al. (2002). Kondo resonance in a single-molecule transistor. *Nature* 417 (6890): 725–729.

60 Schmid, J., Weis, J., Eberl, K., and v. Klitzing, K. (2000). Absence of odd-even parity behavior for Kondo resonances in quantum dots. *Phys. Rev. Lett.* 84 (25): 5824–5827.

61 Simmel, F., Blick, R.H., Kotthaus, J.P. et al. (1999). Anomalous Kondo effect in a quantum dot at nonzero bias. *Phys. Rev. Lett.* 83 (4): 804–807.

62 Pasupathy, A.N., Bialczak, R.C., Martinek, J. et al. (2004). The Kondo effect in the presence of ferromagnetism. *Science* 306 (5693): 86–89.

63 Martinek, J., Utsumi, Y., Imamura, H. et al. (2003). Kondo effect in quantum dots coupled to ferromagnetic leads. *Phys. Rev. Lett.* 91 (12): 127203.

64 Utsumi, Y., Martinek, J., Schön, G. et al. (2005). Nonequilibrium Kondo effect in a quantum dot coupled to ferromagnetic leads. *Phys. Rev. B* 71 (24): 245116.

65 Świrkowicz, R., Wilczyński, M., Wawrzyniak, M., and Barnaś, J. (2006). Kondo effect in quantum dots coupled to ferromagnetic leads with noncollinear magnetizations. *Phys. Rev. B* 73 (19): 193312.

66 Meir, Y., Wingreen, N.S., and Lee, P.A. (1991). Transport through a strongly interacting electron system: theory of periodic conductance oscillations. *Phys. Rev. Lett.* 66 (23): 3048–3051.

67 Meir, Y., Wingreen, N.S., and Lee, P.A. (1993). Low-temperature transport through a quantum dot: the Anderson model out of equilibrium. *Phys. Rev. Lett.* 70 (17): 2601–2604.

68 Świrkowicz, R., Barnaś, J., and Wilczyński, M. (2003). Nonequilibrium Kondo effect in quantum dots. *Phys. Rev. B* 68 (19): 195318.

69 Kashcheyevs, V., Aharony, A., and Entin-Wohlman, O. (2006). Applicability of the equations-of-motion technique for quantum dots. *Phys. Rev. B* 73 (12): 125338.

70 Paulsson, M. and Datta, S. (2003). Thermoelectric effect in molecular electronics. *Phys. Rev. B* 67 (24): 241403.

71 Reddy, P., Jang, S.-Y., Segalman, R.A., and Majumdar, A. (2007). Thermoelectricity in molecular junctions. *Science* 315 (5818): 1568–1571.

72 Cui, L., Miao, R., Wang, K. et al. (2018). Peltier cooling in molecular junctions. *Nat. Nanotechnol.* 13 (2): 122–127.

73 Hicks, L.D. and Dresselhaus, M.S. (1993). Thermoelectric figure of merit of a one-dimensional conductor. *Phys. Rev. B* 47 (24): 16631–16634.

74 Hicks, L.D. and Dresselhaus, M.S. (1993). Effect of quantum-well structures on the thermoelectric figure of merit. *Phys. Rev. B* 47 (19): 12727–12731.

75 Mahan, G.D. and Sofo, J.O. (1996). The best thermoelectric. *Proc. Natl. Acad. Sci. U.S.A.* 93 (15): 7436–7439.

76 Taylor, J., Guo, H., and Wang, J. (2001). Ab initio modeling of quantum transport properties of molecular electronic devices. *Phys. Rev. B* 63 (24): 245407.

77 Brandbyge, M., Mozos, J.-L., Ordejón, P. et al. (2002). Density-functional method for nonequilibrium electron transport. *Phys. Rev. B* 65 (16): 165401.

3

Fabricating Methods and Materials for Nanogap Electrodes

Xianhui Huang[1], Weiqiang Zhang[2], Dong Xiang[2], and Tao Li[1]

[1]Shanghai Jiao Tong University, School of Chemistry and Chemical Engineering, Shanghai Key Laboratory of Electrical Insulation and Thermal Aging, Key Laboratory of Thin Film and Microfabrication (Ministry of Education), 800 Dongchuan Road, Minhang District, 200240, Shanghai, China
[2]Nankai University, Tianjin Key Laboratory of Micro-scale Optical Information Science and Technology, Institute of Modern Optics, 94 Weijin Road, Nankai District, Tianjin 300071, China

3.1 Introduction

A single molecule, even an atom, represents the ultimate limit that we can handle so far. As early as 1974, Aviram and Ratner [1] had predicted that individual molecules someday would be used as circuit elements in devices and the dream of utilizing molecules as functional units in electronic circuits has motivated researchers for years. If this comes true, it would be possible to integrate transistors in the trillions. The molecular-based devices possess unique advantages for electronic applications [2], such as lower cost, lower power dissipation, higher efficiency, ability of self-assembly and recognition, distinct optical and electronic properties, and synthetic tailoring ability by elaborate choice of geometry and composition. A variety of specific electronic functions performed by single molecules, including rectifiers [3, 4], switches [5, 6], and transistors [7–9], have been accordingly designed and reported. All the above aspects render molecules as ideal candidates for the next generation of electronics.

Despite all these advantages, there remain many challenges before molecular computers come to reality. The main obstacle lies in how to integrate nanometer-sized molecules to macroscopic electronic circuits. There are mainly two approaches for wiring molecules between electrodes. One method is to make top contact junctions, which includes scanning probe microscopy (SPM), (scanning tunneling microscopy [STM], and conducting atomic force microscopy [AFM]) [10–21], cross-wire junctions [22–24], mercury drop electrodes [25, 26], and thermally deposited metal films [5]. All devices manufactured by this kind of method can be categorized as "prototype devices," which are very useful for fundamental investigations and have already provided many important results [3–6, 10–24]. However, these devices are far from practical applications, as we cannot imagine a nanometer device carrying a huge SPM system or other systems. The other way utilizes nanogap

Nanogap Electrodes, First Edition. Edited by Tao Li.

electrodes [27–31] to form metal/molecule/metal devices. Compared with the "prototype devices," devices based on nanogap electrodes with conductive metal nanowire circuits and functional molecules inserted in the desired position have the potential ability to realize super integrated circuits, so they are more likely used in practical applications. Moreover, because the nanogap electrodes are fabricated before the molecular components are subsequently inserted, the junction can be characterized with and without molecules in place, which facilitates the distinction of the intrinsic molecule properties. Furthermore, as most nanogap electrodes present a planar configuration, it would be easier for higher density integration and to take the underlying substrate as a gate contact to tune the electrical properties of the molecular components.

In this chapter, we will first discuss the manufacture methods of nanogap electrodes. Typical dimensions of target molecules are well below 5 nm, so fabricating electrodes with separations that are suitable for specific molecules is a very challenging mission because it goes beyond the capability of traditional microfabrication technologies. Also, if the gap is too small, it would be very likely to posit the molecule in a tense and distorted state, resulting in unexpected performances. So, the precise control of the spacing makes it more difficult to prepare such kinds of devices. Despite the obstacles, several effective and creative methods of fabricating nanogap electrodes with controlled spacing have been reported in the last few years, including mechanical break junctions (BJs) [32], electron beam lithography (EBL) [33], electrochemical plating [34], electromigration (EM) [35], focused ion beam (FIB) lithography [31], shadow mask evaporation [8], scanning probe and atomic force microscopy lithography [36], on-wire lithography (OWL) [37], molecular rulers [38], etc. All the above methods have provided promising results and have their own characteristics. Different methods are frequently combined to obtain a desired configuration. For example, EBL is often used to firstly realize nanogap electrodes with space at 10–20 nm, and then other techniques, such as shadow mask evaporation, are implemented to further narrow the gap width to a 1–5 nm scale.

Most of the nanogap electrodes are prepared with metallic materials, since various technologies, especially the well-developed semiconductor manufacturing methods for processing nanogap electrodes are applicable to metals. In addition, the specific metals exhibit desirable chemical or physical properties for constructing molecular junctions, such as gold for self-assembly process. While ordinary metal electrode-based molecular junctions remain operational, other experimental platforms based on nonmetallic materials, such as carbon materials, silicon, and conducting polymers, have also been proposed, which leads to new possibilities for molecular electronics. In detail, the carbon materials possess similar compositions or unit structures to the target molecules, thus ensuring the robust covalent linkage [39, 40]. And the carbon electrode is more stable than some metal electrodes as the carbon atoms are held together by the sp^2 hybridization [41]. The silicon and conducting polymer demonstrate enormous potential in the molecular electronics as well. They not only are abundant and easy processing but also can maintain high electrical conductivity with dimension scale-down [42–45]. Furthermore, the preparation of nanogap electrodes with these nonmetallic materials represents a frontier area in this field. Therefore, the encouraging achievements based on the nonmetallic materials would be demonstrated in Section 3.9.

3.2 Mechanical Controllable Break Junctions

The mechanically controllable break junction (MCBJ) technique, as a handy tool to precisely control the single molecular junction, has been extensively applied in recent years [46]. The schematic of the latest overall MCBJ setup is shown in Figure 3.1a. Viewed from the bottom-up, we observe three sections: a pushing rod for forming nanogap, a flexible substrate (silicon, spring steel, or phosphor bronze) integrated with the nanostructures, and two lateral supports that exert a force to the bendable substrate [49]. A pushing rod maintains contact with the central part of the flexible substrate, on which a micro-fabricated metallic wire is mounted between the two lateral supports. In general, this entire apparatus is driven by a mechanical actuator (a stepping motor or piezo element), and the combination of both has recently been utilized more frequently to enable a fine control of the elongation [50].

In this three-point adjustable bending mechanism as shown in Figure 3.1a, the pushing rod that drives an upward force causes progressive bending of the substrate, which results in continuous stretching of the metal wire. As the metal wire is being elongated, the cross section at the central section is scaled down until its fracture, and then, two metal electrodes at the atomic level with clean surfaces are simultaneously generated. In contrast, as the substrate is relaxed, the inevitable contraction of the metal wire ultimately causes the two electrodes to return to an original wire. The nanogap between two facing electrodes can be accurately regulated during the bending and relaxing process (shown in Figure 3.1b). There is an applied bias voltage within the scope of 80–330 mV to promote the insertion of molecules into the

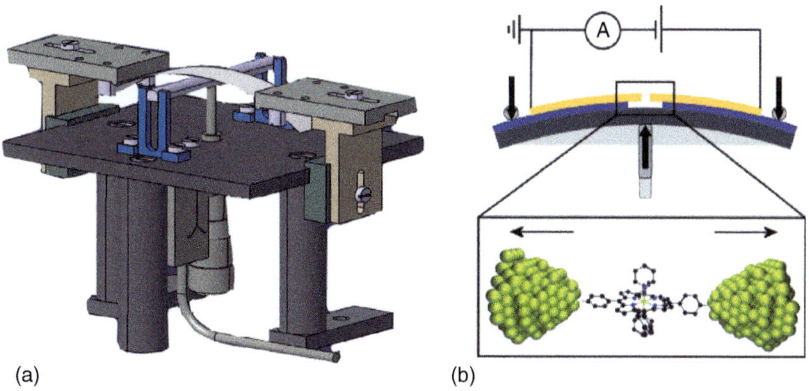

Figure 3.1 Schematic of the MCBJ setup. (a) Side-view schematic of a real MCBJ setup. Source: Based on Xiang et al. [47]. (b) Mechanism of the MCBJ setup. Source: Based on Perrin et al. [48]. Copyright 2011, Beilstein-Institut.

gap. After the molecular junctions are formed, it is important to study the charge transport properties of the single molecular junctions connected with a microscopic circuit [51].

The prerequisite for a high coupling efficiency between the electrodes and molecules is to set the gap size slightly larger than the length of the molecules and to precisely control the size. This requires an incredibly low attenuation factor between the gap size, Δd and the displacement of the pushing rod, Δh. The attenuation factor is defined as $r = \Delta d/\Delta h = 6ts/L^2$, where t is the thickness of the flexible substrate, s is the length of the suspended bridge, and L is the distance between the two lateral supports [52]. Generally, the value of factor $r = 5\times10^{-6}$ can be achieved, which indicates that large displacements of the pushing rod can merely create a little gap changes, i.e. the gap size can be adjusted with picometer precision [53].

Over the years, the constant attention on the MCBJ technique is ascribed to its advantages over other test beds for single-molecule junctions. First, the clean face-to-face electrodes generated by MCBJ suspend the influence of contaminants [54], and the cyclic process for repeatedly breaking and linking the junctions facilitates the collection of more experimental data to decrease data discretization in dynamic configurations [55]. Second, the MCBJ setup is extremely stable in the whole temperature range, and disruptive factors (such as external vibration, thermal drifts, and the discontinuous motion of the pushing rod) can be almost entirely ignored. As a result of the extremely low attenuation ($\sim5\times10^{-6}$) and micro-level control of the pushing rod, the accuracy of mechanical adjustment can approach a picometer scale. Meanwhile, with sophisticated micro-fabrication, the investigation of single molecular junctions can be more easily carried out. Third, the MCBJ technique can be conveniently combined with other systems, such as inelastic electron tunneling spectroscopy (IETS), surface-enhanced Raman scattering, and electrochemical deposition.

Along with the aforementioned significant advantages of the MCBJ technique, a few disadvantages exist in imaging of the junction configuration. In the process of forming single molecular junctions, we cannot use MCBJ technique to obtain the evolution of the tip atoms which is also a challenge for STM and AFM. Another drawback of MCBJ technique is that it is inconvenient to add a gate electrode in a solid state compared to electromigration technique in which the bottom gate electrode can be integrated conveniently. Moreover, the sites where the molecules are inserted remain unknown, which confines the exploration of the underlying influence factor of the conductance. The integration of MCBJ and other techniques will overcome these issues. In addition, it is impossible by MCBJ technique to obtain the surface image as STM and AFM technique. Nevertheless, MCBJ has played a major role in the research of charge transport properties for single molecules.

The fabrication of the MCBJ chip is summarized in Figure 3.2. According to the fabrication methods, there are three main types of MCBJ chips: mechanically cut-based, electrochemically deposition-based [61], and nanofabricated [62, 63]. Based on the structure of the chip, it can be classified as a two-terminal chip [57] or a three-terminal chip [64]. In the three-terminal chip, one of the electrodes

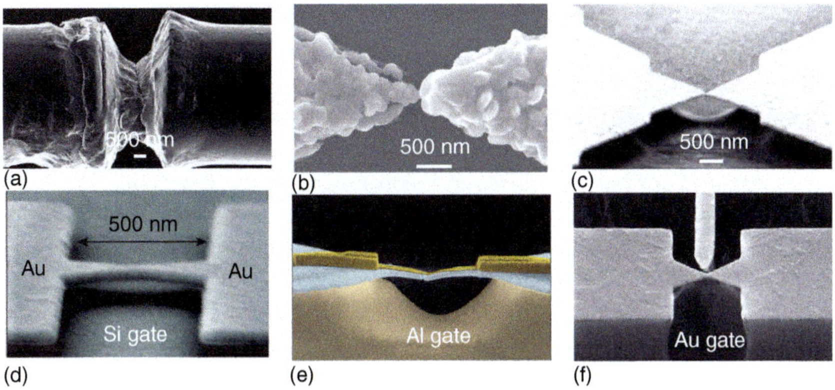

Figure 3.2 Typical MCBJ chips. (a) Mechanical cut to create a notch on the gold wire. (b) Electrochemically deposition-based electrodes. Source: Reproduced with permission from Yi et al. [56]. Copyright 2006, Wiley-VCH. (c) EBL-defined electrodes on flexible metal substrates. Source: Reproduced with permission from Lörtscher et al. [57]. Copyright 2006, Wiley-VCH. (d) Suspended gating (indirectly contacted) above a doped silicon substrate. Source: Reproduced with permission from Parks et al. [58]. Copyright 2010 AAAS. (e) Back gate in direct contact with the source and drain electrodes. Source: Perrin et al. [59]. 2013, Springer Nature. (f) Side gating with a nanometal electrode. Source: Reproduced with permission from Xiang et al. [60]. Copyright 2013, American Chemical Society.

generally operates as the gate electrode by which the electron transport can be modulated, and the other two electrodes are used as the source and the drain electrodes.

An early low-cost macrofabricated chip for an MCBJ experiment is presented in Figure 3.2a. A thin metal wire, normally less than a few millimeters in diameter, was fixed onto an insulated flexible substrate using two drops of glue. A notch was created using a knife through a mechanical cut in the middle of the metal wire to allow it to easily break when subjected to a bend force. These types of chips are cheap and simple to fabricate. The disadvantages of this macrofabricated chip can be described as follows. (i) The large size and the rough surface of the electrode are not quite suitable for a single-molecule investigation. (ii) Contamination may occur because the mechanical cut is typically performed in an ambient environment. (iii) There is a limited attenuation factor, which indicates that the mechanical stability of the macrofabricated chips is not as good as that of the micro-fabricated chip.

The chips for the MCBJ experiments can also be fabricated using electrodeposition on template electrodes that were predefined on the substrate (Figure 3.2b) [56, 65, 66]. The initial electrodes, which have a relatively large gap, are generally fabricated using optical lithographic techniques on the substrates. The gap is then narrowed and fused by depositing specific atoms onto the electrodes. The fused nano-bridge can be reopened by controlling the substrate bend configuration. This strategy, which combines the electrochemical deposition advantages with the MCBJ advantages, e.g. the use of optical lithography instead of EBL, is facile for the fabrication, and the gap size can be changed in situ [67]. The disadvantage of this strategy is that the fabrication process is time-consuming, and contamination is likely due to the electrochemical process.

Recently, advanced EBL and reactive ion etching (RIE) techniques were used to fabricate the MCBJ chip based on a flexible substrate. Figure 3.2c illustrates a typical EBL fabricated MCBJ chip [57]. The chip fabrication process consists of several key steps: (i) an isolating layer (polyimide or silicon dioxide several micrometers thick) is spun onto the substrate, such as thin silicon wafer, spring steel sheet, or bronze sheet; (ii) a standard EBL process is performed to define the metal wire with a constriction in the middle; and (iii) the isolating layer is partly removed by an RIE to obtain a suspended metal bridge.

A significant development of the MCBJ is the introduction of gating control. Ralph and coworkers initially reported the implementation of both electrostatic gating and mechanical adjustability within the same single-molecule device by adding a bottom gate electrode to a mechanically controlled break junction [68]. To fabricate the bottom-gated chip, a thick silicon dioxide film was first grown on top of a doped silicon wafer. Then, an EBL process was performed to pattern the gold electrode structure. Finally, hydrofluoric acid was used to remove the SiO_2 under the gold electrode, thus suspending it above the silicon substrate. In this chip, the doped conductive silicon was used as the gate electrode, as shown in Figure 3.2d. This newly designed chip combines both the ability to mechanically adjust the spacing between the electrodes and the ability to shift the energy levels of the molecules.

van der Zant and coworkers further developed the bottom gate architecture for the independent mechanical and electrostatic tuning of the charge transport [59, 69], as shown in Figure 3.2e. In contrast to previous gated mechanical break junctions, in which the source–drain electrodes are suspended above the gate electrode, and the fabricated source–drain electrodes are in contact with the gate electrode via a thin Al_2O_3 insulator. It was demonstrated that the electrical continuity of the gate electrode was maintained upon the bending of the substrate and successfully achieved mechanical and electrical control of the charge transport through the nanoscale island using such bottom gating. The technique was further applied to the molecular electronics investigation, which clarified the dominant role of the image-charge effects in the single-molecule junctions. A minor fault in this design is that the available gap size range was limited to avoid the partial breakage of the gate dielectric.

Inspired by the bottom gating, Xiang et al. proposed a new design for the MCBJ chips by adding a side-gate electrode near the source–drain electrode [60], as shown in Figure 3.2f. Using the advanced EBL and RIE methods, a suspended metal bridge and a gate electrode in plane were fabricated simultaneously. The Au gate electrode was located a few nanometers away from the suspended metal bridge. Using this noncontact side-gate electrode, the position of the molecular energy level was shifted, and the electron transport characteristics were modulated. The advantage of this design is that the gate electrode can remain unchanged during the operation, i.e. the fracture of the gate electrode can be avoided when one tries to break the metal wire or to obtain a large gap by deeply bending the substrate. The shortcoming in this design is that the gate efficiency is average and strongly depends on the distance between the gate electrode and the molecular junctions.

Now, more and more novel designed MCBJ had been put forward and applied in different research area. For example, Tsutsui and coworkers introduced a novel MCBJ to simultaneously evaluate the conductance and thermopower of atomic and molecular junctions [70]. As shown in Figure 3.3a, a microheater consisting of Pt coils with a 300-nm line space is fabricated using the original structure of microchips. Thus, the device can generate a temperature gradient where an electrode close to the Pt coil is hot due to Joule heating from a variable voltage and the other electrode away from the Pt coil is cold, and simultaneously, the local junction configurations can be adjusted by the MCBJ setup (shown in Figure 3.3b) [70]. Negative thermopower in the Au atomic chain is observed at a low bias voltage, which mirrors the quantum nature. However, there is positive thermopower upon an increased bias voltage, which is ascribed to the disordered Au nanocontacts that result from excessive local heating. Meanwhile, a sensitivity of the thermoelectricity on the atomic contact configurations or the 1,4-BDT molecular junction is observed. The microheater-embedded MCBJ, as an improvement on the MCBJ, is a direct and powerful approach to probe low-dimensional thermoelectric structures using thermopower measurements [74].

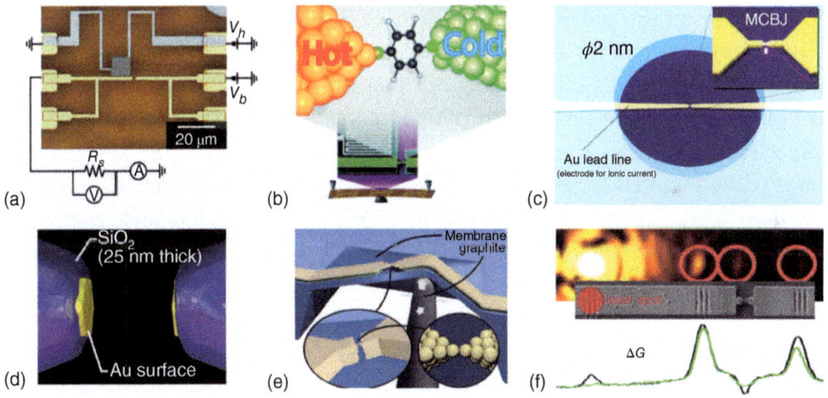

Figure 3.3 Modifications of a basic MCBJ configuration. (a) Optical microscopic image of a microheater-embedded MCBJ consisting of a Pt coil (gray) and a free-standing Au junction (yellow) on a polyimide-coated phosphor bronze substrate. (b) Schematic of a microheater-embedded mechanically controllable break junction with hot and cold electrodes at room temperatures. Source: Morikawa et al. [70]. 2014, Royal Society of chemistry. (c) Schematic illustration of MCBJ electrodes symmetrically placed in a liquid container with a diameter of 2 mm. Source: Doi et al. [71]. 2014, American Chemical Society. (d) The formed Au electrodes are covered with SiO_2 except the fresh breakage surface with an area of approximately 80 nm × 80 nm. Source: Arima et al. [62]. 2014, AIP Publihing LLC. (e) The chips with the membranes are then clamped onto a stretching mechanism consisting of a sample holder which carries the chip and a fine tip controlled by a uniaxial piezo stage. Source: Waitz et al. [72]. 2008, AIP publishing LLC. (f) A sample showing three gratings on the left and the right side of the constriction (top) and light-induced conductance change (bottom). Source: Benner et al. [73]. 2014, American Chemical Society.

To tackle the current noise-jamming problem in electrolyte solutions, as this noise probably affects electrical signals through the junctions, Doi et al. modified the basic MCBJ to determine the mechanisms responsible for the background current [71]. Figure 3.3c shows a scanning electron microscopy (SEM) image of an MCBJ with a nanogap formed by the self-breaking method, and the nanogap is symmetrically immersed in a cylindrical structure with a diameter of 2 mm. Subsequently, NaCl aqueous solutions with various concentrations fill the cylindrical structure, with an external electrical field applied to control ionic transport. The results reveal that the current noise increases as the ionic concentration increased. By solving the time-dependent Nernst–Planck and Poisson equations, it is observed that the applied field and length of the nanogap produce a significant effect on the ionic response. The experimental and theoretical results are conducive to the exploration of the background current and lay a good foundation for follow-up studies. To suppress the effects of electronic noise and ionic current in an electrolyte solution, Taniguchi and coworkers proposed a SiO_2-coated MCBJ method where the electrodes in MCBJ are entirely enclosed with a thin SiO_2 layer [62]. As shown in Figure 3.3d, two SiO_2 layers, respectively, which are under and on top of the electrodes are deposited by a chemical vapor deposition (CVD) based on the e-beam lithography technique.

Besides the contact between pushrod and chips generally is straight-shape, Scheer and coworkers achieved a single-atom contact by fabricating silicon membrane substrate and fine tip (shown in Figure 3.3e). With this method, it allows to control more than one junction on the same chip, detailed explanation can be seen in [72]. Recently, Scheer and coworkers further integrated line gratings into the metallic leads that increase the surface plasmon polarizations (SPPs) excitation efficiency (shown in Figure 3.3f) [73]. They found a clear distance and polarization dependence of the conductance enhancement demonstrating the direct impact of SPPs to the conductance change by both thermal expansion and photo-assisted transport. These innovative structures and results have a huge impact on the design of opto-nanoelectronic devices whose conductance is controllable by the excitation of surface plasmon polaritons [73]. Due to the increased attention on single molecular junctions, there will be more modifications of MCBJ devices in the near future.

The widely application of MCBJ can be found in the references [49, 75]. Here, we only present several updated applications. Recently, Hong and coworkers have conducted a series of research and obtained great achievements by MCBJ techniques. For example, they explored photothermal reaction processes of a photochromic dihydroazulene (DHA)/vinylheptafulvene (VHF) system [76]. To probe the photoreaction process of DHA-6, the in situ ultraviolet (UV) irradiation system was introduced to the MCBJ setup. In the initial state (DHA-6 solution without UV irradiation), all the molecular junctions are formed by trapping the DHA-6 into the nanogaps (DHA-6 junction). With UV irradiation (365 nm), the DHA-6 undergoes a ring-opening process of the five-membered ring to form VHF, as shown in Figure 3.4a. Simultaneously, they obtained corresponding conductance–distance traces. They finally found the reversibility and multiple switching possibilities during the reaction using the well-distinguished conductance states of single-molecule DHA–VHF junctions [76]. Statistical analysis of the break junction experiments provides a quantitative approach for probing the reaction kinetics and reversibility, including the occurrence of isomerization during the reaction. Another research is about the switching of charge transport pathways via delocalization changes in single-molecule metallacycles junctions, as shown in Figure 3.4b. They found that the conductance of the metallacycles junction with phosphonium group is more than 1 order of magnitude higher than that without phosphonium group. Based on X-ray diffraction and UV–vis absorption spectroscopy, they revealed that the attached phosphonium group makes metallacycles more delocalized, which shortened the preferred charge transport pathway and significantly enhanced the single-molecule conductance. At the same time, they reported the feasibility of the conductance distinction of two small molecule isomers (SDPP and SPPO) in single-molecule junctions using the MCBJ technique, as shown in Figure 3.4c. They demonstrated that combination of stimuli-response and quantum interference can be an efficient strategy to enhance isomer recognition and conductance switching in single-molecule junctions.

Utilizing single molecules as electronic functional units meets the increasing demands for the miniaturization of traditional electronic devices. A main challenge to enhance commercial microelectronics with uniform single molecular components is that the performances of single molecular components are strongly influenced by the atomic-scale geometry of the electrodes which leads to unreliability of the devices. Thus, it is critical to clarify the effect of electrodes geometry (extended to the atomic level) on the performance of single molecular junctions and further improve the stability of molecular devices. Zhao et al. present a strategy to generate both atomic-scale sharp and atomic-scale planar electrodes by mechanically stretching/compressing a nanowire [79]. The atomic-scale sharp electrode was fabricated by mechanically one-way stretching a microfabricated gold nanowire, that is, the atomic sharp electrodes were generated automatically during the process of stretching the nanowire. The planar electrode was obtained by cutting a microwire with a scalpel blade followed by an annealing process. It is found that the sensitivity of Raman scattering in tunneling nanogap can be enhanced when atomic-scale planar rather than atomic-scale sharp electrodes are used.

Figure 3.5a shows the schematic of Raman scattering measurement system with two types of electrodes, that is, a pair of atomic sharp electrodes and a pair of atomic planar electrodes. Figure 3.5b,c show the typical surface-enhanced Raman spectroscopy (SERS) spectra obtained from the nanogap area after self-assembling of diaminobenzene (DAB) onto the electrodes surface. It can be found that the SERS intensity decreased when the gap size was increased from 1 to 6 nm for both types of electrodes. Noteworthy, the Raman intensity with planar electrodes drops faster than the one with conical electrodes, which can be clearly observed from the Raman intensity at 1568 cm^{-1} (assigned to benzene ring vibration modes). Figure 3.5d–i show the distribution of the simulated electric field dependent on the gap sizes

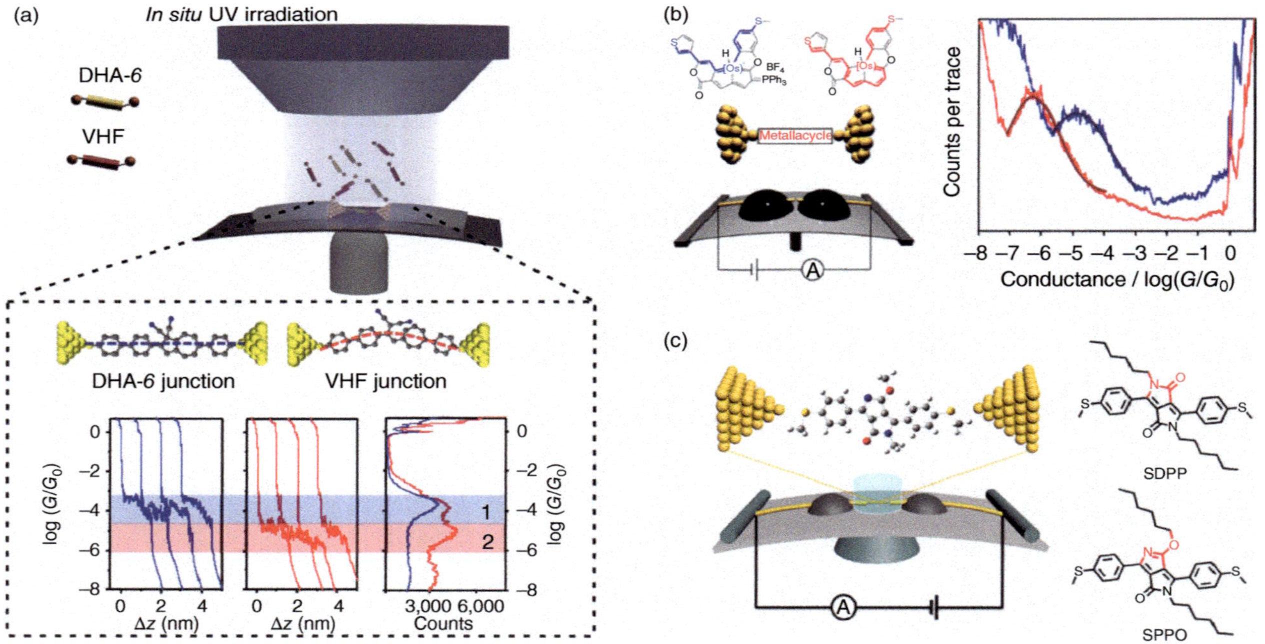

Figure 3.4 Latest developments of original MCBJ techniques. (a) Schematic of the mechanically controllable break junction (MCBJ) measurements during the photo-conversion of dihydroazulene under the in situ ultraviolet (UV) irradiation (top). Typical individual conductance–distance traces recorded in break junction measurements; blue for the DHA-6 junction and red for the VHF junction (bottom). Source: Huang et al. [76]. Licensed under CC BY 4.0.; (b) Single-molecule osmacycle junctions made by MCBJ setup and conductance histograms for two kinds of molecules. Source: Li et al. [77]. 2017, American Chemical Society.; (c) Schematic of MCBJ setup and the two diketopyrrolopyrrole (DPP) isomers. Source: Zhang et al. [78]. 2018, American Chemical society.

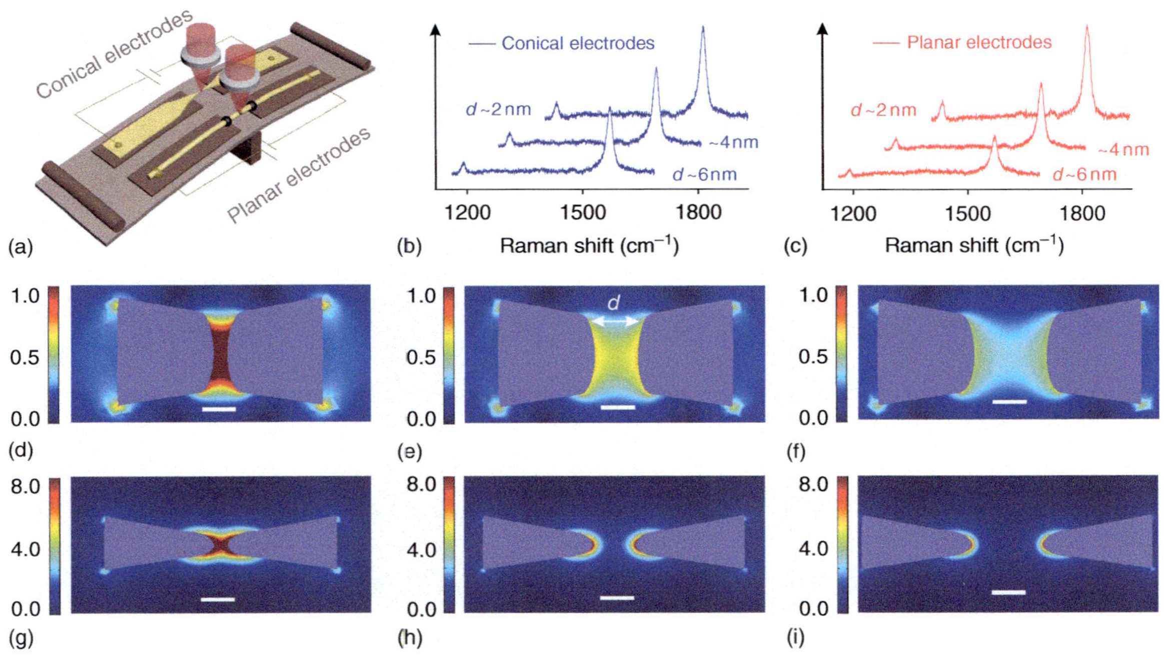

Figure 3.5 Raman scattering with two types of electrodes. (a) Schematic of the experimental setup combined Raman spectrometer with mechanically controllable break junction. Raman spectroscopy of diaminobenzene upon different gap size employing the conical electrodes (b) and planar electrodes (c). (d–f) Distribution of the electric field with a gap size of 1, 3, and 5 nm between plane-shaped electrodes. (g–i) Distribution of the electric field between cone-shaped electrodes with a gap size of 1, 3, and 5 nm, respectively. Scale bar: 2 nm.

for two types of electrodes. This simulation demonstrates that the electric field is considerably reduced when the gap size is increased from 1 to 6 nm for both types of electrodes. Meanwhile, we note that the electric field between planar electrodes decreases faster than the one between conical electrodes. The consistency in simulated field strength and measured Raman intensity indicates that the sensitivity of Raman scattering can be really modulated via local atomic-scale geometries of the electrodes.

Zhao et al. further observed that the conductance of single-molecule junctions can be significantly enhanced as the atomic-scale sharp electrodes are used [79]. The underlying mechanisms for these observations are clarified with the help of density functional theory (DFT) calculation and FEM simulation. These findings not only provide a strategy to improve the stability of single molecular devices but also pave a way to control the electron transport as well as to modulate the optical response for single molecular devices.

3.3 Electrochemical and Chemical Deposition Method

3.3.1 Electroplating and Feedback System

As mentioned above, it is difficult to use the MCBJ method to fabricate an integrated microchip due to the drive apparatus [47]. The electromigration method is unlikely to reproducibly fabricate nanogaps with a precise tunable width [80]. Unlike the electromigration and the MCBJ methods, the electrochemical method, e.g. electrodeposition or electrodissolution, is reversible and device compatible [34, 81, 82]. By electrodepositing metal atoms onto a specific face of the electrodes, the gap between two facing electrodes can be sequentially narrowed. The electrochemical process can be reversed in an electrodissolution mode for controlled etching/dissolution of the metal atoms from a wire/electrode; thus, the gap can be widened. This method requires no special technique or system and can conveniently prepare gaps that range from several angstroms to 10 nm, which is the best scale for fitting desired molecules or nanocrystals for nanodevices. Moreover, it is feasible to fabricate asymmetric nanogap electrodes (e.g. a gold and a platinum finger) and nanogap electrodes with three or more fingers by the electrochemical deposition technology (Figure 3.6) [83], as demonstrated by Kashimura et al., which were both difficult to achieve with conventional lithographic techniques and superior to the MCBJ or SPM techniques discussed above. The simplicity and robustness of the technique provide a promising method for the fabrication of large-scale and highly integrated nanodevices.

The working principle of the electrochemical deposition method is sophisticated [81]. Briefly, an array of original metal electrode pairs were fabricated using the EBL [82, 84] or optical lithography [85] techniques on a chip (such as thermally oxidized Si wafer). The separation between two facing electrodes in each pair was normally larger than dozen of nanometers. It should be noted that the electrodes were typically covered with Si_3N_4 [84], SiO_2 [85], or photoresists [56], except for a window exposing the gap region of the electrodes to the electrolyte for electrochemical

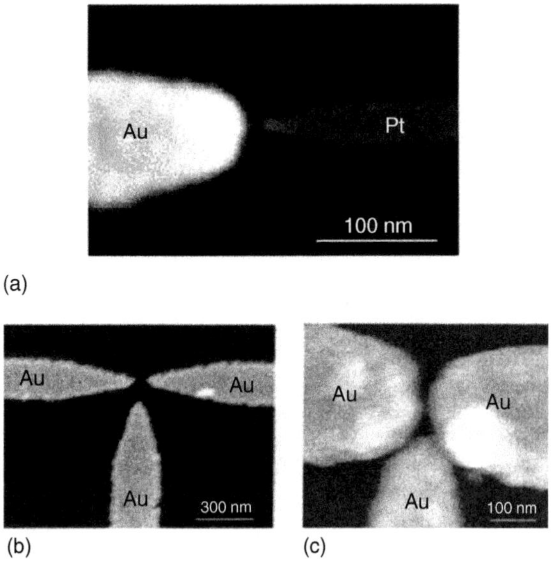

Figure 3.6 (a) Fabrication of nanogap electrodes with a gold and a platinum finger. SEM image of the electrodes with 8-nm separations after electroplating. (b, c) Fabrication of three gold finger nanogap electrodes. (b) SEM images of the electrodes with three sub-10-nm gaps after electroplating. (c) Initial electrodes with three gold fingers before electroplating. Source: Kashimura et al. [83].

deposition and etching. This insulating layer reduced the leakage current due to ionic conduction. When a voltage was applied between the counter electrode (CE) and the working electrode (WE) in the electrolytes, the metal ions were reduced to metal atoms at the surface of the WE. The gap was then reduced due to the electrodeposition of the Au atoms onto the electrodes layer by layer. Using the electrochemical deposition, different metals can be deposited to form junctions of heterogeneous compositions, e.g. the deposition of Cu on Au leads [86].

A key issue of the electrochemical method is the use of a feedback system to precisely control the gap width. There are several strategies using different experimental parameters as feedback signals to control the gap size. (i) Using the electrochemical current and the electrolyte conductance as a feedback indicator to critically monitor the gap width, especially when the gap is narrowed below the sub-micrometer scale [34, 87]. The electrodeposition process was stopped by switching off the potentiostat when the conductance reached a preset value. This provides an accurate and reproducible way to control the gap width below a couple of nanometers. (ii) Using the tunneling current as a feedback indicator to accurately monitor the electrochemical deposition or etching [81, 88]. During the electrodeposition process, the current flow between the two work electrodes was monitored throughout the experiment. Initially, the current was small due to the ionic leakage current; however, it increased rapidly when the gap was small enough to detect a tunneling current across the gap. The tunneling current served as a feedback signal to control the deposition process. This method provides the possibility of achieving a suitable gap fitting different molecules. However, this promising method has a limitation: the gap width range is small; typically, the gap is less than 1 nm because it is based on the detection and the feedback of the tunneling current. (iii) Using the electrode potential (V_{gap}) as the feedback indicator [85, 89]. In the previous report, both of the facing electrodes served as the WEs, and the metal layer was deposited on both of them

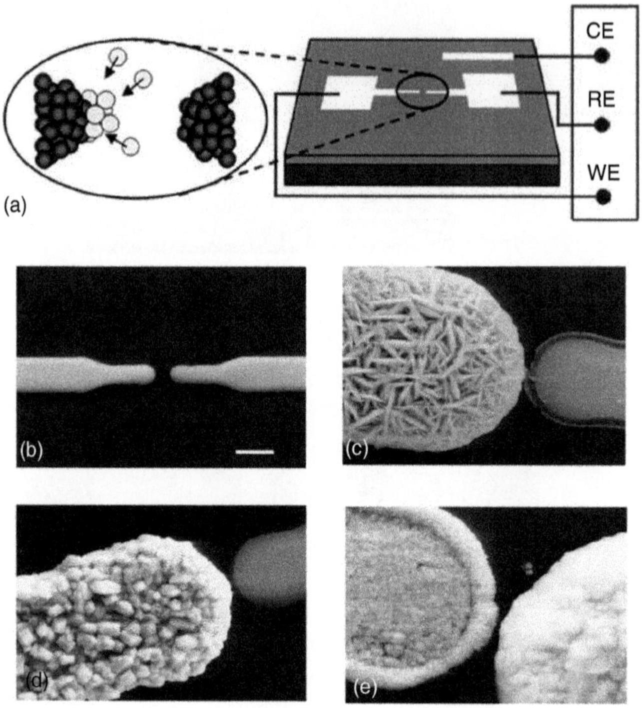

Figure 3.7 Electrochemical deposition (a) Schematic drawing of the experimental setup. A pair of original facing gold electrodes was fabricated on silicon wafers. Three electrodes were used: counter electrode (CE), work electrode (WE), and reference electrode (RE). The gap was reduced by electrochemically depositing metal atoms on the WE. (b) The original electrodes were separated by two micrometers; (c–e) Layer-by-layer metal atom deposition using different constant current densities: 1.0 mA cm^{-2} (c), 0.4 mA cm^{-2} (d), and 0.1 mA cm^{-2} (e). Source: Liu et al. [85].

simultaneously. Tian and coworkers introduced a new design in which the pair of facing electrodes served as the WE and the reference electrode (RE) [85, 89]. Experiments were performed on this new electrode arrangement by applying a controlled current between the WE and the CE, as shown in Figure 3.7a. Metal atoms were deposited layer by layer only on the WE, and the gap width between the WE and the RE was reduced. The value of V_{gap} (between the WE and the RE) decreased and eventually reached zero, thus indicating that the WE and RE electrodes were fully connected. In this way, the monitored V_{gap} can be used as a feedback signal to control the gap width. The disadvantage of this strategy is that the asymmetric deposition will result in asymmetric electrodes, as illustrated in Figure 3.7b–e. (iv) Utilizing high-frequency impedance in the feedbacks system [30, 90]. Liu and coworkers reported a new system that demonstrated an interesting frequency dependency on the final gap size (Figure 3.8). In their design, both electrodes of the gap were used as cathodes to symmetrically deposit the metal. The symmetric design guaranteed that the current signal traveled only in the symmetric circuit loop, and the faradic current was recorded independently on the electrochemical workstation.

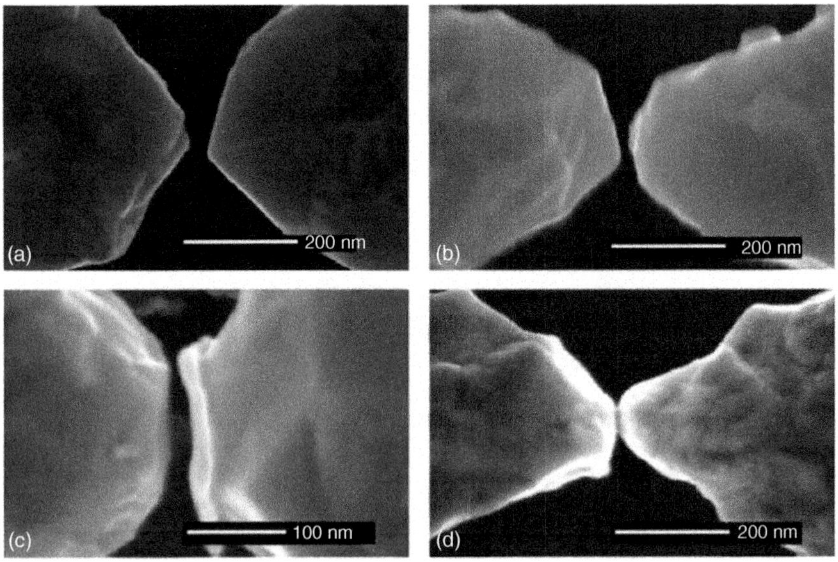

Figure 3.8 SEM images of the samples prepared at 3 kHz with different Δt values: (a) $\Delta t = 9$ seconds, $d = 26$ nm; (b) $\Delta t = 25$ seconds, $d = 16$ nm; (c) $\Delta t = 42$ seconds, $d = 7$ nm; (d) $\Delta t = 62$ seconds, $d = 1$ nm. Source: Qing et al. [30].

The amplitude of the AC voltage (V_m) recorded at one electrode decreased as the gap distance was close enough. The time evolution of V_m in a complete deposition process strongly dependant on the frequencies, thus making it a promising tool to control the gap size [30]. (v) Using a self-termination reaction to control the gap size. Tao et al. achieved nice atom-sized gaps and contacts between electrodes fabricated with a self-terminated electrochemical method based on a built-in self-terminated mechanism. An external resistor (R_{ext}) was connected to one of the electrodes and the final gap width could be predetermined by elaborately choosing $1/R_{ext}$ compared with the conductance quantum ($G_0 = 2e^2/h$). The gap resistance (R_{gap}) decreased with time, and resistances in series (R_{ext} and R_{gap}) shared the total applied bias voltage (V_0). Finally, when R_{gap} was much smaller than R_{ext}, V_{gap} approached 0, which meant almost no applied voltage was used for etching and deposition, the fabrication process was self-terminated. Majima and coworkers demonstrated a self-termination reaction during electroless gold plating (EGP) as well, which provided a uniform gap separation with different initial separations [91, 92]. They designed and implemented a chemical reaction in which the growth of the plating layer in the space between the electrodes was self-terminated (Figure 3.9). Briefly, the probability of the ions and the reducing agent accessing the nanogap decreased as the gap narrowed opposition to other areas. Therefore, the electroless plating rate between the electrodes decreased as the gap separation decreased, even becoming zero when no plating ions were transported into the nanogap. Using this method, nanogap electrodes for nanodevices with a separation of 3.0 nm were simultaneously mass-produced at a yield of 90% [92].

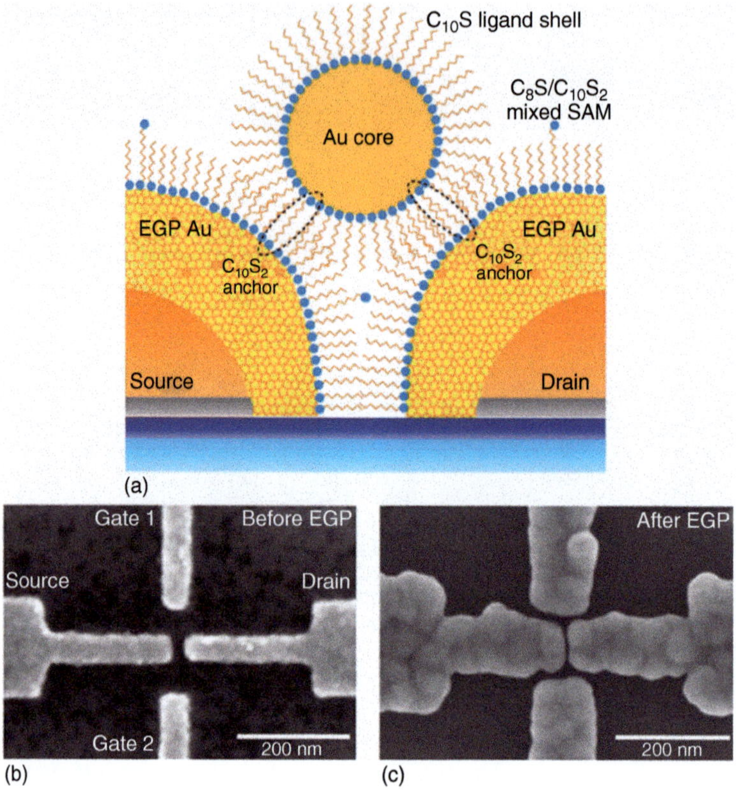

Figure 3.9 (a) Schematic diagram of the single-electron transistor (SET) with a side-view perspective. Typical SEM images of (b) initial gap electrodes before electroless gold plating (EGP) and (c) nanogap electrodes after EGP. Source: Serdio et al. [91]. 2012, Royal society of chemistry.

Recently, Zou et al. [93] demonstrated that facile electrochemical deposition of thin film silicon p–n junction in molten salt for the first time as shown in Figure 3.10. Silicon p–n junction film can be directly produced from inexpensive silicates/silicon oxide precursors through two-step electrodeposition process, which makes this technique attractive for low-cost manufacturing for silicon solar cells. Figure 3.10a shows schematic illustration of the two-step electrodeposition process for fabricating thin film silicon p–n junction. Figure 3.10b,c shows the SEM images of the fabricated typical silicon p–n junction films. The p and n region can be distinguished via a clear boundary. The silicon p–n junction devices were then characterized by I–V measurements, under AM 1.5G, 1 sun illumination as shown in Figure 3.10d.

When performed with gold, the drawback of electroplating lies in the difficulty of obtaining a simple gold electrolyte and special care should be taken to avoid generating highly toxic hydrocyanic acid gases. Accordingly, Umeno and Hirakawa [94] introduced a simple and cost-effective way using an "iodine tincture" as an electrolyte for gold electroplating, which was free of toxic compounds or strong acids. Kervennic et al. [87] found that compared with gold deposition, Pt promised a higher

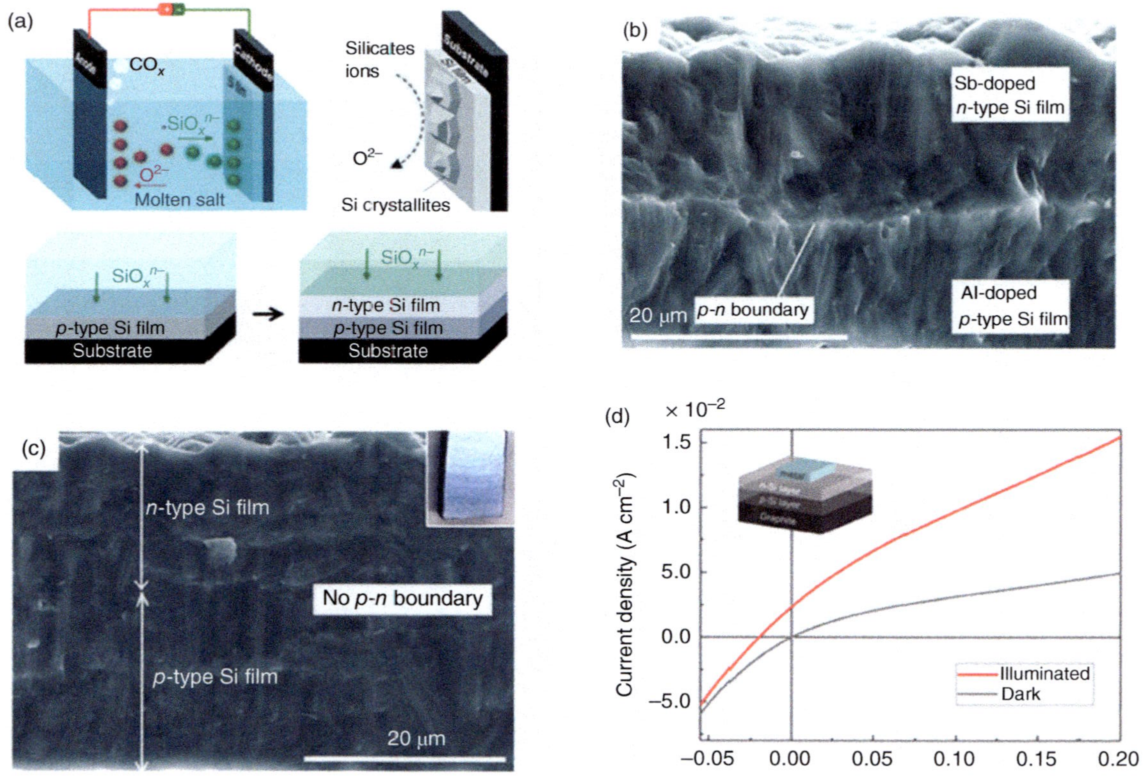

Figure 3.10 (a) Schematic illustration of the two-step electrodeposition process for fabricating thin film silicon p–n junction. (b, c) Typical SEM images of the deposited silicon p–n junction films (b) with and (c) without p–n boundary. (d) Current–voltage characteristics of silicon p–n junction deposited in molten salt. Source: Zou et al. [93]. 2017, American chemical Society.

reproducibility and stability of the electrodes, and pairs of platinum electrodes with a separation between 20 and 3.5 nm were obtained nicely.

The structure of the electrochemically deposited material can be tuned by the deposition current [85], electrode potential [56], electrolyte concentration [85], and plating mode (pulse or DC) [95]. After systematic studies, Liu et al. pointed out that it was essential to select a suitable constant current (I_{dep}) to ensure the deposited layers were smooth and compact rather than rough or loose [85]. They found that a relatively small deposition current led to a more uniform and compact electrode. Yi et al. reported that the morphology of the electrode surface could be modified by the applied electrochemical potential [56], which created nanoelectrodes with rounded surfaces at high over potentials and needle-like surfaces at low over potentials.

Although the electrode shape and the gap size can be accurately controlled by the electrochemical deposition method, this deposition method cannot offer a gap size change in situ; e.g. the gap size cannot be changed to match the length of the target molecules after the molecule is integrated into the gap. To overcome this problem, two strategies were introduced, i.e. deposition combined with the MCBJ [61, 65, 66] and deposition combined with STM [66, 96, 97]. Three groups did the main contributions to the combined deposition method with MCBJ method, i.e. Tian's group, Tao's group, and Mayer's group. It was demonstrated by Tian et al. that the chip fabricated by the deposition method can be further etched to obtain a suspended metal bridge [61]. Using an MCBJ setup, the substrate of the chip can be bent to break the metal bridge mechanically, offering a tunable gap size in situ. In this way, a series of molecule junctions can be fabricated with a high yield, and the properties of the single molecules can be correspondingly investigated [65, 98]. Tian and coworkers also presented an electrochemical strategy to extend the capability of the STM break junctions to create nanowires as well as nanogaps using various metals [96, 99]. The principle of the approach was based on the electrochemical atom on an STM tip. The tip was driven toward the surface to a sufficiently close distance within an extremely short period of time to allow a jump-to-contact occurrence, which led to the transfer of the deposited atoms from the tip to the substrate surface to create a nanoconstriction. Subsequently, the nanoconstriction was gradually stretched and eventually broken to create two sharp nanoelectrodes. This strategy has been successfully proven to measure the conductance of Cu, Pd, Fe, and various single molecules.

3.3.2 Chemical Deposition

Compared with the electroplating method, surface-catalyzed chemical deposition represents a simpler way to obtain narrow gap electrodes pre-patterned by conventional lithography, as it does not even have to carry an external circuit. By localizing a catalyst on the initial electrodes, the metal atoms can be selectively deposited onto the electrode surface assisted by reductive agents. One just need to immerse the chips into a stock solution containing metal ions and a mild reducing agent, and the gap distance depends on the reaction time and reactant concentration, which endows this method with controllability and mass producibility. Gu and

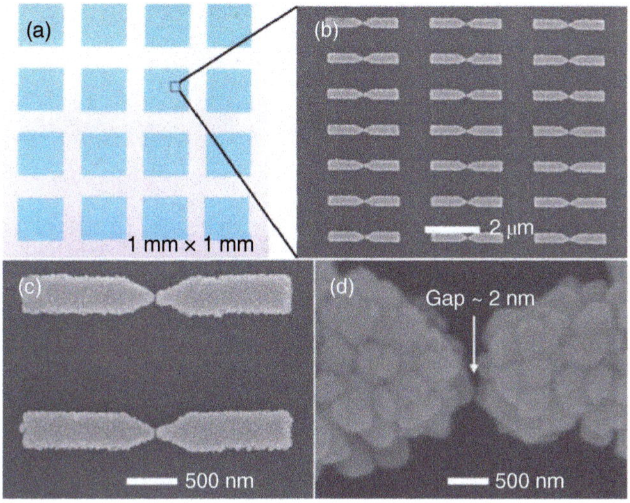

Figure 3.11 (a) Optical and (b) field emission scanning electron microscopy (FESEM) images of an array of nanogaps with sub-5-nm separations after surface-catalyzed chemical deposition. FESEM images of two-fingered nanogap electrodes with magnifications of (c) ×40 000 and (d) ×320 000. Source: Ah et al. [101]. 2006, AIP publishing LLC.

coworkers [100] firstly modified the gold leads with 2-mercaptoethylamine, the amino groups of which bind Pd^{II} particles to catalyze the deposition of metallic copper, which resulted in an electrode nanogap of 45 nm. Yun and coworkers [101] reported the creative fabrication of integrated nanogap electrodes of a few nanometers in separation with a yield over 90% (Figure 3.11). They proposed that the Au deposition rate of the gap edge was smaller than that of the other edges of the electrodes because of the mass transport limitation in the narrow gap and consequently the plating process slows down as the electrodes approach tunneling distances. Multi-fingered nanogap electrodes can also be conveniently prepared by an electroless plating method [92]. If the fabrication process of chemical deposition can be monitored in real time and the main controlling factors, such as reaction time and reactant concentration, are more theoretical rather than empirical, a much better controllability of this method can be expected.

3.4 Oblique Angle Shadow Evaporation

The oblique angle shadow evaporation method was pioneered by Dolan [102] in 1977. This technique was then adopted and refined for the fabrication of nanogap electrodes. Shadow evaporation is often combined with optical and EBL to define metal leads [8, 103–105]. By suspending the mask above the substrate and controlling the direction of the deposition angle to be oblique against the surface normal (Figure 3.12), feature sizes that are smaller than those of the masks can be obtained perfectly as reported by Bjørnholm and his colleagues [8]. Gap lengths between the electrodes of less than 10 nm can be reproducibly fabricated in this way. The gap size

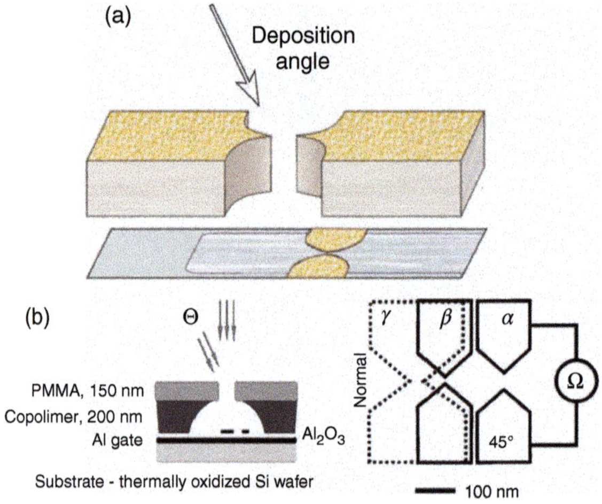

Figure 3.12 (a) Oblique angle deposition with an elevated mask for the preparation of two metal electrodes spaced to a few nanometers. Source: Kubatkin et al. [8]. 2003, Springer nature. (b) Schematic representation of the device preparation procedure. Left: reduction of the effective size of the opening in the resist by angle evaporation. Right: tuning the constriction size: starting from $\Theta = 45°$, where the gap between source and drain electrodes is about 50 nm (contour α), the tilt Θ was decreased stepwise, with a step of about 2°, to reduce the gap about 5 nm every step. After each tilt decrement, about 2 nm of material was deposited and a check was done to see if a measurable tunneling conductance had appeared, which would indicate a gap below 3 nm (contour β). Contour γ is a sketch of the mask shape without its reduction by angle evaporation. Source: Kubatkin et al. [105]. 2004, Elsevier.

can be adjusted by changing stepwise the tilt angle until a desired space is obtained [105], and in situ sample conductance measurements can help to find the proper angle [106], which endows this method with high controllability.

If the substrate is uneven, it can provide a similar effect as an elevated mask and can simplify the fabrication process. Accordingly, double-angle evaporation [107, 108] of thin metallic films was introduced. By simply changing the film thickness, the separation of the electrodes can be adjusted (Figure 3.13) [109]. Furthermore, even without employing conventional lithography methods, nanogap electrodes can also be accurately fabricated. Kawai and coworkers [110] developed a method based on oblique angle shadow evaporation to construct nanogap electrodes without a wet process to measure the electrical characteristics of molecules after they were positioned on the substrate. The gap length between the electrodes was easily controlled under 100 nm and a top contact configuration prevented structural deformation of the molecules. Moreover, this method is suitable to reliably fabricate nanogap arrays for molecular electronics. For example, Sun et al. [111] achieved the fabrication of gap electrode arrays as small as 3 nm in large quantities and applied them to the electrical study of nanocrystals successfully. Frequently, oblique angle shadow evaporation method is conducted at low temperature [8] (e.g. in liquid helium), as under such conditions the grain sizes of the metal deposits are relatively small, which results in the gap width remaining closer to the designed value. In this sense,

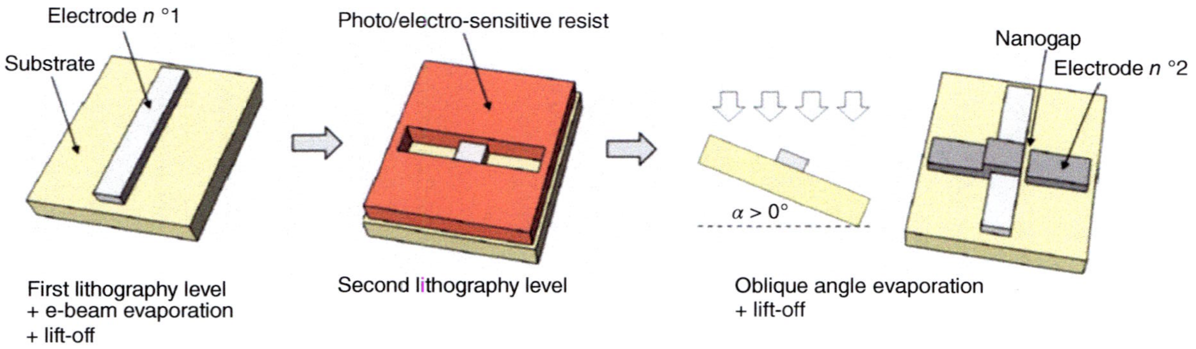

Figure 3.13 Schematic representation of the shadow edge evaporation process. Source: Thomas et al. [109]. 2017, IOP Publishing.

other improvements, such as using metals with smaller granules, decreasing the film thickness, and adjusting the evaporation rate, can be made to enable a narrower gap width distribution [107].

3.5 Electromigration and Electrical Breakdown Method

The phenomenon of electromigration has been well known for over a hundred years. In an applied electric field, the momentum of the moving electrons can be transferred to the metal atom, which results in the gradual movement of the atoms. Electromigration has been considered as the primarily failure mode in microelectronic circuitry for a long time. Recently, it has been well suited to fabricate nanogapped electrodes and molecular-based devices [112]. Generally, a constriction is fabricated in the metal (e.g. gold) wire for nanogap formation. The current density is largest at the constriction part, which favors electromigration. A local reduction in the constriction width increases the local current density, thus leading to the eventual breakage of the nanowire into two distinct (source and drain) electrodes. This process can yield a stable electrode separation of 1 nm with high efficiency and it can be monitored in real time by observing the current–voltage characteristics until only a tunneling signal is present. To incorporate the molecules into the electromigration breakdown junctions (EBJs), two different approaches were developed. One approach is to deposit the molecules onto the electrode surface, after which the breaking process is performed [113]. The other approach first breaks the nanowire and then assembles the molecules into the separated electrodes [114].

In contrast to the MCBJs, the nanogapped junctions formed by electromigration cannot make substantial repeatable measurements with the same junction. Thus, a large number of devices need to be fabricated to examine the statistical behaviors of the molecular junctions [115]. Additionally, two primary issues need to be carefully handled when processing the electromigration procedures. The first is heating in the breakage of a nanowire. The second is the gap size control. When the gap distance is comparable to the molecular length, only then it is possible to bridge the two electrodes using molecules. To control the gap size, several strategies have been introduced, which are addressed below in detail.

Although electromigration is a thermally assisted process that requires Joule heating to start gap formation, excessive heating should be avoided because it can result in undesired melting of the metal, destruction of the molecules, and formation of gold debris (islands) inside the gap [116]. Unintentional metal debris in the gap interferes with the insertion of the molecules of interest and may mask the intrinsic molecular signals [116, 117]. This metal debris can exhibit the same behavior as the molecules, e.g. similar current–voltage characterization, Coulomb blockade and Kondo effect [117–119]. To distinguish the molecular transport from the transport through small gold grains or other imperfections, three strategies are typically used: (i) the lead–molecule coupling in three-terminal devices can be used to realize this distinction because metal grains and molecules exhibit completely different behaviors on the gating voltage [116, 117]; (ii) using specific characterization techniques, e.g. inelastic tunneling spectroscopy, to address the molecular vibration modes to realize the distinction [113, 120]; and (iii) using functional molecules, i.e. the molecules with magnetic properties or light-sensitive molecules will behave differently from that expected for gold grains [121]. To reduce the metal debris in the nanogap, one relatively simple method is to reduce the resistance in series with the junction. By minimizing the total series resistance of the system, the temperature can be kept low during the electromigration process [122]. A single suspended nanostructure (SSN) grown by nanoscale migration across a nanogap is presented by Lee et al. [123]. In their work, a DC current bias, 1 nA, is applied across the metallic gap to induce nanoscale migration of Zn or ZnO. The nanogap formed by atomically sharp tips is fabricated on a silicon-on-insulator (SOI) wafer with a conventional optical lithography and anisotropic wet chemical Si etch. They put forward the application prospects that the nanogap fabrication can be complementary metal oxide semiconductor (CMOS) compatible allowing integration with Si-based electronics. Figure 3.14a shows the device with Ni/ZnO-coated tips forming a nanogap. When the source is applied to the device, the voltage across the device will rise as a function of time to its maximum limiting value. This voltage drop across the nanogap is monitored during the biasing step and the voltage vs. time graph is shown in Figure 3.14c. Figure 3.14b shows SEM image of the device after the DC bias, and Figure 3.14d is the magnified view of the nanogap in Figure 3.14b.

3.5.1 Device Fabrication

The wire with a constriction was typically defined using optical lithography or the EBL. Employing the lithographic technique, different types of metal wires were prepared. Following the electromigration process, different types of nanogaps, such as gold nanogaps [122], zinc nanogaps [123], nickel nanogaps [124], silver nanogaps [125], platinum nanogaps [126], and palladium nanogaps [127], were fabricated. In addition to the standard lithographic method used to fabricate the metal wire, several novel methods were recently reported. (i) Optical lithography combined with FIB lithography. Asghar et al. reported the use of a FIB to "scratch" and remove a thin layer of gold wires predefined by optical lithography. The scratch results in thinning of the metal wire at the designated point, which is unavailable using the standard optical lithography. Combining FIB scratching with electromigration provides an elegant approach to create nanoscale break junctions at an exact location with an extremely narrow distribution of the nanogap sizes [128]. (ii) Electrochemical deposition combined with electromigration. Valladares et al. reported a method to repeatedly generate a nanogap [124]. Briefly, an electrochemical deposition technique was used to bridge two electrodes. During the deposition process, a metal atom was deposited at the predefined electrodes, and the gap width decreased exponentially with time until the electrode gaps were eventually bridged. Once the bridge was formed, the ex situ electromigration technique was used to reopen the nanogap. (iii) Evaporation synchronized with electromigration. Naitoh et al. developed a procedure for the fabrication of Au electrodes using evaporation synchronized with electromigration [129]. The key point is that the electromigration

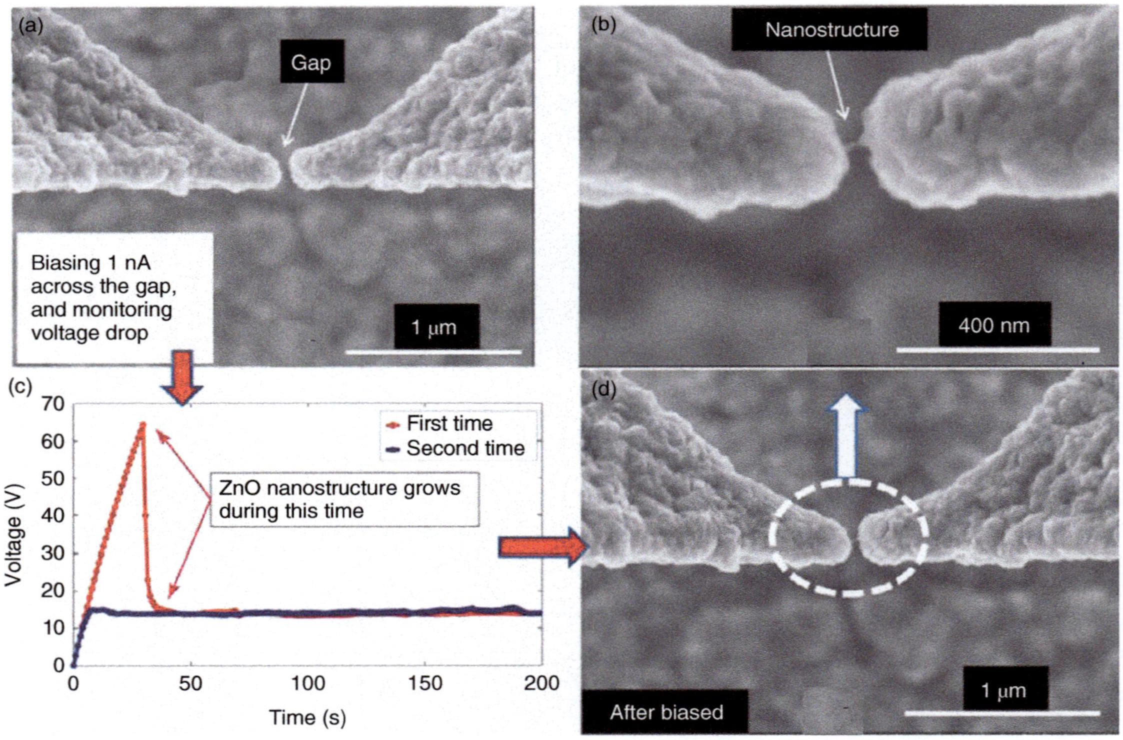

Figure 3.14 (a) The tips forming the nanogap are covered with Ni/ZnO. There is still a gap at this stage. (c) A DC bias of 1 nA is applied across the gap while monitoring the voltage drop. Initially, the voltage drop increases due to an open circuit. After a few seconds, the voltage drop decreases when the SSN grow between the nanogap. (b) SEM image of the device after the DC bias. (d) Magnified view of the nanogap in (b). A single nanostructure bridging the nanogap is visible. Source: Lee et al. [123]. 2011, IOP Publishing.

process was performed simultaneously with the metal evaporation process. It was demonstrated that the application of this method can form nanogaps with a gap size as small as 1 nm, and the gap size can be simply controlled by changing the magnitude of the applied voltage.

In addition to metal materials, nonmetal materials, such as graphene [130, 131] and SWCNTs [39], can use the electromigration technique to fabricate the nanogap. For example, Zhang et al. reported visualizing the electrical breakdown of suspended graphene sheets [131]. Using a pulsed electrical breakdown technique, narrow gaps were formed in suspended single to few-layer graphene devices. It was observed that the broken region started from the center of the graphene sheet and expanded. Upon a further increase in the amplitude of the pulse voltage, the graphene sheet was completely broken but remained freestanding. This observation underscores the potential of all-carbon devices to be integrated with graphene electronics. An additional discussion on molecular devices based on carbon-based electrodes is provided in Section 3.9.2.

Because a gate electrode can be easily fabricated on the substrate using the standard lithographic technique before the electromigration process (Figure 3.15) [29], the EBJ is especially advantageous in creating three-terminal device configurations, in which the molecule energy level can be changed based on the applied gate voltage [118, 121]. The gate coupling efficiency factor (α), which describes the effectiveness

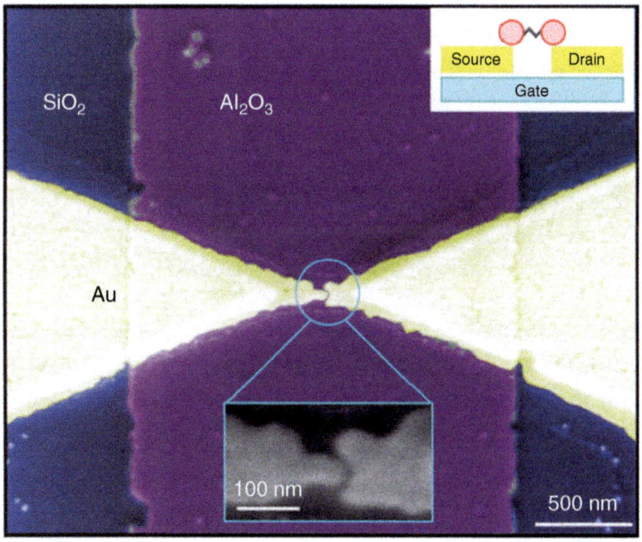

Figure 3.15 SEM image (false color) of the metallic electrodes fabricated by electron beam lithography and the electromigration-induced break junction technique. The image shows two gold electrodes separated by ~1 nm above an aluminum pad, which is covered with a ~3-nm thick layer of aluminum oxide. The whole structure was defined on a silicon wafer. The bright yellow regions correspond to a gold bridge with a thickness of 15 nm and a minimum lateral size of ~100 nm. The paler yellow regions represent portions of the gold electrodes with a thickness of ~100 nm. Inset: schematic diagram of single-molecule transistors that contain individual divanadium molecules. Source: Liang et al. [29].

of the molecular orbital gating, is of key importance to a three-terminal device. For example, if $\alpha = 0.1\ \text{eV V}^{-1}$, then the molecular orbital energy changes by 0.1 eV when 1 V is applied to the gate electrode. To obtain a large gate coupling efficiency, two conditions need to be satisfied: (i) a large local electrostatic field is generated upon the gate voltage [117] and (ii) the molecule resides near the gate metal–dielectric interface to reduce screening by the electrodes [113].

Typically, two types of gate geometries were used. The first type, noted as bottom gating, is commonly used [117, 118]. In this geometry, a gate electrode is fabricated beneath source–drain electrodes that are separated by an insulation layer. Currently, two materials are frequently used for bottom gating: highly doped (HD) Si substrates with a thermally grown SiO_2 layer on the surface and Al strips with a native oxide Al_2O_3 layer on the surface. An example of the Al gate electrodes is provided by Song et al. [113]. Two factors, which are (i) a higher relative dielectric constant of the Al gating and (ii) a thinner insulation layer (Al_2O_3), result in the gating efficiency of aluminum-based devices to be higher than that of silicon-based devices [117].

Remarkably, outstanding achievements have been made, as the electromigration is an advantageous way to prepare three-electrode nanodevices with the substrate as a gate. Park et al. constructed the single-electron transistors based on the colloidal cadmium selenide nanocrystals [112], providing evidence of nanomechanical oscillations of the C_{60} molecule against the gold electrode in a single-C_{60} transistor [7], and observation of a Coulomb blockade and the Kondo effect in single-atom or single-molecule transistors [29, 35]. Recently, van der Zant et al. have achieved a lot in this field, such as electron measurements through single molecular magnets [132], observation of the Kondo effect in gold break junctions in the presence of magnetic impurities [133], research on the temperature dependence of three-terminal molecular junctions based on sulfur end-functionalized tercyclohexylidenes [134], etc.

The second geometry is an in-plane side-gate electrode, in which the gate electrode and the source–drain electrodes are located in the same plane [116]. In this geometry, the electrical field is enhanced by the thickness of the gate electrodes, especially on the gate side compared to a typical back-gate configuration. Generally, to directly visualize the electromigration process and discriminate the trapped molecules/metal clusters from dusts inside the gap, a transmission electron microscopy (TEM) requires the use of electron transparent substrates, which are noncompatible with the most common three-terminal devices that have a back-gate electrode. This side-gate geometry is relatively compatible with the TEM inspection and can overcome this problem. Besides these advantages, the side-gate geometry also face the challenge, that is, the gating potential could be screened by the source–drain electrodes which means that only the molecules facing the gate electrodes can be influenced by the gate electrodes.

3.5.2 Gap Size Control

As mentioned above, controlling the gap size is of key importance to the electromigration method. The earlier electromigration approaches that used a single voltage ramp normally yielded a thermal runaway and a random distribution of the gap sizes. Most of the gap sizes are larger than hundreds of nanometers, resulting in them being unsuitable for the single-molecule investigation. There are two typical strategies, i.e. self-breaking and feedback control, used to control the gap size between the eventually formed electrodes.

Self-breaking: The self-breaking method has recently proven to be an efficient technique for reducing the gap size and avoiding the nanodebris formation [135]. In this method, a nanoscale metal junction is gradually shrunk using a typical electromigration procedure where bias ramps are applied repetitively. When the metal junction was narrowed to the atomic size, the active breaking process was stopped, and the junction was held under a very low field condition. The atom-sized contact then ruptures spontaneously due to the thermal fluctuations, efficiently avoiding excessive heating. Employing this self-break method, O'Neill et al. demonstrated no signs of metal clusters or metal islands inside the gap, a problem that is often observed in the traditional electromigration techniques [135]. Tsutsui et al. demonstrated that 0.5 nm electrode gaps could be achieved using the self-breaking technique assisted by the MCBJ technique [136].

The self-breaking process is relatively easy for gold (Au) electrodes at room temperature due to the high mobility of the Au atoms and the residual stress on the wire after electromigration. However, for other metals, such as Pt, self-breaking is not possible at room temperature due to the higher activation barrier. Fortunately, Prins et al. found that self-breaking did occur at a temperature approximately 100 °C or higher. Based on these results, they developed a method by which they performed electromigration and self-breaking of Pt nanowires at lifted temperatures. Then, the sample was cooled to room temperature to freeze the atomic mobility. This process results in nanogaps that are free of metallic clusters and relatively stable at room temperature for months [126].

Feedback control: A feedback-controlled electromigration (FCE) technique was recently developed, which actively adjusted the applied voltage in response to the changing conductance of the weak link to controllably create the nanoscale junctions [119, 137]. The distinctive advantage of the FCE is that the feedback is used to ensure that the nanogap formation occurs through electromigration at a regulated onset temperature in the constriction region, and a small gap size can thus be obtained. Unlike earlier electromigration approaches that used a single voltage ramp, the FCE typically uses a programmed algorithm to govern the electromigration process [138, 139]. For example, Xiang et al. used a LabVIEW-programmed voltage algorithm to control the electromigration process [138]. The process for their method involved four steps. (i) An initial voltage bias was applied to the metal wire, and the initial wire resistance was measured. (ii) A bias voltage was then increased gently in a ramp, and the wire resistance was measured simultaneously. (iii) When the resistance change ratio exceeded a predefined threshold, the applied voltage was reduced by 5%. (iv) A new reference resistance value was measured again, and the cycle was repeated.

In addition to the resistance as well as the effective conductance, noise can also be used as a feedback signal. Dong and Parviz observed that the measured relative current noise immediately before the nanogap formation increased 2 orders of magnitude higher than the resistance in the first stage of the electromigration process [140]. The considerable change in the current noise during the electromigration process makes it an attractive candidate to be used as a parameter in the control feedback loop to break the wires. They also presented initial results for using noise to control the disassembly process. In a typical two-terminal system, a fixed bias voltage U_0 is applied to the junction resistance as well as the series lead resistance. Consequently, as electromigration begins to shrink the junction, the junction resistance increases, and the voltage fall down to the junction increase. The power dissipated in the junction increases proportionally to the junction resistance, which may cause a thermal runaway and destroy the junctions. To overcome this problem, Schönenberger and coworkers developed a highly reproducible method to fabricate metallic electrodes with a nanometer separation using four terminal devices [141]. The four terminals comprise two symmetric pairs of contacts. In one pair, the bias voltage U_0 is supplied, and in the other pair, the voltage drop U_J over the junction is simultaneously measured. Regardless of the actual value of the junction resistance, the source maintains a constant U_J value. This novel design effectively avoids the thermal instability because the power over the junction will decrease with a fixed U_J and an increased junction resistance.

The limitation of the FCE is that it is time-consuming to fabricate a complex circuitry because the formation of multiple nanogaps requires each junction to undergo the FCE process individually. To overcome this problem, Johnston et al. developed a technique using the FCE that permits the simultaneous fabrication of dozen of nanogaps at room temperature in a single processing step (Figure 3.16) [142]. In their experiments, the formation of parallel nanogaps was achieved by a balanced simultaneous process that used a novel arrangement of nanoscale shorts between the parallel constrictions where the nanogaps are formed. By making the interjunction resistance smaller than the individual junction resistance, they solved the problem that uniform nanogaps could not be formed consistently as a result of uneven power dissipation and therefore successfully achieved parallel electromigration of nanogaps. This technique can be potentially used for constructing complex circuits of molecular-scale electronic devices.

3.5.3 Electromigration Applications

Imaging the electromigration process: Several research groups found that the gaps formed during electromigration may contain metal nanoparticles that might produce artifacts for the molecular signals [119, 133, 143]. Taychatanapat et al. addressed this issue by observing the electromigration process in real time within a SEM [144]. It was found that the amount of series resistance in the electromigration circuit is a critical parameter in controlling the overall morphology of the junction after electromigration. This study provides a direct confirmation that the simple voltage ramp with no added series resistance in the circuit gradually

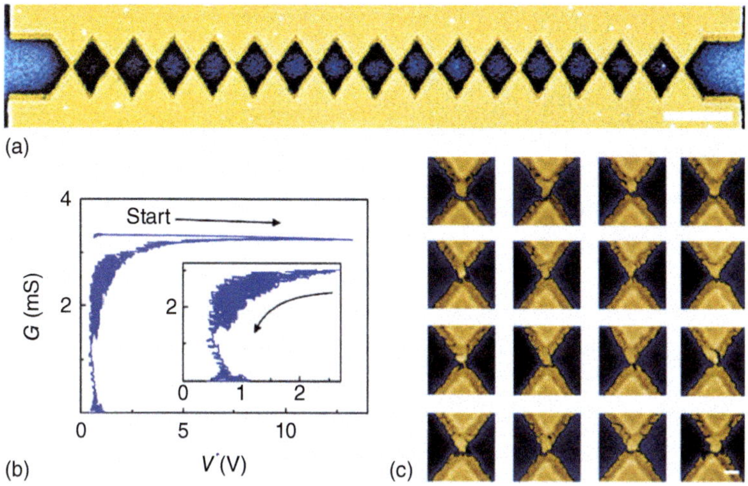

Figure 3.16 Feedback-controlled electromigration (FCE) of a 16-junction array. (a) SEM image of an array of 16 junctions made by electron beam lithography and shadow evaporation. Scale bar is 1 μm. (b) G–V data from FCE of the 16-junction array. (c) SEM images of nanogaps formed by parallel FCE of a 16-junction array that clearly shows the gold removed from the thin overlap junctions. Scale bar is 100 nm. Source: Johnston et al. [142]. 2007, American Chemical Society.

changes the wire morphology without the formation of any visible nanoparticles. Recently, McRae et al. demonstrate the capability to engineer a giant $e-h$ charging energy and conductance asymmetry in ultraclean (single quantum dot [QD] or ballistic) suspended single-walled carbon nanotube (SWCNT) channels whose lengths range down to 14 nm using a feedback-controlled EM procedure [145]. The transistors consist of naked nanotube channels contacted with sections of tube under annealed gold, which acts as an n-doping top gate allowing coherent quantum transport. Figure 3.17a shows a break junction and L_{sus} indicates the length over which the gold is suspended. Figure 3.17b shows the I–V_B EM data for a gold on SWCNT break junction. Figure 3.17c,d shows top-view SEM images and geometry of the suspended devices.

Unfortunately, SEM images of the nanogaps do not have the sub-nanometer resolution required to view the detailed structure of the nanogap; thus, it is a big challenge to view the electromigration process at atom-scale resolution. To address this critical issue, Strachan et al. prepared electromigrated nanogaps on free-standing SiN_x membranes that are transparent to the high-energy electrons and are thus compatible with the TEM at a sub-nanometer resolution [146]. It was determined that the junction pinched away from the edges of the original leads for the samples prepared using the FCE technique. They pointed out that the electrodes were clear of residual debris left from the evaporation procedures, and it was possible to avoid potential parasitic conductance channels through spurious metallic particles using the electromigration samples.

Moreover, Strachan and coworkers used real-time TEM imaging to monitor the dynamics of the FCE process at the sub-nanometer scale, which was not

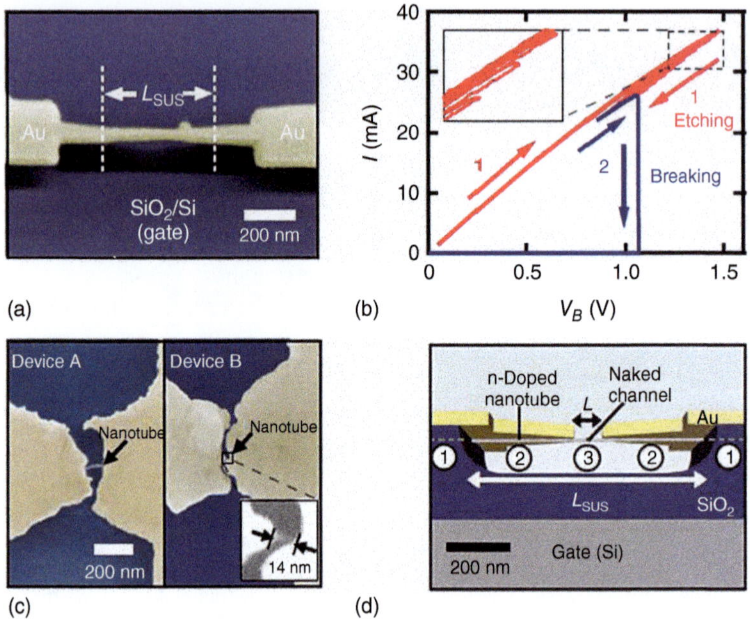

Figure 3.17 (a) Titled SEM image showing a suspended gold break junction fabricated on top of a SWCNT. The break junction is suspended over a length over a length $L_{sus} = 350\,\text{nm}$. The back plane (blue) is used as a global back gate. (b) The $I-V_B$ EM data for a gold-on SWCNT break junction at $T = 4\,\text{K}$, showing the process to (1) narrow the junction and (2) create a naked nanotube channel. (c) SEM images of devices A and B after EM. The naked channels are visible and their lengths are $L_{SEM} = 111$ and 14 nm, respectively. (d) Geometry of our suspended SWCNT transistors. The labels 1, 2, and 3 refer, respectively, to the disordered on-substrate nanotube sections, the n-doped gold-covered suspended SWCNT sections, and the naked SWCNT channel. Source: McRae et al. [145].

accessible to the earlier electromigration investigations [139]. It was determined that well-defined crystal facets appeared during the nanogap formation, indicating a layer-by-layer electromigration process. Furthermore, it was observed that the electromigration process is reversible; e.g. the void can be refilled with gold atoms when the applied voltage is reversed. This finding is clear evidence that the nanogap does not melt during the FCE process, which is contrary to what occurs during the single-voltage ramp procedure.

Although several attempts have been made to develop electron transparent substrates and perform two-terminal transport measurements in addition to the TEM inspection [139, 147, 148], it is a considerable challenge to perform these measurements with bottom-gated devices. Gao et al. first demonstrated a technique to perform three-terminal transport measurements that were compatible with the ex situ TEM inspection [149]. An electron transparent Si_3N_4 membrane was used as the substrate, and a Cr/Au layer was deposited on the backside of the membrane, which served as the back gate. The nanogaps were formed on the membrane using electromigration. The three-terminal electron transport measurements were performed without molecules at cryogenic temperatures, and the Coulomb blockade features were observed. After the transport measurements, the Cr/Au back gate was removed to make the substrate transparent, and the TEM inspection was conducted ex situ again on these samples. The TEM inspection revealed the presence of a few nanoparticles in agreement with the transport measurements. In other words, the sample topology deduced from the differential conductance (dI/dV) plots agreed with the one used for the TEM. This technique can be used to reduce the ambiguous interpretation by providing more information on the sample topology and the structure.

Inelastic electron tunneling spectroscopy: With the I–V characterization, it is difficult to distinguish the molecular transport from the transport through small gold debris or other imperfections [117]. Molecule-specific features (such as vibrational modes or gate coupling) can distinguish the presence of single molecules in the gap from spurious gold dots [114]. The IETS provides not only a characterization technique to identify the molecules in the junctions but also important insights into how the interaction of tunneling charge carriers with molecular vibrational modes influences the overall charge transport characteristics. Most of the IETS studies have been performed based on the enormous number of molecules in the monolayer junctions. At the single-molecule level, Song et al. measured the vibrational signatures of the molecular junctions using the IETS, which was confirmed by the estimated conductance [80]. The Au–S stretch mode was clearly observed in the measurement, and the IETS was extremely sensitive to the chemical structures of the target molecules, suggesting that a reliable metal–molecule contact was formed. The IETS provided unambiguous experimental evidence of the existence of molecules in the nanogap electrodes fabricated using the electromigration technology.

Surface-enhanced Raman spectroscopy: SERS is a useful tool for accessing vibrational and chemical information down to the single-molecule level. The SERS typically relies on plasmon excitations in metal nanostructures to concentrate the incident radiation. Generally, the SERS platforms involve metal nanoparticles to generate the required electromagnetic enhancements. The electromigration technique provides a true nanoscale gap between the extended electrodes instead of discrete sub-wavelength nanoparticles. The ability to fabricate precise gaps on demand combined with the simultaneous electronic transport measurements of the nanogap provides the fingerprint information of the molecules [150]. Ward et al. used the nanoscale gap structures prepared using the electromigration technology to perform simultaneous measurements of the electronic transport and the SERS [151]. It was observed that the conductance changes often correlated with sudden changes in the intensity of different sets of Raman modes. These correlated fluctuations in the tunneling conductance and the SERS intensity could not be explained by the changes in the gap configuration due to the fact that the tunneling conductance depended exponentially on the gap geometry (e.g. conductance could change by a factor of 10 for a 0.1 nm change in the gap separation); however, the electromagnetic enhancement was not as strongly affected by the gap size. The experiments suggested that both the SERS and the conductance changes were most likely due to changes in the conformation and the binding of the individual molecules, indicating a single- or few-molecule SERS sensitivity.

Typically, incident light with an electric field polarized across the gap (along the nanowire) is well known to induce the strongest SERS enhancements [98]. Surprisingly, Herzog et al. observed that for a nanogap located within a nanowire, the greatest enhancement and resulting SERS signal occurred when the electric field of the incident light was polarized along the gap (transverse to the nanowire) [152]. This polarization dependence was interpreted by the striking asymmetries in the nanogap fabricated using the electromigration technique. Due to the intrinsic asymmetries of the nanogap, the dipolar plasmon mode that resonated transversely across the nanowire coupled with dark multi-polar modes, thus causing the intense hybrid plasmon modes to be significantly stronger than that of the longitudinal modes in the structures. These results put light on the understanding of the nature of plasmons in the asymmetric structures and open the possibility of further geometric optimization of the near field and far field. In a further study, the same group developed the Raman technology to address vibrational and electronic heating in situ in nanoscale junctions fabricated using the electromigration technology [153]. When the current flowed through the molecules, the inelastic processes transferred energy from the electrons to the local molecular vibrational modes. In the presence of optical radiation, direct optical absorption and the decay of plasmon excitations produced hot electrons and holes in the metal. These hot electrons and a hot substrate can excite the molecular vibrational modes during the Raman scattering processes. It was demonstrated that the SERS could be used to determine the effective temperature in the molecular junction regime. By measuring the Stokes spectrum, anti-Stokes spectrum, and electrical transport characterization, it was revealed that the mode-specific vibrations could be pumped by both optical excitation and direct current with effective temperatures exceeding several hundred degrees Kelvin. These experiments open a new window to investigate local heating in the molecular junctions [153].

Kondo effect: When a quantum dot (e.g. nanocrystal and nanotube) is attached to the metallic electrodes via the tunneling barriers, the electron transport is dominated by single-electron charging and energy-level quantization [29]. As the coupling to the electrodes increases, higher-order tunneling and correlated electron motion will lead to new phenomena, including the Kondo effect, which results from the exchange interaction between an unpaired localized spin and the conduction electrons in the metallic electrodes. Most of the studies on the Kondo phenomena in quantum dots were typically performed on systems where precise control over the spin degrees of freedom was difficult. Using a chemical design, the spin and the orbital degrees of freedom can be controlled for those molecules (e.g. metal ion complex), thus providing a new system for the Kondo phenomena investigations. Liang et al. first reported the observation of the Kondo effect in single-molecule junctions, where an individual divanadium molecule was trapped in the nanogap by the electromigration method and served as a spin impurity [29]. In the differential conductance maps (dI/dV as a function of the gate voltage V_G and the source–drain voltage V_{SD}), a diamond-shaped pattern was observed, indicating the achievement of a single-electron transistor. More significantly, the Kondo resonance (i.e. a sharp zero-bias conductance peak) was clearly observed, and it was demonstrated that this Kondo resonance could be tuned reversibly using the gate voltage to alter the charge and the spin state of the molecules.

In another example, the Kondo effect in C_{60} single-molecule junctions was investigated [9]. Yu and Natelson used the electromigration technique to fabricate C_{60}-based single-molecule transistors and demonstrated that each of them could act as a single-electron device in the Coulomb blockade regime at low temperatures. Interestingly, sidebands parallel to the zero-bias peak were observed, thus indicating the conduction processes that involved both multi-body correlations and inelastic coupling to the vibrational modes. In the same year, Pasupathy et al. measured Kondo-assisted tunneling using C_{60} molecules in contact with ferromagnetic nickel electrodes fabricated by the electromigration method [154]. It was demonstrated that the C_{60} molecules could be strongly coupled to nickel (Ni) electrodes to exhibit the Kondo effect. More importantly, it was revealed that ferromagnetism could suppress Kondo-assisted tunneling; however, Kondo correlations were still present within the ferromagnetic electrodes. They found that a Kondo peak split was strongly affected by the magnetic moment alignment in the two electrodes under an applied magnetic field, hence proving a new window to modulate the Kondo resonance.

3.6 Molecular Scale Template

3.6.1 Molecular Rulers

The organic molecules whose properties have been extensively investigated in molecular electronics based on the test beds with nanospaced electrodes can in turn serve as templates for the fabrication of nanogap electrodes [38, 155–157]. In this way, the gap dimensions are determined and easily tuned by the length of a certain molecule. Hatzor and Weiss [38] innovated the construction of well-defined nanostructures with mercaptoalkanoic acids as molecular templates (Figure 3.18). Briefly, two Au predetermined patterns were first shaped on the silicon substrate by the EBL. Then the coordinated multilayers of organic molecules were prepared layer by layer on the Au traces. After a second metal deposition and the lift-off of the organic multilayer resist, the nanogaps were formed with separations determined by the thickness of the mercaptoalkanoic acid layer. This molecular ruler resist method has been refined for the fabrication of the nanogap electrodes, as it provides a simple, highly reproducible, and economical way to manufacture metallic electrodes separated on the molecular scale [158–160]. McCarty increased the patterning resolution to only a few nanometers using the combination of conventional photolithographic and molecular lithographic techniques (Figure 3.19) [161]. Initially, the Au film was thermally evaporated onto the silicon wafer, followed by the photolithography and RIE to form the initial metallic electrode. Then the wafer

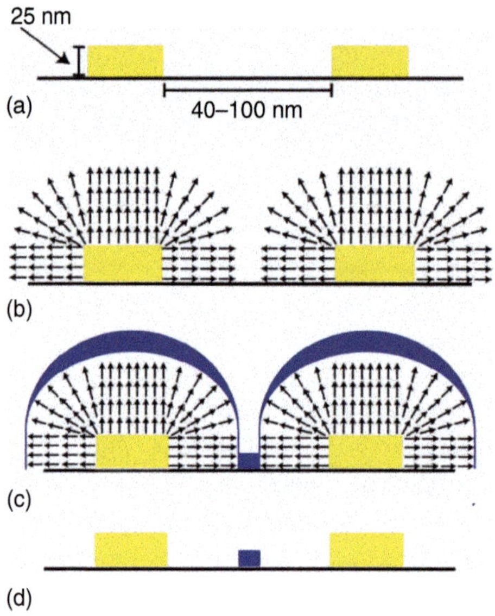

Figure 3.18 Schematic of the scaling-down process. (a) EBL forms the parent structures (yellow rectangles). (b) Layer-by-layer construction of metal–organic resist (2 nm per layer). The arrows represent the mercaptoalkanoic acid (tail is the SH group); Cu^{2+} ions are not depicted. (c) Metal (blue arcs and rectangle) evaporation into the gap (the gap size is determined by the number of organic layers deposited). (d) Lift-off of the organic layer in HCl/dimethylformamide (DMF) solution. The thin Ti layer was omitted for clarity. Source: Puntes [38]. 2001, AAAS.

was immersed in the alkanethiol solution to prepare a dense, self-limiting single layer of molecules selectively on the Au surface. Finally, the second electrode was defined by the lift-off photolithographic process which overlapped with the initial electrode by several micrometers. It was shown that at the overlap, the distance between the two electrodes was determined by the thickness of the molecular resist. To fine tune the separation, multilayered resists could be employed which were connected by the Cu^{2+} ions, for example, a nanogap of about 6 nm was obtained with the 5.6-nm-thick multilayered molecular resist. The worker also demonstrated that the target molecules could be added to, or removed from, patterned molecular junctions. Therefore, this novel technology provides a relatively simple and efficient means for systematic analysis of molecular conduction.

A number of researches have been conducted using the molecular ruler resist method. For instance, it was found that the electronic properties of nanogap electrodes were extremely sensitive to the molecular bridge species for their nanometer-sized spacing [162, 163]. The molecular junction in which the dodecanethiol molecules self-assembled on both surface of the nanogap electrodes could sense the addition of groups such as *N*-hydroxysuccinimide ester (NHS) and phenylalanine by monitoring the conductance change of the system [163]. This demonstration suggested it was possible to perform label-free detection of biological species by the nanogap sensor. Moreover, it is feasible to pattern sub-30-nm relief structures on quartz templates with high precision, accuracy, and throughput for the application in nanoimprint lithography, since the molecular ruler resist method circumvent the proximity effects which limits the EBL technology [164]. Remarkably, Cao et al. [165] have successfully fabricated three-terminal electrode structures in a wafer scale based on the McCarty's concept [161], as the molecular scale field-effect transistors (MSFETs) are key building blocks for molecular electronics. In addition to the mercaptoalkanoic acid molecular resist, Choi and coworkers [157] proposed the DNA-templated molecular lithography to construct sub-30-nm

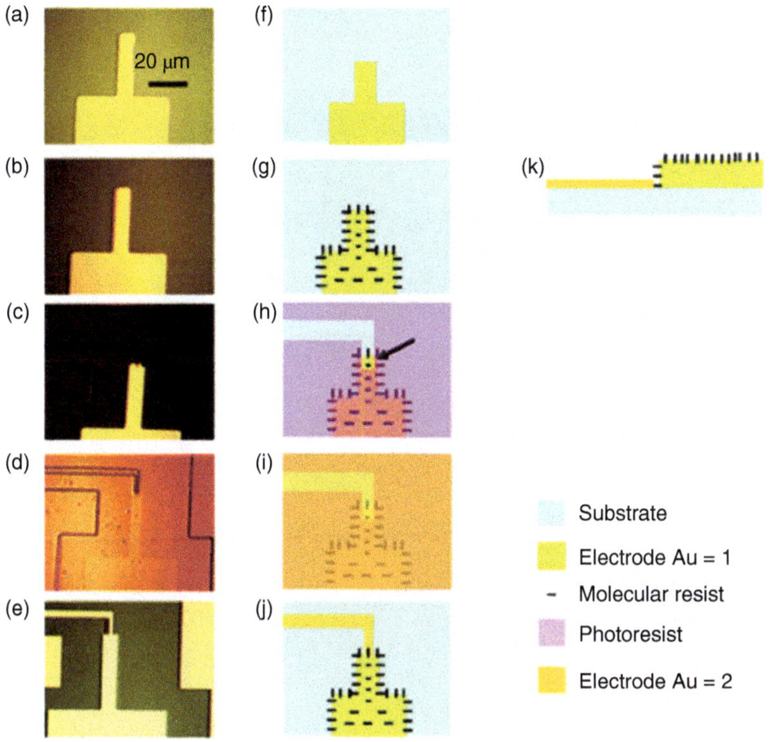

Figure 3.19 Series of optical micrographs (a–e) and cartoon depictions (f–k) showing the steps used to fabricate the electrode structures. (a) and (f) First, a chrome adhesion layer and then a gold layer are thermally evaporated onto the substrate, and photolithography and RIE are used to generate the initial metallic feature. (b) and (g) This structure is modified with a molecular resist. (c) and (h) A second photolithographic process is used to create the shape of the second electrode. The arrow points to the area exposed on the initial electrode. (d) and (i) A second layer of gold is evaporated evenly across the entire substrate. (e) and (j) The resist used in the photolithographic process is removed, effectively stripping the gold deposited on top of it and resulting in the completed electrode pair. (k) Cartoon depiction of a side view showing the separation at the interface of a completed electrode pair. Source: McCarty [161]. 2004, American Chemical Society.

complementary doping-free carbon nanotube field-effect transistors (CNT-FETs). It was demonstrated that the polarity of the CNT-FET device was easily controlled by modulating the Schottky barrier height between the electrode–SWCNT interfaces, thus promoting its practical applications. This simple fabrication method is also suitable for other nanoscale materials to achieve the ultra-miniaturized devices.

On the other hand, Bjørnholm and coworkers introduced a novel approach of the direct synthesis and growth of end-to-end linked gold nanorods by assembling the gold nanoparticle seeds before the rod synthesis, to result in a nanogap (Figure 3.20) [166]. In detail, the citrate-stabilized nanoparticle seeds were first linked with the dithiol-functionalized poly(ethylene glycol) (SH-PEG-SH) linkers, which were then added to the growth solutions containing a gold salt, a mild reducing agent and a surfactant. It was shown that most of the gold nanorods grown from the chemically linked seed particles were end-to-end linked and the size of the resulting nanogaps

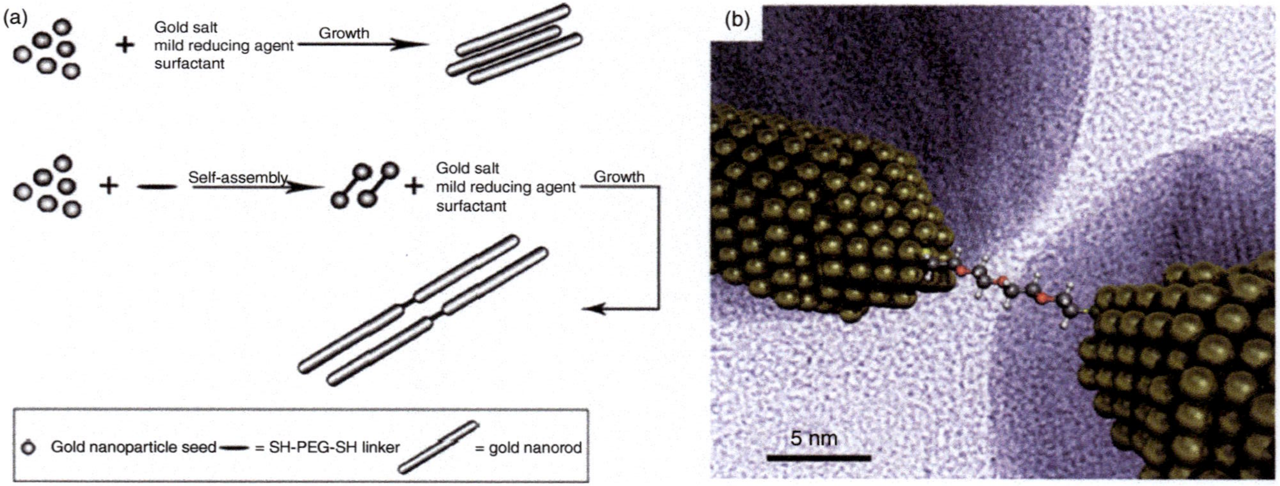

Figure 3.20 (a) Fabrication scheme of gold nanorods. Top scheme: Seed-mediated growth route for synthesis of gold nanorods (AuNRs), which are typically postmodified with desired linkers to create one-dimensional end-to-end linked linear chains as described in the text. Bottom scheme: Our new approach for direct synthesis and growth of end-to-end linked AuNRs using gold nanoparticle seeds that are linked with a SH–PEG–SH linker before addition to the growth solution. (b) High-resolution transmission electron microscopy (TEM) images and schematic view of the linkage between two gold nanorods. The nanogap is 1–2 nm. Source: Jain et al. [166]. 2009, American Chemical Society.

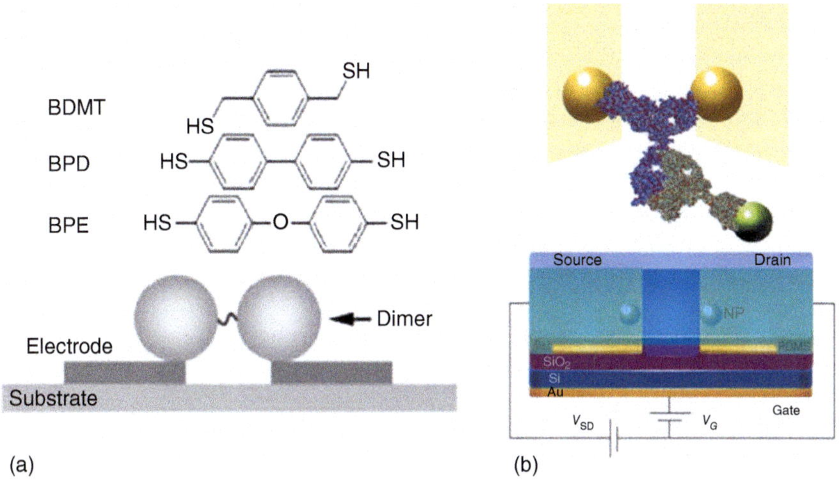

Figure 3.21 (a) The structures of the three molecules: 1,4-benzenedimethanethiol (BDMT), 4,4′-biphenyldithiol (BPD), and bis-(4-mercaptophenyl)-ether (BPE), and the dimer contacting scheme. Source: Dadosh et al. [167]. 2005, Springer nature. (b) Top: Schematic showing one possible molecular arrangement. Bottom: schematic representation of the protein transistor. The device is covered by polydimethylsiloxane (PDMS) to form a liquid channel (100 nm wide and 20 nm deep) and for protection. Source: Chen et al. [168]. 2012, Springer nature.

between the two rods was typically 1–2 nm. Therefore, the electrodes prepared via this bottom-up chemical assembly approach may be more perfectly suited for single-molecule electronics than those fabricated by the top-down way. To realize a functioning single-molecule device, Dadosh et al. [167] have reported an innovative method that was based on synthesizing a dimer structure which consisted of two colloidal gold particles connected by a dithiolated short organic molecule, and electrostatically trapping it between two metal electrodes (Figure 3.21a). The approach avoided the need for fabricating nanometer-sized gaps, as the diameter of the colloidal gold particles was as large as 30 nm. The contacts to the molecules were well defined, and hence the single-molecule devices could be constructed with high certainty, which contributed to the measurement of the conductance of single molecules. Subsequently, Chen et al. [168] prepared the protein transistor made of an antibody molecule and two gold nanoparticles (Figure 3.21b). Typically, a nanogap separated by 10 nm was defined by the EBL technology. Then, the gold nanoparticles with 5-nm diameter were brought onto the edge of the selected electrode pairs with the AFM tip. And the transistor was achieved by the binding of the immunoglobulin to the nanoparticles. The self-assembled way ensured a high reproducibility of the devices, and the protein transistors could be effectively gated with an applied voltage or optical field. Therefore, it is promising to realize the large-scale manufacture of integrated bioelectronic circuits.

Furthermore, Tang et al. [169] proposed the self-assembled nanogaps by assembling the gold nanorods to gold particles which were coated with the self-assembled monolayers (SAMs) of thiol end-capped oligo(phenylenevinylene)s (OPVs). The molecular gap was easily created by the rigid molecule itself during the self-assembly process, and thus the size of the gap was related with the molecular length. Notably, Chiechi and coworker [170] manufactured the metal/molecule/metal tunneling junction by a simple nanoskiving means using the SAMs as templates. The rectangles of gold were deposited onto a silicon wafer through the Teflon mask as the initial electrodes. And after removing the mask, the epoxy polymer was poured over the entire wafer, which was then cured. Subsequently, the gold features were transferred onto the polymer via separating the epoxy from the wafer, followed by the SAMs which served as the templates self-assembling on them. Then the deposition of the second set of gold rectangles were carried out, during which the Teflon mask was placed on the SAM with an offset of 250–500 μm relative to the first gold evaporation. The arduous lithography steps were successfully obviated in the whole processing and the obtained SAM-templated addressable nanogaps were below 3 nm. Meanwhile, this technology circumvented the electrical short problem and thus significantly improved the productivity of the metal/molecule/metal tunneling junction.

3.6.2 Inorganic Films as Templates

The inorganic films can also serve as templates to prepare nanogap electrodes for their precise atomic thickness. For example, Krahne et al. [171] fabricated nanoscale gaps by the gallium arsenide (GaAs) films whose thickness varied between 5 and 10 nm in various samples. In detail, an aluminum gallium arsenide (AlGaAs)/GaAs/AlGaAs quantum well was grown via the molecular beam epitaxy (MBE) technology. Then the 200-nm-high mesa structure was defined through the optical lithography and a standard wet etching process, so that the GaAs layer was exposed on the side of the mesa. Subsequently, the selective wet etching was employed to remove a few tens of nanometers of the GaAs layer from the sloped edges of the mesa structure. And finally, the metal electrodes were thermally evaporated from a direction perpendicular to the plane of the wafer surface. The nanogap was formed exactly where the GaAs layer was removed. It was found that the size of the nanogap was determined by the thickness of the GaAs layer, the surface roughness of the selectively etched AlGaAs/GaAs/AlGaAs interface, and the thickness of the deposited metal layer. This method allowed the simultaneous fabrication of numerous nanogap electrodes on the wafer surface, since the mesa structure and the electron pattern were defined by the optical lithography which was an important step toward low-cost mass production. It is possible to investigate the Coulomb blockade behavior and construct single-electron transistors with these nanojunctions. However, the main disadvantage of this technology lies in the considerably large bulk leakage currents through the semiconductor substrate at ambient conditions which limits the use of the nanojunctions. Notably, Maruccio et al. [172] improved this mesa-gap technique by removing the GaAs gap layer and selectively oxidating the Al-rich layers before the deposition of the metal electrodes. They obtained as best parameters an oxidation time of two hours at 450 °C. After the oxidation, the Al-rich layers were converted into stable native oxide layers which significantly reduced the leakage current and hardly changed the interface roughness. Accordingly, the workers showed the economical fabrication of the complex molecular circuits on a large scale and the resulting gap between electrodes was only a few nanometers. Meanwhile, the transport properties such as the negative differential resistance (NDR) of the molecular tunnel junctions based on the blue-copper protein Azurin were studied for the first time at the single-molecule level in device geometry. Therefore, this molecular circuit demonstrated prospective applications in biosensors and switchable devices.

On the other hand, Luber et al. [173] proposed the high-aspect-ratio nanogap electrodes for addressing the conductance properties of organic molecules (Figure 3.22). In order to prepare the nanogap electrodes, a thin GaAs layer was first embedded in between the two AlGaAs layers by the MBE technology with monolayer precision. Then, the stack structure was cleaved perpendicular to the growth direction, so that an atomically flat crystallographic plane was exposed. Afterward, the thin intermediate GaAs film was selectively recess-etched into this plane with citric acid and H_2O_2, followed by the deposition of the metal layer. The metal film was interrupted at the trench edges and thus formed the two nanogap electrodes. This high-aspect-ratio configuration was potentially suitable for averaging molecular conductance measurements, since the target molecules could assemble across

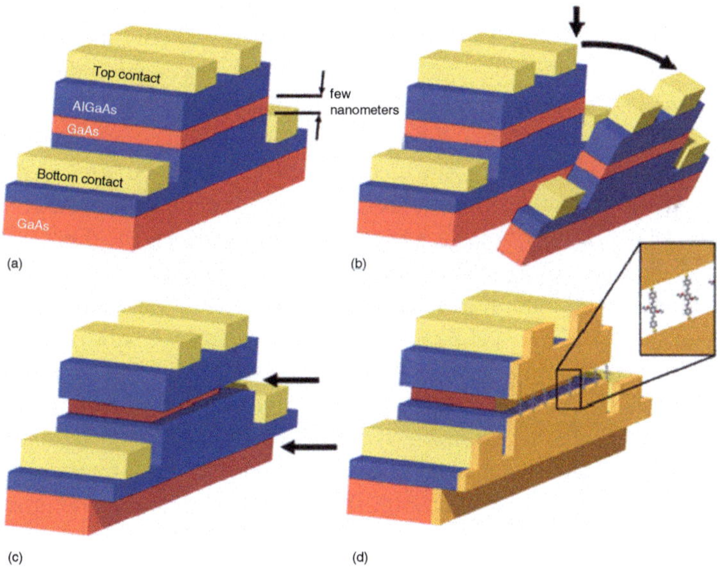

Figure 3.22 Schematic representation of the fabrication process for high-aspect-ratio nanogap electrodes (not to scale). (a) Molecular beam epitaxy (MBE)-grown AlGaAs/GaAs heterostructure after conventional microstructuring with lithographically defined metallic leads. (b) Atomically smooth cleavage along the crystal plane. (c) Selective recess etching of the GaAs layers. (d) Complete device after thin film metallization and subsequent deposition of molecules. The magnified section shows how several molecules bridge the gap. Molecules adsorbed on only one of the electrodes are omitted for clarity. Duplicate metallic leads that connect the bottom and top thin film electrodes allow for a device operational test. Source: Luber et al. [173]. 2007, John Wiley & Sons.

the electrode gap at the desired surface density without interacting. Furthermore, the fabrication protocol [173] was extended toward a three-dimensional surface patterning using the cleaved edge overgrowth technique [174]. A second MBE regrowth step on the in situ cleaved, previously MBE-grown heterostructure side plane was involved during the construction process. Therefore, it was viable to prepare several nanogap electrodes with different lateral widths which had the same electrode gap separation in parallel, which would allow comparing the conductance of a single or few molecules vs. the conductance of a large molecule ensemble average under identical conditions. Moreover, the silicon nanogap electrodes have also been exploited by selectively etching the intermediate layer of silicon oxide. The economical and mass fabrication of the silicon nanogap electrodes is feasible with the standard optical lithography and CMOS process [175, 176].

In addition, Chen et al. [177] constructed the planar array of nanogap electrodes using the sidewall spacer as a sacrificial layer. The size of the nanogap could be precisely controlled by varying the thickness of tetraethylorthosilicate (TEOS) oxide spacer which was prepared through the low-pressure chemical vapor deposition (LPCVD) technology. It was shown that the reproducible nanogap distances from 10 to 100 nm on the 6-in. wafer were easily obtained. Furthermore, the sidewall spacer could also be produced through the anodic oxidation of the patterned metal films [178]. The gap width was easily and accurately regulated by the applied voltage of the anodic oxidation process, for example, when the anodic voltage was 15 V, a 10-nm wide gap was formed. This low-cost and simple fabrication technique is useful in the DNA analysis and to realize the DNA chips, since several different gap widths can be achieved on the one substrate. Remarkably, Tornow and coworkers [179] manufactured the planar nanogap electrodes by the direct high-resolution nanotransfer printing (nTP) method (Figure 3.23). The process started from a solid MBE-grown mold which comprised a few-nanometers-thick GaAs layer embedded in AlGaAs and had an atomically flat bottom surface after being cleaved. Then the thin GaAs layer was recess-etched, featuring a nanoscale groove on the flat cleavage plane. Subsequently, a thin two-layer composite metal film was deposited as transfer material onto the cleavage plane and the MBE-mode was pressed against the substrate. The metal film was transferred onto the planar Si/SiO_2 upon the release of the mode for the stronger adhesion of the chosen transfer material to the activated substrate surface as compared to the mold surface. As the nanogap size was largely predetermined by the chosen configuration of the mode, a separation of about 9 nm between neighboring metal areas was successfully obtained. This hard-to-hard, one-step transfer printing process is simple and fast, which requires neither a flexible buffer layer nor any organic adhesion promoters of additional chemical post-processing.

3.6.3 On-Wire Lithography

The molecular or atomic templates provide simple and mass fabrication processes for the nanogap electrodes; however, the final gap is unlikely to be the exact size of the ruler element on account of a relaxation effect of the metal structures. Remarkably, a more precise template nanolithography method was developed to

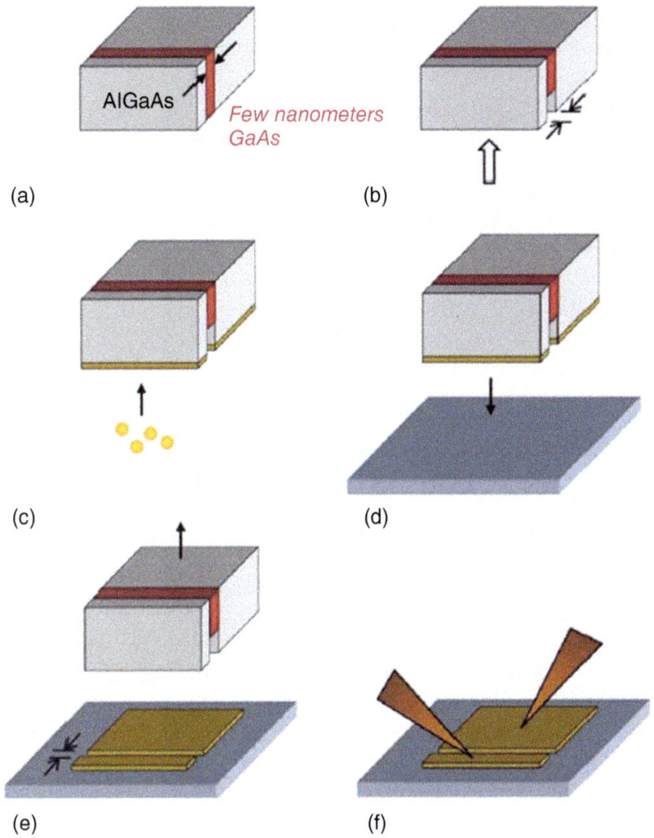

Figure 3.23 Schematic image of nanogap electrode fabrication by direct nTP. (a) MBE-grown mold comprising a few-nanometers-thick GaAs layer embedded in AlGaAs, and having an atomically flat bottom surface after cleaving (not shown). (b) Selective recess etching of GaAs. (c) Deposition of transfer metal thin film on structured mold. (d) Aligning coated mold coplanar to the SiO_2/Si substrate, contacting and pressing it against the same. (e) Release of mold: the metal thin film is transferred, leaving a nanogap between two planar metal layers. (f) Electrical characterization using submicrometer probe needles. Source: Strobel et al. [179]. 2009, John Wiley & Sons.

process one-dimensional wires with sacrificial materials strategically located along the long axis of the wire [180–182]. Typically, the one-dimensional wire, ideally micrometer in length, is synthesized by electrochemically depositing materials in a porous anodic aluminum oxide template. For example, Bao and coworkers [180] have prepared the Au–Ag–Au nanowires of 250 nm in diameter and 6 μm in length, containing Ag segments of 20 μm, 100 nm, and 20 nm respectively. Subsequently, the Ag section was removed using high temperature or chemical etching approaches, to result in a corresponding electrode gap in the Au nanowire. The size of the nanogap was precisely dictated by the thickness of the Ag section, which was easily controlled in the nanowire synthesis process [37]. This technique is termed as OWL, and an attractive advantage is that the entire process

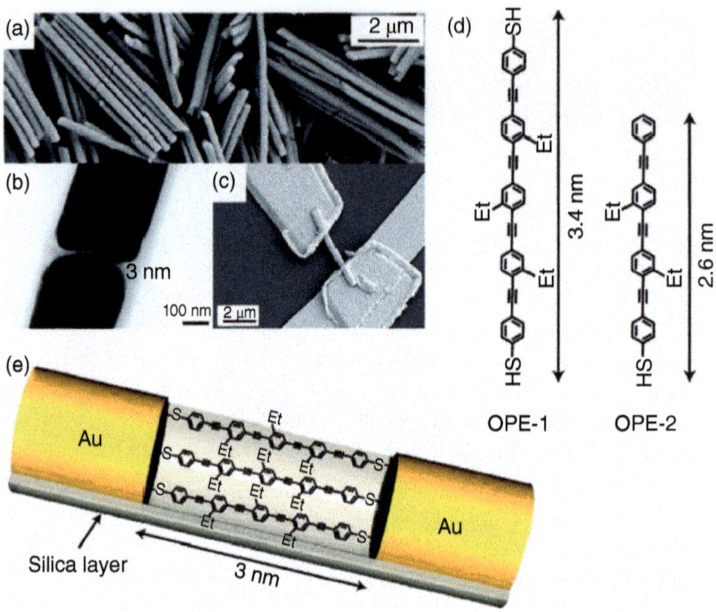

Figure 3.24 (a) SEM image and (b) TEM image of wires with 3.2 ± 0.4 nm gaps fabricated by OWL; (c) SEM image of a device prepared with an OWL-fabricated wire with $a \sim 3$ nm gap; (d) molecular structure of OPE-1 and OPE-2; (e) a diagram of OPE-1 molecules spanning the ~3 nm gap. Source: Chen et al. [184]. Copyright 2008, American Chemical Society.

involves only electrochemical synthesis, heating, or selective chemical etching of the nanowire, all of which are considerably simple without the requirement of sophisticated facilities. It was shown that the nanoscale poly(3-hexylthiophene) (P3HT) organic field-effect transistors (OFETs) were successfully constructed with the 100-nm-gap nanowire electrodes prepared by this technique. In addition, Tian et al. [183] synthesized the Pt–Au–Pt nanowires with 50-nm diameter and 10-μm length, which were then annealed to obtain the sub-10-nm gap nanowire electrodes. The electronic conduction of short double-strand DNA was measured, which demonstrated that the nanowire electrodes were perfectly suited for electron transport studies of organic molecules. Mirkin and coworkers [184, 185] scaled the nanogap down to 2 nm using this high-throughput and highly reproducible OWL method (Figure 3.24). It was feasible to tailor the size of the gap according to the length of the molecular wire; therefore, different molecular transport junctions (MTJs) were fabricated, which may be used for a variety of applications, such as rectifying and sensing. Meanwhile, the Pt/Au heterometallic nanogaps were also realized [181, 186]. When the symmetric thiol-terminated molecules were assembled into such heterometallic nanogap, the resulting MTJs exhibited molecular diode behavior, as the coupling strength between the gold and sulfur (Au–S) was stronger than that of Pt–S. And it was possible to construct a MTJ with the bifunctional molecules, which were capable of self-organizing and orienting within the heterometallic nanogap. Furthermore, multi-modulated MTJs based

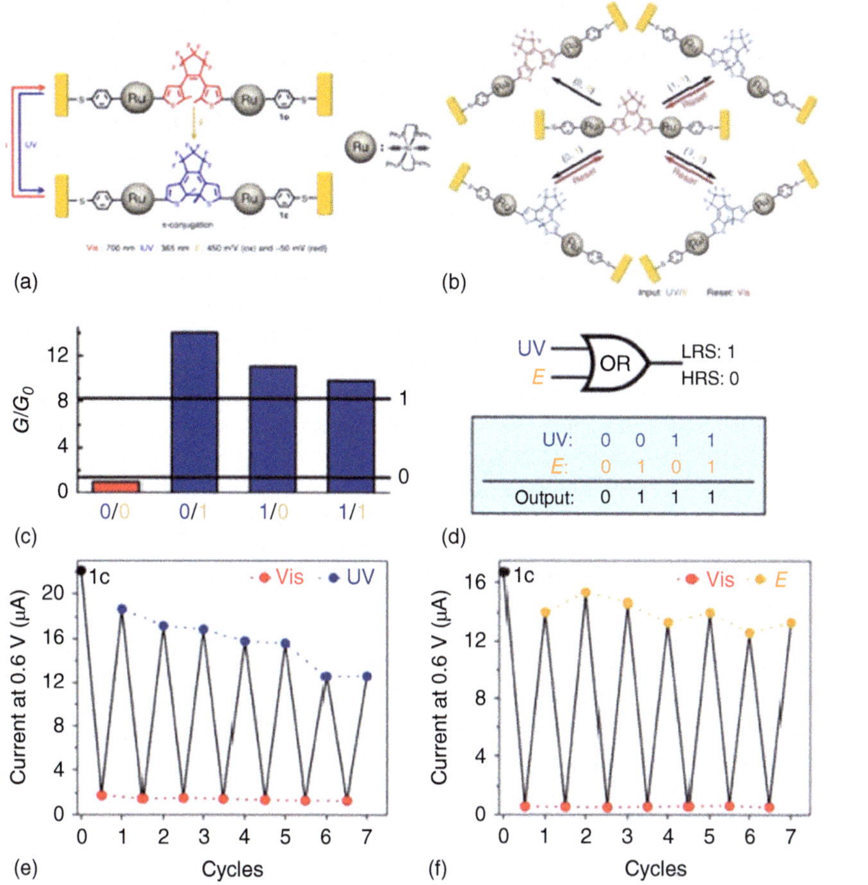

Figure 3.25 Resettable OR logic gate using **1c**-based MTJs. (a) Scheme of molecular isomerization of **1** under external controls. E, electrolysis; UV, UV irradiation; Vis, visible light irradiation. (b) Scheme of **1**-based logic gate: **1** isomerization within nanogap devices under different inputs (UV irradiation and electrolysis). Reset: visible light irradiation (700 nm, 2 hour). (c) Conductance (*G*) changes with different input combinations. (d) The electronic symbol and a truth table of the OR logic gate. (e) and (f) Endurance performance: current response of a **1c**-based nanodevice under alternate modulation of visible light irradiation (700 nm for 2 hour) and (e) UV irradiation (365 nm for 30 minutes) or (f) electrochemical stimuli (oxidation at 450 mV and reduction at −50 mV for 10 minutes). The current values were recorded at 0.6 V bias in vacuum. Source: Meng et al. [188]. Springer nature. Licensed under CC BY 3.0.

on the association of original ruthenium/dithienylethene (DTE) complexes and OWL-fabricated nanogaps with suitable gap sizes were manufactured [187, 188]. This molecular device demonstrated the bi-state switchable conductivity for the first time by judicious control of the orthogonal light irradiation and electrochemical stimuli (Figure 3.25). It was found that the MTJs could be applied as resettable electronic logic gates for Boolean computing, which showed immense potential to develop multifunctional molecular devices.

3.6.4 Nanowire Mask

The carbon nanotubes (CNTs) with nanometer-sized diameters and micrometer lengths exhibit superb mechanical stiffness and chemical stability. Hence, they are considered a good choice of shadow masks to construct nanogap electrodes. Metzger and coworker [189] prepared a reproducible 10-nm nanoditch up to a few micrometers long. Briefly, the multiwalled carbon nanotubes (MWCNTs) or SWCNTs were deposited on a suitable substrate. Then, the metal evaporation and a successive lift-off process were carried out, so that the nanogaps consistent with diameters of the CNTs could be readily obtained. On the other hand, Johnson and coworkers [190] proposed a highly controllable and scalable method to fabricate nanogap arrays using the suspended MWCNT or SWCNT bundle as the shadow mask combined with a trilayer EBL process (Figure 3.26). In detail, the 80-nm-thick e-beam resist was first coated onto an oxidized silicon substrate, followed by the deposition of CNTs sonicated in isopropanol onto the resist layer. Then another 80 nm of e-beam resist was spun on the CNTs layer. Next, the trilayer resist was exposed in a pattern which consisted of a 100–500-nm-wide continuous line crossing, and thus the CNT mask was left free standing above the oxidized silicon substrate. Subsequently, a 25-nm-thick Au film was thermally evaporated; it was found that the CNT bridge could effectively prevent the metal from reaching the area directly underneath it. And finally, to remove the undesirable metal and CNTs, the unexposed resist was dissolved in acetone. This reliable technology not only made the gap dimensions more controlled but also allowed a thicker deposition of metal films and cleaner nanojunctions [191, 192]. It was found that the diameter and spatial placement of the CNTs were the two key factors which contributed to a successful CNT shadow mask fabrication. Initially, the CNTs were dispersed by an ultrasonication method which had the advantages of low cost, scalability, and compatibility with many substrate materials [190]. However, the ultrasonication

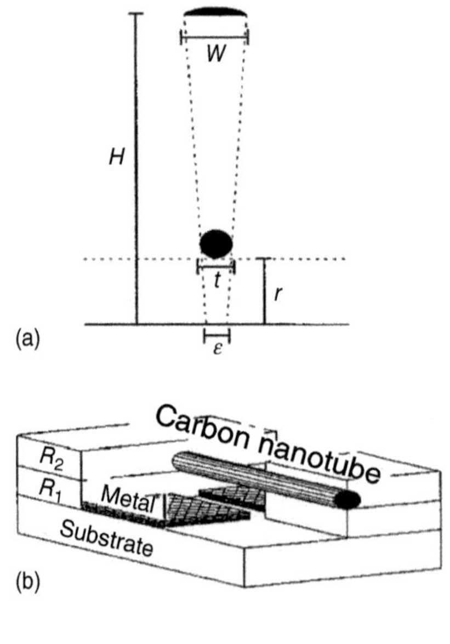

Figure 3.26 (a) In a shadow mask method, the shadow or gap size (ε) depends only on geometrical factors: the mask width or tube diameter (t), the tube-substrate distance (r), the evaporation source distance (H), and width (W). (b) Carbon nanotubes are incorporated between two layers of e-beam resist (R_1 and R_2) to act as a shadow mask preventing metal deposition at one point along a thin wire. The figure illustrates the process after metal deposition. Source: Lefebvre et al. [190]. 2000, AIP publishing LLC.

led to the random placement of the CNTs and may break them into small pieces. Remarkably, a composite electric field-guided assembly (CEGA) technique was developed to precisely position the nanosized gaps by elaborately manipulating the CNTs [193, 194]. In particular, the dielectrophoretic force induced by the gradient of a high-frequency oscillating field attracted the MWCNT across a pair of electrodes. And the DC electric field prevented the deposition of more than one CNT across the same pair of electrodes by an electric field–fluid interaction. Moreover, the CNT mask could be accurately placed by direct mechanical transfer techniques [191, 195].

In addition to the CNT, other organic or inorganic nanowires can also serve as the shadow masks. For example, Jiang et al. [196] presented an "organic ribbon mask" technology in which the single-crystalline nanometer-sized ribbons acted as the template to determine the channel length of the OFETs. The organic ribbon was exactly located and peeled off by a mechanical probe to form the nanogap. Therefore, this technology furnished a cheap, facile, and controllable way to construct single-crystal devices with the short conducting channels. It was demonstrated that the gap between the metal leads could be easily scaled down to submicrometer

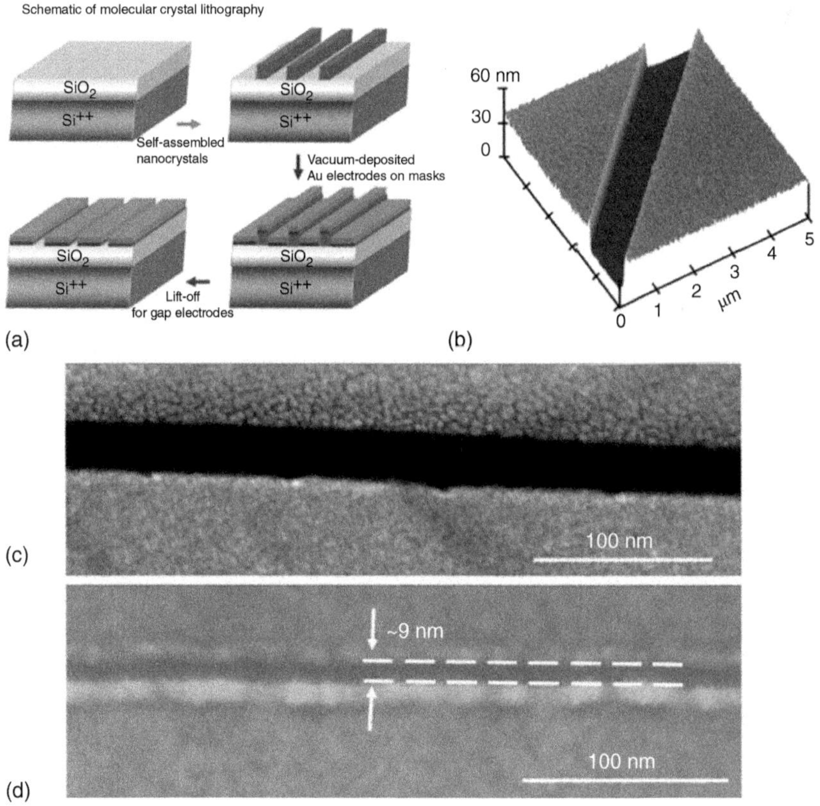

Figure 3.27 (a) Schematic process for fabricating nanogap electrodes by molecular crystal lithography. (b) The electrode pair with clear and smooth edges. (c, d) Example SEM images of obtained nanogap electrodes: ≈30 nm (c) and ≈9 nm (d). Source: Jiang et al. [197]. 2012, John Wiley & Sons.

level, since it depended on the width of the organic ribbon. Afterward, the workers developed a low-cost and simpler method, named molecular crystal lithography, to fabricate nanogap electrodes efficiently (Figure 3.27) [197]. It was notable that the two-dimensional (plate-like) crystals could be self-assembled and stand on the substrate, which simplified the formation of the mask and narrowed the gap down to a few nanometers. Furthermore, Min et al. [198] proposed a simple, reliable way of fabricating large-area nanogap electrodes using the printed semiconducting nanowire arrays as shadow masks. More innovatively, Ongaro et al. [199] developed a thiol-DNA-biotin template method to assemble the protein-functionalized Au electrodes. Typically, the thiol-DNA-biotin template was assembled in solution and deposited on the silicon substrate. Then the modified gold nanoparticles were selectively attached onto the DNA backbone to form the electrode. It was shown that the size of the gap was approximately 10 nm, which was consistent with the dimensions of the proteins located between the DNA backbones. This method opens up the possibility of building molecular devices directly onto molecular frameworks.

3.7 Focused Ion Beam

Ions, compared to electrons, are much heavier and can strike with greater energy at relatively short wavelengths, which makes FIB a powerful technique for creating well-defined nanostructures [200]. A FIB instrument is normally coupled in a system with a SEM (Figure 3.28a), allowing the simultaneous monitoring during the manipulation process. On one hand, FIB enables site-specific removal of materials by the localized impacting ions (usually Ga^+) which eject the surface atoms from the target substrate, referred as FIB milling (Figure 3.28b); on the other hand, FIB can be used for guiding the deposition of a material from the gaseous precursor onto the specimen surface, known as FIB-induced deposition (Figure 3.28c) [200, 201].

For the fabrication of nanogap electrodes, FIB is particularly attractive, benefited from heavy mass of ions, small beam spot size, low lateral exposure, and few proximity effects [200, 202], which enables maskless, high-speed generation of relatively narrow gaps with good reproducibility and yield [203–205].

Table 3.1 lists some representative nanogap electrodes fabricated by standard FIB etching. It shows that Ga-FIB is flexible in the choice of electrode materials, which include the most widely used Au, nonmetal CNTs, and less studied Pd and Al. The minimal gap sizes achieved by these FIB processes range from sub-10 to 50 nm, which was suitable for sandwiching C_{60} clusters [208], individual nanoparticles, and their aggregates [203]. However, further narrowing would be rather difficult due to the limit of the beam spot size. The highest resolution of FIB is 4–5 nm [202], which means that fabrication of nanogaps with sub-5 nm sizes, comparable with the dimensions of organic molecules, is a highly challenging task.

In 2003, Nagase et al. [31] found that Au electrodes with gaps of 5–8 nm were reproducibly formed by etching with ~12 nm size Ga-FIB. Although the gap size was still much larger than the length of typical molecules, they demonstrated for the first time that nanogaps could be prepared with separation much smaller than the beam diameter. This experimental discovery was subsequently supported by numerical

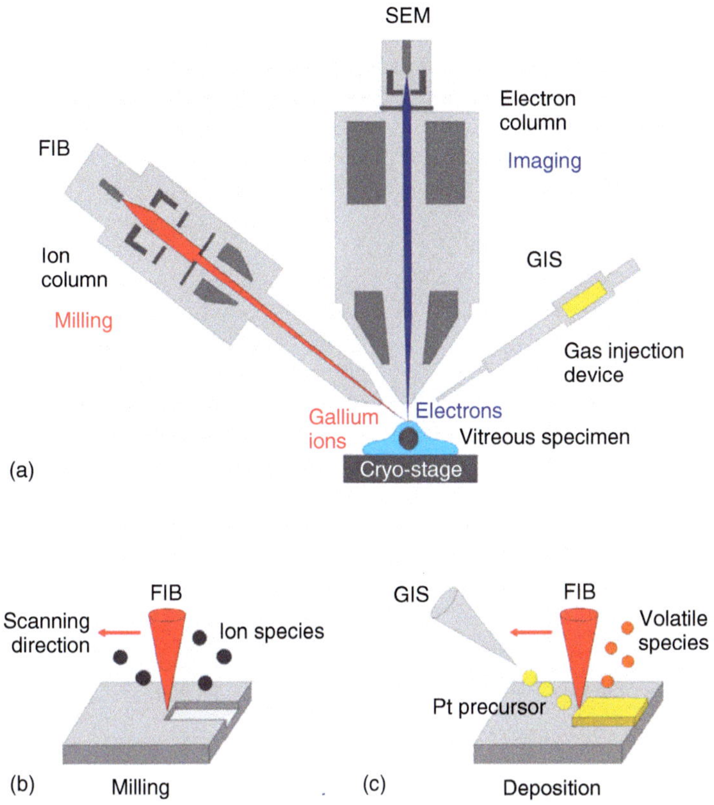

Figure 3.28 Schematic of an FIB system and its working modes. (a) A focused gallium ion beam milling with simultaneous SEM monitoring. The cryo-stage is designated for specimen preparation in structural biology study. (b) Ion beam milling which leads to progressive removal of material. (c) Ion beam-assisted deposition that is used to guide the deposition of material from a gaseous onto the specimen surface [201]. Source: Rigort and Plitzko [201]. 2015, Elsevier.

Table 3.1 Examples of nanogap electrodes fabricated by conventional FIB method.

Electrode material	Ion	Beam spot size (nm)	Minimum gap size (nm)	References
Au	Ga^+	10	20	[203]
Au	Ga^+	6	8	[206]
Au	Ga^+	12	35	[207]
Carbon nanotube	Ga^+	22	50	[208]
Al	Ga^+	10	30	[209]
Pd	Ga^+	Not mentioned	25	[210]

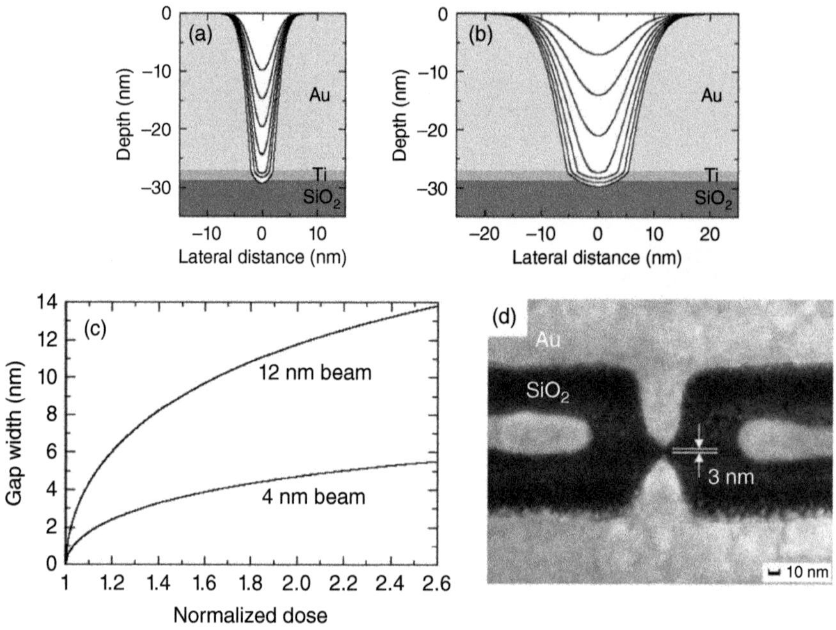

Figure 3.29 (a, b) Calculated cross-sectional profiles for a gap fabricated by Ga-FIB with a beam diameter of 4 nm (a) and 12 nm (b). (c) Gap width as a function of dose normalized to the critical dose to cut films. (d) SEM photograph of a fabricated nanogap [211]. Source: Nagase et al. [211]. 2005, Elsevier.

simulation [211], as shown in Figure 3.29a–c, which revealed, if the beam dose was controlled precisely low enough, very narrow gaps would be formed regardless of the large beam spot used for etching. For example, it was expected that at a normalized dose lower than ~1.1 narrow gap with a width smaller than 2 or 4 nm could be produced using a beam size of 4 or 12 nm, respectively (Figure 3.29c). Based on that, the authors performed careful FIB etching combined with in situ monitoring of gap formation using current flowing as feedback. In this manner, they were able to fabricate ~3-nm Au gaps (Figure 3.29d) with ~12-nm diameter FIB [211]. They further optimized the electrode structure through adding a thin Ti as top layer to prevent excess etching of the underlying Au film by FIB tails, which led to excellent yields of 90% for ~3–6-nm gaps and higher insulating resistance ranging from 10 GΩ to 1 TΩ [204]. In a more recent study, Li et al. [205] observed the gap size evolution by precisely tuning the ion dosage in FIB milling. At a critical dosage when Au gap started to be fully formed, nanogaps as narrow as 2.3 nm were achieved using 7 nm beam spot. These nanogap electrodes exhibited resistance of tens of teraohms; Fowler–Nordheim tunneling effect was observed and Simmon's model was applied to accurately determine the gap size. As indicated by the above studies, low-dose FIB opens an avenue to surpass the resolution limit encountered by the conventional FIB method.

Cui et al. [212] developed a creative strategy of utilizing single grain boundary (GB) junctions for creating ultra-narrow suspended nanogaps by FIB milling (Figure 3.30). GB is the interface between two grains within a polycrystalline

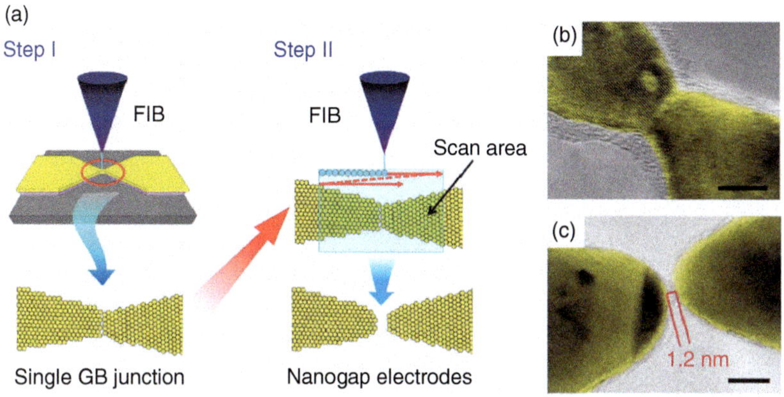

Figure 3.30 (a) Illustration of nanogap fabrication through the break of singe GB junctions: step 1, fabrication of single GB junction; step 2, break of the single GB junction by FIB milling. (b) TEM images of the fabricated single GB junction; (c) TEM images of fabricated nanogap electrodes with 1–2 nm gap width (scale bar: 5 nm) [212]. Source: Cui et al. [212]. 2015, John Wiley & Sons.

material, where the inter-granular fracture takes place easily. The fabrication process involved: (i) producing a suspended single GB junction by dose-controlled FIB milling on a suspended Au nanowire in a designated area, and (ii) formation of a nanogap by careful FIB scanning around the location of the GB junction. Break of such junctions under FIB enabled gap width down to 1–2 nm, benefitting from the preferential removal of the atoms along the GB. Such suspended nanogap electrodes are stable, freestanding, and debris-free, which are suitable for single molecular electronic device construction. In a subsequent research, the authors suggested the use of thermal annealing method, which broke the GB junction and generated sub-5-nm nanogaps through thermal-induced atomic migration at GB, with additional advantages of higher time- and cost-efficiency, better morphology stability, and less structural defect [213].

Besides FIB milling, FIB deposition was proposed as another method to overcome the limitation of the beam size. Shigeto et al. [214] reported the formation of sub-2-nm tungsten nanogaps by FIB deposition on suspended silicon nitride membrane with a 1–2-μm preformed slit (Figure 3.31). During the local deposition, tungsten was deposited not only on the silicon nitride substrate but also on the edge of the slit, resulting in the growth of tungsten electrodes from both sides of the slit. Meanwhile, tunneling current between the electrodes was monitored in situ and used as a good criterion to control the gap separation.

The recent emergence of He-based FIB, implanted in helium ion microscope (HIM), provides a superior tool in high-resolution nanofabrication with an unparalleled beam spot size of <0.5 nm [215, 216]. It would, in principle, allow the cutting preparation of nanogaps with separations below 1 nm. Although such results have not come out yet, He FIB has already manifested its overwhelming advantages over the well-established Ga technique in the preliminary practice. Thiele et al. [217] used helium ion beam lithography to pattern nanogaps into metallic

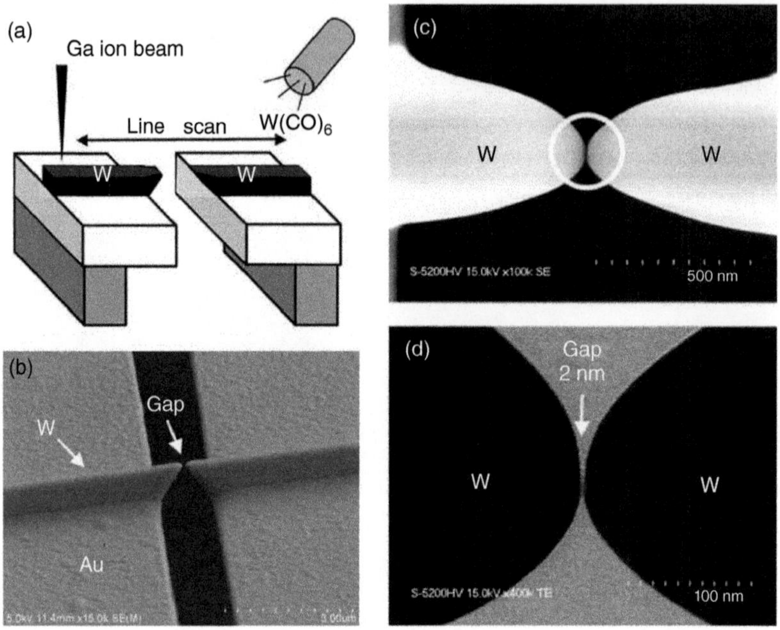

Figure 3.31 (a) Schematic illustration of the gap fabrication process. Tungsten is deposited through the decomposition of a tungsten hexacarbonyl vapor onto the substrate where the Ga ion beam is irradiated. The tungsten is deposited not only on the substrate but also on the edge of the slit. Consequently, the tungsten electrodes grow from both sides of the slit. (b) SEM image obtained for a sample during the fabrication process. (c) SEM image of top view of suspended tungsten electrodes with nanometer-sized gap. The existence of the gap is not clear. (d) Scanning transmission electron microscope (STEM) image of the center part of (c), showing the existence of a 2 nm gap [214]. Source: Shigeto et al. [214]. 2006, Elsevier.

SWCNTs embedded in transistor (Figure 3.32a), obtaining narrow gaps with size of 2.8 ± 0.6 nm (Figure 3.32b). The critical dose for gap formation was determined by in situ voltage-contrast microscopy analysis. The resulting nanogap-containing device had a resistance of hundreds of teraohms, 9 orders of magnitude higher than the pristine SWCNT device. These nanogaps in device were then tested as contacts for oligo(phenylene ethynylene) (OPE) molecular wires of 3.9 nm length, to demonstrate their practical usage in molecular-scale electronics. The first trial of He^+ beam fabrication of Au nanogap, using a slow line scan, was performed by cutting a thin Au nanowire on a carbon TEM grid, and an 8 nm gap width was produced [219]. On the other hand, scientists from Carl Zeiss microscopy reported ~4-nm nanogaps machined in 100-nm-thick Au [220]. Scholder et al. [221] created Au pairs with minimum gap of 3.5 nm by controlling the He milling conditions, from cutting 20-nm-thick Au islands. It was found that for gaps above 10 nm, their final size was predominately defined by the width of the He^+ milled area, while the dose of He^+ need to be carefully adjusted in order to produce narrower gaps. Kollmann et al. [218] demonstrated a combined approach of Ga^+ FIB and milling-based He^+-ion lithography for the fabrication of nanogap antenna arrays. First, arrays of connected

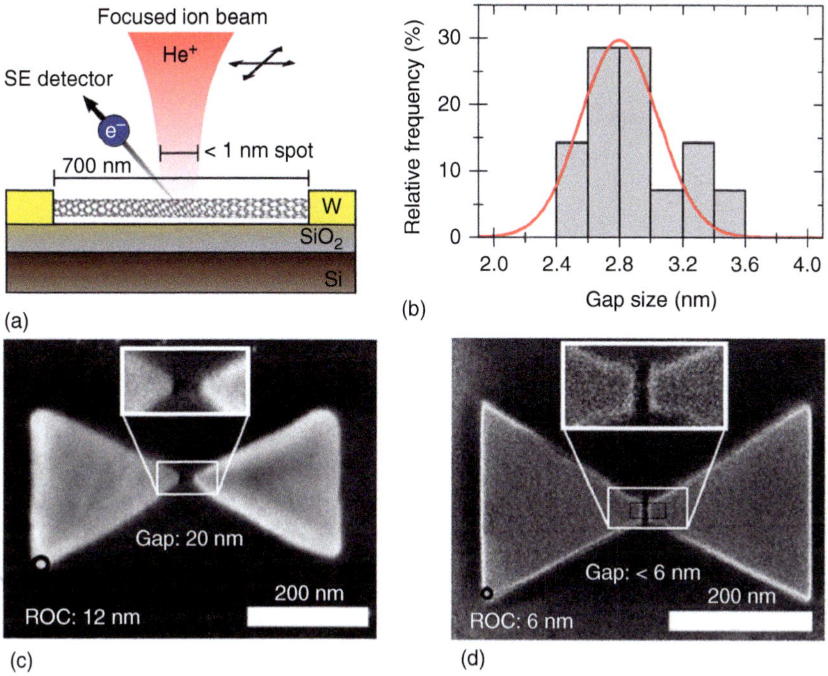

Figure 3.32 (a) Scheme of helium ion sputtering of metallic SWCNTs embedded in device. The nanotubes are deposited on tungsten electrodes and supported by SiO_2/Si substrates. (b) Histogram of 14 nanogap sizes of metallic SWCNTs. (c–d) Images of bowtie nano antennas fabricated by Ga^+-ion (c) and He^+-ion; (d) beam milling, ROC referring to radius of curvature. Source: (a–b) Thiele et al. [217]. 2014, AIP publishing LLC. (c–d) Kollmann et al. [218].

equilateral triangles was created by standard Ga^+ FIB from a 30-nm-thick gold film; then, He^+-ion milling was performed for cutting short lines that separate the triangle and form the nanogaps. By this approach, gap size of less than 6 nm was produced with single-nanometer accuracy and high reproducibility, and the bowtie pairs showed sharp corners and straight edges (Figure 3.32d). As a comparison, if Ga^+ beam was used for the final cut, the resultant gaps were no narrower than 20 nm with rounded corners (Figure 3.32c) after time-consuming steps of focus optimization. They further showed that the superior nonlinear optic performance of the He^+ lithography produced nanogap antennas.

Because of the lighter mass of helium ion, He^+ beam milling is much slower than the Ga treatment, which is considered as a limitation or drawback in nanofabrication. However, the reduced speed can in turn be recognized as mildness and better controllability that are in favor of high-resolution fabrication of nanogap electrodes. In practice, other high-speed techniques, such as Ga FIB and EBL [218, 221], can always be used to remove most of the materials in advance, while He^+ milling is performed to do the most precise work – the final cut. This combination approach keeps full advantage of the sub-nanometer beam size without sacrificing the time efficiency.

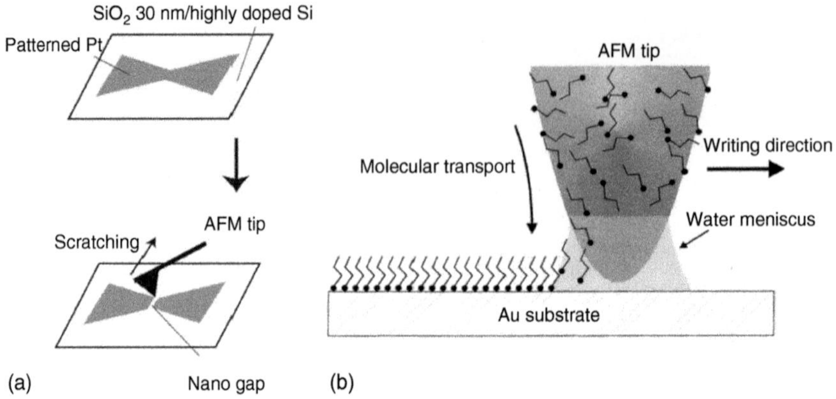

Figure 3.33 (a) Fabrication process of Pt nanogap electrodes. Source: Miyazaki et al. [225]. 2003, IOP Publishing. Copyright (2021) The Japan Society of Applied Physics. (b) Schematic representation of dip-pen nanolithography (DPN). A water meniscus forms between the AFM tip coated with 1-octadecanethiol (ODT) and the Au substrate. The size of the meniscus, which is controlled by relative humidity, affects the ODT transport rate, the effective tip-substrate contact area, and DPN resolution. Source: Piner et al. [226]. 1999, AAAS.

3.8 Scanning Probe Lithography and Conducting Probe-Atomic Force Microscopy

The scanning probe lithography (SPL) describes a set of lithographic methods in which a microscopic or nanoscopic stylus is moved mechanically across a surface to form a pattern [222–224]. There are two main types of SPL. One is a destructive way (Figure 3.33a) which physically, chemically, or electronically deforms the substrate's surface to make up the pattern by providing the substrate with energy [225, 227, 228]. And the other is a constructive way (Figure 3.33b) in which the patterning is performed through directly transferring chemical species to the surface [226, 229]. Since the sharp tips of a scanning probe microscope with a radius smaller than 10 nm can be neatly and controllably manipulated across the substrate, the SPL is quite suited for the fabrication of nanogap electrodes.

3.8.1 Destructive Way

The simplest way is to scratch the pre-patterned metal strip with the scanning probe to form a gap on the order of a few nanometers. The advantages of this method are obviously the precision of alignment, the nondamaging definition process, and the absence of addition etching procedures [227]. Venema et al. [230] showed that it was feasible to cut CNTs in a controlled way by applying a voltage pulse to the tip of a STM. The gap in the tube was typically 5–10 nm. On the other hand, the AFM is another promising tool for the SPL, as it is not restricted to conducting materials. For example, Miyazaki et al. [225] fabricated the Pt nanogap electrodes by a scratching technique using an AFM tip. Briefly, a Pt pattern with a narrow part which was 3-μm wide was prepared via a lift-off method. Then the contact-mode AFM tip was brought down into contact with the surface and scanned across the narrow Pt pattern. Notargiacomo et al. [36] reported on the investigation of direct writing as well, where the AFM probe scratched a metal strip. The material removal was performed in a contact mode by applying an amount of force of several micronewtons, thus producing fine grooves. It was demonstrated that the scratching properties were closely related with the applied force, number of scan cycles, and scan speed [231]. Notably, He et al. [228] successfully constructed the graphene and graphene oxide (GO) nanogap electrodes with gap width below 10 nm by AFM nanolithography.

In general, the direct scratching is an instructive while not a versatile technique, for the limitation and damage of the probe and unsatisfactory material removal. Therefore, the second way which combined the SPL with EBL and wet etching was proposed, avoiding a direct interaction between the probe and the metal to be patterned. Magno and Bennett [227] wrote the patterns on an etch stop layer by an AFM tip, which were then transferred into the underlying layer by a wet etching procedure. In order to narrow the gap, an optimized AFM-assisted mask patterning was suggested (Figure 3.34) [36]. The titanium stripes were first covered with a 60-nm-thick polymeric film, followed by the deposition of a 5-nm thin aluminum sacrificial layer. After the AFM lithography, the gaps opened in the top aluminum layer were transferred to the polymer layer underneath via an oxygen plasma etching process. Then, the aluminum/polymer bilayer acted as a mask for the subsequent titanium etching by the RIE using a $SF_6/CHF_3/O_2$ mixture. Finally, the aluminum and residual polymer film were removed, respectively, through wet etching and oxygen plasma. It was found that the gap as narrow as 40 nm was achieved, since the very thin top metallic layer sustained the whole etching process

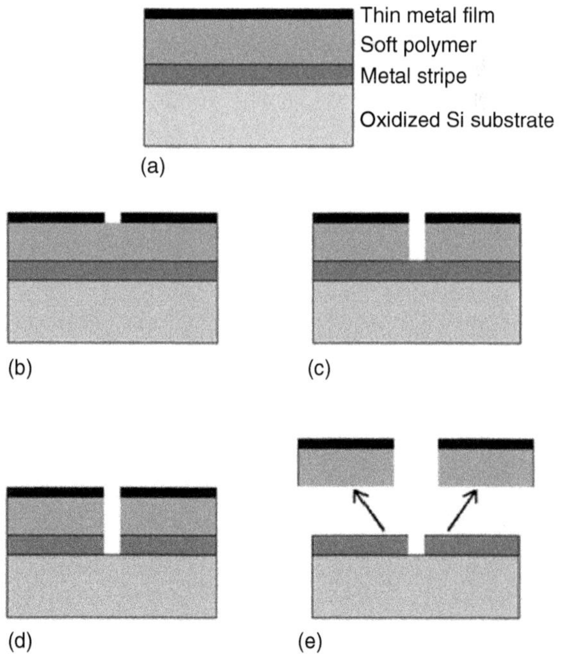

Figure 3.34 Layout of the sample and the process steps: (a) sample multilayer structure; (b) thin metal mask patterning; (c) polymer removal in plasma oxygen; (d) titanium stripe etching; and (e) resulting electrodes after sacrificial layers removal. Source: Notargiacomo et al. [36]. 1999, IOP publishing.

without being appreciably back sputtered and the $SF_6/CHF_3/O_2$ mixture could produce very anisotropic profiles.

In addition to the mechanical scratch lithography, the direct resist exposure shows immense potential for the fabrication of nanogaps in the metal nanowire. Miyazaki et al. [232] utilized the SPL technology to pattern an electron beam negative resist, and the resist patterns could be transferred to the underlying titanium film by wet etching. In this way, Ti nanowires with a desired dimension could be easily obtained. Furthermore, the nanogap was obtained on this line pattern by switching off the voltage bias between the AFM tip and the sample for a short time during the scanning process, to result in an unexposed section which eventually formed the gap.

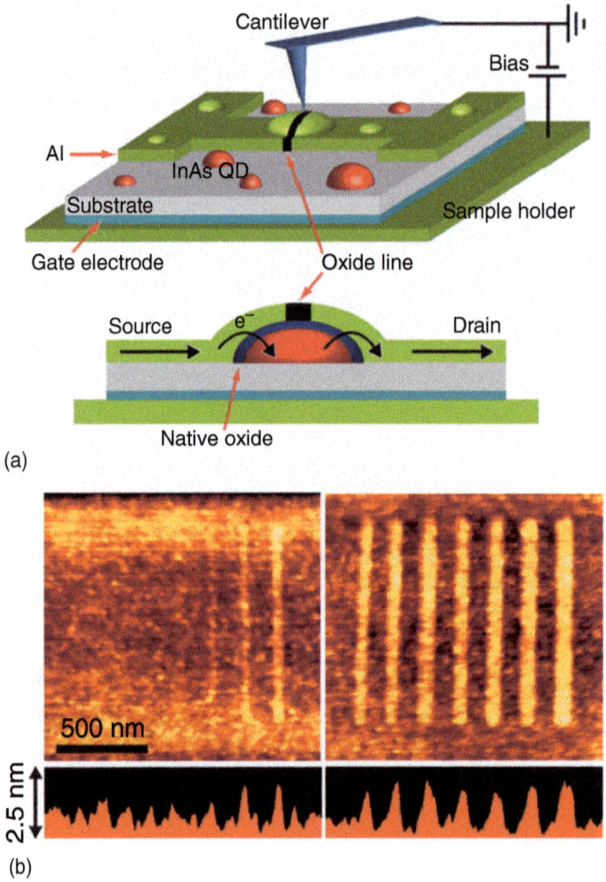

Figure 3.35 (a) Schematic illustration of AFM local oxidation method for fabrication of nanogap electrode. A bias voltage is applied between the cantilever and the sample holder. The figure also shows a cross section of the central part of the device. The native oxide layer of InAs is shown in blue. Electron transport after the fabrication of the nanogap electrode is shown by black arrows. (b) AFM topography image of oxide lines fabricated on the surface of a 5 nm Al/1.5 nm Ti film. A cross section of the topography is also shown. The bias voltage was varied from 1 to 14 V (left to right in the figure) in 1 V steps at a constant scan speed of 200 nm/s. Source: Moriya et al. [233]. 2010, IOP Publishing. Copyright (2021) The Japan Society of Applied Physics.

More innovatively, Moriya et al. [233] employed the AFM local oxidation method to construct the inorganic single-electron transistor in which the nanogap electrodes were aligned with a single self-assembled QD (Figure 3.35). Initially, the 200 nm GaAs/100 nm AlGaAs/300 nm n^+-GaAs layers were deposited on an n^+-GaAs substrate. Then, the self-assembled InAs QDs were grown by the MBE. Subsequently, an Al strip consisting of 5 nm Al/1.5 nm Ti was prepared using the EBL and lift-off technologies. The humidity around the sample was controlled to about 50%. When the cantilever of the AFM which was operated in the tapping mode was scanned under the application of a bias voltage between the Al strip and the cantilever, an oxide line was formed right on top of the QD. This oxide layer suppressed the conduction through the metal, and hence the electron transport became electron tunneling from the left Al lead to the right one via the InAs QD. It was shown that the width of the oxide line could be controlled by adjusting the bias voltage, and a high-quality single-electron transistor was fabricated reliably with this technique.

3.8.2 Constructive Way

As for the constructive type, Mirkin and his colleagues [234] successfully prepared sub-50-nm solid-state nanostructures in a way termed dip-pen nanolithography (DPN). Among these nanostructures, nanowires with deliberately designed gaps were manufactured by initially using DPN to pattern the etch resist 16-mercaptohexadecanoic acid, on the Au/Ti/SiO_x/Si substrate and then removing the exposed gold via the wet etching (Figure 3.36). It was demonstrated that

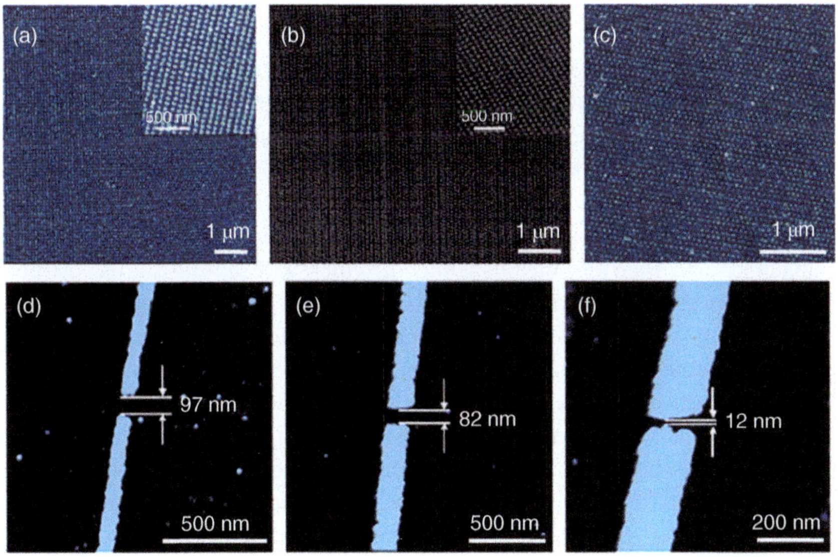

Figure 3.36 Trapping-mode atomic force microscopy (TMAFM) topographic images (a, c) and SEM image (b) of an etched 16-mercaptohexadecanoic acid (MHA)/Au/Ti/SiO_x/Si dot nanoarray. The insets are high-resolution images. (d–f) TMAFM topographic images of etched MHA/Au/Ti/SiO_x/Si nanogaps. Source: Reproduced with permission from Zhang et al. [234]. Copyright 2003 American Chemical Society.

nanogaps as small as 12 nm were obtained. Other than the lithography, fine tips for SPM can also be utilized to manipulate single atoms or molecules on specific substrates to construct the desired configurations [229, 235]. Ho and coworkers [229] investigated the ingenious assembly of a copper(II) phthalocyanine (CuPc) molecule between two Au atomic chains on a NiAl substrate. When the STM tip was close enough to the substrate, the individual Au atoms could be pulled along the trajectory of the tip to build one-dimensional Au chains along the Ni troughs. The CuPc molecule was manipulated in a similar manner, and thus a metal/molecule/metal junction was perfectly fabricated composed of a CuPc molecule and two Au chains. This innovative molecular junction provided a new way for electronic characterization of single molecules and their contacts with metal electrodes.

3.8.3 Conducting Probe-Atomic Force Microscopy

Remarkably, the conducting probe-atomic force microscopy (CP-AFM) provides an efficient and convenient way to fabricate the molecular junction, in which the AFM tip acts as a top electrode [236–238]. Morita and Lindsay [239] obtained the highly reproducible electronic behaviors of single or a few molecules by utilizing the CP-AFM. In detail, the gold-coated mica substrate was first immersed into a monothiol solution to prepare a SAM and then immersed into a dithiol solution to insert the dithiol molecules into the defects in the monothiol matrix. Subsequently, the gold nanoparticles were chemically grafted to the remaining end of the dithiol via treating the sample with the nanoparticle solution. And finally, the electrical circuit was completed by pushing a conducting AFM probe into the gold nanoparticle. It was found that the diameter of the gold particle played an important role in characterizing the molecular properties, since the contact resistance between the AFM probe and the small gold particle would be large enough to allow the nanoparticle to charge independently, which could distort the I–V curves both through the Coulomb blockade and the intrinsic electronic gap in the small particle. The larger nanoparticles were suggested to eliminate these problems, and thus the resulting molecular junctions demonstrated reasonable values for single-molecule conductance and their distance dependence which agreed with the BJ method. This CP-AFM method is a valuable supplement to the BJ technology. In addition, Wei et al. [240] measured the monolayer molecular junctions on the order of a few hundred molecules with this simple and reliable CP-AFM method. In this way, the resulting nanogap electrodes were separated by a well-defined molecular structure, so that the effects of chemical substitutions could be studied expeditiously, which was important for the design of new molecules for functional molecular electronics [241]. Moreover, Wen et al. [242] deposited several layers of graphene onto the conducting AFM tip to reduce the contact effects between the metal tip and molecules such as the potential barrier, interfacial dipoles, and structure disorder. Meanwhile, the graphene improved the endurance, operational stability, and shelf life of the conducting AFM electrode, thus promoting the practical applications of molecular junctions.

The advantage of scanning probe microscope-related technologies stems from their agility and controllability in the range below 100 nm, which is very suitable for research in the laboratory. However, as the scanning area and rate of the probe is limited, it is challenging for the large-scale fabrication.

3.9 Nanogap Electrodes Prepared with Nonmetallic Materials

3.9.1 Introduction

With the rapid development of the CMOS-based integrated circuit technology, it has become more difficult to follow Moore's law to further downscale the size of electronic components both physically and financially [243, 244]. As early as 1974, Aviram and Ratner have predicted that individual molecules someday would be used as active electronic units in devices [1]. In this sense, it is remarkable that chemically identical molecules, with sizes on the order of 1 nm, which could be synthesized in bulk and accomplish a variety of functions, have the potential to partly replace the traditional solid-state device counterparts in the future. And the ultimate goal for shrinking electronic circuits is the realization of molecular-scale electronics, since single molecules constitute the smallest stable structures imaginable [245–247]. If this comes true, it would be possible to integrate transistors in billions. The molecular-based devices exhibit unique advantages for electronic applications, such as lower cost, higher efficiency, ability of self-assembly and recognition, and synthetic tailoring ability by elaborate choice of geometry and composition [2].

To date, molecular electronic devices have been primarily metal/molecule/metal devices utilizing thiolated chemistry on gold electrodes [248]. In molecular electronic junctions, the nanogap electrodes play an important role, as the electrical signal comes in and out of the molecules via contact-coupled electrodes. Most of the nanogap electrodes are prepared with metallic materials, such as gold [29, 32, 249], platinum [87, 250], and titanium [251, 252]. As basic components of modern electronic circuitry, metallic electrodes can be more easily interfaced to the macroscopic world. Furthermore, several effective and creative methods of preparing metallic nanogap electrodes with controlled spacing have been proposed. And the well-developed self-assembly techniques provides a facile and high-yield way to fabricate the molecular devices [253–255]. Based on the metal/molecule/metal junction, it is possible to investigate the molecular properties, and explore the charge transport mechanism; thus, specific functional molecular devices have been accordingly designed, including transistors [7, 8], switches [5, 6], and rectifiers [4].

However, despite the advantages of the metallic materials [256], such as few defects, low cost, and high conductance, there remain many challenges. On the one hand, the atom at the metal electrode surface may rearrange or move due to the local heating and the electric field, which strongly influences the stability of molecular devices [250, 251, 257]. On the other hand, it tends to induce a wide dispersion of conductance values for the poorly defined nature of the contact between the molecules and the metal surface [258]. These obstacles motivate the investigations into the nonmetallic materials, which are the promising candidates for the nanogap electrodes. For example, the carbon-based materials, including CNTs and graphene, can effectively avoid the atom mobility, since the carbon atoms are held together by the sp^2 hybridization [41]. Moreover, there are well-defined covalent bonds between the electrode and molecules, for their similar compositions or unit structures [39, 40]. Remarkably, the carbon-based materials possess strong deformation resistance, high optical transparency, and low electrical resistivity [259]. And considering their dimensional characteristics, it is beneficial to facilitate the fabrication of electrodes and provide desirable point contacts in single-molecule devices [260, 261]. In addition to the carbon-based materials, other nonmetallic materials, such as silicon [42–44] and conducting polymers [45, 262, 263], which exhibit plentiful availability, good stability, high electrical conductance, and easy processing, have also been continuously developed to expand the research regimes of the molecular electronics.

3.9.2 Nanogap Electrodes Made from Carbon Materials

3.9.2.1 Advantages of Carbon Materials

Although noble metals have been extensively applied in the molecular junctions, there are some tricky issues to be solved, for example, the high atomic mobility limits the mechanical stability and the applications at room temperature of the devices [264]. Notably, the carbon materials are promising candidates for nanogap electrodes in molecular electronics. In general, the carbon atoms are held together by the sp^2 hybridization such as fullerenes, CNTs, and graphene, which guarantee higher structural stability [41]. On the one hand, the carbon materials are comparable to metals, for their large fracture stress, strong deformation resistance, excellent thermal conductivity, and high electrical conductance [265]. As well, it is much simpler to prepare nanogap electrodes with the carbon materials than the bulky metals. The carbon-based electrodes are capable to provide point contacts, thus demonstrating a greater possibility to fabricate single-molecule electronic junctions [222, 244]. And the carbon nanogap electrodes enable the gate electrical fields to regulate the transistor channel effectively due to their smaller size [39, 260]. On the other hand, the carbon materials possess similar compositions or unit structures with the organic molecules. Hence, there are reproducible contacts between the carbon electrodes and the molecules by the π–π stacking. Moreover, the edges of carbon electrodes can also be modified with the specific terminal groups, so that the molecules are covalently attached to the electrodes [266]. These well-defined contacts contribute to reduce the wide dispersion of conductance values among the molecular devices. In addition, compared with the conventional indium tin oxide (ITO) material, the carbon-based electrodes with high transparency, superior flexibility, and low resistance are promising in optoelectronic devices [267, 268]. Therefore, the carbon materials are outstanding alternatives to metal electrodes to manufacture reliable solid-state molecular electronic devices.

3.9.2.2 Carbon Nanotubes for Nanogap Electrodes

The CNT which is formed by rolling up graphene sheets seamlessly into tubular shape with fullerence-like end structures is a typical material in carbon families. It has been considered as one of the most perfect nanostructural materials obtained from bottom-up chemical synthesis methods [259, 269, 270]. If the cylinder consists of a single layer of graphene sheet, it is a SWCNT, whereas the MWCNT is a coaxial cylinder that contains multiple layers of graphene sheet. The unique quasi-one-dimensional (1D) structure and extended curved π-bonding configuration endue CNTs with lots of outstanding properties. Mechanically, the Young's modulus reaches 1–2 TPa, illustrating stiff and strong deformation resistance of CNTs [265]. Optically, the CNT thin film with low electrical resistivity exhibits high optical transparency in visible region. And physically, the CNT is either metallic or semiconducting according to its chirality which depends on the lattice vectors. As a semiconductor, the electronic and optical behaviors of individual CNT strongly depend on its bandgap that is inversely proportional to its diameter. The semiconducting CNT can be used as active channel materials for transistors with superb properties such as the high on/off ratio, since it possesses relatively high carrier mobility.

On the other hand, the metallic CNT is an ideal candidate for nanogap electrode materials in molecular electronics, as the metal electrodes are facing more and more challenges. For example, the bonding between molecules and metal electrodes dominating the charge transport behaviors of devices is ill-defined, even for thiolated molecules on the Au electrode, which is a well-developed system. Meanwhile, it is difficult to prepare metal electrodes for the required molecular scale. Remarkably, CNTs could provide desirable point contacts in single-molecule devices. And the diameter of the CNT is almost the same size as the related functional molecules, which could facilitate the fabrication of electrodes. In addition, it is quasi-1D ballistic nanowire with high conductivity and high current carrying capacities [271, 272]. The chemical composition and simple atomic bonding configuration of the CNT guarantee better contacts and efficient carriers transfer between molecules and CNT electrodes [39, 266].

In order to fabricate CNT nanogap electrodes, some strategies have been proposed. In the field of molecular electronics, the electromigration which is based on the current-assisted diffusion process is one of the most reliable methods to prepare metal nanogap electrodes. However, it does not work well for fabricating CNT electrodes, as the strong C—C bonds lead to the CNTs with high current carrying capacities. Fortunately, it was found that when the current was high enough, MWCNTs failed by a layer-by-layer breakdown of individual carbon shells and ultimately nanogaps within MWCNTs were formed (Figure 3.37) [273, 274]. This phenomenon has been extensively utilized to prepare CNT nanogap electrodes, called as the electrical breakdown method [275]. The formation of current-induced defects plays the primary role in the breakdown initiation; meanwhile, the self-heating is a secondary factor. Moreover, the oxygen partial pressure is vital to determine the size of nanogaps within CNTs, for the combustion process of carbon atoms. The electrical breakdown procedure made it possible to intensively study the electrical

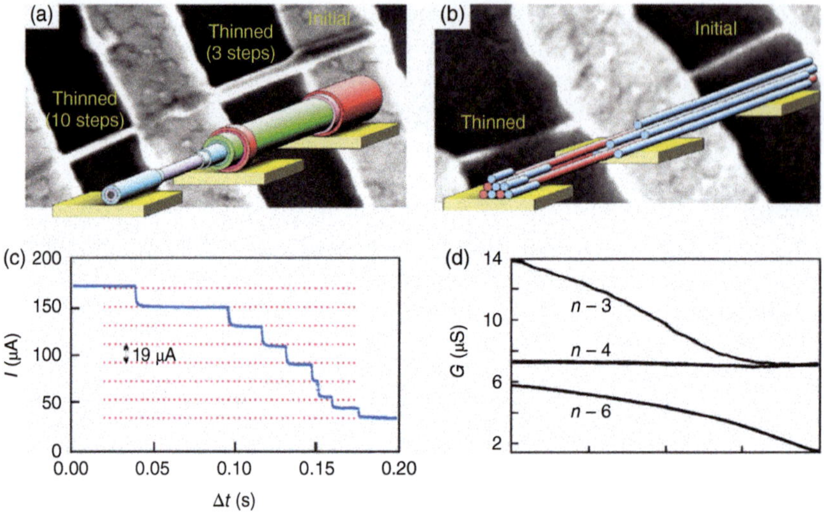

Figure 3.37 (a) Images of partially broken MWCNTs show clear thinning, with a decrease in radius equal to the intershell spacing (0.34 nm) times the number of completed breakdown steps. The two segments of this sample were independently thinned by 3 and 10 shells, as depicted by the color overlays. (b) Similar thinning, as observed in SWCNT ropes. (c) The partial electrical breakdown of a MWCNT at constant voltage stress proceeds in a series of discrete steps corresponding to the loss of individual carbon shells from the MWCNT. Equally spaced dotted lines emphasize the surprisingly regular spacing. (d) The low-bias conductance G (V_G) changes between s and m behaviors due to the changing nature of the outermost shell at each stage of breakdown. Source: Collins et al. [273]. 2001, AAAS.

properties of MWCNTs, since it could eliminate the individual shells one at a time. And it was demonstrated the MWCNTs behaved complex alternating metallic and semiconducting characterizations shell by shell. Tsukagoshi et al. [40, 276] reproducibly prepared MWCNT electrodes with the nanogap less than 50 nm by the reliable electrical breakdown technique. Dai and coworkers [39] suggested that the gap size was associated with the length of the SWCNT between two metal electrodes and they achieved the nanogap whose size was down to 1–3 nm. Liu and coworkers [260] devised a simple and efficient way to manufacture nanogap electrodes with controlled nanogap size on the basis of a combination of the electrical breakdown and electron beam-induced decomposition (EBID) techniques (Figure 3.38). In detail, a high-density current was applied to the CNT to form an initial gap of approximately 10–60 nm, followed by some organic molecules being absorbed in the gap or on the surface of the CNT electrodes. Finally, the EBID of the absorbed molecules was carried out to regulate the nanogap size of CNT electrodes. The electron beam was obtained from the SEM, so that the EBID process could be monitored in real time and in situ, which made it quite easy to accurately control the nanogap size. In total, the combination of these two fabrication processes is valuable in molecular electronics.

The nanogap can also be achieved by the oxidative plasma ablation technique which is quite different from the electrical breakdown. In this technology, the

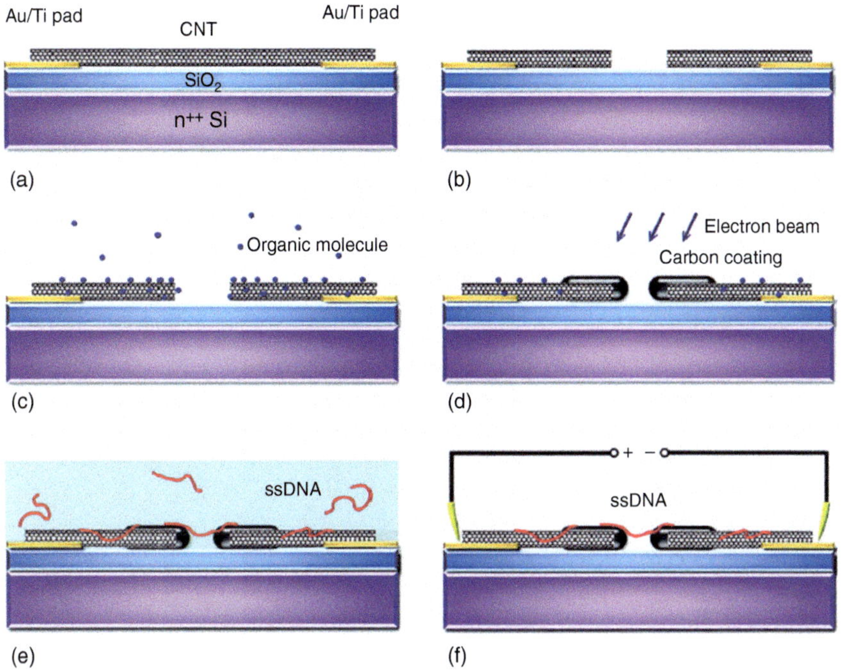

Figure 3.38 Schematic diagram of the process of the fabrication of a CNT electrode with a controlled nanogap (a–d) and the fabrication of a DNA electronic device (e, f). (a) Bridging a CNT between Au/Ti electrodes. (b) Cutting the CNT by current breakdown method. (c) Adsorbing organic molecules on or in the CNT. (d) Irradiating the gap of the CNT by electron beam with in situ observation in SEM. (e) Assembling DNA molecules between the CNT electrodes by immerging in DNA buffer solution. (f) Measuring the electronic properties of the bridged DNA molecules. Source: Wei et al. [260]. 2008, American Chemical Society.

ultrahigh-resolution EBL is applied to open a window on the polymeric mask polymethylmethacrylate (PMMA), and then the oxidative plasma is used to cut SWCNTs through the open window precisely (Figure 3.39) [266]. The oxidation would first occur at a defect on the sidewall of the CNT which either exists before plasma ablation or is induced by etching and then continue to erode the tube. The minimum nanogap size is generally on the order of 10 nm, depending on the resolution of EBL. Longer oxygen plasma etching time leads to higher yield of the cutting tubes, yet decreases the single-molecule devices productivity, since the chemical connections are difficult to carry out in long gaps. The nanogaps whose average size is narrower than the open window in polymeric mask have been prepared with considerable yield of cutting by optimizing the relevant conditions. The oxidative plasma ablation method is more appropriate to construct single-molecule devices. On the one hand, the SWCNTs in diameter are basically the same as the size of the connected molecules. And the separation of two SWCNT electrodes is only a few nanometers. On the other hand, the strongly oxidizing conditions make the cut ends of SWCNTs be capped predominately with carboxylic acids, which provide point contacts to form robust connection between molecules and SWCNT electrodes.

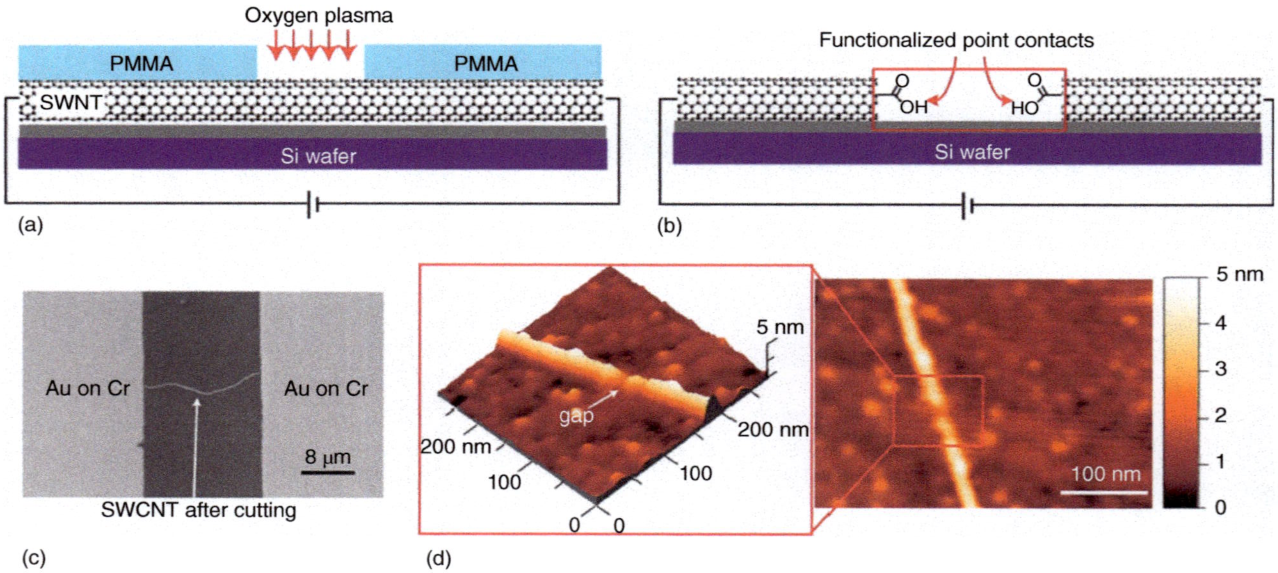

Figure 3.39 (a) Precise cutting of SWCNTs with an oxygen plasma introduced through an opening in a window of PMMA defined with e-beam lithography. (b) Oxidative opening of a tube produces two point contacts functionalized on their ends with carboxylic acids and separated by as little as 2 nm. (c) Scanning electron micrograph of a SWCNT with Au on Cr leads that had been cut using e-beam lithography and an oxygen plasma. (d) AFM image of the gap cut into the SWCNT. (Inset) Height profile of the isolated tubes. The diameter of the SWCNT is 1.6 nm, estimated from the height profile. Source: Guo et al. [266]. 2006, AAAS.

In general, the yield of single-molecule junctions obtained by this technology is typically 10%.

The electron beam etching is prospective in fabricating CNT nanogap electrodes as well [277–279]. The focused electron beam with nanoscale diameter is extremely reliable to prepare small nanogaps reproducibly and the material damage caused by masking and ion implantation can be effectively avoided during the electron beam etching process. Typically, in order to form the nanogap, an electron beam line scan which is provided by the SEM or the TEM is executed across the CNT while injecting the oxygen. The oxygen played an important role in opening a nanogap, as highly reactive oxygen radicals could react with carbon atoms, thus leading to the mass loss of CNTs. The water molecules demonstrated the same effects as oxygen for the similar reactions between high reactive radicals and carbon atoms. Furthermore, compared with the oxygen, the presence of water showed a faster cutting rate at the same partial pressure. However, when the argon was injected instead of oxygen, no mass loss was observed [280, 281]. It was suggested the etching rate was closely related to the primary electron energy, oxygen partial pressure, and the support environment of CNTs. The role of primary electrons was mainly to provide secondary electrodes which contributed to forming active oxygen species, though high energy primary electrons would also damage carbon shells by the direct knock-on [282]. And the CNTs supported by silicon nitride membrane were etched faster than the suspended CNTs, since the former have more secondary electrons that were primarily emitted from silicon nitride membrane, while the secondary electrons of the latter were produced from themselves and the yield of their secondary electrons decreased during the etching process. On the other hand, the nanogap size was essentially determined by the spatial distribution of secondary electrons and the focused electron beam size in this technology. So the size of the CNT nanogap supported by silicon nitride membrane was larger, despite its faster etching rate. The nanogap whose size was as small as 2 nm could be realized with the high-resolution TEM; however, the electron transparent substrate was required, which severely limited the types of CNT-based devices. Noticeably, the SEM could not only prepare the nanogap of about 20 nm but also construct the aligned CNT arrays with similar nanogap size and the same relative gap positions (Figure 3.40) [283].

In addition to the electron beam etching, another technique with the advantages of good reproducibility and maskless nanostructuring is FIB bombardment [203]. It was found that the bombardment demonstrated a high nanogap cutting speed. The nanogap that was as narrow as 50 nm within CNTs has been fabricated with a focused Ga ion beam around 22 nm in diameter [208]. And the focused helium ion beam in a HIM seems more promising in preparing nanogaps less than 10 nm [217]. The size of gaps is normally slightly broader than the original beam diameter for the ion-scattering effect of the surface and the oxygen atoms sputtered from oxide substrate. The ion beam current and the exposure time can be optimized to obtain a narrower nanogap. On the other side, the FIB bombardment has been successfully applied to design a novel nanoelectronic platform (Figure 3.41). In particular, in addition to the nanogap within the SWCNT, a trench in the oxide layer was formed during the ion exposure. The nanotrench suspended the DNA molecule over the

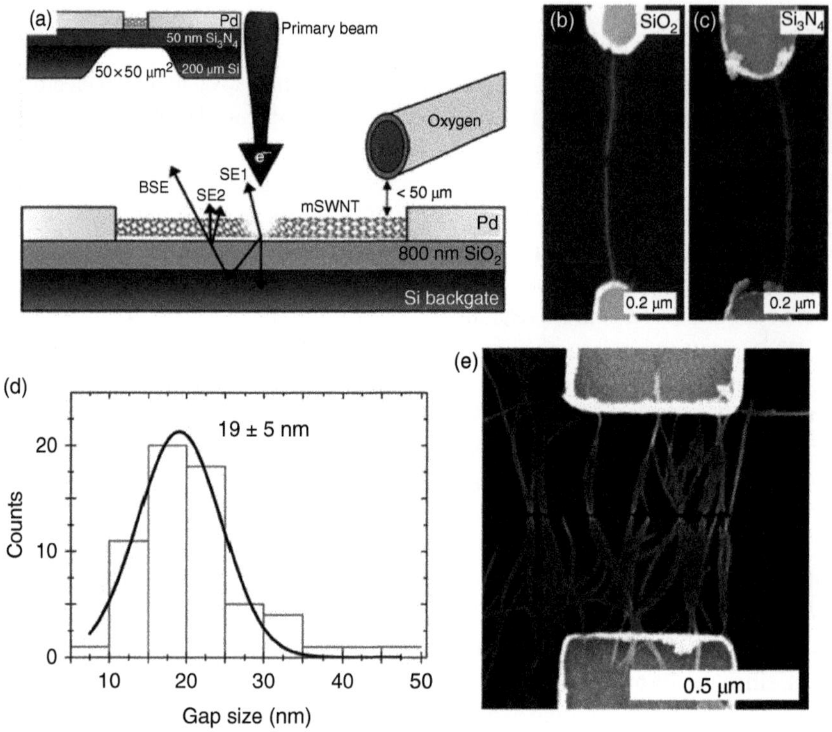

Figure 3.40 (a) Schematic of the electron-beam-induced oxidation setup: primary beam electrons are repeatedly scanned across an mSWCNT in the presence of injected oxygen. The mSWCNTs are wired to Pd electrodes and supported by SiO_2/Si substrates or Si_3N_4 membranes (inset). Indicated are the trajectories of secondary electrons (SE1 and SE2) and backscattered electrons. (b) SE image of an mSWCNT on SiO_2 after gap formation, recorded with $V_G = +10\,V$. (c) SE image of an mSWCNT on a Si_3N_4 membrane after gap formation. (d) Histogram of nanogap sizes of 62 devices on SiO_2/Si substrates. (e) SE image of an mSWCNT thin film device after gap formation. Source: Thiele et al. [283]. 2011, AIP publishing LLC.

oxide surface, so that the interaction between the DNA molecule and the silicon oxide layer was weakened which could cause a large compression of the DNA and disturb its conductivity. Meanwhile, in order to establish robust covalent connections between the molecule and CNT electrodes, the carboxylation of the FIB-cut nanotube edges by the nitric acid treatment is necessary after the bombardment procedure [284]. And the strong electronic coupling resulted from the covalent connection promoted carriers transfer through the whole system without the distribution of Coulomb blockade.

The linkage between the molecule and CNT electrodes significantly affects the performance of molecular devices. Several strategies are proposed to connect molecules with electrodes including π–π stacking, covalent bonding, hydrogen bonding, and so on. The noncovalent π–π stacking is expected to provide a milder molecule–electrode contact, so that the electronic structures of the both are only slightly modified [244, 285]. The pentacene field-effect transistors (FETs) were perfectly constructed by utilizing the electrical breakdown CNTs as the source and

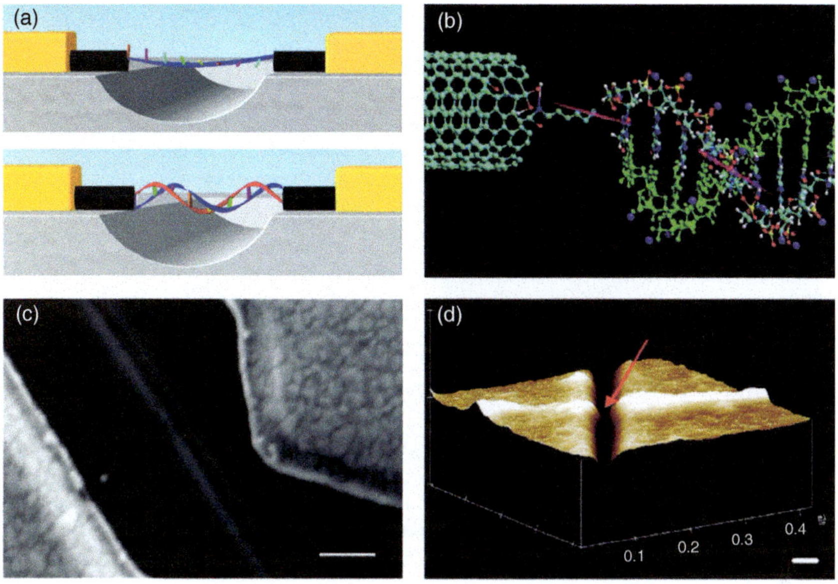

Figure 3.41 (a, b) Schematic illustrations. (a) A covalently attached single-stranded DNA (ssDNA) molecule (upper) or double-stranded DNA (dsDNA) molecule (lower). The nanoelectrodes are separated by a gap of 27 ± 2 nm. The bridging DNA molecule suspends over a trench without touching the silicon dioxide surface. (b) Molecular diagram highlighting the covalent bonding between an amine-terminated ssDNA and a carboxyl-functionalized SWCNT nanoelectrode via a $(\text{-CH2-})_6$ spacer. Charge transport takes place through the stacked base pairs in a helical duplex. (c) SEM image of a pair of FIB-etched SWCNT electrodes, connected to the bonding pads by Ti/Au microcontact leads. Scale bar = 100 nm. (d) High-resolution AFM image of the DNA molecule between the SWCNT electrodes. Scale bar = 30 nm. The red arrow points to the trapped single DNA molecule. Source: Roy [284]. 2008, American Chemical Society.

drain electrodes. During the evaporation, the pentacene film selectively adhered onto the CNT surface for their common aromatic six-membered ring units and then filled the channel. The π–π stacking solved the discontinuity problem of the semiconductor film in organic thin film transistors. Meanwhile, the molecule–electrode interface showed an effective charge transfer, since the conduction-electron wave of the two materials overlapped partially [40]. Furthermore, the FETs with channel in molecular length and width were realized by the SWCNT electrodes. Compared with bulky metal electrodes, the SWCNT electrodes with sharp quasi-1D geometry effectively reduced the screening of the gate electric field, which significantly improved the switching characteristics of devices [39, 260]. In addition, molecules could also be connected to CNT electrodes by being functionalized with the aromatic ring units. Marquardt et al. [286] ingeniously designed a molecule with rod-like structure consisting of two terminal phenanthrene units, two long OPE rods and a central 2,6-dibenzylamino core-substituted naphthalenediimide (NDI) dye to fabricate electroluminescent junctions. The terminal phenanthrene units ensured that the molecule tightly linked with electrical breakdown SWCNT electrodes through the π–π stacking and the OPE rods produced electrostatic attraction in the

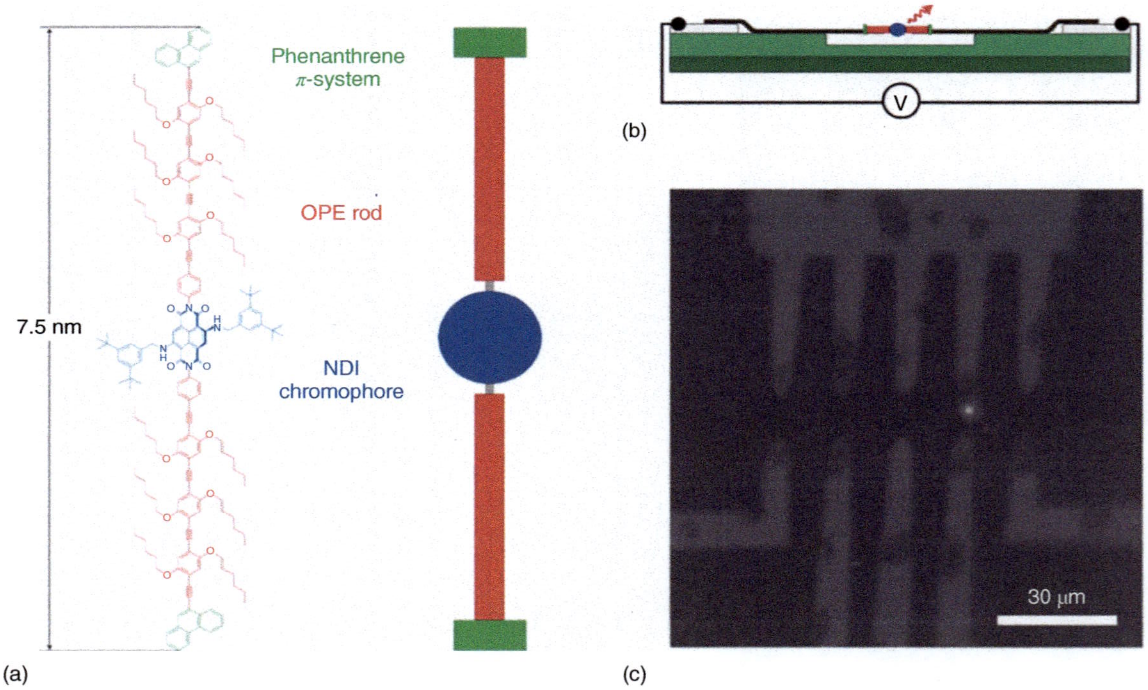

Figure 3.42 Single-molecule electroluminescence. (a) Chemical structure of the molecule, which consists of a central 2,6-dibenzylamino core-substituted NDI chromophore (blue), two OPE rods (red), and phenanthrene anchor units (green). (b) Schematic drawing of the device geometry. (c) Optical micrograph of an image acquired for 20 minutes with one device under a bias of $V = 4\,V$ and without external illumination. A bright electroluminescent spot is visible at the location of the biased carbon nanotube–molecule junction. Source: Marquardt et al. [286]. 2010, Springer nature.

nanogap for their polarizable π-systems. Moreover, the metallic SWCNT electrodes not only showed a ballistic transport property but also suppressed the electron delocalization. Therefore, the charge carriers were able to transport through the junction and lead to the electroluminescence phenomenon. The authors furnished a visible way to prove the charge transfer through the molecular junctions. And it is feasible to achieve diverse molecular light-emitting devices by changing the emission chromophores (Figure 3.42). The π–π stacking increases the possibility of molecule–electrode contact and is less sensitive to the shape and composition of the CNT electrode edges. Accordingly, the measurements of molecular junctions exhibit good reproducibility.

In order to perform specific functions, the single-molecule junctions may confront severe conditions such as the chemical treatments or environmental stimuli. Hence, it is necessary to construct the more robust attachment between the molecules and CNT electrodes [287]. Remarkably, the covalent amide linkage which is achieved by the dehydration reaction has been widely adopted to fabricate sturdy molecular junctions. In detail, the cut ends of the CNT electrodes are modified with carboxylic acid groups through the oxidative plasma ablation or nitric acid treatment. Then, a molecule with amine end groups could be smoothly introduced into the nanogap between the CNT electrodes by the formation of robust amide linkages. Compared with the common Au—S chemical bonds in metal–molecule–metal junctions, the amide linkage exhibits a lot of advantages. Firstly, the amine-containing molecules are easily prepared and more stable than the dithiols which are prone to oxidative oligomerization. Secondly, the CNT electrodes provide a limited number of point contacts, so that the contact geometry and the number of molecular bridges can be exactly controlled. Furthermore, the detrimental stochastic switching and lateral surface mobility of metal grains are effectively avoided. Applying the sturdy covalent amide bonds, Guo et al. have fabricated a series of functional transistors, in which the cut CNTs act as the source and drain electrodes and the doped silicon substrate serves as the back-gate electrode. Considering the electrical conductance of oligoanilines is sensitive to pH, the authors designed an amine-terminated oligoaniline molecule with the defined length to construct molecular-scale devices (Figure 3.43a–c) [266]. It was shown that the SWCNTs were either metallic or semiconducting and the molecular bridges would not change this electrical behavior. The conductance of the device changed by nearly an order of magnitude when alternately immersed in solution of low and high pH, since the protonated molecule was more conductive. This device provided a local monitor for the pH on the basis of one or just a few molecules. Moreover, the molecular junction that can recognize the specific metal ion has also been constructed (Figure 3.43d–e). The junction showed the conductance at the noise limit of the measurement when the ends of cut CNTs were coupled with terpyridine-containing amines. Notably, the Co^{2+} metal ion partially restored the current by connecting the terpyridine ends [288]. However, the performance of this multiple-component molecular device would be limited by the diffusion of the analyte to the receptor and the association constant between the ligand and metal ion. Meanwhile, it was suggested that the pyrrole-based molecular junction was a superb photoswitch device (Figure 3.44a–c) [289]. In solution, the thiophene- and

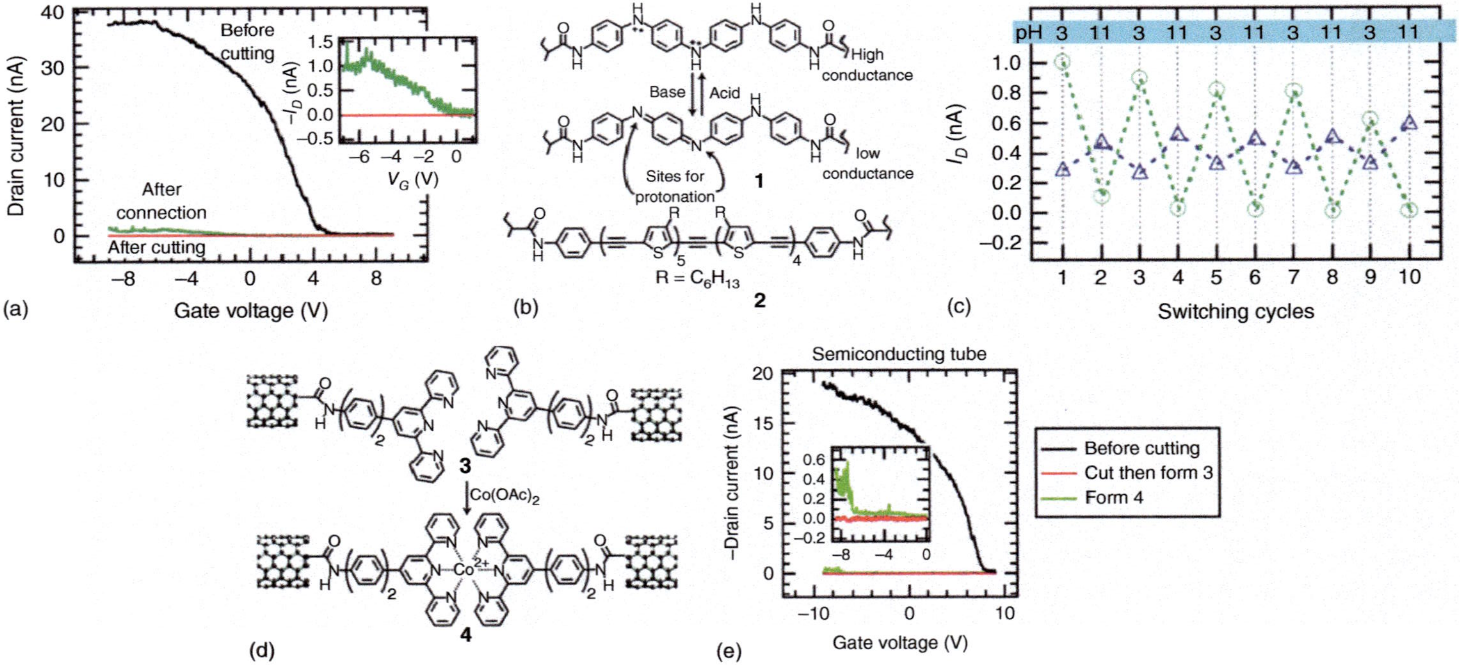

Figure 3.43 (a) Drain current (I_D) in the device as a function of the gate voltage (V_G) at V_{SD} = 50 mV for a semiconducting tube connected with 1. Electrical measurements were made before cutting (black curve), after cutting with oxygen plasma (red curve), and after connection (green curve). Inset is a magnified *y* scale showing the *I*–*V* curves after cutting and connection. (b) Protonation and deprotonation to give high- and low-conductance forms of oligoaniline 1 and the control molecule 2. (c) Green circles show the ON-state resistance for **1** when alternately immersed in solutions of low and high pH. The blue triangles show small changes in ON-state resistance for **2** when alternately immersed in solutions of low and high pH. (d) Terpyridylamine-functionalized SWCNT ends incorporate Co^{2+} ions completing the circuit, and (e) drain current in the device as a function of the gate voltage (V_G). (a–c) Guo et al. [266]. 2006, AAAS. (d–e) Feldman et al. [288]. 2008, American chemical Society.

pyrrole-based molecules can reversibly switch between the open nonconjugated and closed conjugated forms under the appropriate light stimulus. Thus, both of them are expected to manufacture molecular photoswitch devices with the SWCNTs as the source and drain electrodes. When irradiated by the UV light, the molecule linkers would switch into a closed form and open a conjugated pathway between the SWCNT electrodes. Therefore, the current of switch 5 and 6 increasing by 25-fold and more than 5 orders of magnitude, respectively, were observed. Nevertheless, the thiophene-based molecular bridges could not revert back to the open state photochemically, since the energy transferred from the photoexcited molecule to the SWCNT electrodes. Fortunately, the pyrrole-based molecules which demonstrated a closed form with the UV irradiation could recover back to the open form thermally and the on/off cycle of the device could be toggled plenty of times. What's more, the measurement of the conductivity of a single DNA duplex was also realized by covalently wiring the DNA molecule into a SWCNT nanogap. Since the amide linkages were robust enough, it was possible to form the reproducible contact between the DNA molecule and the nanogap electrodes and study the effect of the integrity of the base pair duplex on the DNA charge transport well. This sturdy molecular device could serve as a uniquely powerful reporter to convert the biochemical events into electrical signals (Figure 3.44d–h) [288, 290]. In addition, the covalent bonds between the molecular bridge and SWCNT electrodes also provided a reliable way to construct the complex, selectively reactive heterostructures. The construction was started with a single molecule which bridged the nanogap of electrodes. And the individual binding events could be monitored in real time by sensing the ON-state resistance and threshold voltage changes of the molecular devices [291].

Hydrogen bonding is another efficacious linkage to fabricate molecular junctions, for example, a metalloprotein embedded molecular NDR device has been successfully achieved [292]. The NDR behavior rooted from the reversible active redox reactions of the core Fe ionic species in ferritin molecules which were located into the nanogap of SWCNT electrodes through hydrogen bondings (Figure 3.45). This novel two-terminal planar configuration simplified the complex measurement systems and increased the practical applications of the NDR device. The decent electron conductor peptide unit of ferritin could protect the Fe ionic species from environmental attacks and minimize their dissipation. Meanwhile, the linkages between the ferritin molecules and SWCNT electrodes provided the perfect interfaces to transfer charge carriers. Therefore, the device showed highly reliable and reproducible NDR behaviors and, the NDR peak current density was extremely high. What's more, both positive and negative NDR peaks could be observed for the redox reactions of Fe ionic species.

In addition, the SWCNT electrodes can also be bridged by the organic monolayers, thus forming the monolayer transistors. One of the most outstanding features of these devices is the high yield. The hexabenzocoronene (HBC) derivatives have been extensively studied in this field. The functionalized molecules which assembled laterally into columns were chemically grafted to the silicon oxide surface, so that the device had an exposed active channel. The double attachment at the interface enforced the upright conformation of molecules which was synergistic with the

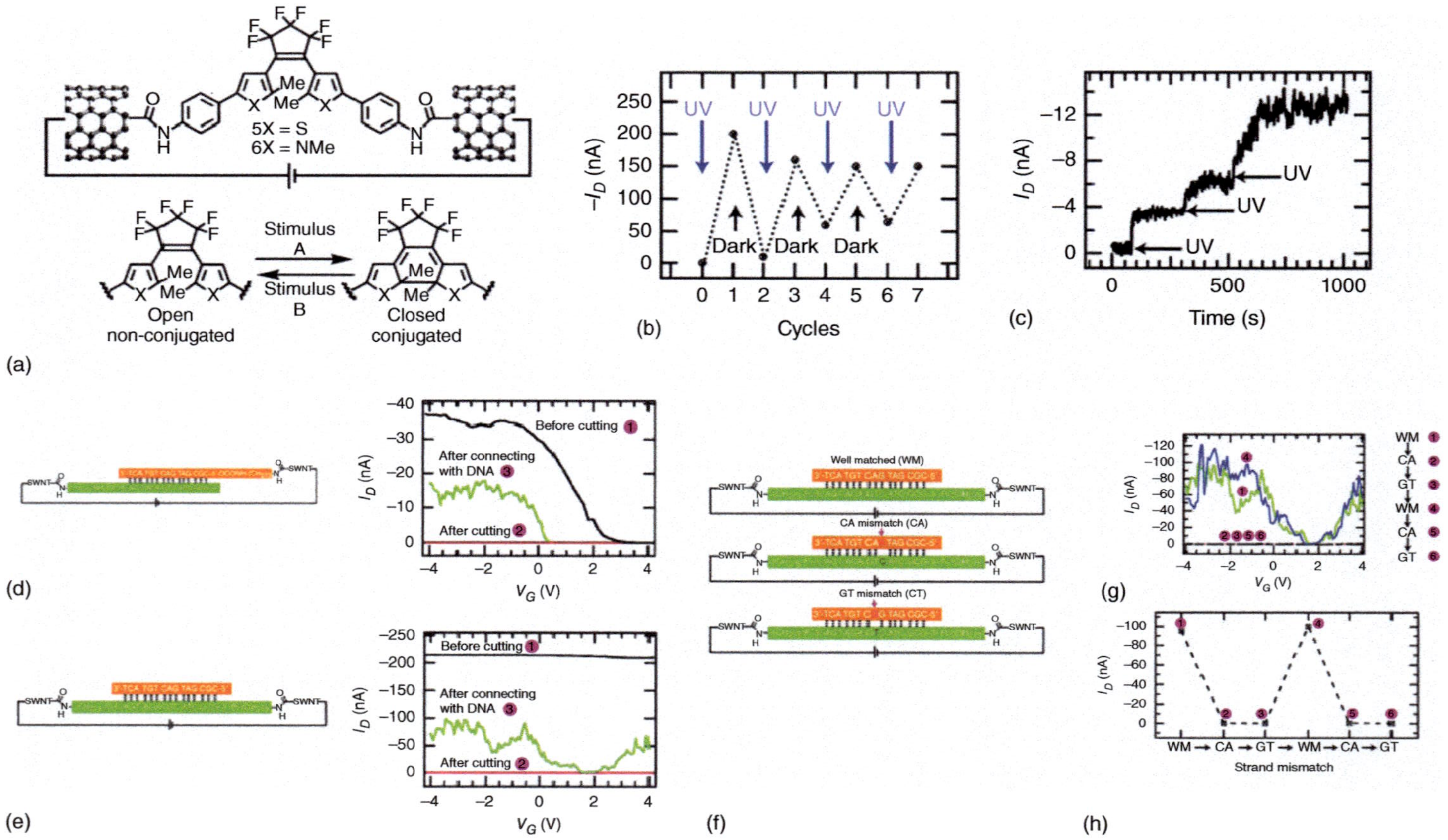

Figure 3.44 (a) Molecular bridges between the ends of an individual SWCNT electrode. (b) Switching between conjugated and nonconjugated molecular structures. (c) Drain current as a function of the switching cycle as the device is irradiated with UV light and then left in the dark for 12 hours. $V_{SD} = -50$ mV and $V_G = -9$ V. (d) Steps in the conductance for a device reconnected with **6** when the UV light is pulsed. $V_{SD} = -50$ mV and $V_G = -1$ V. (d–e) Source–drain current vs. V_G at a constant source–drain voltage (50 mV) for (d) semiconducting SWCNT device with DNA sequence connection and (e) metallic SWCNT device with DNA sequence connection. Guanine, G; cytosine, C; adenine, A; thymine, T. (f) Replacing well-matched (WM) duplexes with CA and GT mismatches. (g) Source–drain current vs. V_G of a SWCNT device taken through the sequence 1–6. The current levels for curves 2, 3, 5, and 6 are ~300 times lower. (h) Source–drain current at $V_G = -3$ V at a constant source–drain voltage (−50 mV) for the sequence 1–6. Source: (a–c) Whalley et al. [289], 2007. American Chemical Society. (d–h) Guo et al. [290], 2008. Springer nature.

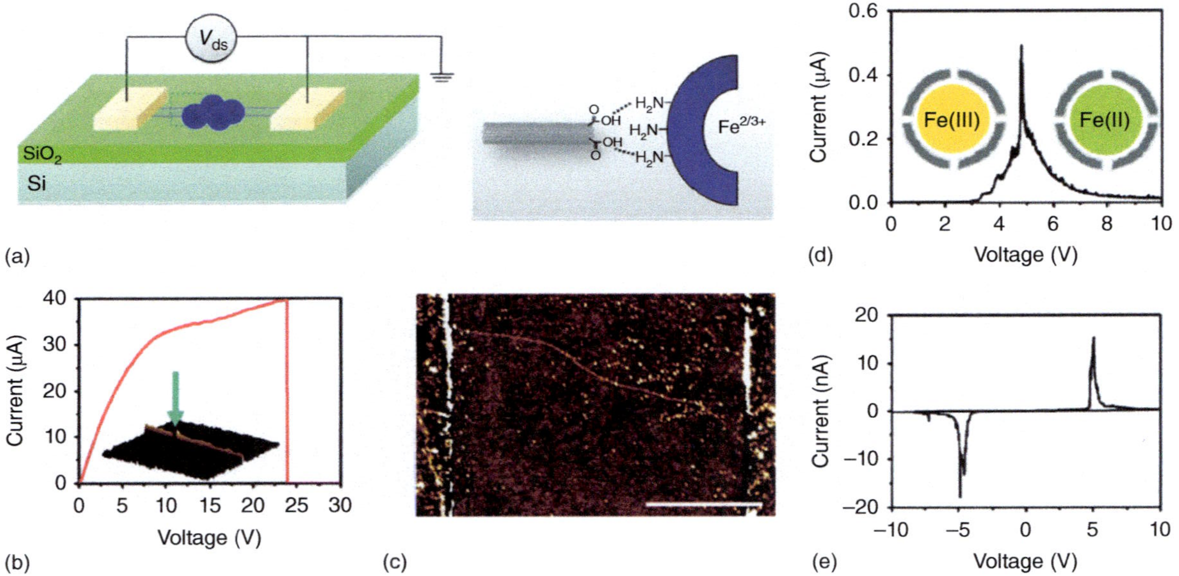

Figure 3.45 (a) A schematic view of the SWCNT–ferritin–SWCNT device. A zoom-in view of the dashed box is represented in the right of panel a. (b) *I–V* curve of cutting metallic SWCNT. AFM image of a created nanogap (arrow) having c. 20 nm gap is shown in the inset. (c) AFM image of a nanotube nanogap device after treating with a ferritin solution ($d_{nanotube}$ = 1.5 nm, scale bar = 1 μm). (d) A representative *I–V* curve of a SWCNT-ferritin-SWCNT device at room temperature. (e) A whole *I–V* cycle sweeping characteristic of the device. Source: Tang et al. [292]. 2007, American Chemical Society.

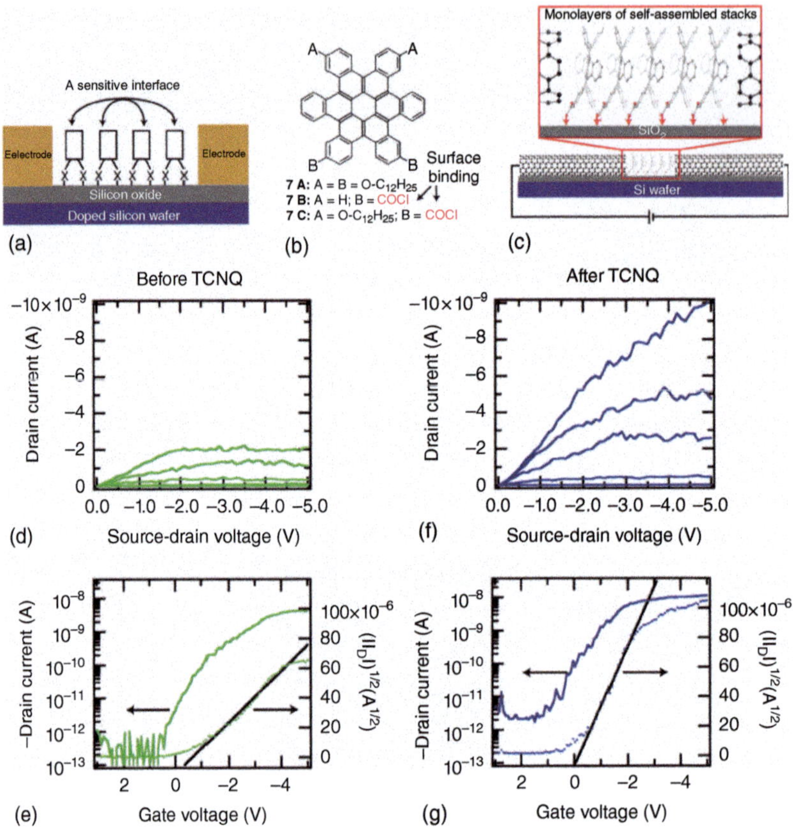

Figure 3.46 (a) An alternative molecular electronics where molecules are attached through the X-group to a primer layer for assembly and probed laterally. The semiconductor–air interface should be very sensitive to its environment. (b) HBC molecules. **7A** assembles into 1D stacks. **7B** and **7C** synthesized with groups to bind them to the surface of silicon oxide. (c) Monolayers of self-assembled stacks being probed with SWCNT electrodes separated by only a few nanometers. (d–g) The device formed by **7C** monolayer on the SiO_2 surface of a silicon wafer with SWCNT electrodes. (d) Transistor output, $V_G = 0$ to −5 V in 1-V steps. (e) Transfer characteristics for the device, $V_D = -2$ V. (f) Transistor output, $V_G = 0$ to −5 V in 1-V steps for the same device measured in (d) after treatment with a tetracyanoquinodimethane (TCNQ) solution. (g) Transfer characteristics for the device, $V_{SD} = -2$ V. Source: Guo et al. [293]. 2006, National Academy of sciences.

inclination of molecules to stack laterally into columns with the good π-overlap. And the quasi-1D SWCNT electrodes guaranteed a more efficient penetration of the gate electric field into the middle of the active channel. Hence, the transistors exhibited the large current modulation and high gate efficiency. Furthermore, the devices were chemoresponsive and thus could be immersed in solutions to test the dopants for the stable chemical attachment between the HBC derivatives in the exposed active channel and the dopants in the solution (Figure 3.46) [293]. Meanwhile, the photoresponsive nanoscale columnar transistors were also perfectly constructed with the tetra(dodecyloxy)hexabenzocoronenes (HBCs) which consisted of two separated π systems: the conductive planar radialene core and the insulating alkoxyphenyl sheath. After the thermal optimization, the molecules could self-organize into ordered columnar liquid crystalline phases, whose orientation was parallel to the nanogap between the SWCNT electrodes. It was demonstrated that the annealing increased the yield of the working devices and improved properties of transistors, since the 1D radialene core that was formed from the ordered columns dominated the charge transport. The SWCNT electrodes and HBCs columns formed a sturdy device which could stand much higher measurement voltage bias. What's more, the transistors were highly photoresponsive as the HBCs columns were sensitive to the visible light and their unique structures restricted the photoexcitations in the cores [294]. Remarkably, the copper phthalocyanine (CuPc) monolayer is another ideal bridge to fabricate ultrasensitive photodetectors (Figure 3.47). On the one hand, the traditional Langmuir–Blodgett (LB) technique is a reliable and environment-friendly self-assembly method to prepare the uniform

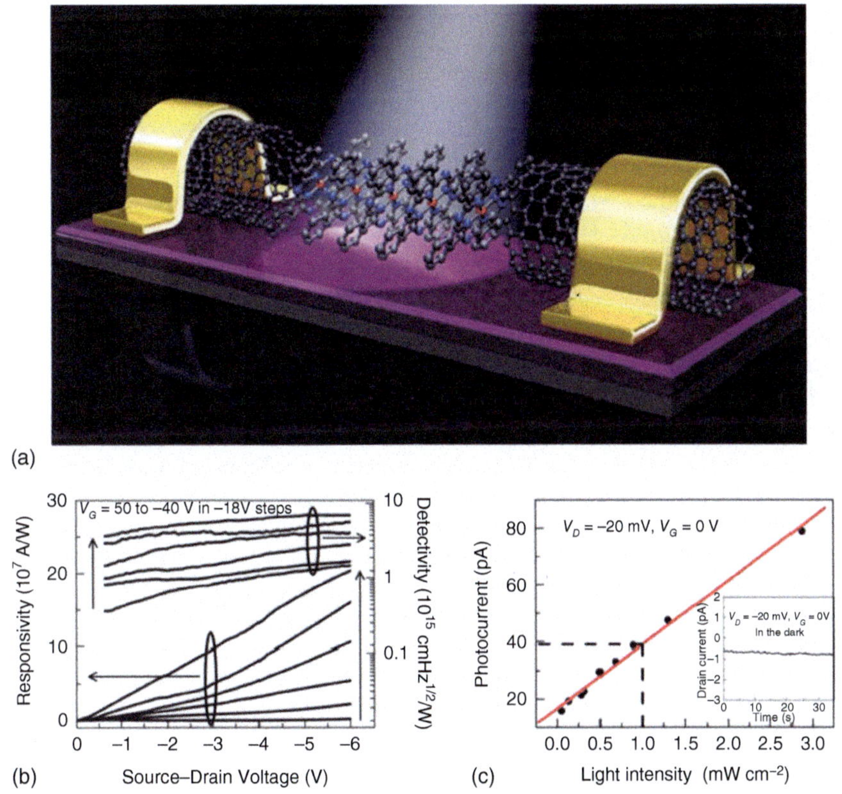

Figure 3.47 (a) Schematic illustration of ultrasensitive water-processed monolayer photodetectors with SWCNTs as point contacts. (b) Bias dependent data of responsivities and detectivities for the device excited using a 620-nm light. V_G = 50 to −40 V in −18 V steps. (c) Linear dynamic range of the device under visible light irradiation. Inset is the dark current of the device. V_G = 0 V and V_D = −20 mV. Source: Liu et al. [295]. 2011, Royal society of chemistry.

monolayers of CuPc which are beneficial to increase the carrier mobility. When the LB monolayer of CuPc is transferred from the water–air interface onto the silicon oxide surface, it is homogenous and high coverage with a columnar structure, in which the molecules are nearly perpendicular to the substrate in a face-to-face closely stacking fashion [295]. On the other hand, the high electric fields in the nanogap avoid the detrimental charge recombination under the light irradiation. And the Schottky barriers between the CuPc LB monolayers and SWCNT electrodes are low enough to allow efficient carrier injection and extraction.

It is viable to fabricate truly single-molecule functional devices based on the CNT nanogap electrodes, since the diameter of the CNT is almost the same size as the connected molecules. However, there are some issues to be solved to improve the CNT-based molecular devices. The connection yield of the nanogap electrodes is too low, which significantly impedes the application of the CNT in molecular electronics. Furthermore, the properties of the SWCNT are closely related with its chirality and diameter, which have an important impact on the performance of the devices, for example, the semiconducting SWCNT electrodes exhibit obvious gate dependence in the transistor, which may mask the gate response of the functional molecules.

3.9.2.3 Graphene

Characteristics of Graphene Graphene whose sp^2-hybrized carbon atoms are arranged into a honeycomb lattice is a zero-bandgap two-dimensional (2D) semimetal material. This nanomaterial has received worldwide attention since its experimental discovery in 2004 [41, 296]. The unique planar shape and chemical structure endow graphene with a collection of fascinating properties [297]. On the one hand, the high Young's modulus and intrinsic strength ensure that it processes good mechanical resistance [298]. Its thermal stability is extremely high owing to the superior thermal conductivity. On the other hand, the graphene with low sheet resistance, high concentration of charge carriers, and giant intrinsic charge carrier mobility demonstrates the ballistic charge transport and strong ambipolar electric field effect [41, 299]. The quantum Hall behavior can be observed even at room temperature, since the charge carriers behave as massless relativistic particles and move with little scattering in graphene [300]. In addition, it is a superbly flexible and stretchable transparent conductor material, and its electronic and optical properties would not be affected by the strong bending or stretching [301]. Therefore, the graphene is promising for next-generation electronics whose properties may exceed those of conventional semiconductors or metals.

The graphene semiconductor shows immense potential for digital electronic device applications. However, the valence and conduction bands of large-area single-layer graphene are cone-shaped and coincide at the K points of the Brillouin zone. In order to obtain the outstanding semiconductor, it is necessary to introduce a bandgap in graphene. Fortunately, several viable methods have been proposed to open and tune the bandgap for the special electronic band structure of graphene. In one approach, the large-area single-layer graphene was patterned into graphene nanoribbon (GNR) by the STM lithography as theoretical studies predicted that the bandgap of the armchair-terminated GNR is a function of its width due to the quantum confinement and edge effects [302]. Li et al. [303] developed a chemical route to produce the sub-10-nm-wide GNR and achieved a graphene FET with on/off ration of about 10^7 at room temperature. Moreover, for the gapless bilayer graphene whose energy bands are parabolic, a perpendicular electric field is useful, as it can break the inversion symmetry of graphene. So the complex FETs with dual-gate structure was extensively suggested [304, 305]. Remarkably, Park et al. [306] put forward a facile dual molecular doping method to avert complicated fabrication steps of the dual-gate structure. Furthermore, Zhou et al. [307] highlighted another promising direction for bandgap engineering of graphene. They ascribed the origin of a finite bandgap to the graphene–substrate interactions which could modify the lattice symmetry in epitaxial graphene.

On the other hand, as a conductor, graphene is one of the excellent candidates for nanogap electrodes in molecules electronics [244, 285]. Its nanogap geometries are quite stable since the strong covalent binding makes the diffusion of carbon atoms in graphene negligible [308]. Only one or a small collection of molecules can be attached to the graphene point contacts, which is a bright way forward in addressing single molecules. Meanwhile, graphene improves the yield of stable molecular junctions compared with SWCNT whose chirality and diameter vary greatly. The 2D flat graphene ensures an enhanced gating efficiency by bringing molecular junctions closer to the gate. In addition, graphene sheets are suitable for nanoscale lithography, so this plane electrode is a potential base for molecular integrated circuits [309]. It is possible to study intrinsic properties of different molecular devices, since contact barriers between molecules and electrodes can be adjusted flexibly by doping graphene with holes or electrons.

Graphene nanogaps can be formed by feedback-controlled electroburning method which exploits the chemical reaction of carbon atoms with oxygen at high temperature. Typically, a voltage ramp is applied to a graphene device while recording the relevant current. In this process, the combustion of graphene channel edges induced by Joule heating at high current densities could lead to negative differential conductance. A feedback loop monitors the current at each voltage step and sweeps the voltage back to zero when the conductance drops by 10%. The repeating of current–voltage ramp allows graphene sheets to narrow down gradually. The device resistance is a sign to determine the formation of electrodes. It was found that the oxygen partial pressure was crucial to generate small nanogaps in graphene during the electroburning process [310]. Prins et al. [261] have utilized this method to fabricate stable gateable molecular junctions with few-layer graphene electrodes whose conductance are largely gate-independent. The size of nanogap between two electrodes was estimated approximately 1–2 nm by fitting the tunneling current with the Simmons model for tunneling. However, there were graphene channels with random widths and shapes in some instances, since lots of current–voltage traces were required to disconnect the graphene. Meanwhile, it is easy to lead to the misshapen or partially burnt graphene channels. Xu et al. [311] added a lithography step which could form a 500-nm-wide constriction in the center of the graphene channel before the electroburning to improve this technology. The nanogaps have

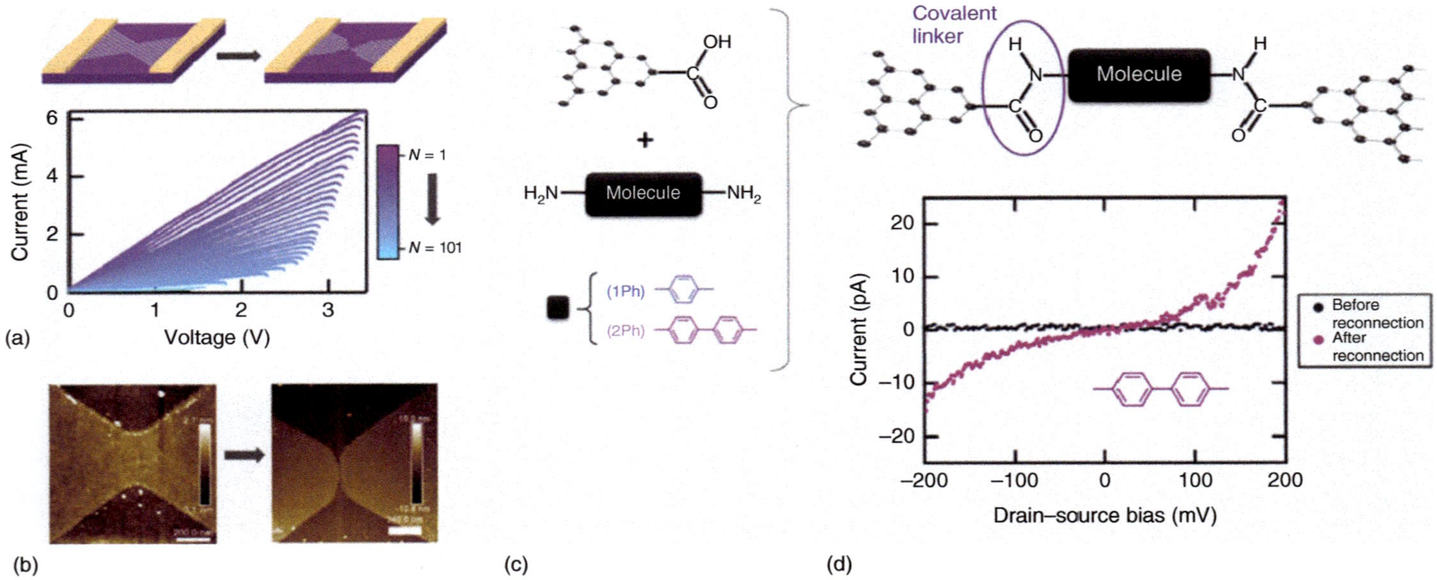

Figure 3.48 Fabrication and covalent reconnection of graphene nanogaps. (a) Gaps are etched in graphene constrictions using a feedback-controlled electroburning process. (b) Atomic force microscopy images of a graphene constriction before and after electroburning, showing the smooth round-off shape of electroburnt graphene electrodes with a gap length minimal in the center. (c) Reaction chemistry of short diamine molecules on the oxidized edges of graphene electrodes, leading to single-molecule circuits connected with covalent amide bonds as the contacts. Two diamine molecules were used in this study, i.e. 1,4-diaminobenzene (1Ph) and 4,4′-diaminobiphenyl (2Ph). (d) Current–voltage characteristics at 0 V gate bias and room temperature, showing negligible current at low bias across the nanogap and a significant increase in conductance after reconnection with a diamine molecule. Source: Xu et al. [311]. 2017, American Chemical Society.

separations of about 0.9 nm as estimated from a Simmons model (Figure 3.48). The preprocessed bowtie-shaped graphene channel resulted in an increasing yield of nanogaps from 30% to 88%, accurately controlled the nanogap position and effectively reduced the critical power which was required to start the electroburning procedure. This modified technology produced a smooth round-off shape at the end of both electroburnt graphene electrodes, so that molecules with different lengths were able to be located into the nanogap of the electrodes. In addition to the lithography technology, the HIM is also suitable to prepare nanoconstriction [312]. In particular, the direct helium ion beam writing provides a cleaner and faster fabrication process. Moreover, compared with gallium ion beam patterning, HIM milling causes less damage to the substrate and allows higher resolution.

Cao et al. [313] proposed an efficient dash-line lithography (DLL) method to construct sturdy single-molecule junctions with indented graphene point electrodes. In detail, a 5-nm-width dash line which consisted of alternating segments and spaces was designed to open windows in the PMMA mask with the ultrahigh-resolution EBL. Then, an oxygen RIE process was applied to selectively etch graphene through the dash-line windows. Nanogaps between indented graphene point contacts were formed by exploiting the gradual etching and undercutting of the PMMA. The size of nanogaps which could be regulated via the etching process was generally less than 10 nm. This DLL method produced graphene point contact arrays on wafer scale and thus simplified the device fabrication procedure greatly. The cut edges of graphene sheets were terminated with carboxylic acid groups for the strongly oxidizing condition induced by oxygen plasma, which contributed to the defined stable contacts between molecules and electrodes through amide linkages. The relationship between the cutting yield and reconnection yield of nanogap electrodes was characterized to optimize device construction conditions. Approximately 50% of reconnection yield has been achieved under optimized conditions which significantly exceeds previous studies using metal or SWCNT leads (Figure 3.49). Furthermore, Gu et al. [314] combined the DLL and electroburning techniques to fabricate graphene electrodes with controllable nanogap size. After opening windows in the PMMA layer by a standard EBL technology, the oxygen RIE cut a graphene sheet into nanoconstrictions which were potential precursors to build electrodes arrays via electroburning process. It was found that the regime of electroburning strongly depended on the feature size of those nanoconstriction units, which furnished new opportunities for future device construction (Figure 3.50). Considering the generation and dispersion of Joule heat, the main hot zone was distributed at the narrow part of nanoconstriction where the resistance was particularly high. In Region 2, the electroburning tended to occur from the edge of graphene that was more reactive owing to abundant oxide defects. However, the hot zone was large and edge atoms were limited in Region 1 as the larger feature size of nanoconstriction. The edge atoms, phonons, and random defects all could promote reactions between oxygen and graphene, inducing a rapid, simultaneous oxidation in the full hot zone. Therefore, compared with Device 1, the Device 2 exhibited a mild, graduated electroburning behavior.

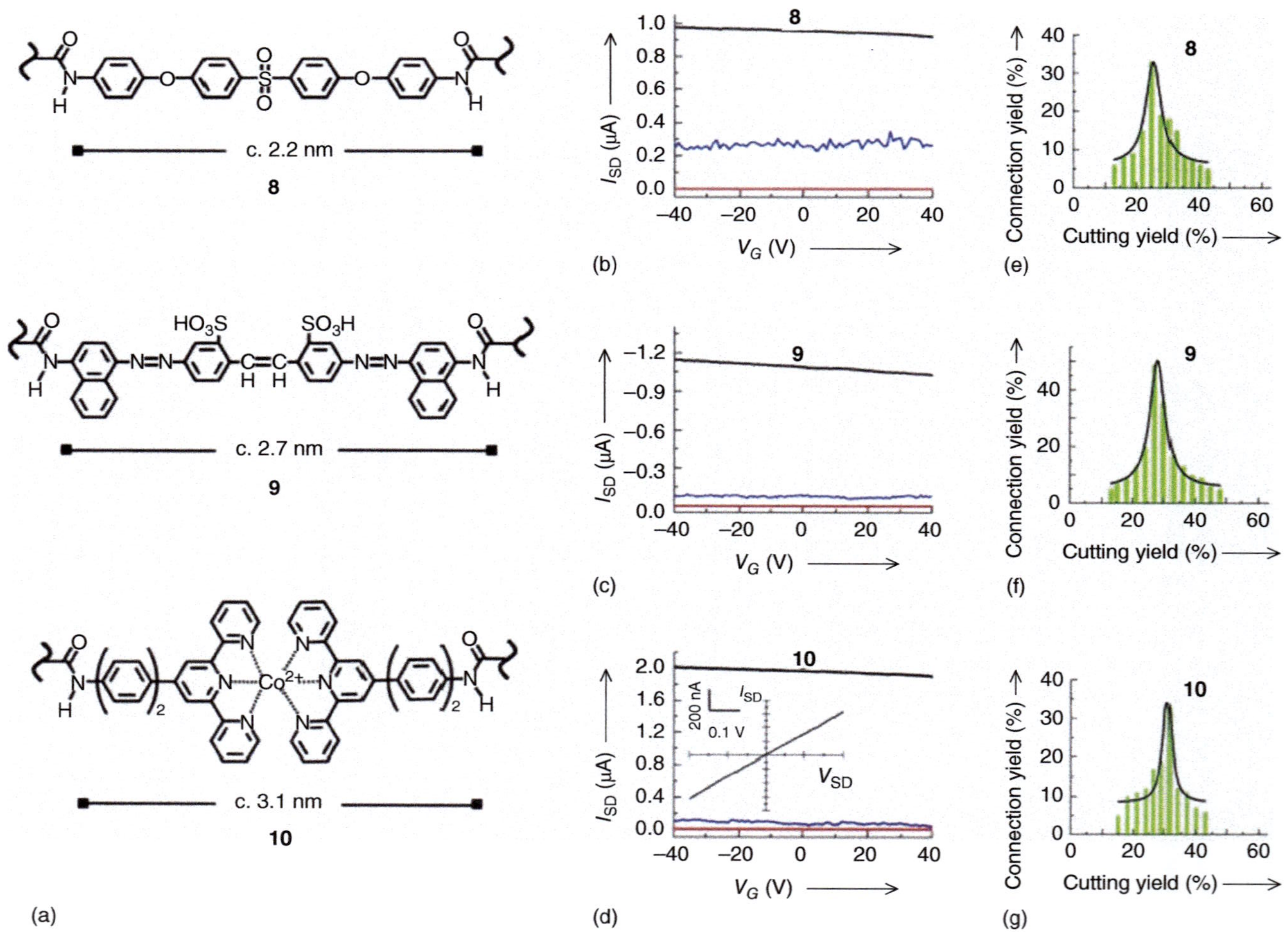

c. 2.2 nm
8
HO3S
SO3H
c. 2.7 nm
9
Co2+
c. 3.1 nm
10
(a)
8
ISD (μA)
VG (V)
(b)
9
ISD (μA)
VG (V)
(c)
10
ISD (μA)
200 nA
0.1 V
ISD
VSD
VG (V)
(d)
8
Connection yield (%)
Cutting yield (%)
(e)
9
Connection yield (%)
Cutting yield (%)
(f)
10
Connection yield (%)
Cutting yield (%)
(g)

Figure 3.49 Reconnection of graphene point contacts by molecules 8–10. (a) Molecular bridges (8–10) spanning single-layer graphene point contacts. (b–d) Device characteristics of the representative devices reconnected by each molecule (b for 8, c for 9, and d for 10) before cutting (black curves, $V_{SD} = -1$ mV), after cutting (red curves, $V_{SD} = -50$ mV), and after connection (blue curves, $V_{SD} = -50$ mV), respectively. Inset: the drain current (I_{SD}) as a function of the S–D voltage (V_{SD}) of the same device after connection at zero gate bias. (e–g) Statistical data of the connection yields as a function of the cutting yields for each molecule (e for 8, f for 9, and g for 10). The simulated curves are shown in black. Source: Cao et al. [313]. 2012, John Wiley & Sons.

Moreover, Zhou et al. suggested that the OWL was relatively suitable for the preparation of graphene ribbons with sub-10-nm gaps (Figure 3.51) [315]. The well-defined metal nanowires etch masks which were synthesized by electrochemically growing alternating Ni and Au segments in an anodic aluminum oxide template took a key role in this technology. In the process, the graphene substrate was first spin-coated with a thin film of PMMA which could secure the nanowires and prevent them from migrating during processing. Then nanowires were drop-cast from an ethanol suspension. In order to form strong adhesion between the nanowire and the substrate, baking the sample on a hot plate is necessary. Subsequently, the Ni sacrificial sections were selectively removed by HCl etch, thus opening windows over the PMMA layer, followed by the etching of the PMMA and graphene layer through an RIE process. It was found that the GNR with 6 nm gap was precisely obtained when the Ni segment was as small as 6 nm, indicating reliable pattern transfer between the wire and the substrate. Therefore, this wet etching approach which is as precise as the EBL technique is more facile. The OWL technology does not require sophisticated nanofabrication instruments and may contribute to rapidly develop a wide variety of molecular devices.

The maskless AFM nanolithography is another attractive technology to fabricate graphene nanogap electrodes for its advantages of efficiency, economy, and easy implementation. Hu and coworkers [228] have successfully tailored both graphene and GO sheets into well-defined nanogap electrodes with a gap width below 10 nm via AFM tips (Figure 3.52). The GO nanogap electrodes were easily formed via an AFM tip scratching on the GO flake which was on a SiO_2/Si substrate, while the graphene tended to fold itself on the SiO_2/Si substrate because of different weak interactions. Fortunately, the authors settled this problem by inserting a thin PMMA layer between the graphene and SiO_2 substrate. The inserted polymer insulator layer could not only prevent the graphene layer from folding but also use as dielectric materials. In addition, graphene electrodes were also constructed on the highly oriented pyrolytic graphite (HOPG). In detail, a large-area HOPG was first patterned into microscale through optical lithography in order to easily transfer the outshoot of the patterned parts later. Then the AFM lithography was utilized to etch patterns into submicroscale or nanoscale. Since only one or several graphene layers on top of the patterned HOPG were transferred to the desired substrate in one time, the same etched nanogap electrodes on HOPG could be used for many times.

The interface between the molecules and graphene electrodes play an important role in constructing robust functional molecular devices [316]. Typically, the

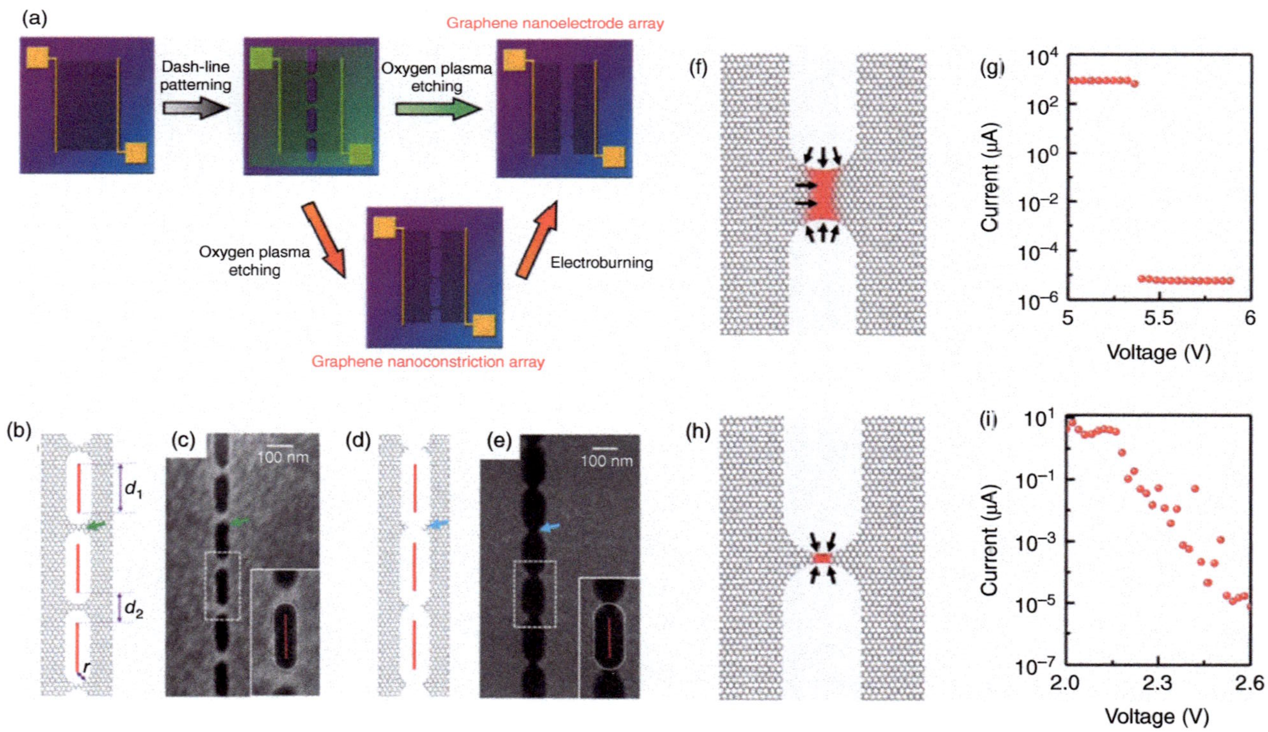

Figure 3.50 (a) Schematic of the fabrication of graphene nanoelectrode arrays. The green and red arrows show the direct cutting and electroburning routes, respectively. (b–e) Schematic and SEM images of graphene nanoconstriction arrays (b, c) and nanogap arrays (d, e) prepared by dash-line lithography. Red segments in (b, d) show the design of the e-beam pattern. The enlarged scales in (c, e) show the shape of the etching holes. (f, h) Schematic of a nanoconstriction unit in Regions 1 (f) and 2 (h), respectively. The red area shows the hot zone. The black arrows show the probable oxidation reactive sites. (g–i) *I*–*V* curves of an electroburning process for Devices 1 (g) and 2 (i). Source: Gu et al. [314]. Copyright 2018 Royal Society of Chemistry.

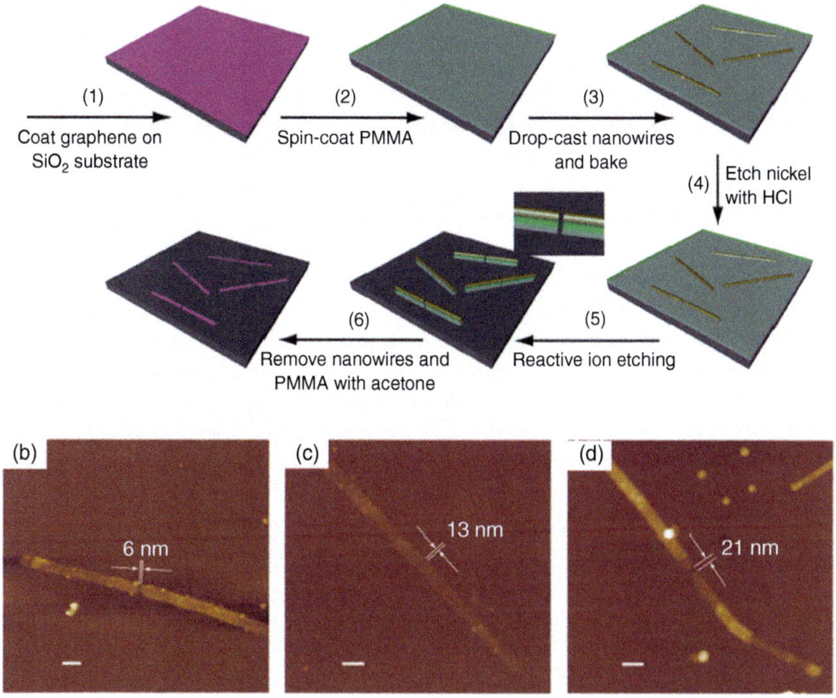

Figure 3.51 (a) Schematic illustration for making graphene nanoribbons with well-defined gaps using OWL-generated, striped nanowires as etch masks. (b–d) AFM images of smaller nanogaps in graphene nanoribbons. (b) A 6-nm gap (RIE time: 40 seconds). (c) A 13-nm gap (RIE time: 50 seconds). (d) A 21-nm gap (RIE time: 60 seconds). Scale bar: 100 nm. The bumps on the graphene are presumed to be residual PMMA, and multilayer characteristics of chemical vapor deposition-grown graphene manifest as plateaus in the ribbon. Source: Zhou et al. [315]. 2012, American Chemical Society.

molecules are attached to electrodes through either chemisorption or physisorption. And in some cases the chemisorption is necessary, as the molecular devices may confront a broad range of chemical treatments. The graphene electrodes prepared via DLL way can be reconnected by those conductive molecules which are terminated with amino groups [317]. The amide linkages between graphene point contacts and molecules guarantee robust molecular junctions, so that it is possible to fabricate devices that are pH sensitive. It was demonstrated the conductance changed by more than 2 orders of magnitude for several switching cycles when the device was alternatively immersed in solution of high and low pH, since the molecules grafted with two sulfonic acid groups could reversibly react with base or acid. Furthermore, the graphene-based single-molecule junction is a superb new-generation test bed due to its easy fabrication and stability. Most of the previous junctions are prepared by the ex situ synthesis of molecular wires which are subsequently inserted into nanogap electrodes. However, molecules are prone to oxidative oligomerization or aggregation, which dose harm to the whole system. In order to avoid these problems, Cao et al. put forward a novel way for the

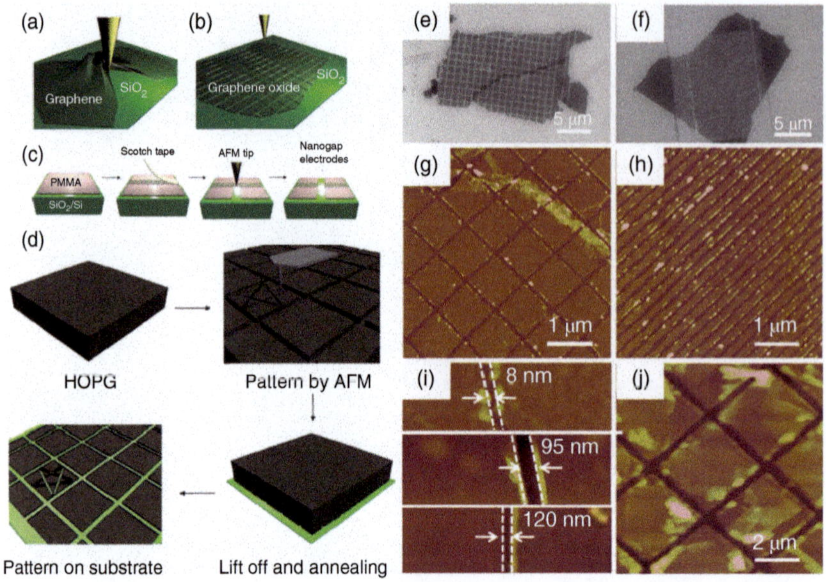

Figure 3.52 (a–c) Schematic diagrams of (a) etching graphene on SiO_2, (b) GO nanogap electrodes on SiO_2, and (c) etching graphene nanogap electrodes on PMMA. (d) Schematic diagrams of etching HOPG by AFM and transferring the patterned graphene nanogap electrodes from HOPG to other substrates. (e–j) SEM images of fabricated nanogap electrodes: (e) squares arrays, (f) linear arrays, AFM image of (g) squares arrays, (h) linear arrays, (i) magnified nanogap electrodes, and (j) magnified nanogap electrode arrays of graphene on PMMA. Source: He et al. [228]. 2010, AIP publishing LLC.

in situ construction of complex molecular wires between two graphene nanogap electrodes by a multi-step reaction sequence. First, the graphene point contacts were initiated with a terpyridyl ligand which contains a single amino group to react with graphene electrodes, while there was no current detected. Then, the primed devices were immersed in a diluted methanol solution of cobalt acetate. In this step, some devices would become conductive, as they were connected through the coordination reaction between metal ions and tridentate aromatic pockets of terpyridyl ligands. The formation of sturdy amide linkage was vital to support the successful coordination chemistry. A longer molecular wire was also prepared via a three-step strategy. After optimizing, the yields of the working devices were approximately 21% and 9% for the two-step and three-step connection reaction, respectively (Figure 3.53). The good stability of both graphene electrodes materials and covalent amide linkages helps to avoid the fluctuation interference problem in metal–molecule–metal junctions and ensure the exploration of the device function from the intrinsic stereoelectronic of the central aromatic chain [318]. It was demonstrated that the thermally induced phenyl twisting in the aromatic chain molecule resulted in different microstates with different degrees of conjugation and thus led to temperature-dependent stochastic conductance switching. This stereoelectronic effect offers a fresh perspective to fabricate practical functional single-molecule devices. The electroburnt graphene nanogaps can be reconnected by molecules

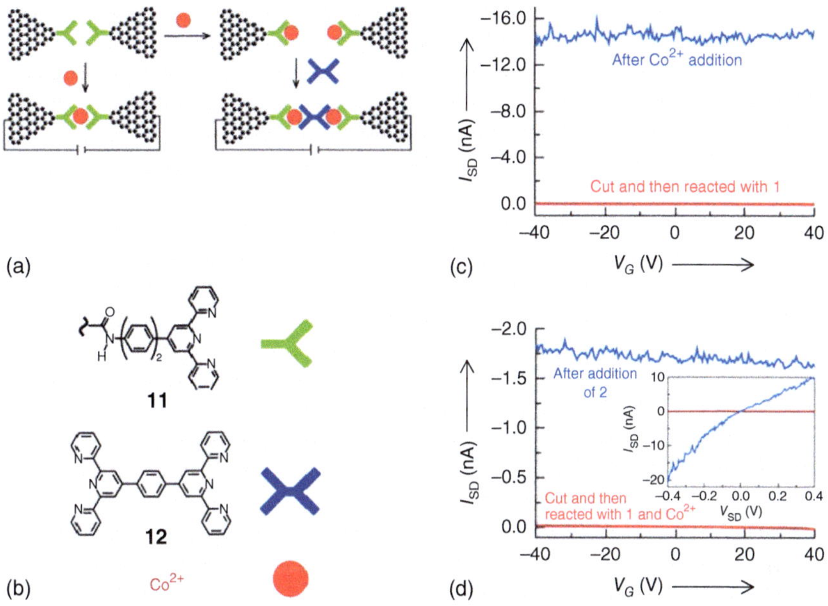

Figure 3.53 In situ construction of molecular junctions. (a) Schematic description of strategies for bridging graphene point contacts through two-step or three-step sequences. (b) Molecules used for reconnection. (c) I_{SD} vs. V_G plot of a device reconnected through the two-step strategy. $V_{SD} = -50$ mV. (d) Device characteristics for a device reconnected through the three-step sequence. $V_{SD} = -50$ mV. Inset shows the I_{SD} vs. V_{SD} plot for the same device after reactions with 11 and cobalt ions (red), and after the addition of 12 (blue) at zero gate bias. Source: Cao et al. [317]. 2013, John Wiley & Sons.

with amine terminal groups as well, considering its formation conditions. Typically, the diamine molecules and 1-ethyl-3-(3-(dimethylamino)propyl)-carbodiimide (EDC) activating agent are introduced simultaneously, and the diamine molecule must interact with the carboxylic acid site which has been transiently activated by the EDC to realize a successful coupling. Xu et al. [311] improved the yield of molecular junctions via optimizing this condensation reaction process. The activated carboxylic acid group was stabilized by the NHS before the introduction of diamine molecules, so the reconnection yield was considerably increased.

In view of the unique honeycomb lattice and 2D planar structure of graphene, π–π stacking is another effective connection. It could lead to a more reproducible bonding for the gentler contact between the molecule and the electrode which preserves a more well-defined geometry. High-performance thin film FETs have been manufactured using single-layer graphene sheets as source–drain electrodes, initiating broad research interest in the field of molecular electronics [308]. Compared with metal electrodes, the graphene has a molecular structure similar to organic semiconductors. Molecules tend to form excellent interface contact with graphene electrodes and self-assemble in the conducting channel benefited from its strong π–π interactions. And the work function of graphite is much lower than that of Au, so there is a reduced hole-injection barrier between the graphene

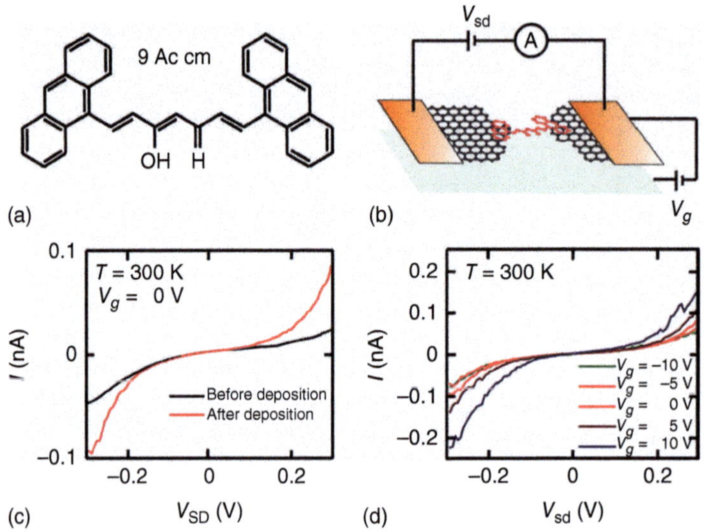

Figure 3.54 (a) Chemical structure of the anthracene terminated curcuminoid wires. (b) Artist's impression of a single 9 Accm molecule bridging a graphene nanogap, representing the ideal transistor configuration. Note that the ambient conditions to which the devices are exposed may result in a thin layer of adsorbates such as water to be present on the surface of the few-layer graphene. (c) I–V characteristics of the nanogapped electrodes before and after being bridged by the 9Accm molecules at 300 K. While the conductance at low bias superimposes with the empty gap characteristic, at higher bias a clear current increase is observed. (d) Dependence of the I–V characteristics of the nanogapped electrodes bridged by 9 Accm molecules on the applied back-gate voltage measured at 300 K. Source: Prins et al. [261]. 2011, American Chemical Society.

and organic layers. Moreover, tight interatomic sp^2 bonds result in quite stiff graphene electrodes, thus making devices sturdy to withstand harsh conditions. These advantages ensured the thin film FETs based on graphene contacts exhibited bulk-like carrier mobility, high on/off current ratio and high reproducibility. As the nanogap size drops to less than 10 nm, more and more molecules with extended π-conjugated terminal groups which could interact strongly with the π-system of graphene electrodes can be designed to fabricate single-molecule devices. The anthracene-functionalized curcuminoid molecule (9 Accm) is one of the outstanding examples (Figure 3.54), whose anthracene group offers a sturdy anchor to the graphene electrodes and curcuminoid wire possesses a high π-electron density to transfer charge carriers. In order to better characterize molecules in the junction, multi-layer graphene sheets were selected which averted the gating of the source and drain electrodes and guaranteed a continuous reservoir of electrons. The molecular devices showed stable gate-modulated transport characteristics at room temperature, and Coulomb-blockaded transport and single-electron tunneling were observed from the conductance map at 10 K [261]. Hybrid molecular junctions were also constructed with superconducting contacts and anthracene-functionalized copper curcuminoid molecules [319]. After molecules deposition, the I–V_b characteristics of devices became linear and the current was strongly dependent on

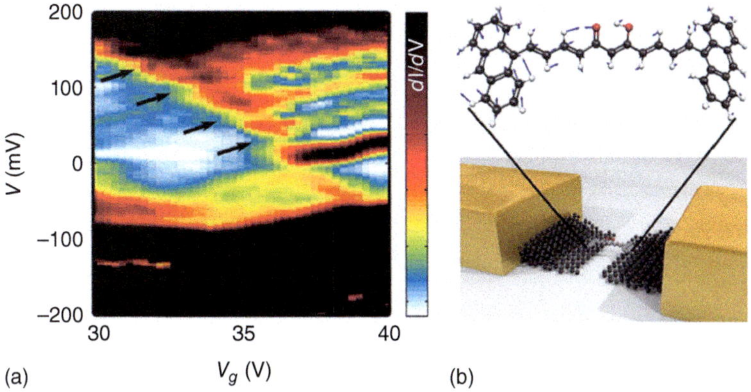

Figure 3.55 (a) *dI/dV* color plot. Several single-electron transport (SET) excitations (marked with black arrows) appear at positive bias at 46(±6), 79(±6), 115(±10), and 150(±10) mV. (b) Schematics of the curcuminoid-based molecule: 9ALCccmoid and artistic representation of a few-layer graphene–curcuminoid junction. The central curcuminoid backbone is completed with a pair of anthracene groups. Color code: C = gray, O = red, H = white. Source: Burzurí et al. [320]. 2016, American chemical society.

gate voltage. Hence, these hybrid molecular devices provided a way to study superconducting effects in single-molecule transport. For the sake of shedding light on π–π stacking strategies in molecular transport, the molecule–electrode contact was expected to be far from graphene edges whose variable shapes and compositions may influence the properties of interface. So another curcuminoid-based molecule (9ALCccmoid) was proposed with an extended transport backbone which ideally brought anchoring groups farther from electrode edges (Figure 3.55) [320]. An intermediate coupling was fulfilled through the modified π–π bonding in the study and thus it was possible to research higher-order processes, for instance, co-tunneling and vibrational excitations. A novel spintronic device made of a single-ion magnet embedded in graphene nanogap has also been successfully fabricated and tested, which implied the graphene platform and π–π stacking were considerably suitable for the design of complex molecular systems [321]. It was predicted that using extended anchor groups on graphene electrodes could significantly suppress single-molecule conductance fluctuations [322–324]. Thus, the tetrabenzofluorene (TBF) and pyrene anchor groups have attracted great attention besides anthracene group. Though a few robust molecular junctions have been realized, there are still some aspects to consider, such as the influence of pyrene connection point on conductance, the effect of nonplanarity of TBF, and the broken conjugation between the cyclopentadiene connection point and molecular wire [325]. The fullerene-based anchoring group which has been applied well in gold single-molecule junctions [326] is also intriguing in graphene-based molecular devices [327, 328]. It was suggested that the fullerene–graphene combination was beneficial to improve experiments on individual single-molecule junctions as their reproducible contacts [329]. Ullmann et al. [330] reported the single-molecule contacts between a fullerene end-capped molecular bridge and epitaxial graphene

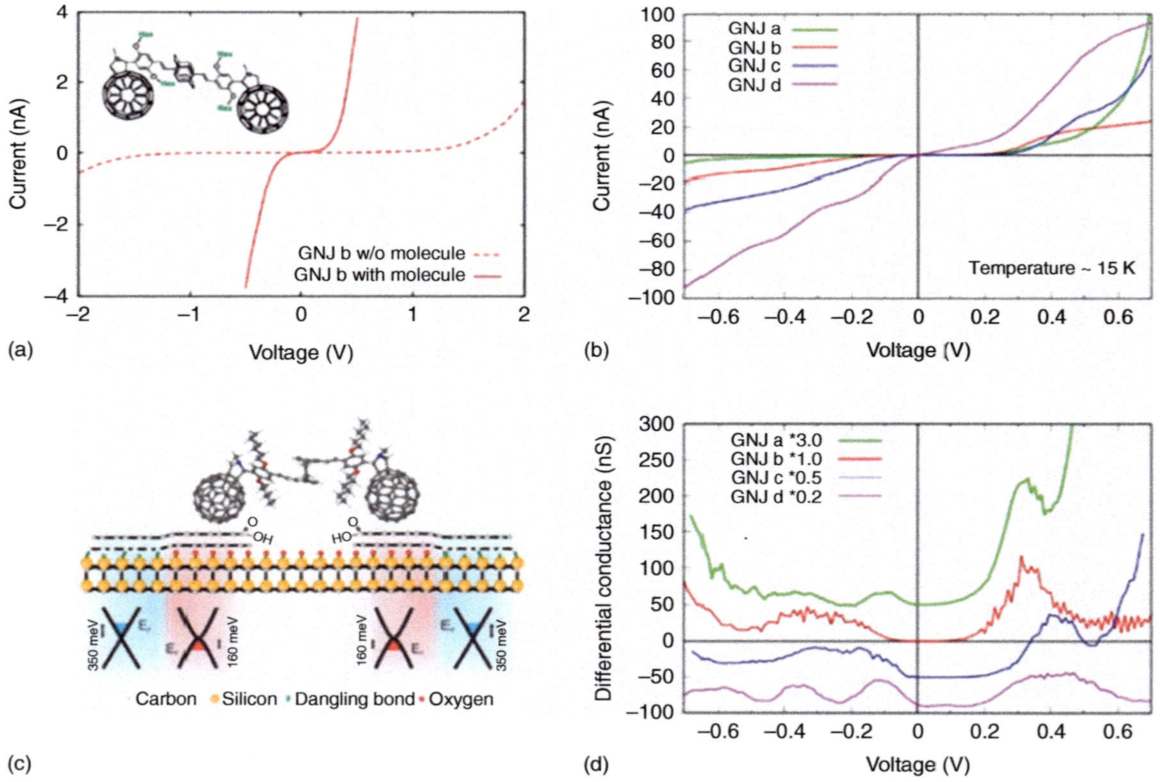

Figure 3.56 (a) C_{60} end-capped molecular wire, used for single-molecule junctions and tunnel characteristics of the same junction before (dashed) and after (solid line) application of molecules, recorded at room temperature. (b, d) Four graphene nanojunctions (GNJ) that display stable nonlinear electrical characteristics at low temperatures, which can be assigned to a single-molecule junction. The differential conductance in (d) is slightly rescaled and shifted by 50 nS each for better visibility. (c) Sketch of a GNJ bridged by a molecule. The graphene is presumably modified at its edge. As a consequence, the fullerene end group is expected to reside on a positively charged bilayer graphene area. Source: Ullmann et al. [330]. 2015, American Chemical society.

electrodes (Figure 3.56). Extremely similar transport properties were observed when using graphene and Au electrodes, indicating that the molecule itself dominated the charge transport in the system. So the asymmetric current–voltage characteristics and switching behavior were considered to be associated with different conformers of the molecule.

In addition to the graphene–molecule–graphene junctions, other robust graphene-based molecular junctions were also studied. Using graphene tip electrodes, sturdy graphene–molecule–Au junctions have been fabricated (Figure 3.57) [242]. The Au AFM tip was coated with more than four-layer graphene by CVD technology. This process took place at high temperature, so that nearly all of contaminants on the Au tip that had a strong effect on the resistance could be removed. And there were good contacts between the graphene tip electrode and organic molecules. Therefore, the graphene–molecule–Au junctions showed more reproducible electronic behaviors and higher yields when compared with Au–molecule–Au junctions. The graphene electrode also exhibited good endurance, operational stability, and long shelf life. This technique provides a simple and reliable way to better understand molecules properties and exploit their applications in molecular electronics.

Graphene is a promising bottom electrode as well, for its availability of relatively defect-free wafer scale and excellent characteristics. Yang and coworkers constructed a series of Au–molecule–graphene junctions in which the graphene acted as the bottom electrode to measure the electrical properties of single molecules via the STM-based matrix isolation $I(s)$ method [331]. It was found that the graphene offered the superior electrode performance. And the single-molecule junctions demonstrated a phase-coherent tunneling mechanism when various lengths of dicarboxylic-terminated alkanes were located within the nanogap. The Au–molecule–graphene junction exhibited a lower attenuation factor compared with the Au–molecule–Au junction because of the difference in Fermi energies of two junctions, relative to the frontier orbitals of the molecules. Further research have indicated the graphene-based nonsymmetric junctions and appropriate anchoring groups are crucial to low decay constants and high conductance junctions at longer molecular lengths, since the electronic structure of graphene and the weak coupling at the molecule–graphene interface disturbed the electrostatic equilibrium and promoted the charge transfer through molecule–Au interface (Figure 3.58) [332].

Application of Graphene Nanogap Electrodes in Optoelectronic Nanodevices The optoelectronic nanodevice whose electrical conductivity can be regulated by external stimuli is a vital component of modern computer circuits. The application of graphene nanogap electrodes provides this field with a broader development prospect for their outstanding characteristics [49, 223, 333]. In order to realize practical applications of the optoelectronic molecular devices, several kinds of photosensitive molecules have been investigated, such as azobenzenes, diarylethene, and P3HT. An azobenzene molecule consisting of two phenyl rings linked by a N=N double bond possesses two stable trans and cis isomer states. The trans isomer

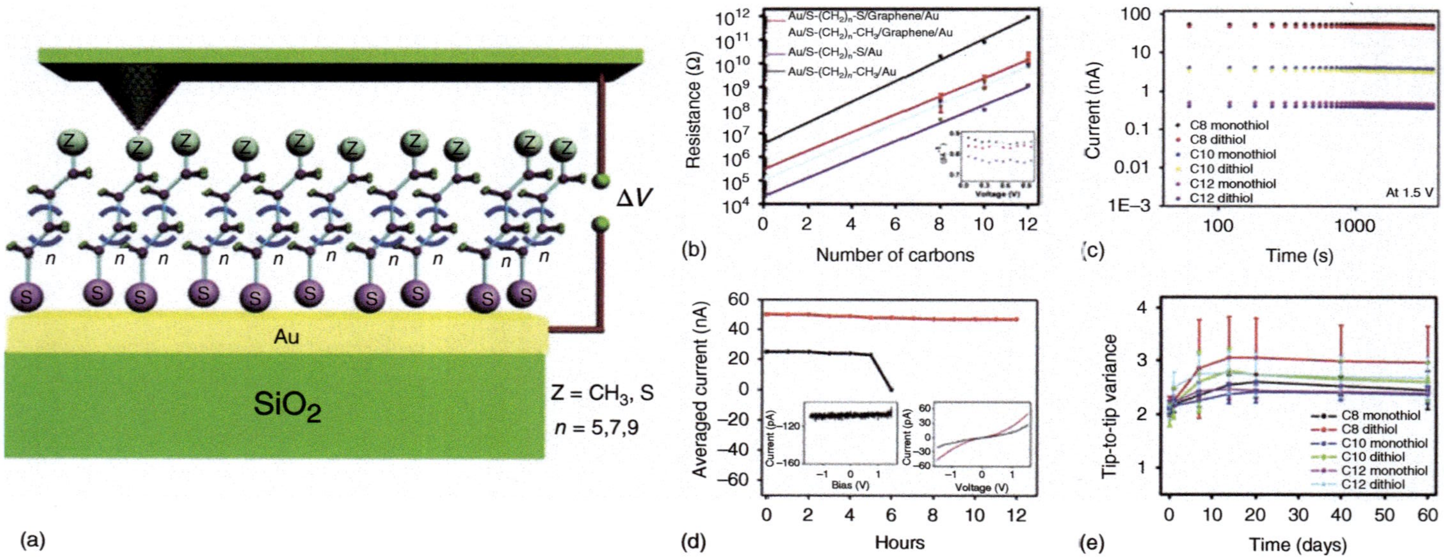

Figure 3.57 (a) Schematic representation of a molecular tunnel junction formed using conducting probe-atomic force microscopy (CP-AFM). (b) Plot of average junction resistance vs. self-assembled monolayer (SAM) chain length (number of carbon atoms). The inset shows the β values of the corresponding junctions as a function of the applied bias. (c) The operational stability of molecular junctions using graphene tips at 1.5 V. (d) The test of endurance of Au and graphene tips. The left inset shows the $I-V$ data for octanemonothiol junctions with Au tips after continuously scanning for six hours. The right inset shows the $I-V$ data for octanemonothiol junctions with Au (black) and graphene (red) tips. (e) Plot of the tip-to-tip variance as a function of the storage time of graphene tips in air. Source: Wen et al. [242]. John Wiley & Sons.

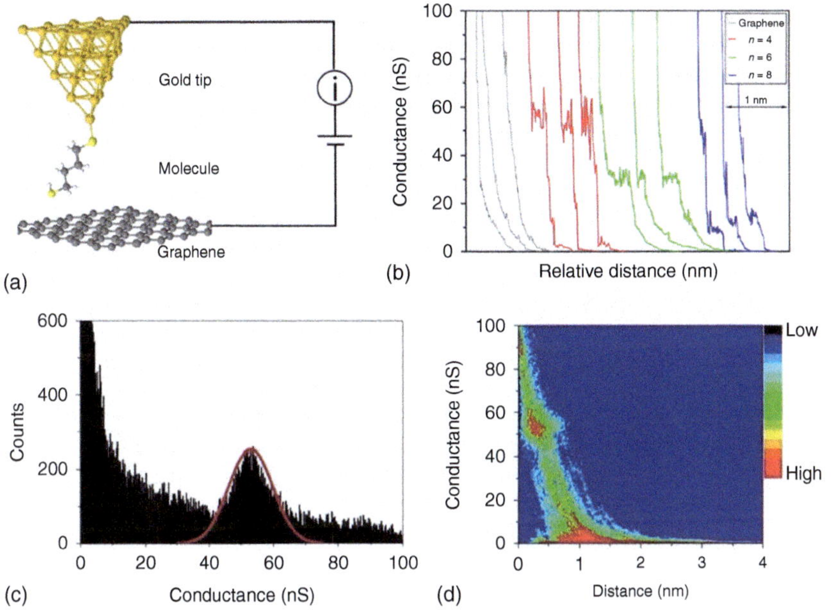

Figure 3.58 (a) Schematic diagram of the molecular junction formed in this study. (b) Typical *I*(*s*) curves of bare graphene (gray, without molecular junctions formed), gold-1,4-butanedithiol-graphene junctions (red), gold-1,6-hexanedithiol-graphene junctions (green), and gold-1,8-octanedithiol-graphene junctions (blue). (c) 1D histograms of single-molecule conductance of gold-1,4-butanedithiol-graphene hybrid junctions with the Gaussian fit. (d) The corresponding 2D histogram with a sensitivity indicator of the conductance counts. Source: Zhang et al. [332]. Licensed under CCBY 4.0.

has a nearly planar conformation, while the cis isomer takes a bent structure. The trans-to-cis and cis-to-trans isomerization could be realized under the UV and visible light irradiation, respectively. Some theoretical studies have suggested that the trans isomer has a smaller energy gap (1.98 eV) of the highest occupied molecular orbital (HOMO) and the lowest unoccupied molecular orbital (LUMO), which is lower about 0.2 eV in energy. Meanwhile, the dipole moment would change from 0 (trans) to 3 Debye (cis) during the transformation process [334, 335]. These performances indicate that some different properties of azobenzene derivatives may be obtained via light illumination. And the graphene point contacts offer a reliable platform to study their electronic transport characteristics [317]. Under optimized conditions, the maximum reconnection yield of the azobenzene derivative could reach 50%, corresponding to a cutting yield of about 28%. It was found the device exhibited two distinct conductive states under UV and visible light irradiation, which was agreed with the theoretical prediction. And a series of UV and visible light irradiations showed that both states could switch stably in a reversible manner. It is the first example of a reversible light-driven single-molecule switch.

The diarylethene is another superb candidate for photoswitching molecular junctions due to its excellent thermal stability and high fatigue resistance [336]. It can be reversibly transformed between two isomers with open/closed conformations.

The nonconjugated ring-open form is achieved with visible light illumination and the conjugated ring-closed form is under irradiation in the UV region. And there are negligible changes in molecular length when the two isomers switch. However, some potential properties shown in solution disappear once the diarylethene derivatives are attached to the solid-state molecular-scale electrodes. In particular, when the diarylethene derivative was wired to gold electrodes via the Au—S bonds, only the transition from the closed and conducting isomer to the open and nonconducting isomer could be observed for the quenching effect of the excited state of the open form in the presence of the Au electrode [337]. And the opposite unidirectional photoswitching occurred in the carbon-based molecular junctions, which was ascribed to the energy transfer from the photoexcited molecule to the electrodes [289]. These results strongly illustrate that the molecule–electrode coupling strength has a great influence on the device performance. It has been proved feasible to adjust the energy-level alignments at the molecule–electrode interface by modifying diarylethene backbones with rationally designed side and anchor groups. Jia et al. [338] effectively reduced the molecule–graphene electrode coupling strength by incorporating three methylene groups into each side of the diarylethene backbone. The narrow resonance half-widths of the transmission peaks showed the interfacial graphene–molecule coupling was on the order of 1 meV in both open and closed forms, which is considerably weaker than those in previous studies. When UV and visible light were toggled back and forth, the sturdy single-molecule junction with the diarylethene linker covalently sandwiched between graphene point contacts could reversibly switch in a reproducible way. The on/off ratio of this molecular device was approximately 100, demonstrating unprecedented levels of accuracy, even after running more than 100 cycles or over one-year storage (Figure 3.59).

It is quite viable to manufacture high-performance photosensitive nanoscale transistors based on 2D graphene electrodes, since there is a good interface contact between the organic semiconductors and graphene by π–π stacking. Given the dielectric thickness of 300 nm, the channel length of transistors was designed to be 100–150 nm, which could guarantee an efficient field effect from the global back gate. A common photoactive organic semiconductor (P3HT) sensitive to visible light was deposited into the channel through a dip-coating method to prepare the thin film transistors [308]. The real-time drain current was measured by switching the light on and off at room temperature and in ambient atmosphere when the device was held at $V_D = V_G = -2$ V. The light increased the photoexcited states of the P3HT molecules and thus improved the current level. Plenty of measurement cycles were carried out since the photocurrent in the transistor was reversible and stable. The photoinduced carriers reached the maximum density within only a very short time, and the calculated responsivity of the P3HT transistor was found as high as about 8.3 AW^{-1} (Figure 3.60a–c). CuPc is also photoactive to visible light and suitable organic semiconductor material for optoelectronic devices. Hu and coworkers have fabricated nanoscale photoswitchers with the channel length below 10 nm by the AFM nanolithography. After the CuPc was vacuum deposited in graphene nanogaps, the device showed obvious photoresponse characteristics:

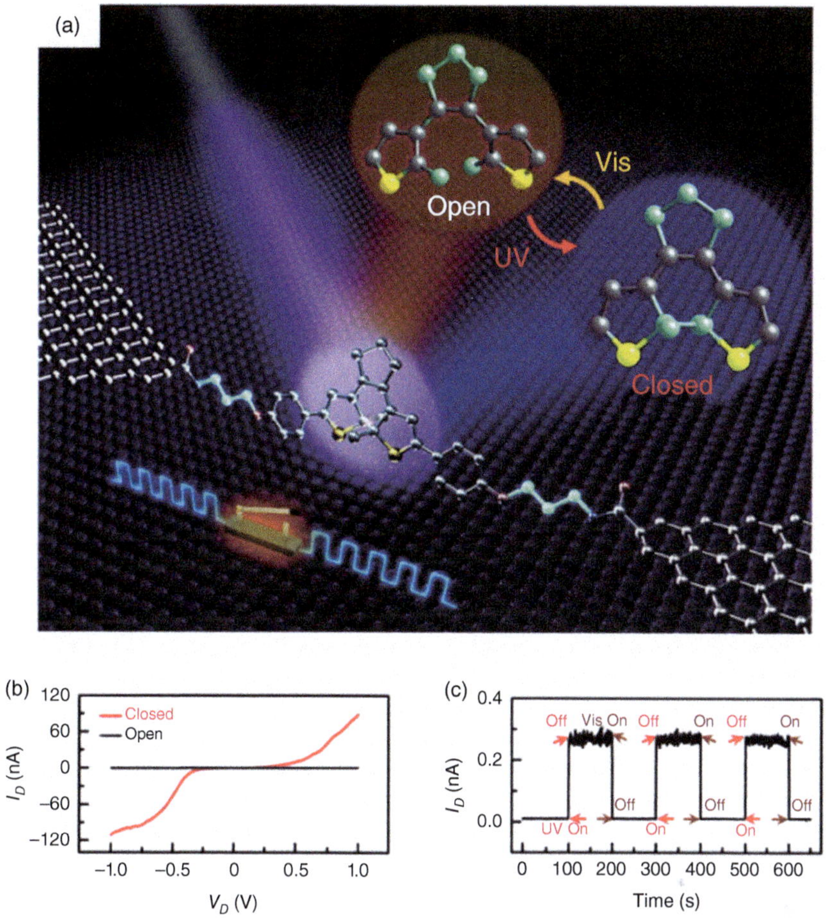

Figure 3.59 (a) Schematic of a graphene–diarylethene–graphene junction that highlights the expansion of the molecular bridge by methylene groups. (b–c) Reversible photoswitching of graphene–diarylethene–graphene junctions. (b) I–V characteristics of individual diarylethenes in open (black line) and closed (red line) format gate voltage $V_G = 0$ V. V_D, drain voltage; I_D, drain current. (c) Real-time measurement of the current through a diarylethene molecule that reversibly switches between the closed and open forms, upon exposure to ultraviolet (UV) and visible (Vis) radiation, respectively. $V_D = 100$ mV and $V_G = 0$ V. Source: Jia et al. [338]. 2016, AAAS.

high current state under light condition and low current state under dark condition. And when the back-gate bias was applied, the on/off ratio of this CuPc-based thin film transistor was up to 800. These excellent performances illustrated prospective applications of the graphene nanogap electrodes in optoelectronic nanodevices [228]. The performance of transistors is significantly influenced by the organic semiconductor layer. So LB technology was adopted to prepare dense and well-ordered SAMs, thus improving the molecular organization [339]. When using the CuPc LB monolayer and 2D planar graphene contacts, the device yield was almost 100% and the properties were highly reproducible. Although the semiconductor layer was

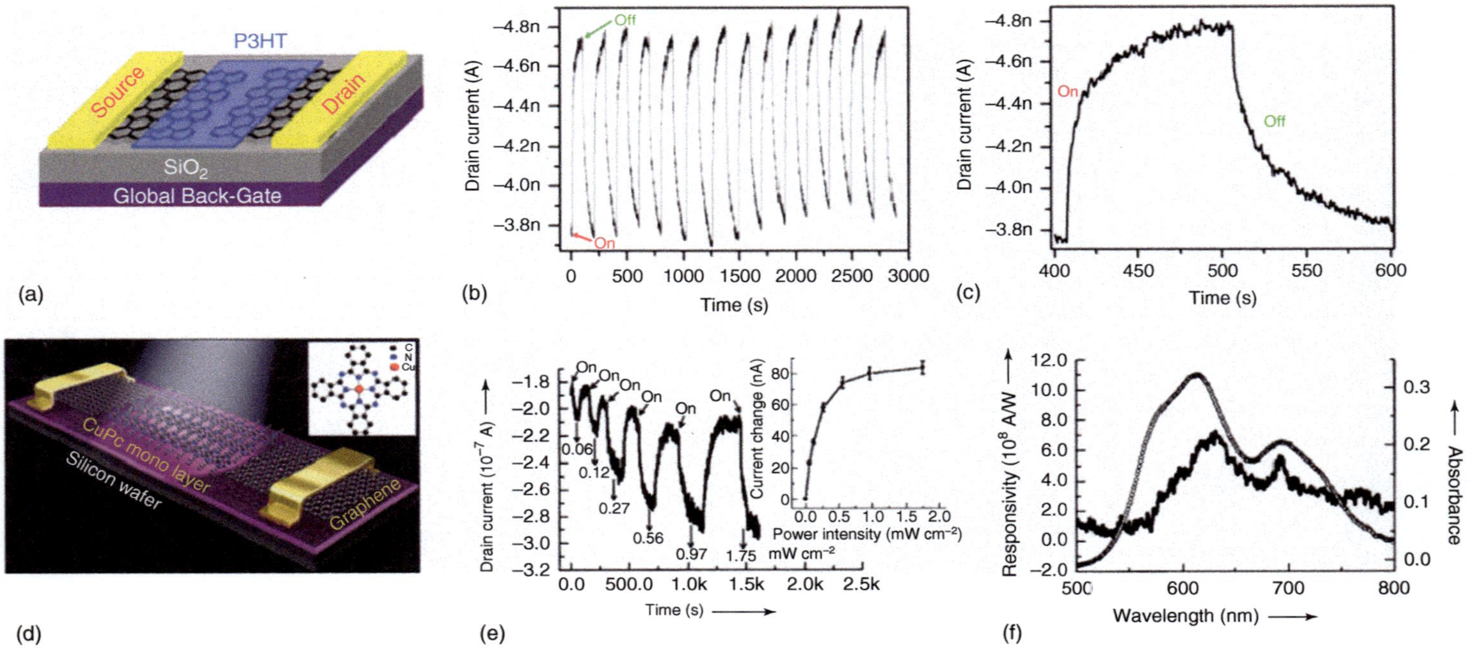

Figure 3.60 (a) Device structure of P3HT transistors made by dip-coating. Only the nanogaps between graphene sheets are covered by P3HT. (b) A time trace of the drain current for the device, showing the reversible photocurrent while the device was held at −2 V source–drain bias and −2 V gate bias by switching on/off light. (c) A representative full switching cycle showing the quick saturation process in drain current of the device under light irradiation. (d) The structure of the CuPc monolayer transistor device with metal electrodes protected by a 50-nm layer of silicon dioxide. Inset: the molecular structure of copper phthalocyanine (CuPc). (e) The time dependence of the drain current I_D as lights of different power are switched on and off. Inset: power dependence of the changes in I_D. $V_G = 0$ V and $V_D = -8$ V. (f) Comparison of the wavelength-dependent spectrum (– –) with the UV/Vis absorption spectrum (○) of CuPc thin films. Light was scanned from 500 to 800 nm in 5 nm steps with each wavelength left on for five seconds. All wavelengths used were adjusted to be constant ($I_{light} = 30\,\mu W\,cm^{-2}$). $V_G = 0$ V and $V_D = -6$ V; $W = 2\,\mu m$, $L = 50$ nm. Source: (a–c) Cao et al. [308]. 2009, John Wiley & Sons. (d–f) Cao et al. [339]. 2010, John Wiley & Sons.

only 1.3 nm thick, the on/off current ratio (over 10^6) and carrier mobility (as high as $0.04\,cm^2\,V^{-1}\,s^{-1}$) of monolayer transistors were comparable to those of bulk-like FETs [340]. On the other hand, the device had a quick response to the light with different power. In detail, two peaks appeared in the responsivity spectrum; the high one at about 625 nm was ascribed to the π–π^* transition of CuPc aggregated species arranged through face-to-face stacking and the weak one is attributed to the π–π^* transition of CuPc monomers. The monolayer transistors form the basis for new kinds of tunable photodetectors and environment sensors (Figure 3.60d–f).

Compared with a single molecule, the SAM can guarantee a more robust and reproducible molecular device [241]. Since the photoswitching phenomenon is resulted from the response of molecules to the light, it is considerably reasonable to fabricate high-performance two-terminal molecular monolayer junctions by using transparent electrodes. To the materials, graphene with atomic thickness and high conductivity is one of the most ideal transparent electrode candidates. Wang et al. [341] have manufactured reliable solid-state molecular junctions utilizing multilayer graphene films as the top electrode. The graphene film could contact with SAMs closely as a result of its high flexibility and stretchability, ensuring a better contact conductance and a high yield of the working molecular devices. Afterward, the two-terminal device with a photoswitchable molecular monolayer anchored between highly flexible and transparent graphene electrodes were reported [342]. In this device, the aryl azobenzene derivative molecules were covalently grafted to the graphene bottom electrode through a dediazoniation process which contributed to the formation of a large-area uniform molecular monolayer. And then the graphene top electrode was transferred onto the other side of the molecular monolayer, resulting in a physical contact between them (Figure 3.61). The UV–vis absorption spectra of azobenzene derivative molecules in solution illustrated the transition near peaks of the wavelength for 360 and 430 nm. Meanwhile, considering the difference of the length in the trans and cis isomers, there was a shorter tunneling distance in cis isomer molecular junctions. Therefore, it was found that the molecular monolayer device transitioned from the low conductance state to the high conductance state with UV light irradiation and would recover the low conductance state under visible illumination. The two conductance states were almost identical and reproducible, which suggested highly stable photoswitchable molecular devices could be realized using outstanding graphene electrodes. In addition, the graphene–SAM–Au junctions in which graphene acted as the top transparent electrode have also been studied [343]. When the device was exposed to UV light, the current increased and resonance oscillations were observed simultaneously (Figure 3.62). The Raman spectroscopy demonstrated that these resonance oscillations originated from quasi-bound states formed in localized regions in the graphene/SAM system, which implied a possibility of in situ reversibly modifying the electrical and quantum properties of Dirac fermions in graphene by functionalizing the substrate. Manipulating the molecule conformations by irradiating may provide a convenient way to program carrier dynamics and transport in graphene nanomaterials.

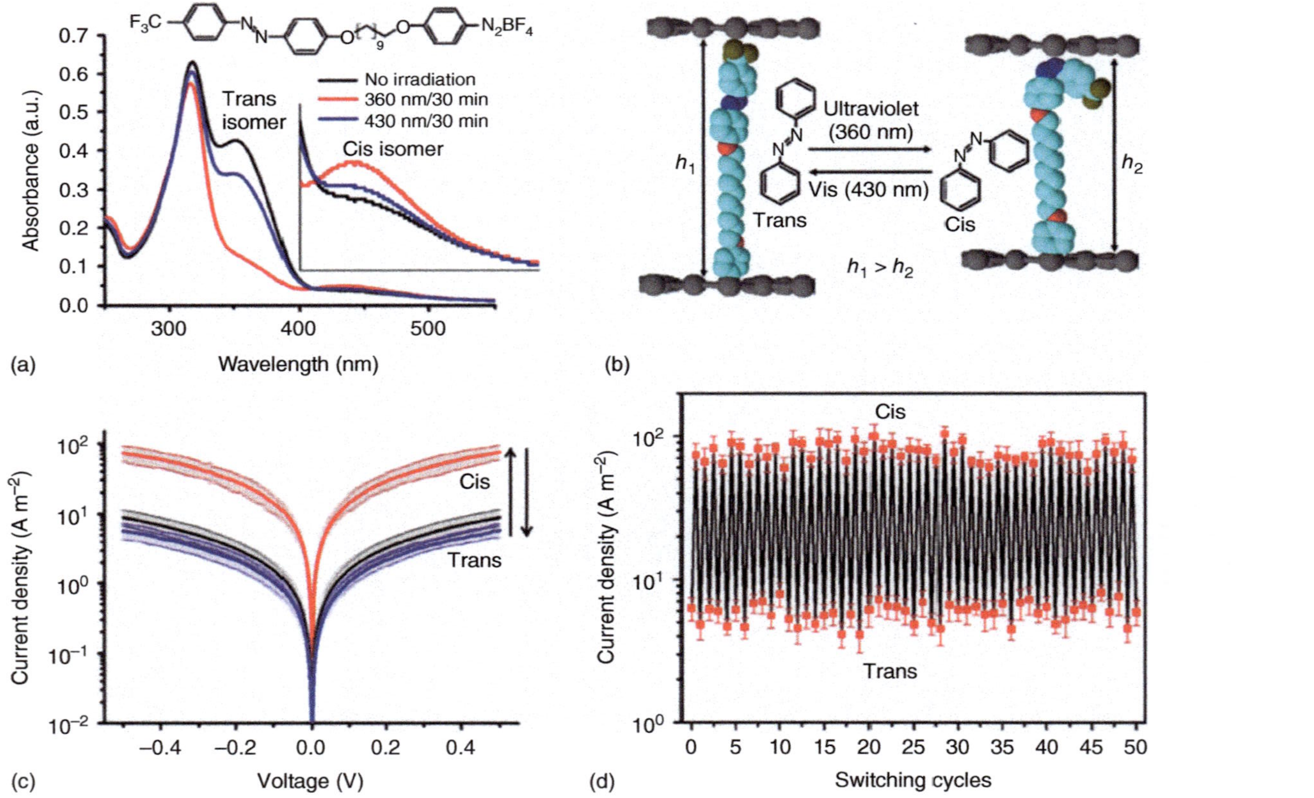

Figure 3.61 Photoswitching of graphene–aryl azobenzene monolayer–graphene devices. (a) Chemical structure of aryl azobenzene and ultraviolet–Vis spectra in solution regarding ultraviolet–Vis irradiation. Inset shows the zoomed-in spectra at a range from 400 to 550 nm. (b) Molecular tunneling barriers (with h1 and h2 being the vertical distance between the two graphene electrodes) corresponding to conformational changes in aryl azobenzene molecules with light irradiation. (c) Photo-induced current density–voltage plots on a log scale over a low voltage range for the graphene–aryl azobenzene monolayer–graphene devices. (d) Current density plots for each isomer state over 50 cycles of reversible photoswitching (induced by ultraviolet–Vis irradiation per 45 minutes); the plots were generated from average values obtained at 0.3 V from over five devices (the error bars denote the s.d. values obtained from over 20 data points for each state). Source: Seo et al. [342]. 2013, Springer nature.

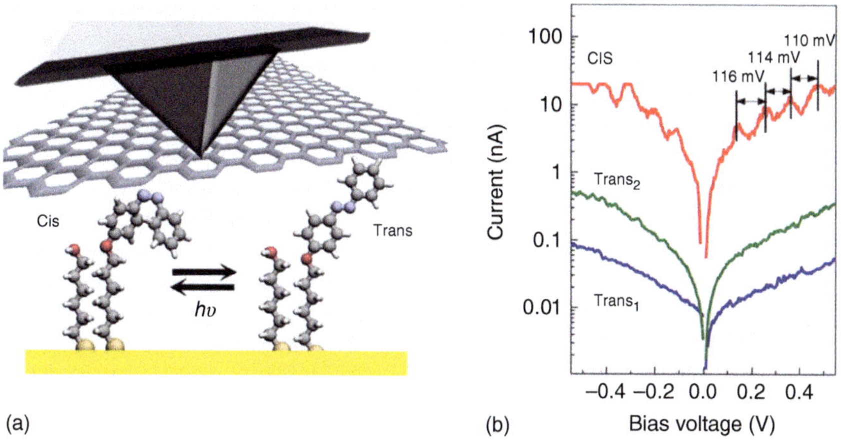

Figure 3.62 (a) Schematic representation of conductive AFM measurement on the graphene–SAM–Au hybrid with the azobenzene molecule in trans (elongated form) and cis (bent) configurations. (b) Typical *I*–*V* characteristics in trans configuration before illumination (blue trace), in *cis* configurations after switching with UV–light (red trace) and after switching back to the trans configuration (green trace) following white light illumination. Source: Margapoti et al. [343]. 2014, American Chemical Society.

It is possible to prepare molecular active components and tune their properties in organic devices; however, the subsequent techniques are necessary to characterize the spatial orientation of both active layer's component and their interfaces with atomic precision. For the device optimization, an atomically precise layout of the constituting components is critical, yet it is rarely reproducible. Fortunately, 2D self-assembled networks of supramolecules provide a potential route to spatially organize different building blocks, allowing the preparation of synthetics in molecular accuracy [344, 345]. It was shown that the atomically flat and transparent graphene-passivated H–C(100) diamond (GHD) was capable of supporting the biomolecular 2D self-assembly. So Wieghold et al. successfully manufactured a special optoelectronic device using the GHD as the bottom electrode. The active network of junctions consisted of a terrylene diimide (TDI) derivative and melamine. The gallium alloys were utilized as the top contact, for maintaining the molecular integrity at the interface. The TDI derivative was designed to response to a 710-nm wavelength of light. After irradiation, the photocurrents of 0.5 nA and open-circuit voltages of 270 mV were observed. The junction demonstrated a 3 order of magnitude increase in the short-circuit current with respect to the dark current (Figure 3.63). This work provided potential feasibility to fabricate optoelectronic device via bottom-up way with molecular resolution.

The pristine graphene prepared by mechanical exfoliation [311, 346] or CVD technology [347] has been extensively applied in molecular electronics. However, it generally suffers from either low-yield or high-cost issues. It is also a challenge to manipulate the ultrathin graphene films as well as to incorporate them into practical electronic devices. Remarkably, the reduced graphene oxide (rGO) which could be prepared via a low-cost solution-processing route and used in large-area

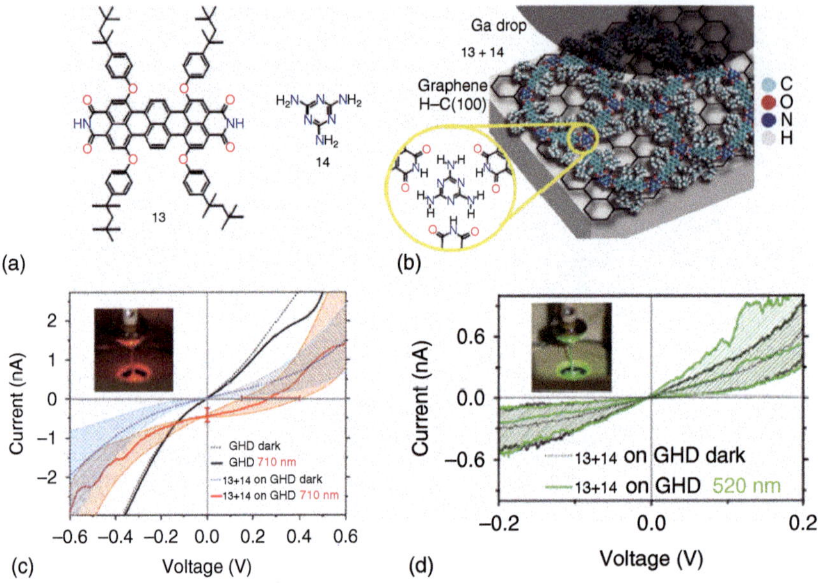

Figure 3.63 (a) Structure of TDI tetracarboxylic acid derivative 13 and melamine 14. (b) Schematic drawing of the photoresponse device setup including the ideal representation of the 13 + 14 mixture yielding a $[13_3 14_2]_n$ hexagonal supramolecular network via hydrogen bonds (yellow circle). (c) Current–voltage characteristics of pristine GHD (black lines) and of 13 + 14 on GHD before (blue lines) and after (red lines) irradiation at $\lambda = 710$ nm. (d) Current–voltage characteristics before (black lines) and after (green lines) $\lambda = 520$ nm photon irradiation. Inset shows the back illumination geometry with $\lambda = 710$ nm in (c) and $\lambda = 520$ nm in (d). Approach parameters: $I_t = 2$ nA and $V_t = 100$ mV. Striped areas indicate the maximum and minimum currents observed in more than 10 curves in a single junction, while error bars indicate the standard deviation for five junctions in three samples and are reported in the main text as $I_{SC} = (0.5 \pm 0.2)$ nA, $V_{oc} = (270 \pm 120)$ mV. Source: Wieghold et al. [344]. Copyright 2016 Nature Publishing Group.

device fabrication promotes the graphene field [223]. Meanwhile, the hydrogen bonding and van der Waals interactions between individual GO sheets are strong enough to maintain robust films which can be transferred onto target substrates without the polymer support. Therefore, several studies have been carried out to non-destructively fabricate molecular junctions which exhibit high yield, good operational stability, and long shelf life [267, 268, 348–351]. In the junctions, the rGO films can perform as conductive connections between junctions in series; on the other hand, they act as the interlayer which are able to effectively protect the SAMs from destructive metal contacts [268]. Moreover, a new type of light switchable solid-state device based on SAMs of DHA/VHF molecules was first successfully manufactured, since the rGO films showed high optical transparency, superior flexibility, and extremely low resistance [267]. The DHA/VHF, a potential photochromic system in optoelectronics, was first introduced by Daub et al. in the 1980s [352]. DHA undergoes a ring-opening reaction to transform to VHF with the UV light irradiation, and then VHF can recover back to DHA through a heat-induced ring-closure reaction. The absorption maximum of VHFs could be availably adjusted by substituent groups. In general, the photoreactions of DHA derivatives to the corresponding VHFs show a higher quantum yields when compared with the trans to cis isomerization of azobenzenes. The double-junction geometry with well-defined electrode area and spacing averted the demand for an additional top metal contact. In addition, the transparent rGO film and unique double-junction layout permitted an in situ photoswitching process of SAMs, and their charge transport properties could be measured simultaneously. The SAM of DHA displayed a higher conductivity than that of VHF by an average current on/off ratio of 5–7, which was ascribed to the intrinsic structural change of DHA/VHF isomers. So bidirectional conductance switching was realized via the alternative UV illumination and thermal annealing (Figure 3.64). This study highlights the immense potential of graphene materials for molecular electronic applications.

3.9.2.4 Silicon Nanogap Electrodes

The silicon not only plays an irreplaceable pivotal role in the CMOS technology but also demonstrates enormous potential as the electrode material in molecular electronics. Physically, the interfacial energy alignment between the silicon conduction/valence bands and the molecular orbitals can be tailored by changing the type and concentration of dopants, so the freedom degrees in the device design will increase [42, 353, 354]. Chemically, organic monolayers covalently bonding to silicon are robust, which can protect the silicon surface from oxidation and extreme conditions [43, 355]. And technologically, the existing powerful semiconductor industry can be effectively utilized to develop integrated molecular junctions or hybrid devices.

The photolithographic technologies and selective etching methods which have been developed fairly well in the traditional semiconductor fields make it reliable and facile to prepare silicon nanogap electrodes. Meanwhile, the ever-developing silicon-related chemistry furnishes plenty of means to construct the robust silicon–molecule linkage such as Si—O—C or direct Si—C bonds [356]. These aspects are promising for the development of silicon electrodes in the molecular electronics. The sub-10-nm fluid channels in poly-Si have been successfully achieved using a reversed spacer lithography method (Figure 3.65) [357]. Being different from the traditional lithography process, the spacer lithography technique determined the feature size by the thickness of the deposited materials, so the LPCVD of an oxide layer was the critical step in this approach. The obtained structure could be utilized to manufacture the nanogap capacitor arrays providing label-free DNA hybridization detection. In detail, the SAM of 3-aminopropyl triethoxysilane was first formed on the silicon surface which enhanced the efficiency of immobilization of DNAs, and then single-stranded DNA was linked to the amine group on the SAM. Furthermore, the CMOS processing and chemical etching could also be utilized to prepare silicon nanogaps, and the gap size was closely related with the thickness of the gate dielectric layer [175]. The silicon electrodes were modified with the 4-*tert*-butoxycarbonylsulfanyl-benzenediazonium tetrafluoroborate for the subsequent formation of the stable molecular layers. The yield of the functional devices was found to be 80–90%. This mass production way promoted the practical

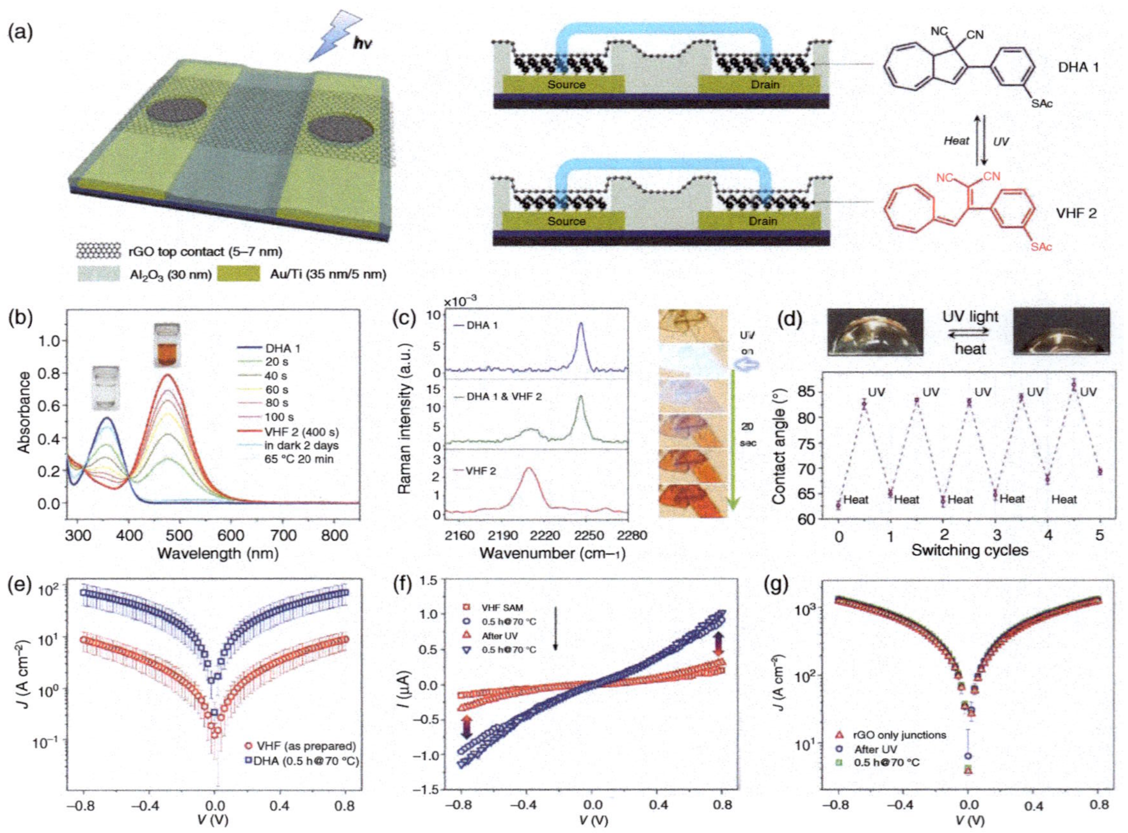
(a)
hν
rGO top contact (5–7 nm)
Al2O3 (30 nm)
Au/Ti (35 nm/5 nm)
Source
Drain
NC
CN
DHA 1
SAc
Heat
UV
VHF 2
(b)
Absorbance
Wavelength (nm)
DHA 1
20 s
40 s
60 s
80 s
100 s
VHF 2 (400 s)
in dark 2 days
65 °C 20 min
(c)
Raman intensity (a.u.)
Wavenumber (cm−1)
DHA 1
DHA 1 & VHF 2
VHF 2
UV
on
20
sec
(d)
UV light
heat
Contact angle (°)
Switching cycles
(e)
J (A cm−2)
V (V)
VHF (as prepared)
DHA (0.5 h@70 °C)
(f)
I (μA)
VHF SAM
0.5 h@70 °C
After UV
(g)
rGO only junctions

Figure 3.64 Light switchable solid-state molecular junctions with rGO thin films as transparent top contacts. (a) Schematic view of molecular test bed with rGO thin film as transparent top contact and schematic cross section of the molecular junction and molecular structure of DHA 1 and VHF 2 illustrating the thermo-optical switching. (b) UV–vis absorption spectroscopy of DHA 1 and VHF 2 in methyl cyanide. (c) (Right) Raman spectra of solidstate powders composed of only DHA 1, a mixture of DHA 1 and VHF 2, and only VHF 2. (Left) Upon UV exposure, the color of the powders changed from transparent to deep red. (d) Optical images of water droplets on surface of DHA 1 (left) and VHF 2 (right) SAMs, and reversible wettability transitions as a function of alternating thermal treatment and UV irradiation. (e) *J–V* characteristics of the VHF 2 (as prepared) and DHA 1 (70 °C for 0.5 hours) isomers self-assembled in the molecular junctions. The data are averaged from at least 24 junctions of the same batch. (f) Representative bidirectional conductance switching characteristics induced by alternating heating and UV irradiation. (g) *J–V* characteristics of junctions without SAMs (rGO only) and with a SAM of C12 alkane thiols, respectively. Source: Li et al. [267]. 2013, John Wiley & Sons.

applications of organic/nanoparticle-film-based sensors. In order to fabricate more reliable molecular junctions, a truly metal-free, silicon-based device was suggested (Figure 3.66) [358]. The fullerene films were grafted into the hydrogen-passivated 7-nm silicon nanogap by a simple solution-based deposition technique. The bipolar memory devices with a high on/off ratio and good stability were realized by the mesitylene- and 1-methylnaphthalene-based grafting of fullerene films, as the silicon–fullerene interface was capable of trapping charge carriers effectively. Since it is difficult to match the molecular length to the width of the electrode gap, the in situ synthesis of conjugated wires within the nanogap may be more feasible to bridge the top and bottom electrodes [176]. The 4-ethynylbenzaldehyde molecules which could protect the electrode from oxidation and act as a template for reacting with amino-terminated units to bridge the nanogap were covalently attached to the silicon surface. It was found that the current increased by at least 4 orders of magnitude after the gap was connected with the coupling linkers. The surface roughness and an uneven oxide layer thickness provided the compatibility of the gap width for bridging, and thus the number of connecting molecules could be controlled by the reacting time. In addition, it was possible to manufacture the real single-molecule junctions through reducing the length of coupling linkers from 5.85 to 2.8 nm (Figure 3.67) [359].

However, the interface between the molecules and silicon top electrode and the uncertain number of bridging molecules are huge challenges for the study of junctions based on the Si/SiO_2/Si structure. Remarkably, among various silicon-based molecular junctions, the Si/molecule/metal monolayer junction is more perspective for the research and practical applications. On the one hand, the self-assembly technology greatly simplifies the fabrication process of Si/molecule/metal junctions, since the molecules tend to self-assemble on the target substrates. In self-assembled molecular monolayer junctions, several devices could be achieved on a single substrate simultaneously and the difficult alignment which appears in the single-molecule junctions would not occur. In addition, many complementary spectroscopic measurements can furnish independent structural and thermodynamic information of Si/organic monolayer/metal junctions by detecting the SAMs,

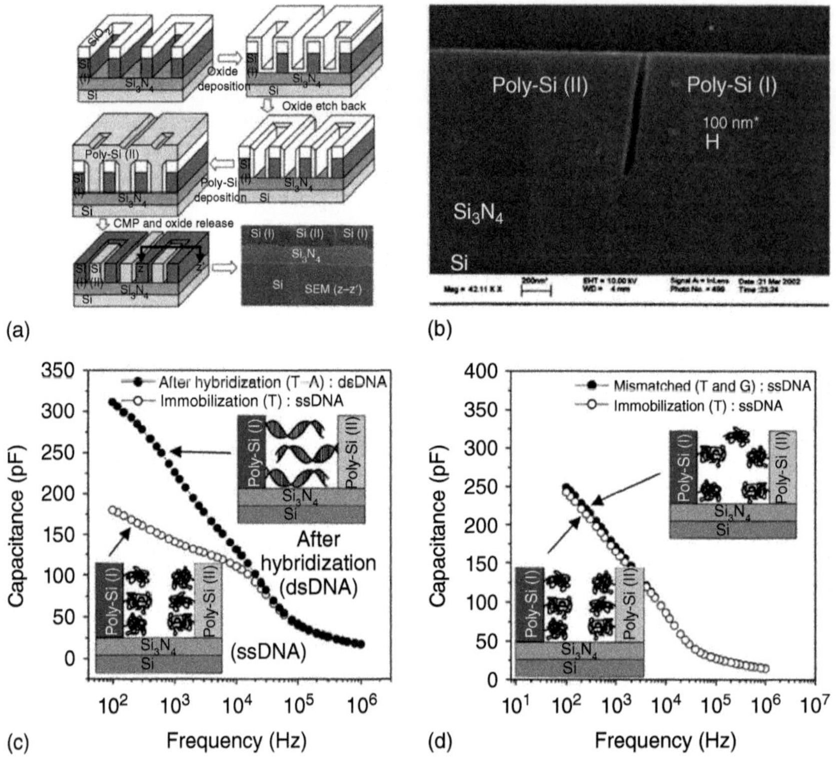

Figure 3.65 (a) Process flow of nanogaps. (b) Cross-sectional SEM photographs. The gap width is determined by the deposited oxide film thickness. (c–d) Measured capacitance vs. input frequency after immobilization and hybridization. (c) Capacitance increase as the frequency decreases in the conjugated base pairs (T–A) and (d) there is no significant difference of capacitance in the nonconjugated base pairs (T–G). Source: Choi et al. [357]. Copyright 2003 American Vacuum Society.

which is significant to understand the charge transport mechanisms and rationally design practical devices [360].

In the 1940s, Zisman and coworkers originally reported the self-assembly process of the long-chain hydrocarbons carrying polar groups on appropriate surfaces [361]. Later in 1980, Sagiv [362] first unequivocally demonstrated the production of the oriented organosilane monolayers on an oxidized silicon substrate by the direct adsorption from solution, since then the field of SAMs on SiO_2 has developed exponentially [363]. Meanwhile, the atom-resolved STM, a key enabling tool in nanoscale science, provided a powerful way of characterizing the relationship between the silicon substrate and molecules visually, chemically, and electrically [364]. Soon afterward, the densely packed organic monolayers directly covalently bonded to silicon surface without the oxide layer promised a more practical application for the molecular electronics [365, 366].

The self-assembling molecules usually contain three parts: (i) the head group, bonds the molecules to the substrates covalently; (ii) the spacer unit, assists in the

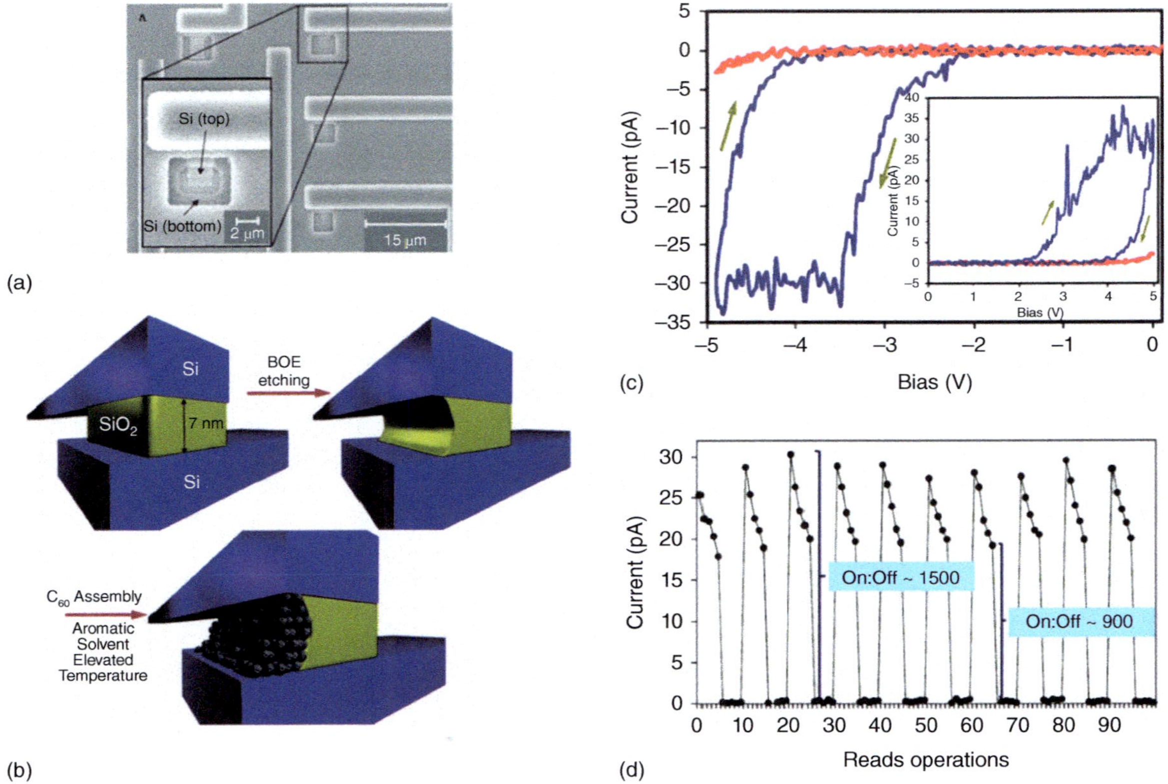

Figure 3.66 (a) SEM image of the nanogap devices from a top view. (b) Schematic of a pristine device, the device after buffered oxide etch (BOE) etching, and the device after C_{60} deposition. (c) 1-methylnaphthalene-deposited devices. The red trace in each graph is the OFF-state of the device, prior to a sweep or pulse of the opposite bias. The blue trace in each graph is the ON-state for the device, after a sweep into the opposite bias direction, which follows a higher-conductivity path until the device is discharged at high bias and returns to a low-conductivity OFF-state. (d) Read operations from 10 device cycles under improved conditions in which one cycle corresponds to one write (−6 V, 10 seconds), five reads (4 V, 1 μs each), one erase (6 V, 10 seconds), and five reads (4 V, 1 μs each). Source: Corley et al. [358]. 2010, American Chemical Society.

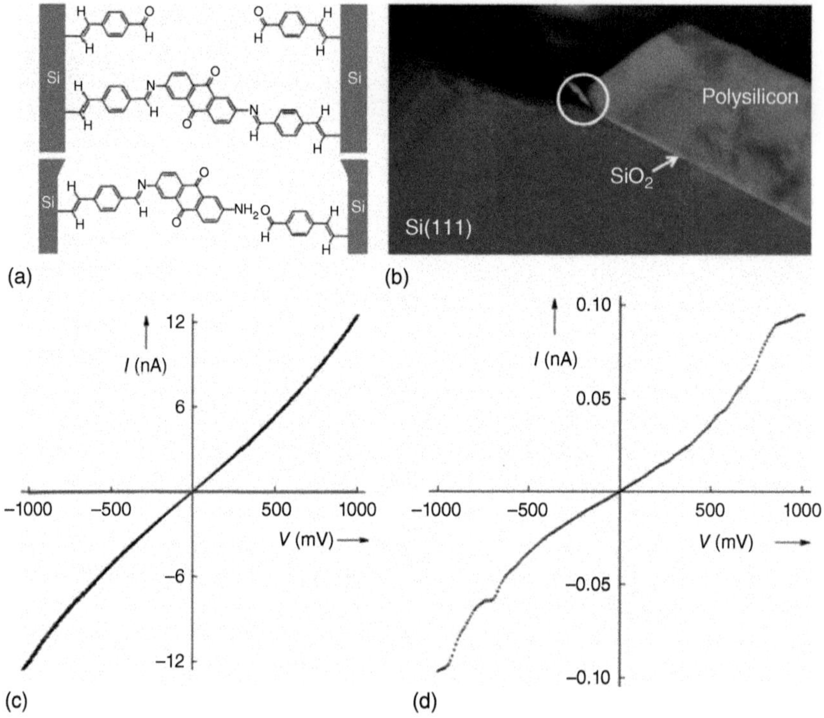

Figure 3.67 (a) Reaction at the silicon surface: Molecular structures of the chemisorbed species and a 2.8-nm-long molecular bridge formed by grafting the diethylacetal derivative of 4-ethynylbenzaldehyde and reacting the grafted molecules with 2,6-diaminoanthra-9,10-quinone at ambient temperature and an asymmetric nonbridging molecule. (b) Scanning electron microscopy image of an $Si/SiO_2/Si$ cross section. The oxide layer is etched along its outer edge to provide a nanosized electrode gap, which is illustrated within the circle by the apex of the V-shaped undercut. (c) Symmetrical $I-V$ characteristics of a bridged $Si/SiO_2/Si$ nanogap device (black) and altered behavior following immersion in an acidified solution to disconnect the anthraquinone linker unit (red). (d) $I-V$ curve of an $Si/SiO_2/Si$ nanogap device. Anomalous distortions in the electrical characteristics are attributed to a nonbridging arrangement, with altered contacting distances induced by rotation about the exocyclic single bonds of the molecule as the voltage is swept. Source: Ashwell et al. [359]. 2011, John Wiley & Sons.

formation of ordered monolayers; and (iii) the surface group, functionalizes the SAMs. One primary mechanism of the SAMs formation on a silicon oxide surface is the silanization reaction and the silanes could be $R(CH_2)_nSiX_3$ (X = Cl, OMe, Oet, and diazo-compounds). Typically, the hydroxylated silicon oxide surface is hydrophilic, which is covered with an ultrathin water layer, so that the silanes get physisorbed on it. Then, the molecular head groups hydrolyze into highly polar trihydroxysilanes and these polar groups are subsequently covalently grafted to the silicon oxide surface through a condensation reaction. The last process is spontaneously driven by the van der Waals interactions among the spacer units to form a compact monolayer [367]. However, the existence of the oxide layer is not favorable in some cases, such as requiring high reproducibility of the molecular devices and studying the transport properties through molecules, so depositing organic SAMs directly on the oxide-free Si substrate is of paramount importance. In general, the highly reactive bare Si(100) or Si(111) surface is hydrogen-terminated or halogen-terminated. Then, the Si radicals which are extremely critical in the formation of SAMs could be activated via many ways such as hydrosilylation catalysts, UV irradiation, and Lewis acid catalysts. Differing from silanes which are usually dissolved in alkanes or carbon tetrachlorides when SAMs are grafted to the silicon oxide surface, the electronic transport quality alkyl monolayers on the oxide-free Si substrate are primarily prepared from the neat melt. It was worth noting that adsorbing alkyl molecules from the dilute solution to produce high-quality SAMs increased the type selectivity of the molecules [368]. Furthermore, the supercritical carbon dioxide ($SCCO_2$) could also be applied as a reaction medium to self-assemble monolayers on the oxide-free Si substrate. It was found that the supercritical medium provided a simple and large-scale manufacturing method and allowed the formation of monolayers with varying reactive surface groups [369].

In decades, various kinds of Si/molecule/metal monolayer junctions have been proposed, which are stimulated by extending the silicon technology and studying the molecular properties more efficiently. Compared with general metal/molecule/metal junctions, the Si/molecule/metal junctions promise the diversity of molecular electronic devices and the convenience of in situ measurements. Nevertheless, the penetration of the top metal electrode, which may cause a short circuit and be extremely harmful to the yield of devices, is one of the major problems in the construction of junctions. Fortunately, a lot of efforts have been made to solve this issue. It was proved that the degree of the metal penetration was closely related to the Si substrate/molecule interfacial chemistry properties and the type of metals [370]. The resulting metal/octadecanetrichlorosilane (OTS) monolayers/SiO_2/Si junctions had a better continuous metal film than the metal/alkoxy monolayers/Si junctions, when using the evaporative Au or Al as top electrodes. Besides, Ti cannot be applied as the top electrode in some specific situations for the stronger interaction of Ti with OTS monolayers than the Au or Al with them. Meanwhile, although evaporating Au as the top electrode is a mature and relatively simple method to assemble molecular junctions, it is very easy to disturb the molecular array and lead to electrical shorts. One effective solution is to modify the organic monolayers surface with the thiol groups which can form stable covalent bond with Au, before the evaporation of the top electrode. Therefore, the Au atoms are more likely to adhere to the monolayers surface than penetrate into them during the evaporation process (Figure 3.68) [371]. On the other hand, a number of superb metallization methods, such as "soft metallization," nTP [372], or lift-off float-on process [373], have also been considered to impede the filament. Detailed information of these methods is provided in Section 6.3. The mercury, a metal in liquid form at room temperature and atmospheric pressure, has demonstrated great success to make reliable and reproducible electrical contacts with the monolayers for its atomically flat surface and the strong electrical coupling with alkanethiolate monolayers. And the mercury with high surface tension hardly permeates through organic monolayers and does not corrode the silicon

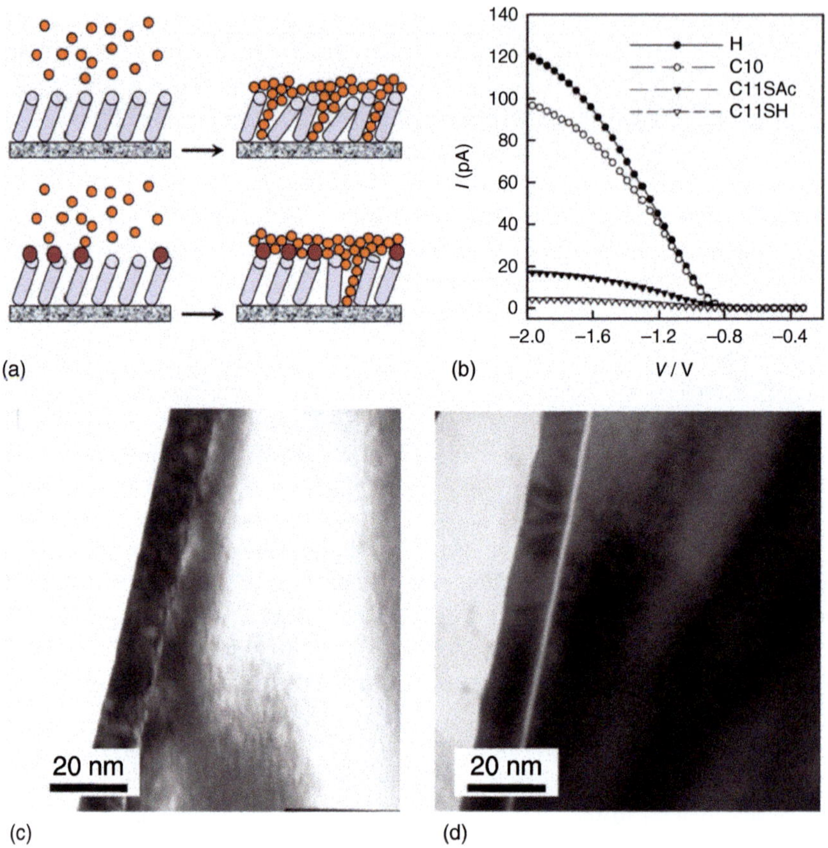

Figure 3.68 (a) Hypothetical illustration of gold penetration upon the formation of diode junctions on silicon modified with *n*-alkyl monolayers and thiol-terminated monolayers. (b) Averaged ballistic electron emission microscopy spectra for diode junctions prepared on hydrogen-terminated and organically modified silicon with evaporated gold pads as top contacts. (c–d) Cross-sectional TEM images of the Au/monolayer/Si(111) diode junctions: (c) Au/C10-Si and (d) Au/HSC11-Si. Source: Kuikka et al. [371]. Copyright 2008 American Chemical Society.

substrate [374]. Notably, the flip-chip lamination provides a nondestructive novel route to construct lots of Si/molecule/metal junctions with high quality in parallel under mild conditions, which is a valuable complement to the "soft metallization" content. This method can absolutely avoid the presence of filament and guarantee the formation of more orderly and nearly defect-free monolayers since it is more flexible for the molecules to move on the Au substrate than on the Si substrate (Figure 3.69) [375, 376].

In order to design integrated molecular devices rationally, it is critical to grasp the mechanisms of charge transport across Si/molecule/metal junctions. In general, there are two transport barriers existing simultaneously in the Si/molecule/metal monolayer junctions: a tunnel barrier closely related to the length of insulating σ-bonded molecules; a Schottky barrier near the interface between the Si substrate

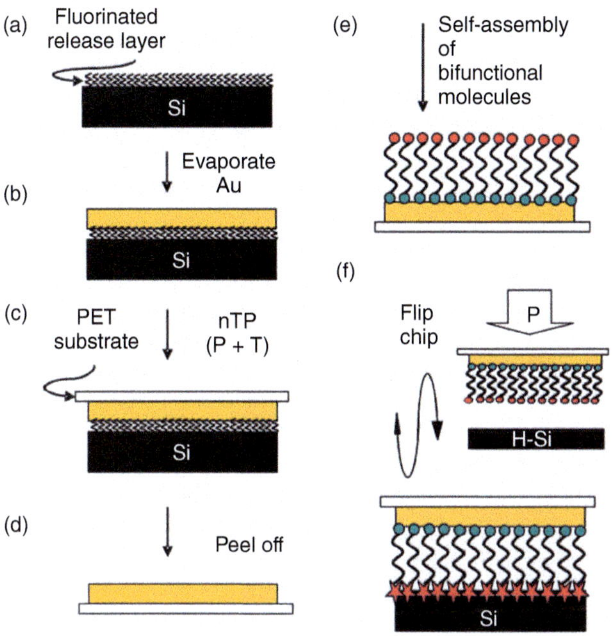

Figure 3.69 Schematic of flip-chip lamination process to form metalmolecule–silicon molecular junctions preserving the integrity of the molecules. Evaporated gold (b) is lifted off of a fluorinated release layer onto a polyethylene terephthalate (PET) substrate using nanotransfer printing (nTP) (c) to reveal the ultrasmooth Au underside (d). Bifunctional molecules are self-assembled onto ultrasmooth gold (uSAu) forming a dense monolayer with the functional group exposed (e). Finally, the two electrodes are laminated together with nTP causing bonding between the exposed functional group and H–Si (111) (f). Source: Coll et al. [375]. 2009, American Chemical Society.

and molecules, formed by band bending of the Si energy levels [377]. The transport properties of junctions are significantly influenced by the dipole moment, length, and conjugated degree of molecules [378]. For example, surface band bending, barrier height, interface density of states, and ideality factor in the p-Si/molecule/Hg junctions can be effectively tuned by covalently connecting conjugated organic molecules with different dipole moments to Si electrodes (Figure 3.70a) [379]. In some situations, the magnitude and direction of the dipole moment of molecules would dominate the conductance in the p^+-Si/molecule/Hg junctions and thus significantly affect the symmetry of current–voltage (I–V) curves. Considering the directional relationship between the electric field of the dipole moment of conjugated molecules and the applied electric field, the junctions may present reversal rectification behaviors (Figure 3.70b,c) [380]. In addition, the molecule–electrode electrical coupling is also crucial to carrier transport across Si/molecule/metal junctions. When alkyl chains are connected to both electrodes by covalent bonds [382], carriers can effectively tunnel across junctions through these chemical bonds. And external slight disturbances have little effect on the tunneling current. However, if alkyl chains are only covalently grafted on one side of the electrode [374, 377], the

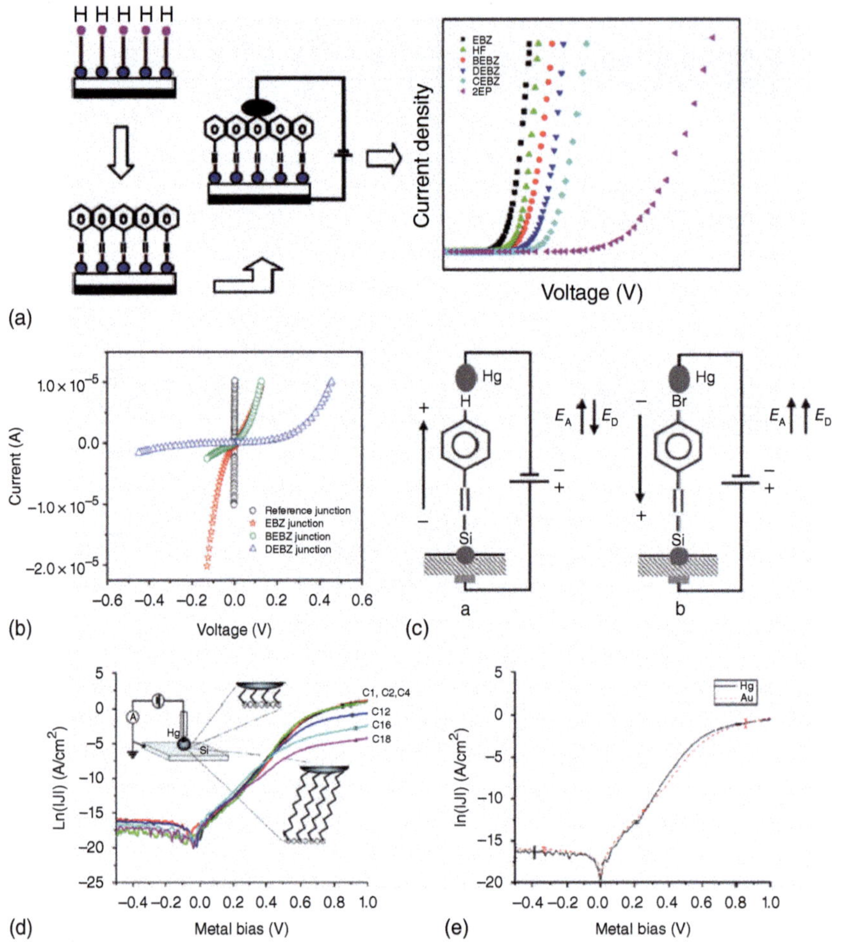

Figure 3.70 The expected configuration of organic molecules upon binding silicon surface and current–voltage (*I*–*V*) curves recorded on various metal–molecules–silicon junctions, using soft mercury contact (ethynylbenzene [EBZ], 4-bromo-1-ethynylbenzene [BEBZ], 1,4-diethynylbenzene [DEBZ], 3-chloro-1-ethynylbenzene [CEBZ], and 2-ethynylpyridine [2EP]). (b) Hg–molecules–p$^+$ (Si) junctions showing variations in the current–voltage behavior according to the dipole moments of molecules. A reference junction is also shown for the comparison. (c) The directions of molecular dipole moment E_D and applied electric field E_A of EBZ and BEBZ molecular junctions. The parallel and antiparallel orientation of these fields controls the current transport through molecular junctions. (d) ln(|*J*|)–*V* curves of (n–Si)–C_nH_{2n+1}/Hg ($n = 1, 2, 4, 12, 16,$ and 18). Bias is applied to the Hg; Si is grounded. Results are logarithmic averages of at least seven different junctions with a scan rate of 20 mV s^{-1}. The error bars represent standard deviations, which are typically less than 5% of the measured current. The insets give "artist" views of the junctions for "short" and "long" monolayers (not to scale!). (e) Comparison of junctions with Hg and Au contacts to (n-Si)–$C_{12}H_{25}$. Source: (a) Hiremath et al. [379]. 2008, American Chemical Society. (b, c) Hiremath et al. [380]. 2010, Royal society of chemistry. (d, e) Yaffe et al. [381]. 2009, American Chemical Society.

tunneling current depends not only on the covalent bonds but also on the molecular chain length which determines the probability of through-space tunneling. In the event that neither side of the electrodes is covalently connected with the molecules, the alkyl monolayer can act as an insulating film whose thickness hardly affect the tunneling current. These findings are closely related with the presence of the Schottky barrier in the molecular junction, as the Si electrode surface that contacts with molecules can be accumulated, depleted, or inverted. In the state of accumulation or depletion, charge transport is dominated by majority carrier, while the former presents near-symmetric I–V characteristics that strongly depend on the molecular length and the latter shows a general diode behavior. When the Si electrode is depleted in the n-Si/alkyl molecule/Hg junction, the current varies sharply at different directional bias. And the carrier transport which is almost independent on the molecular length at reverse and low forward bias is limited by thermionic emission over the Schottky barrier in the Si substrate. In inversion, however, the probability of carriers transport across insulating monolayers, even for the C18, is much higher than that of finding minority carriers at the interface. Therefore, the interface is minority carrier controlled via generation and recombination at low bias (Figure 3.70d,e) [381, 383].

The performance of Si/molecule/metal junctions can be regulated in several ways, such as altering the doping level of the Si substrate or decorating molecules with different dipole moments. Using the highly doped Si, further studies on the molecule–electrode electrical coupling have been realized. The junctions with alkenyl monolayers chemically bonded to the highly doped Si electrode exhibited a higher conductivity than that with alkyl monolayers, since the double bond near the Si surface enhanced the electrical coupling between the organic monolayer and Si electrode. Meanwhile, the moderately doped Si provided the feasibility for adjusting the rectification characteristics of the molecular diode (Figure 3.71) [354]. From the molecular perspective, irradiating alkyl monolayers could introduce new electronic states in the HOMO–LUMO gap of alkyl molecules via the formation of C=C bonds or C–C crosslink in monolayers and improve the conductance of the junction [355]. Considering the compatibility with the existing advanced CMOS technology, an enclosed, planer, and silicon-based molecular electronic device has been fabricated to investigate the electrical properties. In order to separately study the factors affecting the charge transport, such as the bias, silicon doping, and molecular dipole, unconjugated alkyl molecules which displayed simple quantum mechanical tunneling were assembled in this enclosed device (Figure 3.72) [384]. It was shown that the small local electric field formed by the dipole of molecular monolayers not only influenced the effective Schottky barrier but also changed the shape and height of the molecular barrier. Moreover, the chemical bonds that connected the Si electrode and molecules, for instance Si—C bond, Si—O bond, and Si—S bond [356, 385], also played a key role in modifying the characteristics of the junctions, as they significantly impacted the density of states and electrical coupling of the molecule–electrode interface.

A great number of molecular devices based on the Si substrate with different functions, such as memory, storage, and rectification, have been constructed to

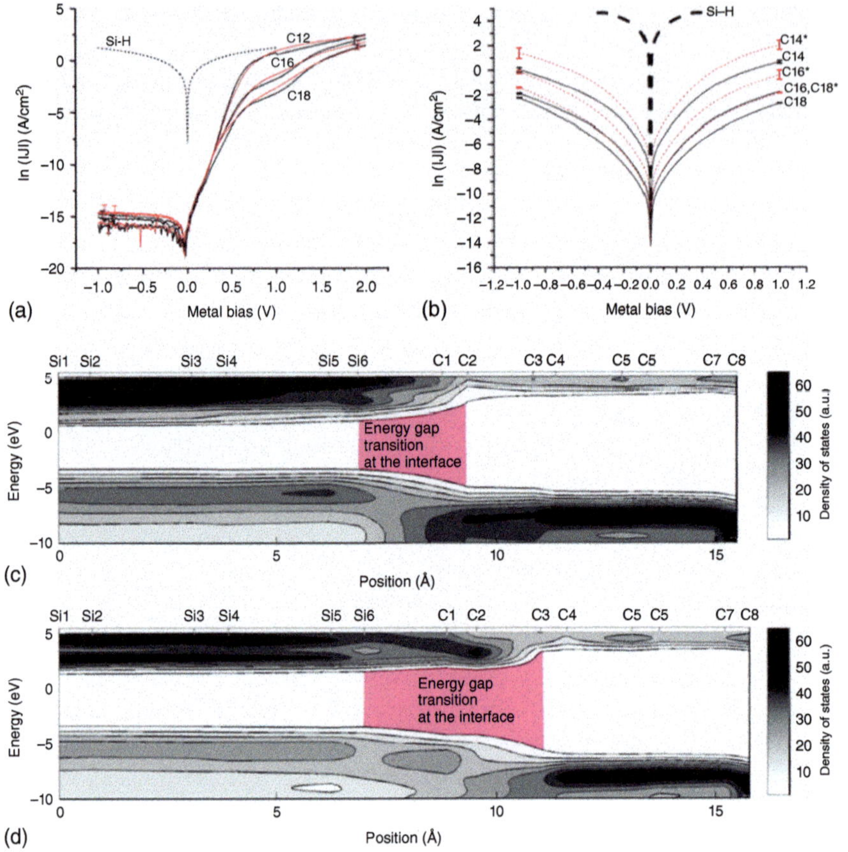

Figure 3.71 (a) ln(|J|)−V curves of moderately doped (MD) n-Si/organic molecular monolayer/Hg of 12, 16, and 18 carbon chain alkyl (black) and alkenyl (red) monolayers, together with the ln(J)−V curve of a freshly etched Si−H metal/organic molecule/semiconductor (MOMS) (dashed). Bias is applied to the Hg, and the Si is grounded. Scan rate: 20 mV s^{-1}. Results are logarithmic averages of at least seven different junctions. The error bars represent the standard deviations, which are typically less than 5% of the measured current. (b) ln(|J|)−V curves of HD n-Si/monolayer/Hg of both alkenes (black-solid) and alkynes (red-dashed lines) of several lengths (14, 16, and 18 carbon chains) and of freshly etched n−Si−H/Hg (----). (c, d) Contour maps of the local density of states for (c) alkyl and (d) alkenyl monolayers on Si(111). The interfacial transition region is emphasized in red. Source: Yaffe et al. [354]. 2010, American Chemical Society.

promote the field of the molecular electronics. The Si/molecule/metal junctions are promising to miniaturize the information processing systems and make the molecular electronics compatible with the powerful silicon technology [44, 386–388]. The molecular memory devices are based on the redox-active molecules which furnish different charged states at the distinct voltages. When an oxidizing voltage is applied, the molecules would lose electrons, resulting in positively charged organic monolayers. While a reducing voltage returns molecules to neutral state via the electron transfer process. The oxidation process is equivalent to writing the information and the reduction process is equivalent to reading out the information.

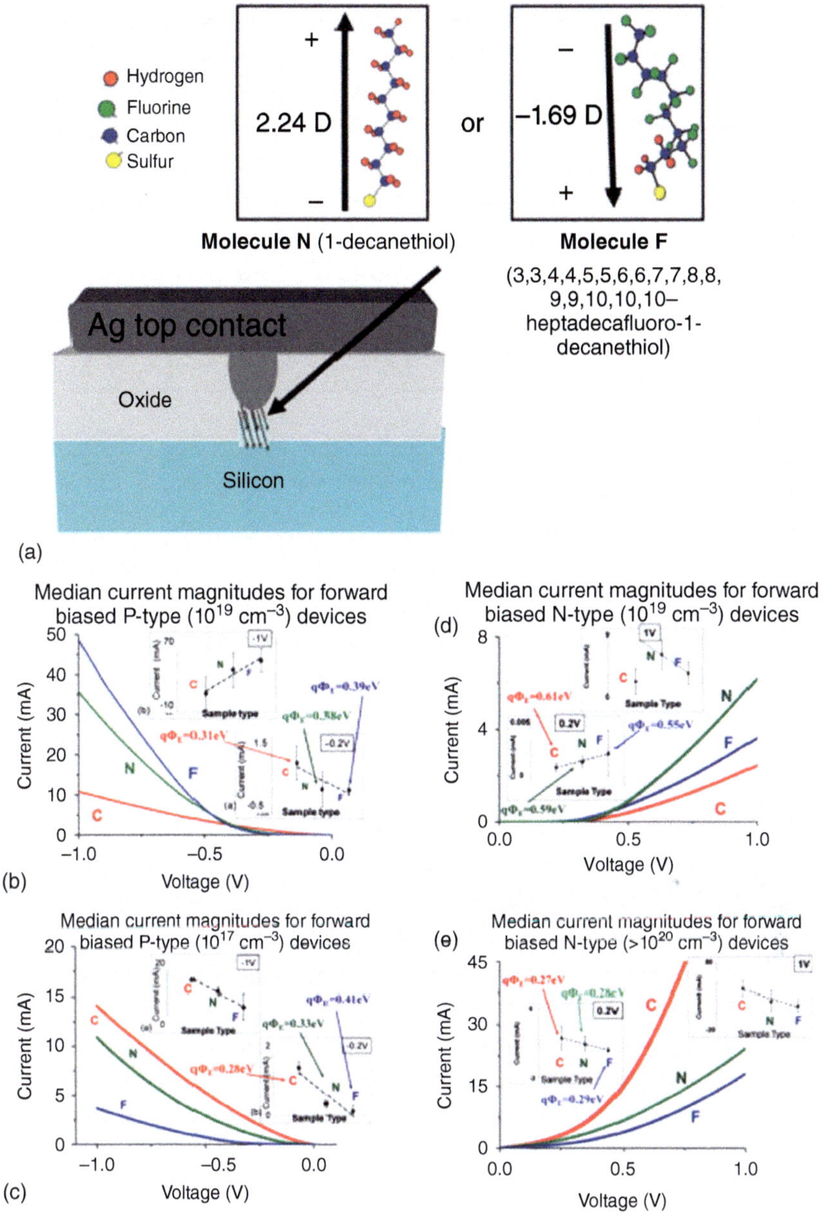
Hydrogen
Fluorine
Carbon
Sulfur
+
2.24 D
–
or
–
–1.69 D
+
Molecule N (1-decanethiol)
Molecule F
(3,3,4,4,5,5,6,6,7,7,8,8,9,9,10,10,10–heptadecafluoro-1-decanethiol)
Ag top contact
Oxide
Silicon
(a)
Median current magnitudes for forward biased P-type (10^{19} cm^{-3}) devices
Current (mA)
Voltage (V)
(b)
Median current magnitudes for forward biased N-type (10^{19} cm^{-3}) devices
(d)
Median current magnitudes for forward biased P-type (10^{17} cm^{-3}) devices
(c)
Median current magnitudes for forward biased N-type (>10^{20} cm^{-3}) devices
(e)

Figure 3.72 (a) Device junction consisting of one of two different molecules assembled in a 5 μm × 5 μm well through an approximately 150-μm-thick silicon dioxide layer on a silicon substrate. (b, c) *I*–*V* curves representing the median electrical characterization of forward-biased devices assembled on p-type Si with hydrogen-terminated silicon junctions (C), nonfluorinated molecules in the junctions (N), and fluorinated molecules in the junctions (F). The insets show the median current values, including standard deviation, at relatively low bias (−0.2 V) and relatively high bias (−1.0 V). Also labeled on the insets are the effective Schottky barrier ($q\Phi_E$) values for each molecular sample. (d–e) *I*–*V* curves representing the median electrical characterization of forward-biased devices assembled on n-type Si with hydrogen-terminated silicon junctions (C), nonfluorinated molecules in the junctions (N), and fluorinated molecules in the junctions (F). The insets show the median current values, including standard deviation, at relatively low bias (0.2 V) and relatively high bias (1.0 V). Also labeled on the insets are the $q\Phi_E$ values for each molecular sample. Source: Gergel-Hackett et al. [384]. 2010, American Chemical Society.

The charge retention time of the molecular memory devices can be modified by introducing a tunneling barrier between the redox-active molecules and Si electrode. Ferrocene-containing molecules are attractive to apply in the memory devices for their low writing and erasing voltages, and changeable charge storage time (Figure 3.73a,b) [389]. However, considering the applications of molecular-based memory devices, the molecular components need to withstand various rigorous conditions, including the high temperature during the manufacturing and operating processes, and a large number of read–write cycles. It was found that the porphyrin-based molecules which were adhered to the Si electrode via covalent Si–O linkages satisfied the above requirements perfectly [390]. Monomeric porphyrins and triple-decker architectures behaving two and four accessible cationic states, respectively, are excellent information storage media. The voltammetric response of the monomeric porphyrin monolayers was identical whether platforms were treated at high temperature or not, significantly indicating the thermal stability of porphyrin monolayers. And the delocalization of the positive charge on the molecules produced by carbon and nitrogen atoms in the macrocycle promoted the read–write cycling endurance of devices (Figure 3.73c–e). Furthermore, as active charge-trapping elements, porphyrins could form stable π-cation radicals and exhibit long charge storage time. Their multiple redox states were accessible at relatively low voltages, which diminished the power consumption. Although the two architectures in which porphyrin molecules tethered to the Si substrate via Si—C bonds or Si—S bonds performed comparable physical properties, the Si–C linkage was more fascinating for the less susceptibility to hydrosis and good stability. In order to promote the practical application of molecular-based memory devices, a series of Si/porphyrin monolayer/metal junctions with different evaporative metals were systematically characterized. Tripodal alkyl groups chemically linking porphyrin molecules to the Si substrate ensured that the monolayers were robust to the top metal deposition. The Au was more likely to penetrate into porphyrin monolayers than the Ag or Cu even with the relatively gentle electron beam deposition way. However, ultraclean conditions were required during Cu deposition to avoid its oxidation. So it appeared that Ag was more viable to form a reliable electrical contact by the electron beam deposition (Figure 3.74) [391].

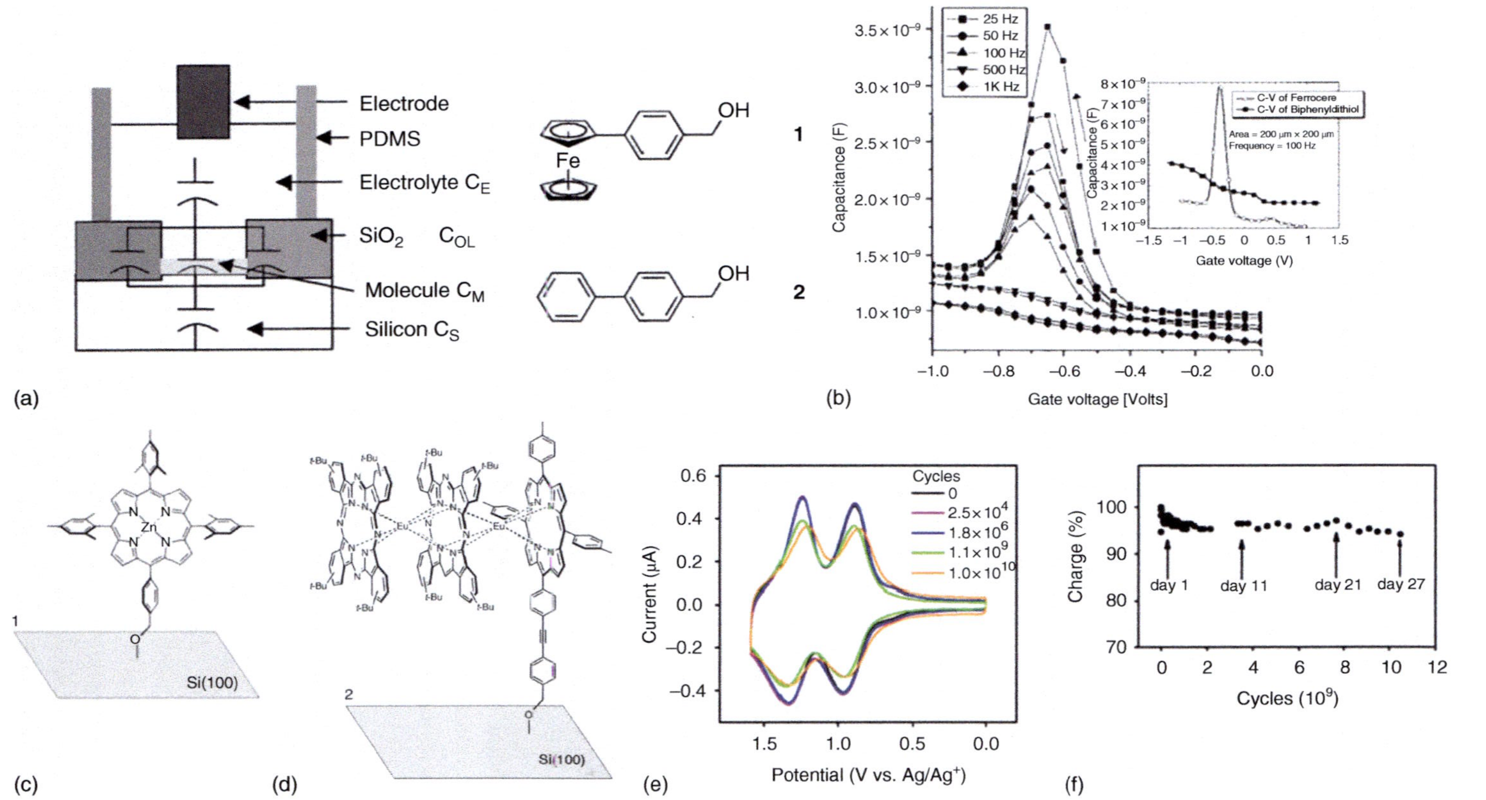

Figure 3.73 (a) Schematic of the electrolyte-molecule-silicon capacitor with a simplified equivalent circuit and the structure of the ferrocene derivative (**1**) and the nonredox-active analog (**2**). In each case, SAM formation affords a covalent attachment between the oxygen of the linker and the silicon surface. (b) Capacitance–voltage characteristics of the electrolyte-molecule-silicon capacitor with redox-active ferrocene monolayers measured at 25, 50, 100, 500, and 1000 Hz. Inset shows the lack of $C-V$ peaks in the nonredox-active monolayers measured at 100 Hz. (c, d) Structure of the Si-tethered porphyrins. (e) Voltammetric response of a monolayer of porphyrin **1** after 0, 2.5×10^4, 1.8×10^6, 1.1×10^9, and 1.0×10^{10} cycles. The cycling process was carried out under inert atmosphere conditions via application of a 100-μs-square-wave pulse of amplitude 1.2 V. (f) Plot of the integrated voltammetric signal (corresponding to the charge in the monolayer) as a function of the number of cycles. Source: (a, b) Li et al. [389]. 2002, AIP publishing LLC. (c–f) Liu et al. [390]. 2003, AAAS.

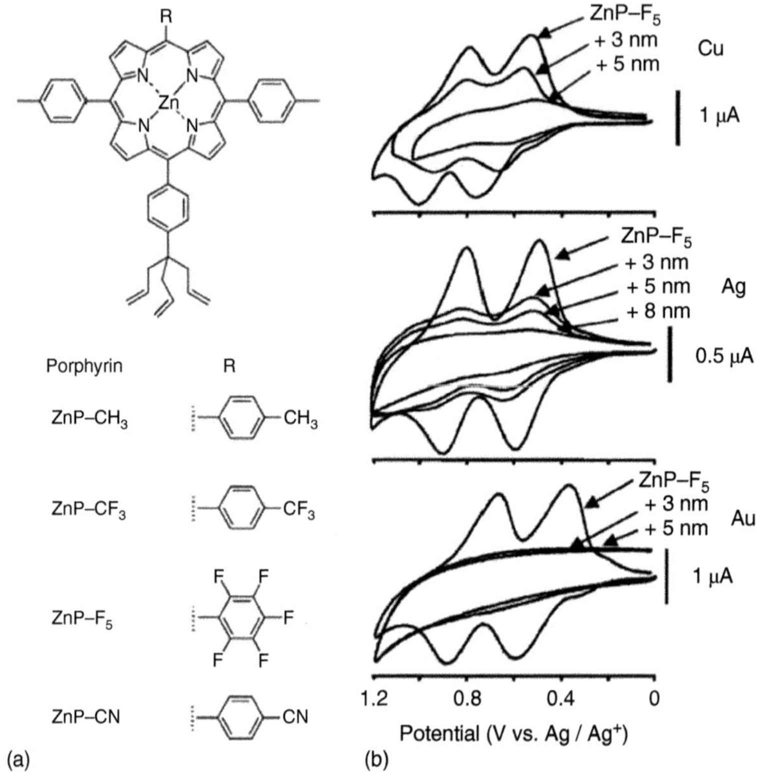

Figure 3.74 (a) A series of porphyrins in monolayers covalently attached to Si(100) substrates. (b) Representative cyclic voltammograms (100 V s^{-1}) of the metal/ZnP-F_5/Si junctions: Cu (top), Ag (middle), and Au (bottom). The cyclic voltammogram of the ZnP-F_5 monolayer on Si(100) prior to metal deposition is shown in each panel for reference. Source: Anariba et al. [391]. 2008, American Chemical Society.

On the other hand, the Si/molecule/metal junctions exhibit immense potential in constructing rectification molecular devices. Since 1974, the first seminal paper on molecular rectifying diode was proposed by Aviram and Ratner, which was based on donor–acceptor pairs linked by σ-bonded tunneling bridges, the field has witnessed spectacular development [1]. Typically, a rectifier is a device with a large number of carriers flowing at one bias while restricted carriers passing through at the opposite bias [392, 393]. The LB films and SAMs of organic molecules on solid substrates have been extensively adopted to construct molecular rectifying diodes [394, 395]. Further experimental results demonstrated that the donor–acceptor pair was not always necessary. The architectures containing σ-bonded alkane chains and π-conjugated components may perform a more consistent and a greater degree of rectification [367, 396]. In addition, compared with the metal/molecule/metal junction, the Si/molecule/metal junction formed by grafting σ–π systems on the Si substrate through an alkyl spacer afforded more possibilities and even better performance for molecular rectifying diodes [397]. In the σ–π system, σ carbon atoms influence the tunneling possibility through alkyl chains and π-conjugated

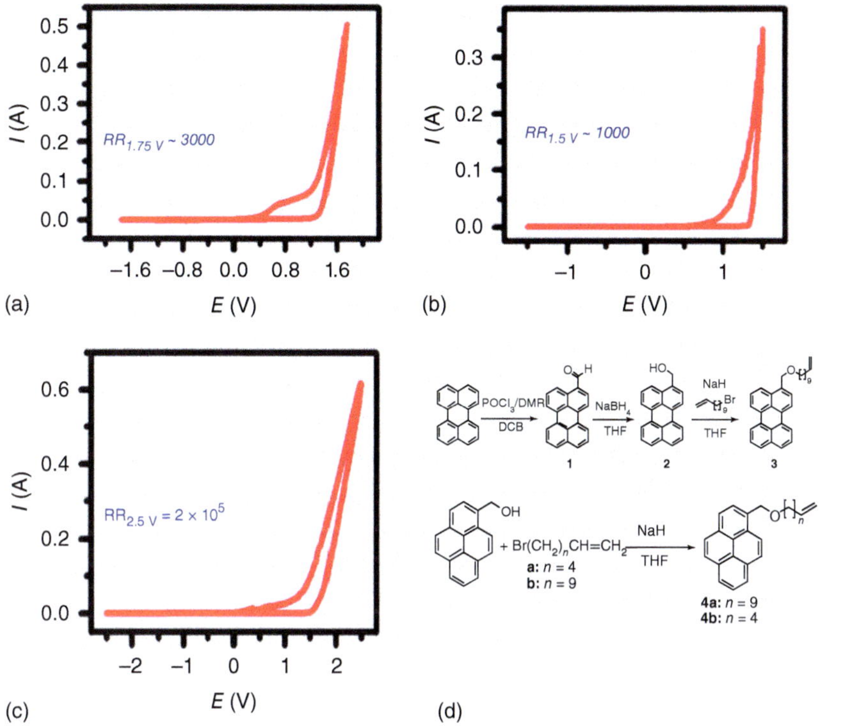

Figure 3.75 *I–V* characteristics of Hg/molecule-grafted monolayer/Si(n^{++}) devices. (a) 3; (b) 4a; and (c) 4b. The *I–V* characteristics of the device samples made from the same molecules were similar, and typical curves are shown. (d) The related molecules. Source: Garg et al. [398]. 2015, Royal Society of chemistry.

components impact the resonant tunneling process for the controllable energy levels. The rectification is attributed to the resonant tunneling since molecular orbitals of π-conjugated components are closer to Fermi energies of electrodes. When only alkyl chains were utilized in place of the σ–π systems, no obvious rectification phenomenon took place. Therefore, the σ–π system is an important adjustable part that affects the performance of devices. Differing from molecules attached to native oxide through silanization, a denser monolayer could be resulted by directly adhering σ–π systems to the H-terminated Si substrate, and thus the diode showed a higher rectification ratio (Figure 3.75) [398]. Meanwhile, the internal dipole of molecules is also of paramount importance to determine rectification ratio [360].

Furthermore, it is feasible to investigate the NDR molecular devices based on the silicon substrate. Several experiments have demonstrated the STM is a powerful tool in nanoscale science for analyzing charge transport behaviors of various junctions at the single-molecule level, such as coulomb blockade, current rectification, Kondo resonance, and NDR [399–402]. The current rectification and NDR have been extensively observed in the Si/molecule/metal junction. It is predicted that the STM field could move the location of molecular orbitals into the energy band gap of the Si substrate and thereafter interrupt the carriers transfer, resulting in NDR peaks in

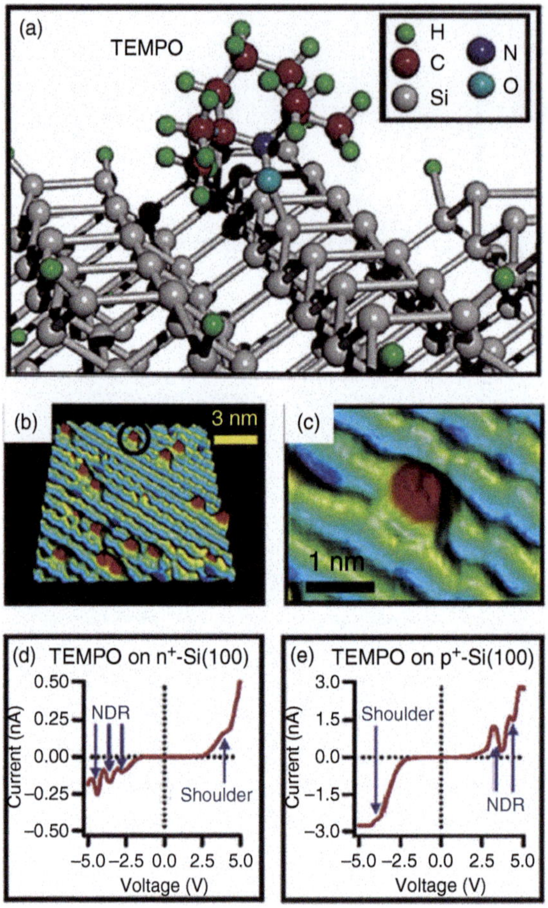

Figure 3.76 Styrene on the Si(100)-2 × 1 surface. (a) Schematic representation of an individual styrene molecule bound to a truncated Si(100) surface following molecular mechanics optimization. (b) STM topography image of isolated styrene molecules bound to the degenerately n-type Si(100)-2 × 1 surface. The black circle identifies a single-styrene molecule chosen for STM spectroscopy. (c) STM image of the circled styrene molecule prior to current–voltage characterization. (d) Current–voltage curve measured on the clean Si(100)-2 × 1 surface within the vicinity of the isolated styrene molecules. (e) Current–voltage curve measured over the isolated styrene molecule. Negative sample bias reveals two distinct NDR events, whereas electron stimulated desorption (ESD) is detected at positive sample bias. Source: Guisinger et al. [399]. 2004, American Chemical Society.

the *I*–*V* curves. A heavily doped substrate and high measurement bias which make sure that the molecular orbitals could slip into the Si energy gap are necessary for the observation of the NDR. In addition, the NDR was reported to be highly substrate doping-type-dependent and specifically, NDRs appeared at positive bias for the degenerately doped p-Si and at negative bias for the degenerately doped n-Si (Figure 3.76). The high packing density of the monolayers would suppress the NDR behavior compared with the isolated molecules [403].

Although the powerful semiconductor technology promotes the development of the silicon nanogap electrode in molecular electronics, it is necessary to improve its chemical stability and conductance to provide the molecular devices with excellent mechanical and electronic performances. On the other hand, in order to fabricate practical silicon-based single-molecule devices, the platforms which have been proposed on the basis of the CMOS techniques need to be further simplified. Moreover, modifying the way which can determine the number of the molecules in the $Si/SiO_2/Si$ structure is essential.

3.9.2.5 Other Materials

In addition to the materials mentioned above, other nonmetallic materials which exhibit good stability, plentiful availability, high electrical conductance, and easy processing have also been continuously developed to expand the research regimes of the molecular electronics. Kinoshita and coworkers introduced a method to prepare the carbon films through pyrolyzing photoresist [45, 262]. It was found the atomically smooth carbon films with superior conductivity could be obtained by optimizing the conditions such as the type of photoresist and the pyrolysis temperature. This pyrolyzed photoresist film (PPF) is a prospective electrode in the molecular monolayer junctions for its flat and sp^2-hybridized carbon surface. Molecules can be grafted to the PPF substrate through the C—C bonds and then form densely packed monolayers. The PPF/molecule/top–electrode monolayer junction furnished a reliable platform to study the charge transport mechanisms which are important in the molecular electronics and the basis to regulate the performance of the devices. However, the preparation of top contact electrode is vital and challenging. For example, a great number of PPF/molecule/Cu junctions have been reported [404–406]. It was found that when the bias was applied, the Cu top electrode was prone to electro-oxidation, thus forming conducting filaments in molecular junctions which could completely dominate the properties of the device. Meanwhile, as another widely used electrode material, although Au is electrochemically stable, it is easy to penetrate into monolayers by the direct deposition [407]. Inserting an interlayer between monolayer and Au top electrode can validly avoid penetration [268, 408]. However, the direct vapor deposition is still imminently demanded, since it is concise and compatible with the existing semiconductor processing technology. Fortunately, the electron beam-deposited carbon film (e-C) shows the immense potential as the top electrode for its greater stability and lower tendency for electromigration (Figure 3.77) [409–411]. Furthermore, in order to increase the applications of PPE–molecule–top-electrode monolayer junctions, Fairman et al. [412] proposed a versatile, simple, and modular method to functionalize the PPE surface via the plasma iodination.

Fabricating practical single-molecule integrated circuits is one of the foremost goals in molecular electronics. In single-molecule electronics, each molecule could perform the basic functions of electronics such as storage, rectification, and amplification, which makes it possible to develop cheaper, more-ecological, and higher-performance alternatives to traditional silicon-based devices. In order to realize the single-molecule integrated circuit, there are some obstacles to overcome.

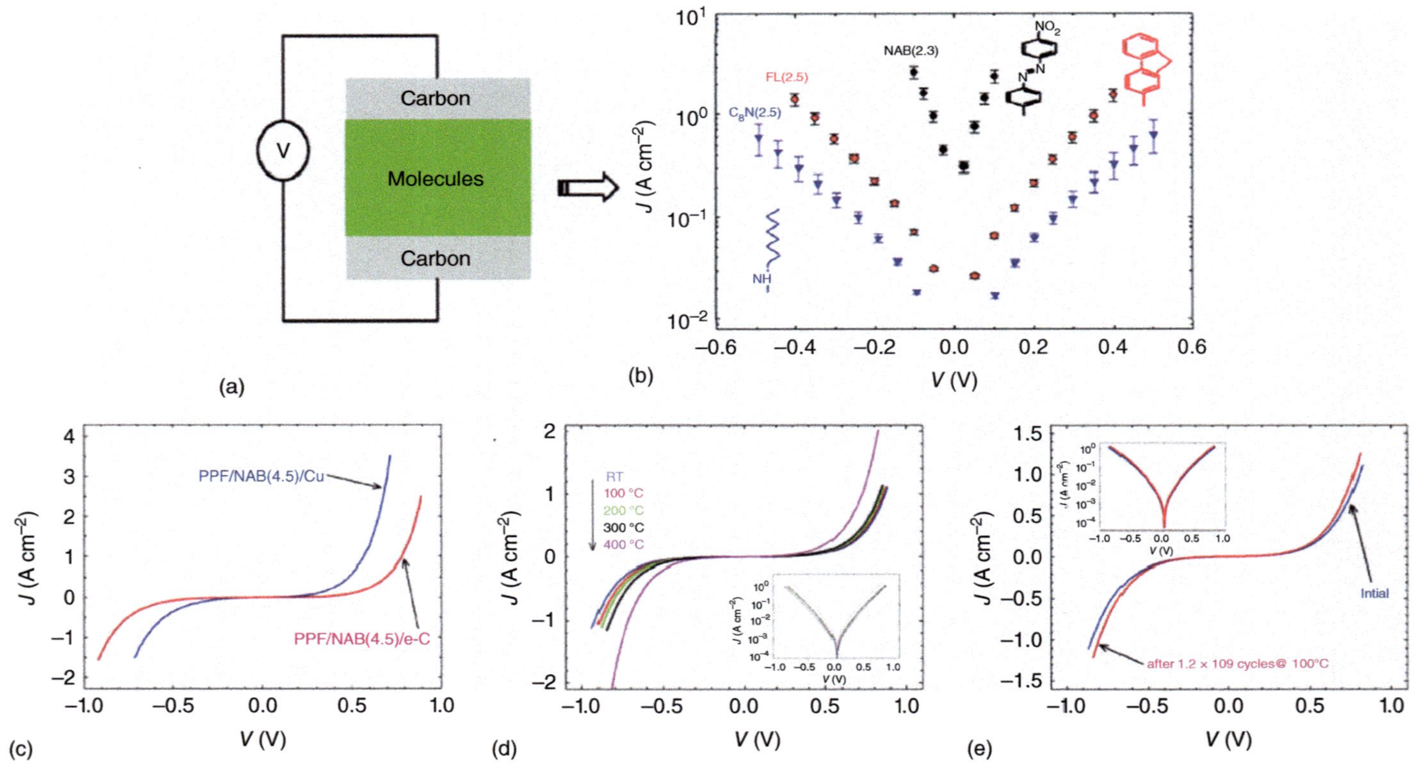

Figure 3.77 (a) Schematic representation of a carbon–molecule–carbon molecular junction. (b) Overlay of J–V curves for PPF–molecule–e-C junctions fabricated with three different molecular layers with similar thickness: nitroazobenzene (NAB) (2.3 nm), fluorene (FL) (2.5 nm), and C_8N (2.3 nm). (c) Comparison of top contact materials: Cu vs. e-C. (d) J–V curves of a PPF–NAB (4.5)–e-C junction heated in a vacuum ($2–4 \times 10^{-6}$ Torr) at progressively elevated temperatures for 30 min at each temperature. Each J–V curve was taken in lab ambient after the sample was cooled to room temperature. (e) J–V curves of a PPF–NAB (4.5)–e-C junction before and after triangle wave (1 V, 5 kHz) cycling up to 1.2×10^9 cycles at 100 °C over the course of~68 hours in air. The insets display the same data plotted in a semilog scale. Source: Yan et al. [409]. 2011, American Chemical Society.

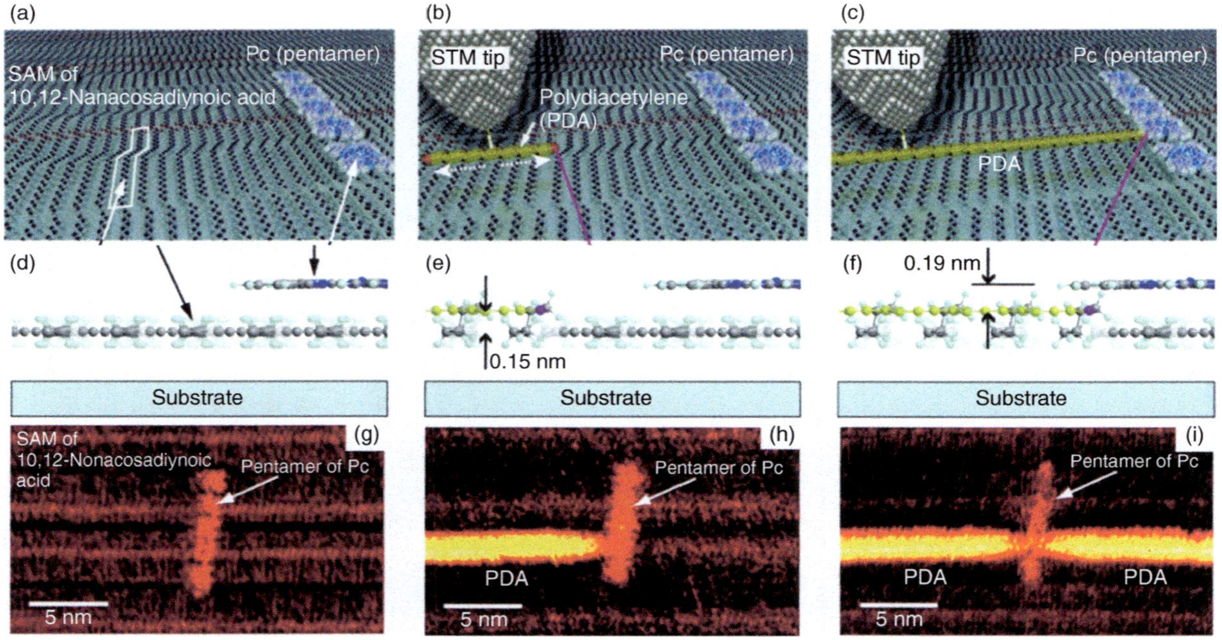

Figure 3.78 (a) Illustration of phthalocyanine molecules adsorbed to a SAM of 10,12-nonacosadiynoic acid molecules. (b) Chain polymerization is initiated using an STM tip. (c) The polymer chain reaches the adsorbed molecule, and spontaneous chemical bond formation occurs. (d–f) Side views of the models. The PDA backbone (yellow spheres) protrudes by 0.15 nm. (g) STM image of an H_2Pc pentamer on a SAM of 10,12-nonacosadiynoic acid. (h) STM image obtained after initiating chain polymerization. Fabricated PDA chain, which is seen as a bright line, is connected to an H_2Pc molecule. (i) STM image showing two PDA chains connected to a single H_2Pc molecule. Source: Okawa et al. [263]. 2011, American Chemical Society.

For example, it is difficult to reduce the total size of the molecular circuits for the wide metal wires. Therefore, new viable architectural schemes are needed for the circuit design, in which creating conductive nanowires at the designated positions and ensuring chemical bonding between the nanowires and functional molecules are able to be realized. Notably, compared with the metal wires, conductive polymer nanowires have an extremely small size in width, which completely meets the requirements of molecular circuits [413, 414]. Accordingly, Okawa et al. [263] reported a novel "chemical soldering" approach and successfully constructed the single-molecule junctions. In detail, the monomer molecules of the diacetylene were deposited on a HOPG surface to prepare a SAM, on which the functional phthalocyanine (Pc) molecules were then adhered. The diacetylene moiety beneath the STM tip was stimulated through a pulsed bias voltage, followed by the chain polymerization of the diacetylene molecules, leading to a single polydiacetylene (PDA) chain along the row. The obtained PDA chain had a lift-up structure and reactive chemical specials at the propagating end, so that it could connect with the encountered molecule through the covalent bonds (Figure 3.78). A resonant tunneling diode stemming from two conductive PDA nanowires bridged with a single phthalocyanine molecule has been perfectly fabricated, which indicated that the "chemical soldering" method was significantly vital for the development of the future single-molecule electronics.

3.10 Summary and Outlook

With the rapid development of molecular nanoelectronics, nanodevices as the next generation of circuit elements with lower power consumption, higher speed, and a higher level of integration have attracted more and more attention. How to conveniently and efficiently fabricate nanogap electrodes is a key step for the construction of nanodevices. Here, we review some representative methods for the fabrication of nanogap electrodes, the application of nanogap electrodes in devices, and the analysis of material properties. All these methods have provided promising results and have their own characteristics. The combination of different methods is frequently adopted to obtain the desired configuration to integrate molecules or other nanometer-sized components into nanogap electrodes, which then function as switches, rectifiers, memories, and transistors, etc.

The nonmetallic materials demonstrate immense potential for the development of molecular electronics. It is more viable to characterize the molecular properties precisely, since the smaller size of the electrode materials makes sure the construction of the truly single-molecule devices with high yield. Furthermore, the molecular devices are expected to realize the mass production and the commercial application due to the powerful semiconductor technology. The nonmetallic materials could continuously and creatively update the architecture and operation of the molecular devices and thus connect the field of molecular electronics with the practical application.

Despite this progress, there remain many challenges in the research of nanogap electrodes such as how to controllably and reliably manufacture electrodes with a nanogap in large numbers and how to precisely and unambiguously graft the desired molecules into the gap electrodes. Moreover, it is difficult to identify the exact number of molecules inserted between the gap electrodes, and the charge transport mechanism through the molecules remains inadequately understood. In addition, the design and synthesis of target molecules with expected electronic properties and proper energy levels is also a long-term, arduous task. All the challenges require further deep and systematical research on nanogap electrodes, both in fundamental understanding and practical embodiment.

References

1 Aviram, A. and Ratner, M.A. (1974). *Chem. Phys. Lett.* 29: 277.
2 Heath, J.R. and Ratner, M.A. (2003). *Phys. Today* 56: 43.
3 Kornyshev, A.A., Kuznetsov, A.M., and Ulstrup, J. (2006). *Proc. Natl. Acad. Sci. U.S.A.* 103: 6799.
4 Metzger, R.M., Chen, B., Höpfner, U. et al. (1997). *J. Am. Chem. Soc.* 119: 10455.
5 Collier, C.P., Mattersteig, G., Wong, E.W. et al. (2000). *Science* 289: 1172.
6 Blum, A.S., Kushmerick, J.G., Long, D.P. et al. (2005). *Nat. Mater.* 4: 167.
7 Park, H., Park, J., Lim, A.K.L. et al. (2000). *Nature* 407: 57.
8 Kubatkin, S., Danilov, A., Hjort, M. et al. (2003). *Nature* 425: 698.
9 Yu, L.H. and Natelson, D. (2004). *Nano Lett.* 4: 79.
10 Andres, R.P., Bein, T., Dorogi, M. et al. (1996). *Science* 272: 1323.
11 Tian, W., Hong, S., Reifenberger, R. et al. (1997). *Phys. Rev. Lett.* 79: 2530.
12 Dhirani, A., Lin, P.H., Guyot-Sionnest, P. et al. (1997). *J. Chem. Phys.* 106: 5249.
13 Venkataraman, L., Klare, J.E., Nuckolls, C. et al. (2006). *Nature* 442: 904.
14 Venkataraman, L., Klare, J.E., Tam, I.W. et al. (2006). *Nano Lett.* 6: 458.
15 Fan, F.F., Yang, J., Cai, L. et al. (2002). *J. Am. Chem. Soc.* 124: 5550.
16 Xu, B. and Tao, N.J. (2003). *Science* 301: 1221.
17 Xiao, X., Nagahara, L.A., Rawlett, A.M., and Tao, N. (2005). *J. Am. Chem. Soc.* 127: 9235.
18 Xiao, X. and Tao, N.J. (2004). *Nano Lett.* 4: 267.
19 Wu, S.W., Ogawa, N., and Ho, W. (2006). *Science* 312: 1362.
20 Lewis, P.A., Inman, C.E., Maya, F. et al. (2005). *J. Am. Chem. Soc.* 127: 17421.
21 He, J., Fu, Q., Lindsay, S. et al. (2006). *J. Am. Chem. Soc.* 128: 14828.
22 Holt, D.B., Yang, J.C., Naciri, J. et al. (2002). *Phys. Rev. Lett.* 89: 86802.
23 Kushmerick, J.G., Holt, D.B., Pollack, S.K. et al. (2002). *J. Am. Chem. Soc.* 124: 10654.
24 Kushmerick, J.G., Naciri, J., Yang, J.C., and Shashidhar, R. (2003). *Nano Lett.* 3: 897.
25 Rampi, M.A., Schueller, O.J.A., and Whitesides, G.M. (1998). *Appl. Phys. Lett.* 72: 1781.

26 Rampi, M.A. and Whitesides, G.M. (2002). *Chem. Phys.* 281: 373.
27 Bourgoin, J.P., Palacin, S., Esteve, D. et al. (1999). *Phys. Rev. B* 59: 12505.
28 Saifullah, M.S.M., Ondarcuhu, T., Koltsov, D.K. et al. (2002). *Nanotechnology* 13: 659.
29 Liang, W., Shores, M.P., Bockrath, M. et al. (2002). *Nature* 417: 725.
30 Qing, Q., Chen, F., Li, P. et al. (2005). *Angew. Chem. Int. Ed.* 44: 7771.
31 Nagase, T., Kubota, T., and Mashiko, S. (2003). *Thin Solid Films* 438–439: 374.
32 Reed, M.A., Zhou, C., Muller, C.J. et al. (1997). *Science* 278: 252.
33 Chen, W., Ahmed, H., and Nakazoto, K. (1995). *Appl. Phys. Lett.* 66: 3383.
34 Morpurgo, A.F., Marcus, C.M., and Robinson, D.B. (1999). *Appl. Phys. Lett.* 74: 2084.
35 Park, J., Pasupathy, A.N., Goldsmith, J.I. et al. (2002). *Nature* 417: 722.
36 Notargiacomo, A., Foglietti, V., Cianci, E. et al. (1999). *Nanotechnology* 10: 458.
37 Qin, L., Park, S., Huang, L., and Mirkin, C.A. (2005). *Science* 309: 113.
38 Hatzor, A. and Weiss, P. (2001). *Science* 291: 1019.
39 Qi, P., Javey, A., Rolandi, M. et al. (2004). *J. Am. Chem. Soc.* 126: 11774.
40 Tsukagoshi, K., Yagi, I., and Aoyagi, Y. (2004). *Appl. Phys. Lett.* 85: 1021.
41 Novoselov, K.S., Geim, A.K., Morozov, S.V. et al. (2004). *Science* 306: 666.
42 He, T., Ding, H., Peor, N. et al. (2008). *J. Am. Chem. Soc.* 130: 1699.
43 Vilan, A., Yaffe, O., Biller, A. et al. (2010). *Adv. Mater.* 22: 140.
44 Roth, K.M., Yasseri, A.A., Liu, Z. et al. (2003). *J. Am. Chem. Soc.* 125: 505.
45 Kostecki, R., Schnyder, B., Alliata, D. et al. (2001). *Thin Solid Films* 396: 36.
46 Karimi, M.A., Bahoosh, S.G., Herz, M. et al. (2016). *Nano Lett.* 16: 1803.
47 Xiang, D., Jeong, H., Lee, T., and Mayer, D. (2013). *Adv. Mater.* 25: 4845.
48 Perrin, M.L., Martin, C.A., Prins, F. et al. (2011). *Beilstein J. Nanotechnol.* 2: 714.
49 Xiang, D., Wang, X., Jia, C. et al. (2016). *Chem. Rev.* 116: 4318.
50 Alwan, M., Candoni, N., Dumas, P., and Klein, H.R. (2013). *Eur. Phys. J. B* 86: 243.
51 Ienaga, K., Nakashima, N., Inagaki, Y. et al. (2012). *Appl. Phys. Lett.* 101: 123114.
52 Ruben, M., Landa, A., Lörtscher, E. et al. (2008). *Small* 4: 2229.
53 Xiang, D., Lee, T., Kim, Y. et al. (2014). *Nanoscale* 6: 13396.
54 Frisenda, R., Parlato, L., Barra, M. et al. (2015). *Nanoscale Res. Lett.* 10: 305.
55 Huang, C., Rudnev, A.V., Hong, W., and Wandlowski, T. (2015). *Chem. Soc. Rev.* 44: 889.
56 Yi, Z., Banzet, M., Offenhäusser, A., and Mayer, D. (2010). *Phys. Status Solidi RRL* 4: 73.
57 Lörtscher, E., Ciszek, J.W., Tour, J., and Riel, H. (2006). *Small* 2: 973.
58 Parks, J.J., Champagne, A.R., Costi, T.A. et al. (2010). *Science* 328: 1370.
59 Perrin, M.L., Verzijl, C.J.O., Martin, C.A. et al. (2013). *Nat. Nanotechnol.* 8: 282.
60 Xiang, D., Jeong, H., Kim, D. et al. (2013). *Nano Lett.* 13: 2809.
61 Tian, J.H., Yang, Y., Liu, B. et al. (2010). *Nanotechnology* 21: 274012.
62 Arima, A., Tsutsui, M., Morikawa, T. et al. (2014). *J. Appl. Phys.* 115: 114310.

63 van der Giessen, E., van der Molen, S.J., Dulic, D. et al. (2005). *Phys. Rev. B* 71: 35313.
64 Champagne, A.R., Hutchison, G.R., Flores-Torres, S. et al. (2007). *Phys. Rev. Lett.* 99: 26601.
65 Yi, Z., Trellenkamp, S., Offenhäusser, A., and Mayer, D. (2010). *Chem. Commun.* 46: 8014.
66 Tian, J., Yang, Y., Zhou, X. et al. (2010). *Chem. Phys. Chem.* 11: 2745.
67 Wen, H., Yang, Y., Zhou, X. et al. (2013). *Chem. Sci.* 4: 2471.
68 Champagne, A.R., Pasupathy, A.N., and Ralph, D.C. (2005). *Nano Lett.* 5: 305.
69 Martin, C.A., van Ruitenbeek, J.M., and van der Zant, H.S.J. (2010). *Nanotechnology* 21: 265201.
70 Morikawa, T., Arima, A., Tsutsui, M., and Taniguchi, M. (2014). *Nanoscale* 6: 8235.
71 Doi, K., Tsutsui, M., Ohshiro, T. et al. (2014). *J. Phys. Chem. C* 118: 3758.
72 Waitz, R., Schecker, O., and Scheer, E. (2008). *Rev. Sci. Instrum.* 79: 93901.
73 Benner, D., Boneberg, J., Nürnberger, P. et al. (2014). *Nano Lett.* 14: 5218.
74 Tsutsui, M., Morikawa, T., He, Y. et al. (2015). *Sci. Rep.-UK* 5: 11519.
75 Wang, L., Wang, L., Zhang, L., and Xiang, D. (2017). *Top. Curr. Chem.* 375: 61.
76 Huang, C., Jevric, M., Borges, A. et al. (2017). *Nat. Commun.* 8: 15436.
77 Li, R., Lu, Z., Cai, Y. et al. (2017). *J. Am. Chem. Soc.* 139: 14344.
78 Zhang, Y., Chen, L., Zhang, Z. et al. (2018). *J. Am. Chem. Soc.* 140: 6531.
79 Zhao, Z., Liu, R., Mayer, D. et al. (2018). *Small* 14: 1703815.
80 Song, H., Kim, Y., Ku, J. et al. (2009). *Appl. Phys. Lett.* 94: 103110.
81 Li, C.Z., He, H.X., and Tao, N.J. (2000). *Appl. Phys. Lett.* 77: 3995.
82 Wu, Y.X., Hong, W.J., Akiyama, T. et al. (2013). *Nanotechnology* 24: 235302.
83 Kashimura, Y., Nakashima, H., Furukawa, K., and Torimitsu, K. (2003). *Thin Solid Films* 438–439: 317.
84 Li, X.L., He, H.X., Xu, B.Q. et al. (2004). *Surf. Sci.* 573: 1.
85 Liu, B., Xiang, J., Tian, J. et al. (2005). *Electrochim. Acta* 50: 3041.
86 Mészáros, G., Kronholz, S., Karthäuser, S. et al. (2007). *Appl. Phys. A* 87: 569.
87 Kervennic, Y.V., Van der Zant, H.S.J., Morpurgo, A.F. et al. (2002). *Appl. Phys. Lett.* 80: 321.
88 He, H.X., Boussaad, S., Xu, B.Q. et al. (2002). *J. Electroanal. Chem.* 522: 167.
89 Xiang, J., Liu, B., Wu, S. et al. (2005). *Angew. Chem.* 117: 1291.
90 Chen, F., Qing, Q., Ren, L. et al. (2005). *Appl. Phys. Lett.* 86: 123105.
91 Serdio, V.V.M., Azuma, Y., Takeshita, S. et al. (2012). *Nanoscale* 4: 7161.
92 Yasutake, Y., Kono, K., Kanehara, M. et al. (2007). *Appl. Phys. Lett.* 91: 203107.
93 Zou, X., Ji, L., Yang, X. et al. (2017). *J. Am. Chem. Soc.* 139: 16060.
94 Umeno, A. and Hirakawa, K. (2005). *Appl. Phys. Lett.* 86: 143103.
95 Lee, Y., Ahn, S.K., and Roh, Y. (2005). *Surf. Coat. Technol.* 193: 137.
96 Zhou, X., Wei, Y., Liu, L. et al. (2008). *J. Am. Chem. Soc.* 130: 13228.
97 Kiguchi, M. and Murakoshi, K. (2006). *Appl. Phys. Lett.* 88: 253112.
98 Tian, J., Liu, B., Li, X. et al. (2006). *J. Am. Chem. Soc.* 128: 14748.
99 Zhou, X., Liang, J., Chen, Z., and Mao, B. (2011). *Electrochem. Commun.* 13: 407.

100 Huang, L., Xu, L., Zhang, H., and Gu, N. (2002). *Chem. Commun.*: 72.

101 Ah, C.S., Yun, Y.J., Lee, J.S. et al. (2006). *Appl. Phys. Lett.* 88: 133116.

102 Dolan, G.J. (1977). *Appl. Phys. Lett.* 31: 337.

103 Klein, D.L., McEuen, P.L., Katari, J.E.B. et al. (1996). *Appl. Phys. Lett.* 68: 2574.

104 Naitoh, Y., Liang, T.T., Azehara, H., and Mizutani, W. (2005). *Jpn. J. Appl. Phys.* 44: L472.

105 Kubatkin, S., Danilov, A., Hjort, M. et al. (2004). *Curr. Appl. Phys.* 4: 554.

106 Kubatkin, S.E., Danilov, A.V., Olin, H., and Cleason, T. (2000). *J. Low Temp. Phys.* 118: 307.

107 Kanda, A., Wada, M., Hamamoto, Y., and Ootuka, Y. (2005). *Physica E* 29: 707.

108 Philipp, G., Weimann, T., Hinze, P. et al. (1999). *Microelectron. Eng.* 46: 157.

109 Thomas, D., Puyoo, E., Le Berre, M. et al. (2017). *Nanotechnology* 28: 455602.

110 Otsuka, Y., Naitoh, Y., Matsumoto, T. et al. (2004). *Nanotechnology* 15: 1639.

111 Sun, L.F., Chin, S.N., Marx, E. et al. (2005). *Nanotechnology* 16: 631.

112 Park, H., Lim, A.K.L., Alivisatos, A.P. et al. (1999). *Appl. Phys. Lett.* 75: 301.

113 Song, H., Kim, Y., Jang, Y.H. et al. (2009). *Nature* 462: 1039.

114 van der Zant, H.S.J., Osorio, E.A., Poot, M., and O'Neill, K. (2006). *Phys. Status Solidi B* 243: 3408.

115 Kim, T.W., Wang, G., Lee, H., and Lee, T. (2007). *Nanotechnology* 18: 315204.

116 Mangin, A., Anthore, A., Della Rocca, M.L. et al. (2009). *J. Appl. Phys.* 105: 14313.

117 van der Zant, H.S.J., Kervennic, Y., Poot, M. et al. (2006). *Faraday Discuss.* 131: 347.

118 Ueno, S., Tomoda, Y., Kume, W. et al. (2012). *Appl. Surf. Sci.* 258: 2153.

119 Houck, A.A., Labaziewicz, J., Chan, E.K. et al. (2005). *Nano Lett.* 5: 1685.

120 Kushmerick, J.G., Lazorcik, J., Patterson, C.H. et al. (2004). *Nano Lett.* 4: 639.

121 Henderson, J.J., Ramsey, C.M., Del Barco, E. et al. (2007). *J. Appl. Phys.* 101: 9E.

122 Trouwborst, M.L., van der Molen, S.J., and van Wees, B.J. (2006). *J. Appl. Phys.* 99: 114316.

123 Lee, C.H., Han, J.H., Schneider, S.C., and Josse, F. (2011). *Nanotechnology* 22: 405301.

124 Valladares, L.D.L.S., Felix, L.L., Dominguez, A.B. et al. (2010). *Nanotechnology* 21: 445304.

125 Baik, J.M., Lee, S.J., and Moskovits, M. (2009). *Nano Lett.* 9: 672.

126 Prins, F., Hayashi, T., de Vos Van Steenwijk, B.J.A. et al. (2009). *Appl. Phys. Lett.* 94: 123108.

127 Arzubiaga, L., Golmar, F., Llopis, R. et al. (2013). *Appl. Phys. Lett.* 102: 193103.

128 Asghar, W., Ramachandran, P.P., Adewumi, A. et al. (2010). *J. Manuf. Sci. Eng.* 132: 30911.

129 Naitoh, Y., Ohata, T., Matsushita, R. et al. (2013). *ACS Appl. Mater. Interfaces* 5: 12869.

130 Lu, Y., Goldsmith, B., Strachan, D.R. et al. (2010). *Small* 6: 2748.

131 Zhang, H., Bao, W., Zhao, Z. et al. (2012). *Nano Lett.* 12: 1772.

132 de Groot, Z., Folk, J.A., van der Zant, H.S.J. et al. (2006). *Phys. Rev. Lett.* 96: 206801.
133 de Groot, Z., Folk, J.A., Kouwenhoven, L.P. et al. (2006). *Phys. Rev. Lett.* 96: 17205.
134 Poot, M., Osorio, E., O'Neill, K. et al. (2006). *Nano Lett.* 6: 1031.
135 O'Neill, K., Osorio, E.A., and van der Zant, H.S.J. (2007). *Appl. Phys. Lett.* 90: 133109.
136 Tsutsui, M., Taniguchi, M., and Kawai, T. (2008). *Appl. Phys. Lett.* 93: 163115.
137 Strachan, D.R., Smith, D.E., Johnston, D.E. et al. (2005). *Appl. Phys. Lett.* 86: 43109.
138 Xiang, C., Kim, J.Y., and Penner, R.M. (2009). *Nano Lett.* 9: 2133.
139 Johnston, D.E., Guiton, B.S., Datta, S.S. et al. (2008). *Phys. Rev. Lett.* 100: 56805.
140 Dong, J.C. and Parviz, B.A. (2006). *Nanotechnology* 17: 5124.
141 Wu, Z.M., Steinacher, M., Huber, R. et al. (2007). *Appl. Phys. Lett.* 91: 53118.
142 Johnston, D.E., Strachan, D.R., and Johnson, A.T.C. (2007). *Nano Lett.* 7: 2774.
143 Sordan, R., Balasubramanian, K., Burghard, M., and Kern, K. (2005). *Appl. Phys. Lett.* 87: 13106.
144 Taychatanapat, T., Bolotin, K.I., Kuemmeth, F., and Ralph, D.C. (2007). *Nano Lett.* 7: 652.
145 McRae, A.C., Tayari, V., Porter, J.M., and Champagne, A.R. (2017). *Nat. Commun.* 8: 15491.
146 Strachan, D.R., Smith, D.E., Fischbein, M.D. et al. (2006). *Nano Lett.* 6: 441.
147 Rudneva, M., Gao, B., Prins, F. et al. (2013). *Microsc. Microanal.* 19: 43.
148 Fischbein, M.D. and Drndić, M. (2007). *Nano Lett.* 7: 1329.
149 Gao, B., Osorio, E.A., Gaven, K.B., and van der Zant, H.S.J. (2009). *Nanotechnology* 20: 415207.
150 Natelson, D., Li, Y., and Herzog, J.B. (2013). *Phys. Chem. Chem. Phys.* 15: 5262.
151 Ward, D.R., Halas, N.J., Ciszek, J.W. et al. (2008). *Nano Lett.* 8: 919.
152 Herzog, J.B., Knight, M.W., Li, Y. et al. (2013). *Nano Lett.* 13: 1359.
153 Ward, D.R., Corley, D.A., Tour, J.M., and Natelson, D. (2010). *Nat. Nanotechnol.* 6: 33.
154 Pasupathy, A.N., Bialczak, R.C., Martinek, J. et al. (2004). *Science* 306: 86.
155 Kwon, N., Kim, K., and Chung, I. (2010). *J. Vac. Sci. Technol., A* 28: 730.
156 Arielly, R., Vadai, M., Kardash, D. et al. (2014). *J. Am. Chem. Soc.* 136: 2674.
157 Kim, K., Kim, J., Huang, X. et al. (2008). *Small* 4: 1959.
158 Anderson, M.E., Mihok, M., Tanaka, H. et al. (2006). *Adv. Mater.* 18: 1020.
159 Anderson, M.E., Srinivasan, C., Jayaraman, R. et al. (2005). *Microelectron. Eng.* 78–79: 248.
160 Negishi, R., Hasegawa, T., Terabe, K. et al. (2006). *Appl. Phys. Lett.* 88: 223111.
161 McCarty, G.S. (2004). *Nano Lett.* 4: 1391.
162 McCarty, G.S., Moody, B., and Zachek, M.K. (2010). *J. Electroanal. Chem.* 643: 9.
163 McCarty, G.S. (2006). *J. Appl. Phys.* 99: 64701.
164 Srinivasan, C., Hohman, J.N., Anderson, M.E. et al. (2008). *Appl. Phys. Lett.* 93: 83123.

165 Cao, L., Chen, S., Wei, D. et al. (2010). *J. Mater. Chem.* 20: 2305.

166 Jain, T., Westerlund, F., Johnson, E. et al. (2009). *ACS Nano* 3: 828.

167 Dadosh, T., Gordin, Y., Krahne, R. et al. (2005). *Nature* 436: 677.

168 Chen, Y., Hong, M., and Huang, G.S. (2012). *Nat. Nanotechnol.* 7: 197.

169 Tang, Q., Tong, Y., Jain, T. et al. (2009). *Nanotechnology* 20: 245205.

170 Pourhossein, P. and Chiechi, R.C. (2012). *ACS Nano* 6: 5566.

171 Krahne, R., Yacoby, A., Shtrikman, H. et al. (2002). *Appl. Phys. Lett.* 81: 730.

172 Maruccio, G., Marzo, P., Krahne, R. et al. (2007). *Small* 3: 1184.

173 Luber, S.M., Zhang, F., Lingitz, S. et al. (2007). *Small* 3: 285.

174 Luber, S.M., Bichler, M., Abstreiter, G., and Tornow, M. (2011). *Nanotechnology* 22: 65301.

175 Howell, S.W., Dirk, S.M., Childs, K. et al. (2005). *Nanotechnology* 16: 754.

176 Ashwell, G.J., Phillips, L.J., Robinson, B.J. et al. (2010). *ACS Nano* 4: 7401.

177 Chen, C.C., Sheu, J.T., Chiang, S.L., and Sheu, M.L. (2006). *Jpn. J. Appl. Phys.* 45: 5531.

178 Hashioka, S., Saito, M., Tamiya, E., and Matsumura, H. (2003). *J. Vac. Sci. Technol., B* 21: 2937.

179 Strobel, S., Harrer, S., Penso Blanco, G. et al. (2009). *Small* 5: 579.

180 Liu, S., Tok, J.B.H., and Bao, Z. (2005). *Nano Lett.* 5: 1071.

181 Chen, X., Yeganeh, S., Qin, L. et al. (2009). *Nano Lett.* 9: 3974.

182 Meng, F., Sana, B., Li, Y. et al. (2014). *Small* 10: 277.

183 Tian, X., Li, J., and Xu, D. (2010). *Electrochem. Commun.* 12: 1081.

184 Chen, X., Jeon, Y., Jang, J. et al. (2008). *J. Am. Chem. Soc.* 130: 8166.

185 Chen, X., Braunschweig, A.B., Wiester, M.J. et al. (2009). *Angew. Chem. Int. Ed.* 48: 5178.

186 Osberg, K.D., Schmucker, A.L., Senesi, A.J., and Mirkin, C.A. (2011). *Nano Lett.* 11: 820.

187 Meng, F., Hervault, Y., Norel, L. et al. (2012). *Chem. Sci.* 3: 3113.

188 Meng, F., Hervault, Y., Shao, Q. et al. (2014). *Nat. Commun.* 5: 3023.

189 Xu, T. and Metzger, R.M. (2002). *Nano Lett.* 2: 1061.

190 Lefebvre, J., Radosavljević, M., and Johnson, A.T. (2000). *Appl. Phys. Lett.* 76: 3828.

191 De Poortere, E.P., Stormer, H.L., Huang, L.M. et al. (2006). *Appl. Phys. Lett.* 88: 143124.

192 De Poortere, E.P., Stormer, H.L., Huang, L.M. et al. (2006). *J. Vac. Sci. Technol., B* 24: 3213.

193 Chung, J. and Lee, J. (2003). *Sens. Actuators, A* 104: 229.

194 Chung, J., Lee, K., and Lee, J. (2003). *Nano Lett.* 3: 1029.

195 Huang, X.M.H., Caldwell, R., Huang, L. et al. (2005). *Nano Lett.* 5: 1515.

196 Jiang, L., Gao, J., Wang, E. et al. (2008). *Adv. Mater.* 20: 2735.

197 Jiang, L., Dong, H., Meng, Q. et al. (2012). *Adv. Mater.* 24: 694.

198 Min, S., Kim, T., Kim, B.J. et al. (2013). *Nat. Commun.* 4: 1773.

199 Ongaro, A., Griffin, F., Nagle, L. et al. (2004). *Adv. Mater.* 16: 1799.

200 Tseng, A.A. (2005). *Small* 1: 924.

201 Rigort, A. and Plitzko, J.M. (2015). *Arch. Biochem. Biophys.* 581: 122.

202 Cui, Z. (2017). *Nanofabrication: Principles, Capabilities and Limits*, 401. Cham, Switzerland: Springer.

203 Blom, T., Welch, K., Strømme, M. et al. (2007). *Nanotechnology* 18: 285301.

204 Nagase, T., Gamo, K., Kubota, T., and Mashiko, S. (2006). *Thin Solid Films* 499: 279.

205 Li, H., Wani, I.H., Hayat, A. et al. (2015). *Appl. Phys. Lett.* 107: 103108.

206 Gazzadi, G.C., Angeli, E., Facci, P., and Frabboni, S. (2006). *Appl. Phys. Lett.* 89: 173112.

207 Han, G., Weber, D., Neubrech, F. et al. (2011). *Nanotechnology* 22: 275202.

208 Horiuchi, K., Kato, T., Hashii, S. et al. (2005). *Appl. Phys. Lett.* 86: 153108.

209 Bhattacharjee, S. and Chowdhury, T. (2009). *Appl. Phys. Lett.* 95: 61501.

210 Tsai, C., Chen, K., Pan, F. et al. (2008). *J. Electrochem. Soc.* 155: J361.

211 Nagase, T., Gamo, K., Kubota, T., and Mashiko, S. (2005). *Microelectron. Eng.* 78–79: 253.

212 Cui, A., Liu, Z., Dong, H. et al. (2015). *Adv. Mater.* 27: 3002.

213 Cui, A., Liu, Z., Dong, H. et al. (2015). *Sci. China Mater.* 58: 769.

214 Shigeto, K., Kawamura, M., Kasumov, A.Y. et al. (2006). *Microelectron. Eng.* 83: 1471.

215 Hill, R., Notte, J.A., and Scipioni, L. (2012). *Scanning Helium Ion Microscopy*, 65. Amsterdam: Elsevier.

216 Scipioni, L., Stern, L.A., Notte, J. et al. (2008). *Adv. Mater. Processes* 166: 27.

217 Thiele, C., Vieker, H., Beyer, A. et al. (2014). *Appl. Phys. Lett.* 104: 103102.

218 Kollmann, H., Piao, X., Esmann, M. et al. (2014). *Nano Lett.* 14: 4778.

219 Maas, D., van Veldhoven, E., Chen, P. et al. (2010). Nanofabrication with a helium ion microscope. Metrology, Inspection, and Process Control for Microlithography XXIV. *Proceedings of SPIE*. International Society for Optics and Photonics, Volume 7638, 763811.

220 Elswick, D., Ananth, M., Stern, L. et al. (2013). *Microsc. Microanal.* 19: 1304.

221 Scholder, O., Jefimovs, K., Shorubalko, I. et al. (2013). *Nanotechnology* 24: 395301.

222 Li, T., Hu, W., and Zhu, D. (2010). *Adv. Mater.* 22: 286.

223 Zhang, X. and Li, T. (2017). *Chin. Chem. Lett.* 28: 2058.

224 Zhang, Z. and Li, T. (2016). *Chin. Chem. Lett.* 27: 1209.

225 Miyazaki, T., Kobayashi, K., Ishida, K. et al. (2003). *Jpn. J. Appl. Phys.* 42: 4173.

226 Piner, R.D., Zhu, J., Xu, F. et al. (1999). *Science* 283: 661.

227 Magno, R. and Bennett, B.R. (1997). *Appl. Phys. Lett.* 70: 1855.

228 He, Y., Dong, H., Li, T. et al. (2010). *Appl. Phys. Lett.* 97: 133301.

229 Nazin, G.V., Qiu, X.H., and Ho, W. (2003). *Science* 302: 77.

230 Venema, L.C., Wildöer, J.W.G., Tuinstra, H.L.J.T. et al. (1997). *Appl. Phys. Lett.* 71: 2629.

231 Miyashita, K., Nishimura, S., Toyofuku, T., and Shirakashi, J. (2009). *J. Vac. Sci. Technol., B* 27: 953.

232 Miyazaki, T., Kobayashi, K., Horiuchi, T. et al. (2001). *Jpn. J. Appl. Phys.* 40: 4365.

233 Moriya, R., Kobayashi, H., Shibata, K. et al. (2010). *Appl. Phys. Express* 3: 35001.

234 Zhang, H., Chung, S., and Mirkin, C.A. (2003). *Nano Lett.* 3: 43.
235 Repp, J., Meyer, G., Paavilainen, S. et al. (2006). *Science* 312: 1196.
236 Lee, T., Wang, W., Klemic, J.F. et al. (2004). *J. Phys. Chem. B* 108: 8742.
237 Engelkes, V.B., Beebe, J.M., and Frisbie, C.D. (2005). *J. Phys. Chem. B* 109: 16801.
238 Cui, X.D., Zarate, X., Tomfohr, J. et al. (2002). *Nanotechnology* 13: 5.
239 Morita, T. and Lindsay, S. (2007). *J. Am. Chem. Soc.* 129: 7262.
240 Wei, Z., Li, T., Jennum, K. et al. (2012). *Langmuir* 28: 4016.
241 Wei, Z., Wang, X., Borges, A. et al. (2014). *Langmuir* 30: 14868.
242 Wen, Y., Chen, J., Guo, Y. et al. (2012). *Adv. Mater.* 24: 3482.
243 Moore, G.E. (1965). *Electronics* 38: 114.
244 Cui, A., Dong, H., and Hu, W. (2015). *Small* 11: 6115.
245 Bryce, M.R. and Petty, M.C. (1995). *Nature* 374: 771.
246 Ratner, M. (2013). *Nat. Nanotechnol.* 8: 378.
247 Tour, J.M., Reinerth, W.A., Jones, L. et al. (2006). *Ann. N.Y. Acad. Sci.* 852: 197.
248 Salomon, A., Cahen, D., Lindsay, S. et al. (2003). *Adv. Mater.* 15: 1881.
249 Xu, B., Zhang, P., Li, X., and Tao, N. (2004). *Nano Lett.* 4: 1105.
250 Lau, C.N., Stewart, D.R., Williams, R.S., and Bockrath, M. (2004). *Nano Lett.* 4: 569.
251 Stewart, D.R., Ohlberg, D.A.A., Beck, P.A. et al. (2004). *Nano Lett.* 4: 133.
252 Kumagai, S., Yoshii, S., Matsukawa, N. et al. (2009). *Appl. Phys. Lett.* 94: 83103.
253 Mann, B. and Kuhn, H. (1971). *J. Appl. Phys.* 42: 4398.
254 Polymeropoulos, E.E. and Sagiv, J. (1978). *J. Chem. Phys.* 69: 1836.
255 Kagan, C.R. and Ratner, M.A. (2004). *MRS Bull.* 29: 376.
256 Vuillaume, D. (2008). *C.R. Phys.* 9: 78.
257 Anaya, A., Korotkov, A.L., Bowman, M. et al. (2003). *J. Appl. Phys.* 93: 3501.
258 Ramachandran, G.K., Hopson, T.J., Rawlett, A.M. et al. (2003). *Science* 300: 1413.
259 Ajayan, P.M. and Ebbesen, T.W. (1997). *Rep. Prog. Phys.* 60: 1025.
260 Wei, D., Liu, Y., Cao, L. et al. (2008). *Nano Lett.* 8: 1625.
261 Prins, F., Barreiro, A., Ruitenberg, J.W. et al. (2011). *Nano Lett.* 11: 4607.
262 Kostecki, R., Song, X., and Kinoshita, K. (2002). *Electrochem. Solid-State Lett.* 5: E29.
263 Okawa, Y., Mandal, S.K., Hu, C. et al. (2011). *J. Am. Chem. Soc.* 133: 8227.
264 Ulman, A. (1996). *Chem. Rev.* 96: 1533.
265 Yu, M.F., Lourie, O., Dyer, M.J. et al. (2000). *Science* 287: 637.
266 Guo, X., Small, J.P., Klare, J.E. et al. (2006). *Science* 311: 356.
267 Li, T., Jevric, M., Hauptmann, J.R. et al. (2013). *Adv. Mater.* 25: 4164.
268 Li, T., Hauptmann, J.R., Wei, Z. et al. (2012). *Adv. Mater.* 24: 1333.
269 Iijima, S. (1991). *Nature* 354: 56.
270 Dai, H. (2002). *Acc. Chem. Res.* 35: 1035.
271 Yao, Z., Kane, C.L., and Dekker, C. (2000). *Phys. Rev. Lett.* 84: 2941.
272 Javey, A., Guo, J., Paulsson, M. et al. (2004). *Phys. Rev. Lett.* 92: 106804.
273 Collins, P.G., Arnold, M.S., and Avouris, P. (2001). *Science* 292: 706.
274 Collins, P.G., Hersam, M., Arnold, M. et al. (2001). *Phys. Rev. Lett.* 86: 3128.

275 Naitoh, Y., Yanagi, K., Suga, H. et al. (2009). *Appl. Phys Express* 2: 35008.
276 Yagi, I. (2004). *Microelectron. Eng.* 73–74: 675.
277 Toth, M. (2014). *Appl. Phys. A* 117: 1623.
278 Liu, P., Arai, F., and Fukuda, T. (2006). *Appl. Phys. Lett.* 89: 113104.
279 Yuzvinsky, T.D., Mickelson, W., Aloni, S. et al. (2006). *Nano Lett.* 6: 2718.
280 Yuzvinsky, T.D., Fennimore, A.M., Mickelson, W. et al. (2005). *Appl. Phys. Lett.* 86: 53109.
281 Spinney, P.S., Howitt, D.G., Collins, S.D., and Smith, R.L. (2009). *Nanotechnology* 20: 465301.
282 Yoshida, H., Tomita, Y., Soma, K., and Takeda, S. (2017). *Nanotechnology* 28: 195301.
283 Thiele, C., Engel, M., Hennrich, F. et al. (2011). *Appl. Phys. Lett.* 99: 173105.
284 Roy, S., Vedala, H., Roy, A.D. et al. (2008). *Nano Lett.* 8: 26.
285 Perrin, M.L., Burzuri, E., and van der Zant, H.S.J. (2015). *Chem. Soc. Rev.* 44: 902.
286 Marquardt, C.W., Grunder, S., Błaszczyk, A. et al. (2010). *Nat. Nanotechnol.* 5: 863.
287 Chen, X., Guo, Z., Yang, G. et al. (2010). *Mater. Today* 13: 28.
288 Feldman, A.K., Steigerwald, M.L., Guo, X., and Nuckolls, C. (2008). *Acc. Chem. Res.* 41: 1731.
289 Whalley, A.C., Steigerwald, M.L., Guo, X., and Nuckolls, C. (2007). *J. Am. Chem. Soc.* 129: 12590.
290 Guo, X., Gorodetsky, A.A., Hone, J. et al. (2008). *Nat. Nanotechnol.* 3: 163.
291 Guo, X., Whalley, A., Klare, J.E. et al. (2007). *Nano Lett.* 7: 1119.
292 Tang, Q., Moon, H.K., Lee, Y. et al. (2007). *J. Am. Chem. Soc.* 129: 11018.
293 Guo, X., Myers, M., Xiao, S. et al. (2006). *Proc. Natl. Acad. Sci. U.S.A.* 103: 11452.
294 Guo, X., Xiao, S., Myers, M. et al. (2009). *Proc. Natl. Acad. Sci. U.S.A.* 106: 691.
295 Liu, S., Wei, Z., Cao, Y. et al. (2011). *Chem. Sci.* 2: 796.
296 Wang, Q.H. and Hersam, M.C. (2009). *Nat. Chem.* 1: 206.
297 Pinzon, J.R., Villalta-Cerdas, A., and Echegoyen, L. (2012). *Top. Curr. Chem.* 312: 127.
298 Booth, T.J., Blake, P., Nair, R.R. et al. (2008). *Nano Lett.* 8: 2442.
299 Fang, T., Konar, A., Xing, H., and Jena, D. (2007). *Appl. Phys. Lett.* 91: 92109.
300 Novoselov, K.S., Jiang, Z., Zhang, Y. et al. (2007). *Science* 315: 1379.
301 Kim, K.S., Zhao, Y., Jang, H. et al. (2009). *Nature* 457: 706.
302 Tapasztó, L., Dobrik, G., Lambin, P., and Biró, L.P. (2008). *Nat. Nanotechnol.* 3: 397.
303 Li, X., Wang, X., Zhang, L. et al. (2008). *Science* 319: 1229.
304 Taychatanapat, T. and Jarillo-Herrero, P. (2010). *Phys. Rev. Lett.* 105: 166601.
305 Xia, F., Farmer, D.B., Lin, Y., and Avouris, P. (2010). *Nano Lett.* 10: 715.
306 Park, J., Jo, S.B., Yu, Y. et al. (2012). *Adv. Mater.* 24: 407.
307 Zhou, S.Y., Gweon, G.H., Fedorov, A.V. et al. (2007). *Nat. Mater.* 6: 770.
308 Cao, Y., Liu, S., Shen, Q. et al. (2009). *Adv. Funct. Mater.* 19: 2743.

309 Bundesmann, J., Liu, M., Richter, K., and Ryndyk, D.A. (2012). *Phys. Rev. B* 86: 195425.

310 Nef, C., Pósa, L., Makk, P. et al. (2014). *Nanoscale* 6: 7249.

311 Xu, Q., Scuri, G., Mathewson, C. et al. (2017). *Nano Lett.* 17: 5335.

312 Bell, D.C., Lemme, M.C., Stern, L.A. et al. (2009). *Nanotechnology* 20: 455301.

313 Cao, Y., Dong, S., Liu, S. et al. (2012). *Angew. Chem. Int. Ed.* 51: 12228.

314 Gu, C., Su, D., Jia, C. et al. (2018). *RSC Adv.* 8: 6814.

315 Zhou, X., Shade, C.M., Schmucker, A.L. et al. (2012). *Nano Lett.* 12: 4734.

316 Jia, C. and Guo, X. (2013). *Chem. Soc. Rev.* 42: 5642.

317 Cao, Y., Dong, S., Liu, S. et al. (2013). *Angew. Chem. Int. Ed.* 52: 3906.

318 Xin, N., Wang, J., Jia, C. et al. (2017). *Nano Lett.* 17: 856.

319 Island, J.O., Holovchenko, A., Koole, M. et al. (2014). *J. Phys. Condens. Matter* 26: 474205.

320 Burzurí, E., Island, J.O., Díaz-Torres, R. et al. (2016). *ACS Nano* 10: 2521.

321 Lumetti, S., Candini, A., Godfrin, C. et al. (2016). *Dalton Trans.* 45: 16570.

322 Ferradás, R., Carrascal, D., Ferrer, J., and García-Suárez, V.M. (2013). *Phys. Rev. B* 87: 235425.

323 Li, Y., Tu, X., Wang, H. et al. (2015). *J. Chem. Phys.* 142: 164701.

324 Lambert, C.J. and Péterfalvi, C.G. (2012). *Phys. Rev. B* 86: 85443.

325 Sadeghi, H., Sangtarash, S., and Lambert, C. (2017). *Nano Lett.* 17: 4611.

326 Zhu, H., Li, T., Zhang, Y. et al. (2010). *Adv. Mater.* 22: 1645.

327 Leitherer, S., Coto, P.B., Ullmann, K. et al. (2017). *Nanoscale* 9: 7217.

328 Martin, C.A., Ding, D., Sørensen, J.K. et al. (2008). *J. Am. Chem. Soc.* 130: 13198.

329 Berland, K., Hyldgaard, P., Kubatkin, S. et al. (2011). *Phys. Rev. B* 84: 155451.

330 Ullmann, K., Coto, P.B., Leitherer, S. et al. (2015). *Nano Lett.* 15: 3512.

331 Liu, L., Zhang, Q., Tao, S. et al. (2016). *Nanoscale* 8: 14507.

332 Zhang, Q., Liu, L., Tao, S. et al. (2016). *Nano Lett.* 16: 6534.

333 Jeong, H., Kim, D., Xiang, D., and Lee, T. (2017). *ACS Nano* 11: 6511.

334 Del, V.M., Gutierrez, R., Tejedor, C., and Cuniberti, G. (2007). *Nat. Nanotechnol.* 2: 176.

335 Zhang, C., Du, M.H., Cheng, H.P. et al. (2004). *Phys. Rev. Lett.* 92: 158301.

336 Pshenichnyuk, I.A., Coto, P.B., Leitherer, S., and Thoss, M. (2013). *J. Phys. Chem. Lett.* 4: 809.

337 Frisbie, C.D. (2016). *Science* 352: 1394.

338 Jia, C., Migliore, A., Xin, N. et al. (2016). *Science* 352: 1443.

339 Cao, Y., Wei, Z., Liu, S. et al. (2010). *Angew. Chem. Int. Ed.* 49: 6319.

340 Feng, L., Dong, H., Li, Q. et al. (2017). *Sci. China Mater.* 60: 75.

341 Wang, G., Kim, Y., Choe, M. et al. (2011). *Adv. Mater.* 23: 755.

342 Seo, S., Min, M., Lee, S.M., and Lee, H. (2013). *Nat. Commun.* 4: 1920.

343 Margapoti, E., Strobel, P., Asmar, M.M. et al. (2014). *Nano Lett.* 14: 6823.

344 Wieghold, S., Li, J., Simon, P. et al. (2016). *Nat. Commun.* 7: 10700.

345 MacLeod, J.M., Lipton-Duffin, J.A., Cui, D. et al. (2015). *Langmuir* 31: 7016.

346 Kang, J., Shin, D., Bae, S., and Hong, B.H. (2012). *Nanoscale* 4: 5527.

347 Lee, W.H., Suk, J.W., Lee, J. et al. (2012). *ACS Nano* 6: 1284.

348 Wang, Z., Dong, H., Li, T. et al. (2015). *Nat. Commun.* 6: 7478.
349 Hauptmann, J.R., Li, T., Petersen, S. et al. (2012). *Phys. Chem. Chem. Phys.* 14: 14277.
350 Yang, J., Huo, N., Li, Y. et al. (2015). *Adv. Electron. Mater.* 1: 1500267.
351 Min, M., Seo, S., Lee, S.M., and Lee, H. (2013). *Adv. Mater.* 25: 7045.
352 Daub, J., Knöchel, T., and Mannschreck, A. (1984). *Angew. Chem. Int. Ed. Engl.* 23: 960.
353 Hunger, R., Jaegermann, W., Merson, A. et al. (2006). *J. Phys. Chem. B* 110: 15432.
354 Yaffe, O., Scheres, L., Segev, L. et al. (2010). *J. Phys. Chem. C* 114: 10270.
355 Seitz, O., Vilan, A., Cohen, H. et al. (2008). *Adv. Funct. Mater.* 18: 2102.
356 Yaffe, O., Ely, T., Har-Lavan, R. et al. (2013). *J. Phys. Chem. C* 117: 22351.
357 Choi, Y., Lee, J.S., Zhu, J. et al. (2003). *J. Vac. Sci. Technol., B* 21: 2951.
358 Corley, D.A., He, T., and Tour, J.M. (2010). *ACS Nano* 4: 1879.
359 Ashwell, G.J., Phillips, L.J., Robinson, B.J. et al. (2011). *Angew. Chem. Int. Ed.* 50: 8722.
360 Lamport, Z.A., Broadnax, A.D., Harrison, D. et al. (2016). *Sci. Rep.-UK* 6: 38092.
361 Bigelow, W.C., Pickett, D.L., and Zisman, W.A. (1946). *J. Colloid Sci.* 1: 513.
362 Sagiv, J. (1980). *J. Am. Chem. Soc.* 102: 92.
363 Mazo, R. and Sagiv, J. (1983). *J. Colloid Interface Sci.* 100: 465.
364 Wolkow, R. and Avouris, P. (1988). *Phys. Rev. Lett.* 60: 1049.
365 Linford, M.R. and Chidsey, C.E. (1993). *J. Am. Chem. Soc.* 115: 12631.
366 Linford, M.R., Fenter, P., Eisenberger, P.M., and Chidsey, C.E. (1995). *J. Am. Chem. Soc.* 117: 3145.
367 Aswal, D.K., Lenfant, S., Guerin, D. et al. (2006). *Anal. Chim. Acta* 568: 84.
368 Lavi, A., Cohen, H., Bendikov, T. et al. (2011). *Phys. Chem. Chem. Phys.* 13: 1293.
369 Puniredd, S.R., Jayaraman, S., Yeong, S.H. et al. (2013). *J. Phys. Chem. Lett.* 4: 1397.
370 Richter, C.A., Hacker, C.A., and Richter, L.J. (2005). *J. Phys. Chem. Lett. B* 109: 21836.
371 Kuikka, M.A., Li, W., Kavanagh, K.L., and Yu, H. (2008). *J. Phys. Chem. Lett. C* 112: 9081.
372 Loo, Y., Willett, R.L., Baldwin, K.W., and Rogers, J.A. (2002). *J. Am. Chem. Soc.* 124: 7654.
373 Shimizu, K.T., Fabbri, J.D., Jelincic, J.J., and Melosh, N.A. (2006). *Adv. Mater.* 18: 1499.
374 Selzer, Y., Salomon, A., and Cahen, D. (2002). *J. Phys. Chem. B* 106: 10432.
375 Coll, M., Miller, L.H., Richter, L.J. et al. (2009). *J. Am. Chem. Soc.* 131: 12451.
376 Coll, M., Gergel-Hackett, N., Richter, C.A., and Hacker, C.A. (2011). *J. Phys. Chem. Lett. C* 115: 24353.
377 Salomon, A., Boecking, T., Chan, C.K. et al. (2005). *Phys. Rev. Lett.* 95: 266807.
378 Scott, A., Janes, D.B., Risko, C., and Ratner, M.A. (2007). *Appl. Phys. Lett.* 91: 33508.

379 Hiremath, R.K., Rabinal, M.K., Mulimani, B.G., and Khazi, I.M. (2008). *Langmuir* 24: 11300.

380 Hiremath, R.K., Rabinal, M.H.K., and Mulimani, B.G. (2010). *Phys. Chem. Chem. Phys.* 12: 2564.

381 Yaffe, O., Scheres, L., Puniredd, S.R. et al. (2009). *Nano Lett.* 9: 2390.

382 Böcking, T., Salomon, A., Cahen, D., and Gooding, J.J. (2007). *Langmuir* 23: 3236.

383 Salomon, A., Boecking, T., Seitz, O. et al. (2007). *Adv. Mater.* 19: 445.

384 Gergel-Hackett, N., Aguilar, I., and Richter, C.A. (2010). *J. Phys. Chem. Lett. C* 114: 21708.

385 Bora, A., Pathak, A., Liao, K.C. et al. (2013). *Appl. Phys. Lett.* 102: 241602.

386 Zhou, C., Deshpande, M.R., Reed, M.A. et al. (1997). *Appl. Phys. Lett.* 71: 611.

387 Wei, L., Syomin, D., Loewe, R.S. et al. (2005). *J. Phys. Chem. Lett. B* 109: 6323.

388 Gryko, D., Li, J., Diers, J.R. et al. (2001). *J. Mater. Chem.* 11: 1162.

389 Li, Q., Mathur, G., Homsi, M. et al. (2002). *Appl. Phys. Lett.* 81: 1494.

390 Liu, Z., Yasseri, A.A., Lindsey, J.S., and Bocian, D.F. (2003). *Science* 302: 1543.

391 Anariba, F., Tiznado, H., Diers, J.R. et al. (2008). *J. Phys. Chem. C* 112: 9474.

392 Garg, K., Majumder, C., Gupta, S.K. et al. (2016). *Chem. Sci.* 7: 1548.

393 Liu, Y. and Yu, H. (2003). *J. Phys. Chem. B* 107: 7803.

394 Metzger, R.M. (2003). *Chem. Rev.* 103: 3803.

395 Schreiber, F. (2000). *Prog. Surf. Sci.* 65: 151.

396 Nerngchamnong, N., Yuan, L., Qi, D. et al. (2013). *Nat. Nanotechnol.* 8: 113.

397 Lenfant, S., Krzeminski, C., Delerue, C. et al. (2003). *Nano Lett.* 3: 741.

398 Garg, K., Majumder, C., Nayak, S.K. et al. (2015). *Phys. Chem. Chem. Phys.* 17: 1891.

399 Guisinger, N.P., Greene, M.E., Basu, R. et al. (2004). *Nano Lett.* 4: 55.

400 Liang, G.C., Ghosh, A.W., Hersam, M.C. et al. (2005). *Phys. Rev. B* 72: 125305.

401 Robert, A.W. (2001). *Jpn. J. Appl. Phys.* 40: 4378.

402 Wolkow, R.A. and Dickie, A.J. (2008). *Phys. Rev. B* 77: 115305.

403 Davidovits, J.V., Rondelez, F., Vuillaume, D., and Boulas, C. (1996). *Phys. Rev. Lett.* 76: 4797.

404 Bergren, A.J., McCreery, R.L., Stoyanov, S.R. et al. (2010). *J. Phys. Chem. C* 114: 15806.

405 Bergren, A.J., Harris, K.D., Deng, F., and McCreery, R.L. (2008). *J. Phys. Condens. Matter* 20: 374117.

406 Sayed, S.Y., Fereiro, J.A., Yan, H. et al. (2012). *Proc. Natl. Acad. Sci. U.S.A.* 109: 11498.

407 Zhu, Z., Daniel, T.A., Maitani, M. et al. (2006). *J. Am. Chem. Soc.* 128: 13710.

408 Akkerman, H.B., Blom, P.W.M., de Leeuw, D.M., and de Boer, B. (2006). *Nature* 441: 69.

409 Yan, H., Bergren, A.J., and McCreery, R.L. (2011). *J. Am. Chem. Soc.* 133: 19168.

410 Morteza Najarian, A., Szeto, B., Tefashe, U.M., and McCreery, R.L. (2016). *ACS Nano* 10: 8918.

411 Sayed, S.Y., Bayat, A., Kondratenko, M. et al. (2013). *J. Am. Chem. Soc.* 135: 12972.
412 Fairman, C., Chockalingam, M., Liu, G. et al. (2013). *Langmuir* 30: 332.
413 Ouchi, M., Badi, N., Lutz, J., and Sawamoto, M. (2011). *Nat. Chem.* 3: 917.
414 Ariga, K., Mori, T., and Hill, J.P. (2013). *Langmuir* 29: 8459.

4

Characterization Methods and Analytical Techniques for Nanogap Junction

Baili Li, Ziyan Wang, Bin Han, and Xi Yu

Department of Chemistry, Tianjin Key Laboratory of Molecular Optoelectronic Sciences, School of Science, Tianjin University, No. 92 Weijin Road, Tianjin 300072, P. R. China

The realization of the single-molecule electronic device requires well characterization of the junction, more specifically, the active molecule–electrode region. Actually, one of the great challenges in nanogap molecule electronics is to demonstrate that the electrical properties obtained originate explicitly from the charge transport through the molecule and to correlate the molecule–electrode structure to the functions of the junction. The most direct characterization of the junction is, obviously, the current–voltage (I–V) measurement of the device. Combined with physical models of charge transport, the I–V character analysis has been used to extract key parameters of electronic structure and charge transport, like the energy level and coupling of the molecule to the electrode, etc. More advanced spectroscopy methods and analytical techniques based on electronical measurement have also been developed, like the inelastic tunneling spectroscopy (IETS) and transition voltage spectroscopy (TVS), etc., by which, charge transport pathway, molecule–electrode interface structure, and vibronic effect can be obtained. In addition to the electrical methods, thanks to the strong plasmonic field, therefore enhancement effect in the metal gap region, the optical spectroscopies have worked as powerful tools with ultra-high sensitivity in revealing detailed chemical structure information of the junction.

In this chapter, we will introduce briefly the most common and popular characterization techniques, both electrical and optical techniques, and analytical methods used in the molecular electronics study.

4.1 Current–Voltage Analysis

The charge transport in nanogap molecular junction can generally be categorized into two mechanisms: coherent (super-exchange) tunneling and incoherent hopping, with quite distinct I–V behaviors [1]. The coherent tunneling is a one-step quantum tunneling through the molecule which can be formulated by quantum

Nanogap Electrodes, First Edition. Edited by Tao Li.

transport based on the transmittance of running wave across the energy barrier of molecule. On the other hand, hopping is a step-wise transport from one electrode to the molecule (maybe more than one step on the molecule) and then to the other electrode, which can be regarded as consecutive redox reactions with kinetics based on rate equation method. The analysis of the transport mechanism of the junction is commonly done by either examining the I–V properties against certain experimental variables, like temperature, molecule length, etc., or fitting the experimental data to the theoretical I–V relation of certain physical model.

4.1.1 Coherent Tunneling Transport

In the coherent tunneling scenario, temperature has little effect on the Fermi distribution of the electrons in the electrode. Therefore, coherent tunneling is basically temperature-independent. While for incoherent hopping, charge transport takes place with the thermo-activation of the electron, so conductance increase with temperature. Therefore, the temperature-dependent behavior is the first criteria for distinguishing between tunneling and hopping mechanism.

In a simplified tunneling model, charge transport through molecule is regarded as quantum tunneling through a rectangular energy barrier with a linear drop in the potential when bias applied. Ignoring the temperature effect, the I–V relation is shown by Simmon as [2]

$$I = \frac{e}{4\pi^2\hbar d^2}\left(\Phi - \frac{eV}{2}\right)\left\{\exp\left[-\frac{2(2m)^{1/2}}{\hbar}\alpha\left(\Phi - \frac{eV}{2}\right)^{1/2}d\right] - \left(\Phi + \frac{eV}{2}\right)\exp\left[-\frac{2(2m)^{1/2}}{\hbar}\alpha\left(\Phi + \frac{eV}{2}\right)^{1/2}d\right]\right\} \quad (4.1)$$

where, Φ, d, m, and V are the barrier height, width, electron mass, and applied voltage, respectively; α is a unitless parameter that accounts for adjusting the effective mass of the tunneling electrons and modifying the rectangular barrier model. $\alpha = 1$ corresponds to tunneling through the rectangular barrier with bare electron (Figure 4.1a). Fitting the experimental I–V result to the equation, one can obtain key physical parameters for the charge transport, most importantly the barrier height and tunneling length.

Simplification of the I–V relation by Simmon's model can be done by dividing it into high-bias and low-bias regime, where in high-bias regime, the I–V relation reads:

$$I = \left(\frac{e}{4\pi^2\hbar d^2}\right)\left(\Phi - \frac{eV}{2}\right)\exp\left[-\frac{2(2m)^{1/2}}{\hbar}\alpha\left(\Phi - \frac{eV}{2}\right)^{1/2}d\right] \quad (4.2)$$

and in low-bias regime:

$$I = \left(\frac{(2m\Phi)^{1/2}e^2\alpha}{\hbar d^2}\right)V\exp\left[-\frac{2(2m)^{1/2}}{\hbar}\alpha(\Phi)^{1/2}d\right] \quad (4.3)$$

It can be found out that as the barrier thickness increased, the tunneling current will experience an exponential decrease. This exponential length dependence

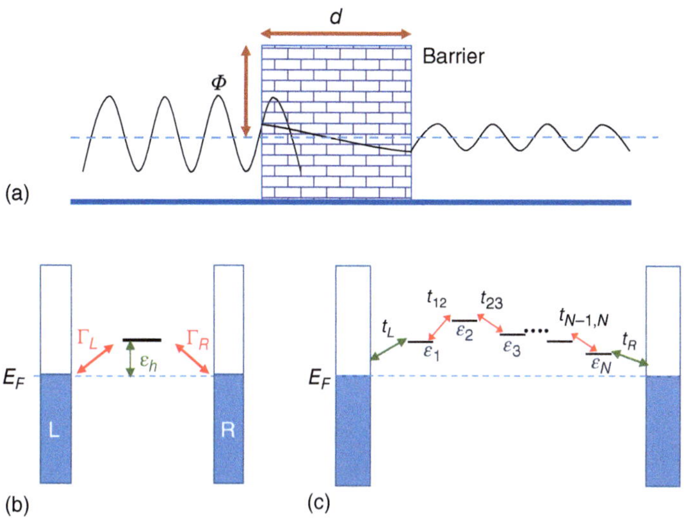

Figure 4.1 The schematic of the tunneling process. (a) The case when a wave tunnels through a energy barrier of width d and height Φ. (b) Single-level model with energy offset ε_h and coupling strength $\Gamma_{L/R}$. (c) Multi-level model in the case of molecules with multi-electronic states.

is commonly used in experimental analysis and takes a simplified form as:

$$I = I_0 e^{-\beta d} \tag{4.4}$$

where $\beta = 2(2m\Phi/\hbar^2)^{1/2}$ is defined as the tunneling decay coefficient and I_0 is the hypothetical contact current across the molecule–electrode interface.

The exponential decay of the conductance with molecule length has been used as another criteria for tunneling transport. The decay constant is related to the tunneling barrier; the higher the barrier, the bigger the decay constant. In a typical molecular junction, tunneling barrier can be regarded as the energy offset between the molecule highest occupied molecular orbital (HOMO) or lowest unoccupied molecular orbital (LUMO) level relative to the Fermi level of the electrode. So conjugate molecules usually have a higher conductance and lower decay coefficient than alkane molecules due to smaller energy offset. Experimental investigation-based different junction structures have found similar tunneling decay coefficient, which is, for alkyl chains, 0.7–0.9 and 0.2–0.5 Å^{-1} for π conjugated molecules [3].

Reed and coworkers first showed the use of Simmon's model in analyzing tunneling mechanism through alkane monothiol self-assembled monolayer (SAM) [4] They fabricated nanopore-based molecular junction (as shown in Figure 4.2a and studied alkanethiol molecule of different length, C8, C12, and C16. Temperature-dependent I–V measurement of C12 in the range from 300 to 80 K was conducted, and no significant temperature dependence was observed, indicating that the conduction mechanism is tunneling (Figure 4.2b). Using Simmon's model, they fitted the measured I–V data (C12, circular symbols in Figure 4.2c) using barrier height $\Phi = 1.42 \pm 0.04$ eV and $\alpha = 0.65 \pm 0.01$ as fitting parameters (solid

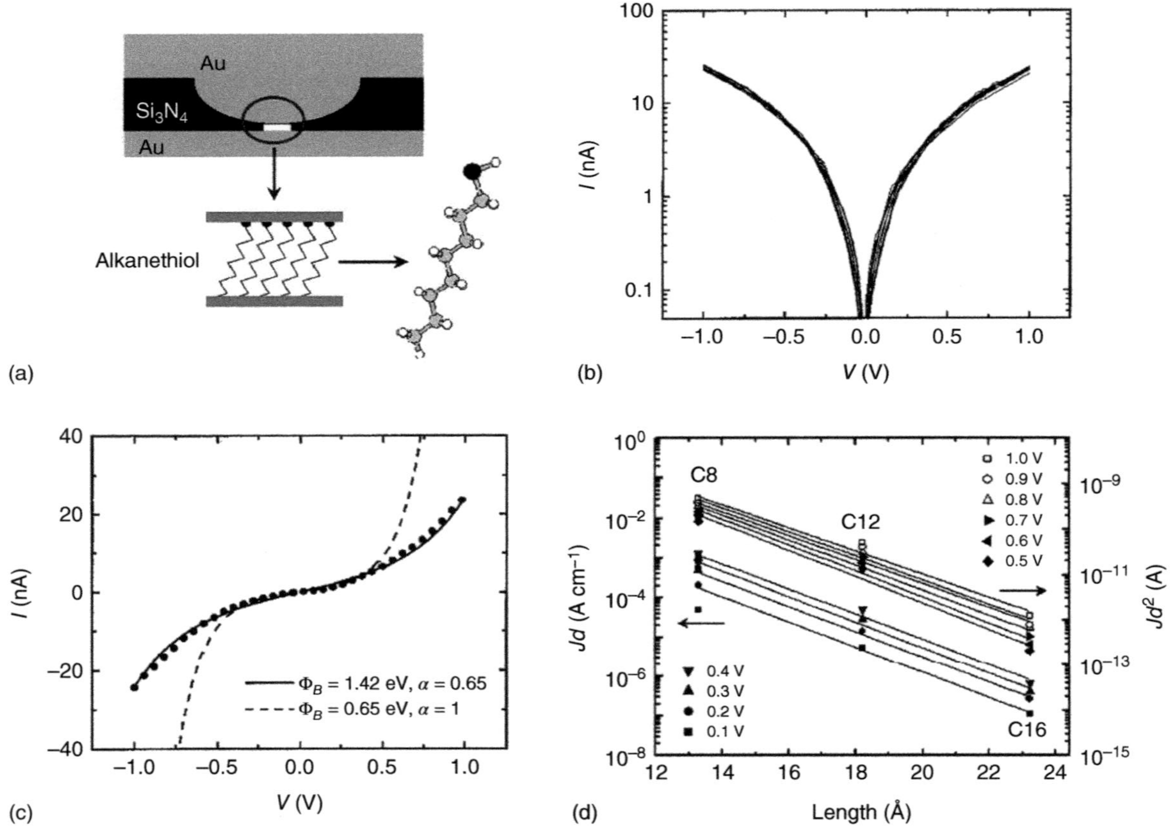

Figure 4.2 (a) Schematics of the nanogap junction device structure. (b) Temperature-dependent log(I)–V curves from 300 to 80 K with 20 K steps of C12. (c) The C12 I–V data that were measured (circular symbols) and calculated (solid curve). Dashed lines show the calculated I–V from a simple rectangular barrier. (d) Linear fittings of log curves that tunneling current densities J times molecular length d at low bias and d^2 at high bias (symbols) *vs.* molecular lengths [4].

line in Figure 4.2c). Similarly, the parameters $\Phi = 1.40 \pm 0.03$ eV and $\alpha = 0.68 \pm 0.01$ could be obtained for C16.

On the other hand, the length-dependent conductance both at low bias, $I \propto (1/d)\exp(-\beta_0 d)$, and high bias, $I \propto (1/d^2)\exp(-\beta_V d)$, were also demonstrated (shown in Figure 4.2d). The tunneling barrier, $\Phi = 1.32 \pm 0.18$ eV and $\alpha = 0.63 \pm 0.03$ for C12, obtained by the decay constant in the length-dependent measurement is in good agreement with the values from I–V curve fitting, indicating a good self-consistency and success of the Simon's model.

Further research in experiment and theoretical analysis, however, pointed out limitation of the Simmon's model due to over-simplified physical picture using rectangular potential barrier and the inherent mathematical difficulty in fitting of the I–V equation [5, 6]. Alternative transport formula has been derived from advanced quantum transport using Landauer's formula and Green's function technique. The simplest case is to consider one electronic state in the metal junction interacting with electrodes [7]. The I–V relation in this single-level model (ignoring again the temperature effect) becomes

$$I(V) = \frac{2e}{h}\frac{4\Gamma_L\Gamma_R}{\Gamma}\left[\arctan\left(\frac{\varepsilon_h + eV/2}{\Gamma}\right) + \arctan\left(\frac{\varepsilon_h - eV/2}{\Gamma}\right)\right] \tag{4.5}$$

where $\Gamma_{L/R}$ is the scattering rates, referring to the electron injection probability from the left (L) or right (R) electrode into the molecule, and Γ is the sum of the them; ε_h is the energy offset between the electronic state and Fermi level of the electrode (Figure 4.1b).

At low bias, this equation can be further simplified as:

$$I = GV\frac{{\varepsilon_h}^2}{{\varepsilon_h}^2 - (eV/2)^2} \tag{4.6}$$

where G is the conductance at zero bias:

$$G = G_0\frac{\Gamma_L\Gamma_R}{{\varepsilon_h}^2} \tag{4.7}$$

with $G_0 = 2e^2/h = 77.48$ μS the conductance quantum.

The single-level model has achieved great success in explaining many experimental systems of molecular junctions. Erbe and coworkers has reported that mechanically controlled break junction (MCBJ) of single toluene molecule modified with different anchoring groups (thiol, nitro, and cyano) [8]. I–V curves can be fitted exactly with the single-level model. Figure 4.3a–c show the symmetric parts of experimental I–V characteristics and fitted curves for three molecules of different anchoring groups. As comparison, these curves could not be so precisely fitted by models like Simmon's model. As seen in Figure 4.3, the anchoring groups that bound to the molecule play an important role in the scattering rate (in the case of $\Gamma_L = \Gamma_R$).

Another example of successful application of single-level model has been provided by Frisbie's group in fitting I–V curves measured from conducting probe atomic force microscopy (CP-AFM) of junctions based on oligophenylene dithiol connected to several types of metal (Au, Ag, and Pt) electrodes [9]. Tunneling parameters including energy offset ε_h and molecule–electrode coupling strength Γ can be extracted

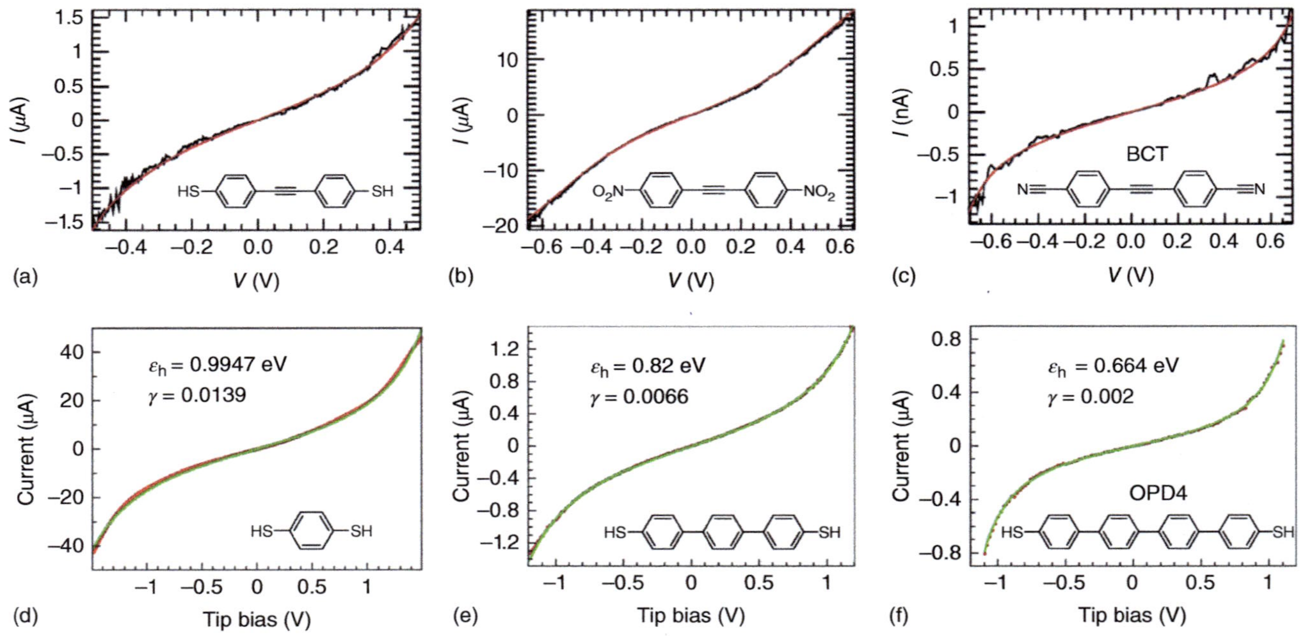

Figure 4.3 The experimentally measured *I*–*V* curves (black line) and fitted by the single-level model (red line) of molecules linked via thiol (BTT) (a), nitro (BNT) (b), and cyano (BCT) (c) anchoring groups and OPD1 (d), OPD3 (e), OPD4 (f) connected to Ag electrodes, respectively. Source: Zotti *et al.* [8]. 2010, John Wiley & Sons.

from well-fitted I–V curves as shown in Figure 4.3, and ε_h agreed with the energy offset between HOMO level of the molecule and Fermi level of the electrode measured by ultraviolet photoelectron spectroscopy (UPS) measurement.

For a long molecule of multi-electronic state, the one level model can be expanded, and the conductance at low bias in this case become

$$G = \frac{e^2}{h} \frac{\Gamma_{L1}(E)\Gamma_{NR}(E)}{\Gamma^2} \left(\frac{\Gamma}{\varepsilon_h}\right)^{2N} \tag{4.8}$$

where $\Gamma_{L1}(E)$ and $\Gamma_{NR}(E)$ are the scattering rates referring to the electron injection probability from the left (L) or right (R) electrode into the molecule first and Nth energy level. N is the number of energy levels in the junction (Figure 4.1c).

It can be easily found that the conductance is again exponentially dependent on length (N), in accordance with the Simmon's model. The tunneling decay coefficient, β, in this model, however, did not depend only on energy gap but also related to coupling between energy levels of the molecule.

It is worth noting that the single-level model generally works well for long molecule. The assumption here is that the multi states of the molecule in the junction, by their coupling with each other, can reorganized into one close packed narrow energy band. So, the single-level model here reflects the collective behavior of the energy band. The tunneling decay is then accounted for by the decrease of the coupling to the electrodes, accompanied by smaller energy barrier. Unfortunately, a uniform physical model to coordinate the single- and multi-level picture is still not available, though Whitesides and coworkers have done some preliminary exploration [10].

The single-level model and Simon's model are popular analyzing methods due to their simplicity and indeed work well in many cases. On the other hand, however, the simplified physical picture set their limitation for molecules with complex electronic structures. People should realize that the two models generally can only apply to molecules with uniform chemical structure (so that uniform electronic structure) or those with only one electronic state close to the Fermi level of the electrodes. Improper use of the model will for sure be misleading. Therefore, it is always important to check electronic structure of the molecule, sometimes it should also include the electrode, by using for example, ab initio calculation.

4.1.2 Transition Voltage Spectroscopy

TVS represents another popular way of analyzing the I–V relation. Beebe *et al.* found that when replotting the I–V data into $\ln(I/V^2)$ against $1/V$, an inflection behavior was fond. This was regarded as the transition from direct tunneling to Fowler–Nordheim tunneling, when the energy barrier is changed from trapezoidal to triangular shape at high bias. Based on Simon's equation, the I–V relation at low bias is (according to Eq. (4.3))

$$I \propto V \exp\left[-\frac{2d\sqrt{m\Phi}}{\hbar}\right] \tag{4.9}$$

and at high-bias Fowler–Nordheim tunneling:

$$I \propto V^2 \exp\left[-\frac{4d\sqrt{2m\Phi^3}}{3\hbar eV}\right] \tag{4.10}$$

It can be seen that $\ln(I/V^2)$ is proportional to $\ln(1/V)$ in low bias, while linear related with $1/V$ in the high-voltage *F–N* regime. Therefore, the cross point of the two *I–V* relations in *F–N* plot is regarded as the transition voltage from direct tunneling to *F–N* tunneling, *i.e.* the energy barrier height.

Experimental studies by Beebe *et al.* have found seemingly evidence that TVS can indicate the energy barrier of tunneling junction [11]. They measured a series of conjugated monothiol molecules by either CP-AFM or crossed-wired junction. The V_{trans} obtained by *F–N* plotting of the *I–V* measurement showed a linear correlation to energy offset between electrode Fermi energy and molecule HOMO from UPS measurement, as shown in Figure 4.4. Thereafter, TVS became a popular technique in exploring the transport barrier [14].

However, careful theoretical studies demonstrate more complex situation and call for caution in use of TVS. Araidai and coworkers investigated the inflection in the *F–N* plot using two types of quantum transport theories. They found that the inflection does not necessarily mean the transition between the direct and *F–N* tunneling. Rather, it takes place when the molecular level responsible for electric currents approaches to the edge of the bias window. Vilan analyzed the *F–N* plot mathematically by Taylor expansion and found that the inflection is actually a mathematical sign of nonlinearity of the *I–V* plot, and V_{trans} actually does not have true physical significance. On the other hand, she found the V_{trans} can, albeit in a model-dependent manner, be related to energy barriers. This is because the nonlinearity of the *I–V* relation is indeed due to the increased transmission probability, and so the conductance, at higher bias when the energy offset (barrier) between the electrode Fermi level and molecular electronic state become smaller. Baldea derived an approximate relation between V_{trans} and energy offset based on the above-mentioned single-level model [12], which is

$$eV_{trans} = 2\varepsilon_h/\sqrt{3} \tag{4.11}$$

Xie and coworkers found experimentally the energy offset obtained in this way match pretty well with the UPS measurement, indicating the TVS can still work as an effective method in extracting meaningful physical information of the junction. Of course, as mentioned above, both Simon's model and single-level model face the risk of over-simplified physical picture relative to complex molecule–electrode electronic states, and the exact relation between energy offset and V_{trans} should not be robust for all the molecular junctions. Therefore, the energy offset value obtained from TVS should always be used with great caution.

On the other hand, however, considering the close correlation between energy barrier, nonlinearity of the *I–V* and V_{trans}, TVS still can work well in indicating the relative change of the barrier in a controlled experiment. For example, Jia *et al.* used TVS to show the energy offset difference for a diarylethene junction in its closed and open form [13]. In another gated nanogap junction, V_{trans} was found by Song *et al.*

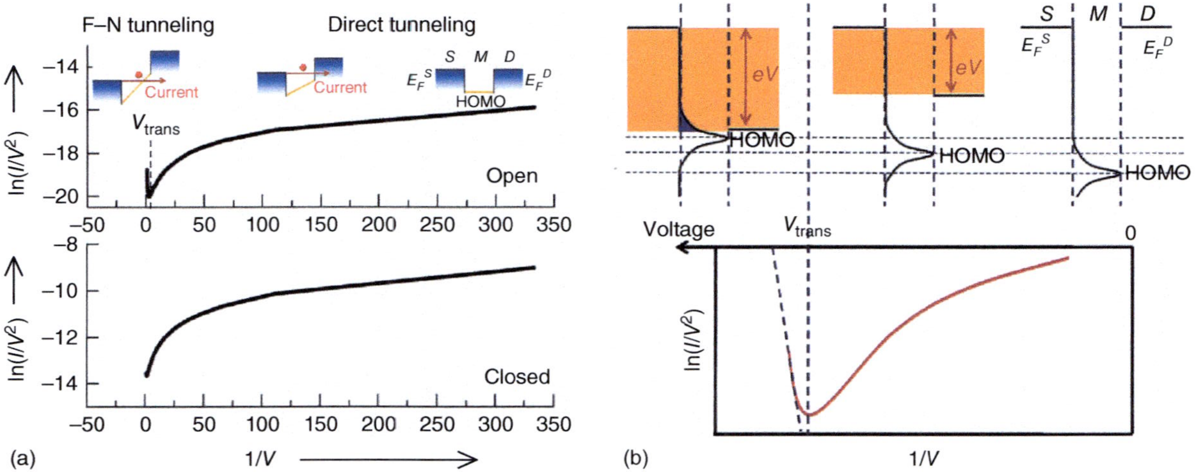

Figure 4.4 (a) F–N plots of the I–V curves for graphene–diarylethene molecular junctions with open (top) and closed (bottom) states. Inset: barrier model to qualitatively explain the inflection of the F–N plots. (b) The resonant tunneling model to explain the inflection of the F–N curves [13].

to change the linearly with gating voltage, suggesting a barrier height modulation by gating the energy level of the molecule and electrode. We believe these could be the proper way of using TVS [49].

Another development of the I–V analysis beyond TVS method is the so-called "law of corresponding states" (LCS), a universal law that describes the I–V characteristics of molecular junctions by Bâldea *et al.* It is based on the fact that (i) V_{trans} in the F–N plot is nothing but a scaling factor and can be related to ε_h by a constant in a single-level model, and (ii) single level actually work pretty well for most of the tunneling transport [15]. In the single-level model, the voltage scale $eV_{trans} = 2|\varepsilon_h|/\sqrt{3}$ and a current scale $I_{trans} = G_1|\varepsilon_h|\sqrt{3}/e$ are introduced into this simplified current voltage equation. V_{trans} stands for the threshold voltage that separates the I–V curves into two regimes, a linear regime at low bias $V < V_{trans}$ and a nonlinear regime at high bias $V > V_{trans}$. Accordingly, I_{trans} represents the current at bias V_{trans}. Inserting these tow scales, one can get the universal law in terms of the reduced quantities $I_R = I/I_{trans}$ and $V_R = V/V_{trans}$ as follows:

$$I_R = \frac{2V_R}{3 - {V_R}^2} \tag{4.12}$$

This simplified equation describes the universal LCS of the electron tunneling through the molecular junctions.

Besides the concise form of this equation, the successfully application of the universal law in fitting I–V data obtained from experiments to theoretical simulations has been proved by several works. As an example, Daniel Frisbie reported their measured I–V data from conductive AFM experiments in different molecules based on molecular junctions can be well fitted to numerical generated curves using the universal law. To show the broad use of this universal law, a reproduced figure summarized the accurate fitting of reduced current–voltage characteristics by several works of scanning tunneling microscopy (STM) junctions, see Figure 4.5. As the comparison between experimental data and theoretical generated curves shows great similarity, the importance of this universal law is obvious. In their recent paper [20], they studied the effect of heteroatom substitution on transport in alkanedithiol-based junction and universal behavior was also found in these junctions.

4.1.3 Incoherent Transport

The incoherent hopping represents another major charge transport behavior. In contrast to the one-step coherent tunneling, hopping is accompanied by the relaxation of the charge, and a complete loss of the phase coherency, in the transport process by molecule vibration and environmental dielectric response.

The relaxation of the charge in the transport depends on the relative timescale of the electron transversal time τ and nuclear dynamics/environment dielectric response time [21]. The traversal time τ in tunneling can be obtained based on energy–time uncertainty principle of quantum mechanics as $\hbar/\varepsilon_h$. On the other hand, the time for an electron to enter into and escape from molecule is related to

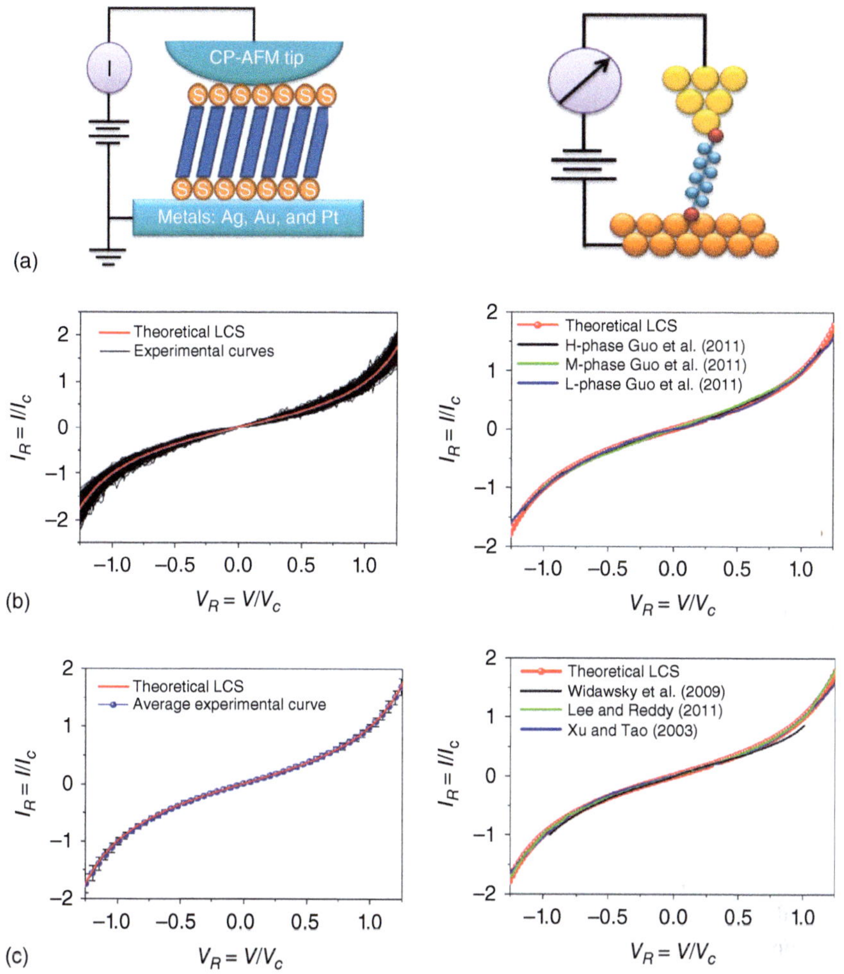

Figure 4.5 (a) Schematic image of CP-AFM setup. (b and c) Reduced $I_R - V_R$ curves of theoretical (red) and ~570 experimental (black) results measured for 24 different types of molecular junctions of OPD*n* and C*n*DT molecules linked to metal electrodes (Ag, Au, and Pt). (d) Schematic image of STM set-up. (e) Averages of thousands of $I_R - V_R$ curves for high (H) (black line), medium (M) (green line), and low (L) (blue line) conductance of octanedithiol STM junctions and theoretical fitted curve (red line). (f) $I_R - V_R$ curves for 4,4′-diaminostilbene (black line), C8DT (green line), and 4,4′-bipyridine (blue line) linked to gold electrodes and theoretical fitted curve (red line). Reproduced experimental data can be found in the Ref. [15–19]. Source: Modified from Bâldea *et al.* [15].

the coupling energy Γ by $\hbar/\Gamma$. A unified expression for the total traversal time of the electron is

$$\tau = \hbar/\sqrt{\varepsilon_h{}^2 + \Gamma^2} \tag{4.13}$$

When transversal time is short, for example, in most of the alkane bridging molecule junction where energy barrier is big, the transversal time for a 10 Å long and 1 eV high barrier is ~0.2 fs, the vibration and environment effect can be neglected, and the transport is generally coherent. In the cases of small energy gap (such as conjugate structure), weak coupling between molecule and electrode, or when molecule is long, electron will spend more time on the molecule and interact with the vibration of the molecule and environment, well enough for it to lose its energy to environment and get trapped (a relaxation process). Now the transport has to proceed by reactivating the electron to overcome the trap, which means it becomes hopping transport. So it is not surprising to observe hopping transport in junction with long conjugate molecule.

In the hopping transport, as mentioned above, charge is trapped temporarily by molecule deformation and environmental polarization, and its de-trapping and transport have to go through a thermal re-activation. As a result, the conductance in hoping is temperature-dependent and follows an Arrhenius relation given by:

$$G \propto \exp\left(\frac{-E_a}{k_B T}\right) \tag{4.14}$$

where k_B is the Boltzmann constant, T is temperature, and E_a is the hopping activation energy. As we have shown above, the coherent tunneling is generally temperature-independent, so temperature dependency is a good indicator of the two transport mechanisms.

The first temperature-dependent transport in single molecular junction was reported by Selzer *et al.* using 1-nitro-2,5-di(phenylethynyl-4′-mercapto) benzene bridging the nanogap fabricated by electromigration (Figure 4.6) [22]. The current decrease exponentially with $1/T$ at high temperature and the activation energy at low bias was found to agree with the energy barrier of internal conformation change (relative rotation of the benzene ring), which works as the hopping barrier. Meanwhile, they observed a transition of the conductance from temperature dependence to independence at low temperature, and the transition temperature becomes lower as the bias increases. They attributed this to the fact that with increasing bias, the vibrational temperature of the molecule(s) increases, and the transition to incoherent tunneling is promoted at a lower bath temperature.

Another criterium is the length dependence of the conductance. Different from the exponential decay with molecule length in coherent tunneling, conductance decreases linearly with length in hopping transport. An intuitive understanding of this phenomena is to regard the overall hopping process as a chain of serially connected resistors following the basic circuit law, where tunneling between every two hopping sites works as a resistor and the hopping sites as connector.

Moreover, a transition from tunneling to hopping transport, indicated by the temperature and length dependency change of the conductance, can often be observed when the length of the molecule increases gradually.

For example, Frisbie and coworkers grew conjugated oligophenyleneimine (OPI) molecule wires of different lengths (ranging from 1.5 to 7.3 nm) stepwise on Au substrate [23]. The I–V characterization of different length wires were performed using either CP-AFM or Au–molecule–EGaIn (eutectic indium gallium alloy). The semi-log plot of resistance *vs.* molecular length measured using CP-AFM is shown

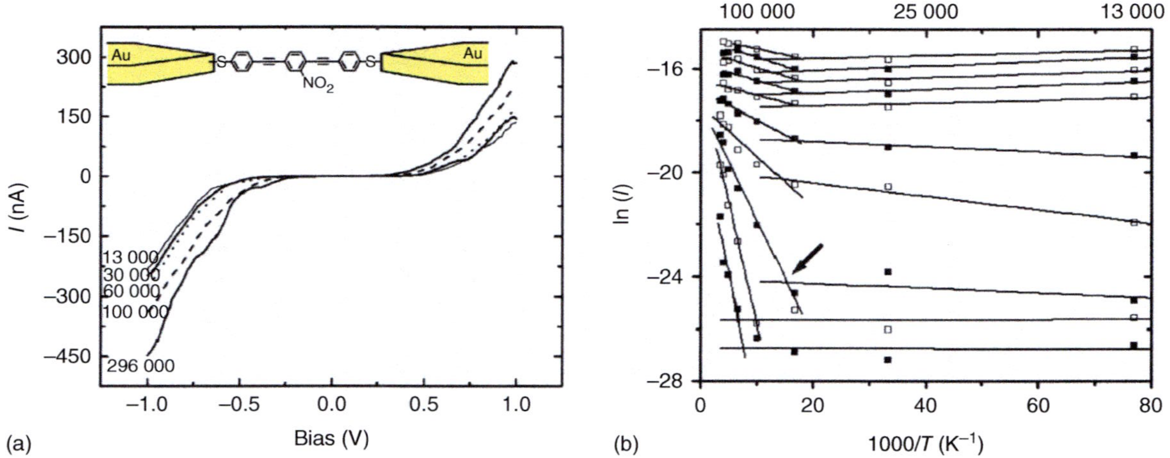

Figure 4.6 (a) Temperature-dependent I–V curves of the 1-nitro-2,5-di(phenylethynyl-4′-mercapto)benzene junction. Inset is the schematic of the junction. (b) Arrhenius plots of ln(I) *vs.* 1/T at different bias voltages showing a transition in conductance from temperature independent at low temperature range and temperature dependent at high temperature. The lowest bias is 0.1 V and the increment between curves is 0.1 V. The arrow shows the turning point from temperature-independent to thermal-activated process. Source: Selzer *et al.* [22]. 2004, American Chemical Society.

in Figure 4.7a, and an obvious transition can be observed. The conductance of OPI of <4 nm decreased exponentially with length, and the decay constant β was found to be 3.0 ± 0.08 and $2.84 \pm 0.13\,nm^{-1}$ for CP-AFM and EGaIn configuration, respectively. For wires >4 nm, the plot of resistance *vs.* molecular length showed a linear relationship and the much smaller β (0.9 ± 0.07 and $1.01 \pm 0.08\,nm^{-1}$) than those for short molecules from the semi-log plot. In addition, the long wires present a strong temperature-dependent resistant in contrast to short wires, which tells that the transport mechanism changes from coherent nonresonant tunneling for short wires to thermally activated hopping for long wires (shown in Figure 4.7b). Similar phenomenon also was observed in other molecules, and Figure 4.7c–f showed the results of conjugated oligo-imine (ONI) wires measured by CP-AFM [24].

The charge transport dynamics and I–V character in hopping transport is generally obtained by solving the kinetic rate equations. The full relaxation of the charge on molecule actually implies that molecule in the transport process is reduced (when it is electron transport) or oxidized (when it is hole transport), and the trapping and de-trapping process can be regarded as a continuous redox cycle of the molecule in the overall transport process, as shown in Figure 4.8.

So kinetics that is used for a consecutive chemical reaction can be borrowed to account for the hopping charge transport. Using the steady-state approximation, *i.e.* assume the ratio of reduced/oxidized states in the junction does not vary as time in the steady current flow, the net current becomes

$$I = e\frac{k^L_{o/r}k^R_{r/o} - k^L_{r/o}k^R_{o/r}}{k_{r/o} + k_{o/r}} \tag{4.15}$$

where $k^L_{o/r}$, $k^L_{r/o}$, is the transfer rate of electron entering (reducing) or leaving (oxidizing) the molecule from/to left electrode, $k^R_{r/o}$, $k^R_{o/r}$ is the rate of electron leaving or entering molecule from/to the right electrode, and $k_{o/r} = k^L_{o/r} + k^R_{o/r}$ is the net reduction rate (oxidized state to reduced state transition rate), while $k_{r/o} = k^L_{r/o} + k^R_{r/o}$ is the net oxidation rate.

Here, the elementary reaction is the charge transfer from electrode to molecule and from molecule to the final electrode. The rate k_{et} is described by the well-known Marcus theory:

$$k_{et} = \int d\varepsilon\, \Gamma(\varepsilon) f(\varepsilon) \frac{1}{\sqrt{4\pi\lambda k_B T}} \exp\left[-\frac{(\varepsilon - \varepsilon_A - \lambda)^2}{4\lambda k_B T}\right] \tag{4.16}$$

where $\Gamma(\varepsilon)$ is the coupling between molecule and electrode, ε_A is the energy level of the redox molecule, $f(\varepsilon)$ is the Fermi distribution of the electron in electrode, and λ is the reorganization energy.

The Fermi distribution depends on the electrochemical potential μ and applied electro-potential φ by:

$$f(\varepsilon, \varphi) = 1/\left(1 + \exp\left[\frac{\varepsilon - \mu + e\varphi}{k_B T}\right]\right) \tag{4.17}$$

The transport equation for hopping can be then obtained by combining Eqs. (4.15–4.17), and the final formula of the transport is much more complex than

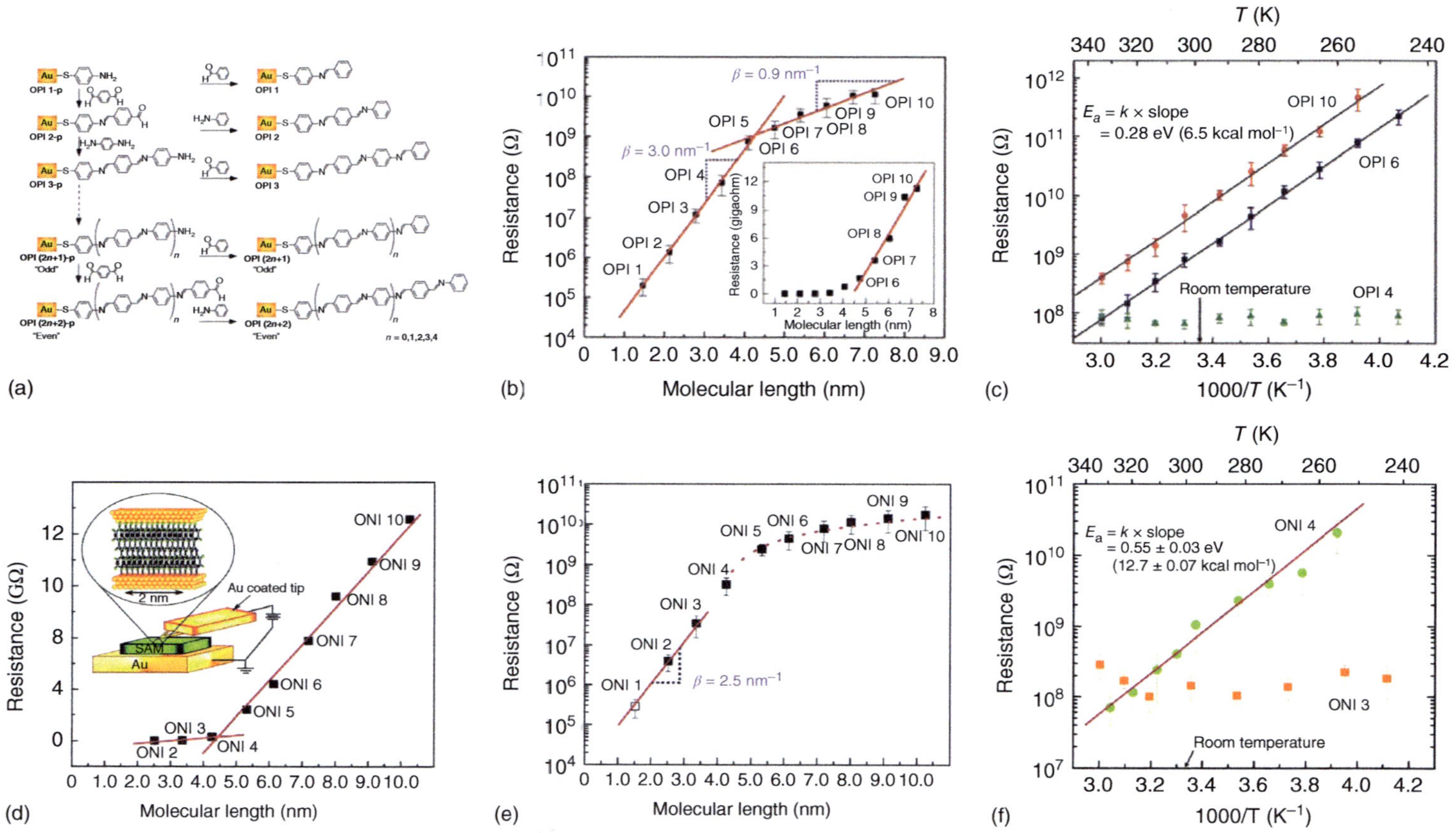

Figure 4.7 (a) The synthetic route of various length OPI wires [23] (b) Semi-log plot of *R vs. L* for the OPI junctions measured by CP-AFM. (Inset) A linear plot of *R vs. L* [23] (c) Arrhenius plot for OPI 4 (short wire), OPI 6, and OPI 10 (long wires). Straight lines are linear fits to the data. Source: Ho Choi *et al.* [23] (d) The plot of *R vs. L* for the ONI junctions measured by CP-AFM. Straight lines are linear fits to the data. Inset: The schematic diagram of CP-AFM with an ONI monolayer [24] (e) The semi-log plot of *R vs. L* for the ONI junctions measured by CP-AFM [24] (f) Arrhenius plot for ONI 3 (short wire) and ONI 4 (long wire). Straight lines are linear fits to the data. Source: Choi *et al.* [24]. 2010, American Chemical Society.

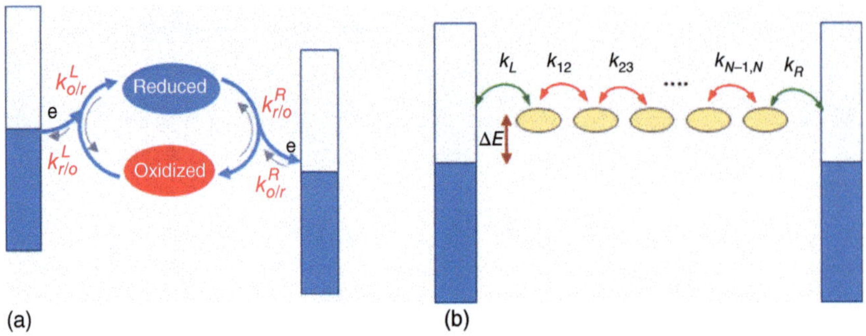

Figure 4.8 Model of charge and electron transport process through a molecule with redox center (a) and schematic of multi-step hopping process (b).

tunneling. We refer the reader to works by Ulstrap and coworkers [25] or Nitzan and coworker [26] for the detailed discussion of it.

In the situation of multiple step hopping, and consider the situation that transfer rates k between redox states are the same, the current becomes [27]

$$I = e\frac{e^{-\Delta E/k_BT}}{[1/k_L + 1/k_R + (N-1)/k]}\left[e^{\frac{eV}{k_BT}}f_L - f_R\right] \tag{4.18}$$

where $k_{L/R}$ is the transfer rate of electron entering/leaving the molecule, k is the transfer rate of electron tunneling between each energy levels of the molecule, and $\Delta E = \varepsilon_A - \mu$ is the energy barrier. f_L and f_R are the Fermi distribution of the electron in the left and right electrode, respectively.

The corresponding linear conductance can be expressed as:

$$G = \frac{e^2}{k_BT}\frac{e^{-\Delta E/k_BT}}{[1/k_L + 1/k_R + (N-1)/k]} \tag{4.19}$$

Equation (4.19) is the most useful formula for hopping transport. It tells that conductance in hopping transport decrease linearly as the number of the redox sites, *i.e.* the length of the molecule, and also have an Arrhenius-type temperature dependence.

The full current-bias formula in a hopping transport is rarely used to fit the I–V experimental results as that was done is coherent tunneling. The complex relation between current and bias and the mathematical difficulty might be the main reason. Instead, the dependence of the low-bias conductance on temperature, and gating voltage, which can change the energy barrier between chemical potential of the electrode and redox potential of the redox molecule, are more studied.

Li *et al.* systematically studied transport properties of a series of redox-active PTCDI molecules as a function of electrochemical gate voltage and temperature in different solvents, using STM break junction [28]. They found that a maximum transport current occurs near the redox potential, and the conductance is temperature-dependent with activation energy being gating voltage-dependent. The results can be well interpreted by the model of two-step sequential charge transfer process (hopping process) developed by Kuznetsov and Ulstrup, also mentioned above by Eq. (4.16), which predict a temperature-dependent conductance and

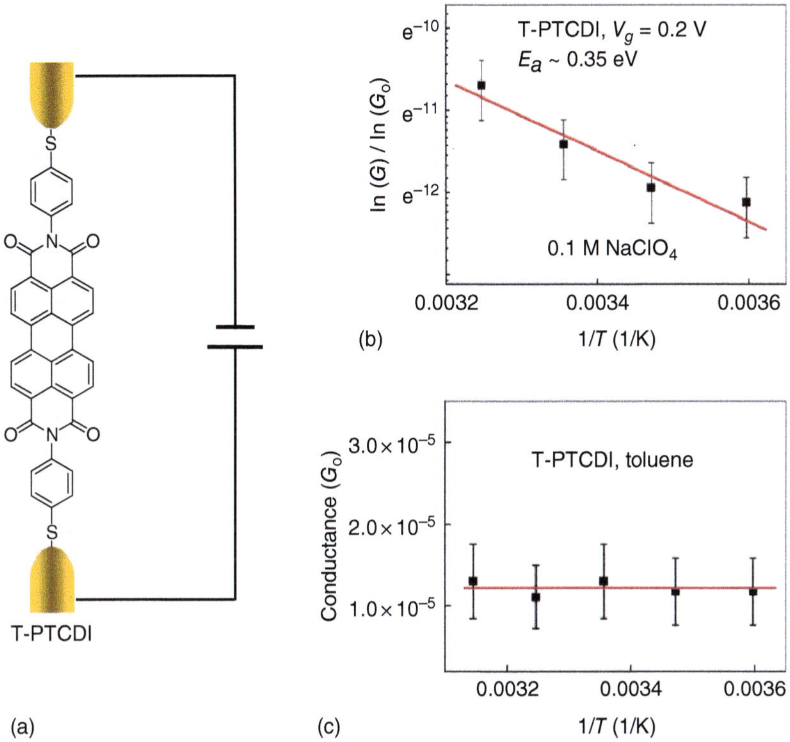

Figure 4.9 (a) Schematic setup of *I–V* measurement of T-PTCDI molecule. (b) Conductance *vs.* temperature for T-PTCDI at gate voltage of 0.2 V. (c) Conductance *vs.* temperature of T-PTCDI measured in toluene. The bias was 0.1 V. Source: Li *et al.* [28]. 2007, American Chemical Society.

gating voltage-dependent activation energy (Figure 4.9). Another interesting result they found is that the temperature dependence of the transport disappears when the solvent was changed from water to toluene. As we have discussed at the beginning of this section, the condition for hopping transport is that electron (or hole) has to relax either by molecule vibration or by solvent polarization when transports through molecule. Therefore, it do make sense that the hopping transport disappeared in the nonpolar toluene solvent.

Before closing this section, it deserves to discuss a bit again TVS for the hopping transport. As we have mentioned, the inflection in *F–N* plot is nothing but a mathematical result of a nonlinear relation. In the coherent tunneling, it can be used to indicate the relative change of the barrier in a control experiment. The hopping *I–V* relation, as indicated above, is more complex than tunneling and quite nonlinear. It is basically exponentially dependent on the voltage. So the V_{trans} is even less related to any specific physical parameter for the transport. So generally speaking, if the transport mechanism is found to be hopping, either by temperature or by length-dependent measurement, TVS may not be applied. Artés in his research on a protein azurin junction with two-step redox-mediated charge transport studied the TVS of the junction and found that the V_{trans} strongly depends on one coupling parameter Γ as well as charge transfer free energy [29].

The current–voltage character of the molecular junction, though can provide plenty of key information about the charge transport as we have discussed above, faced a serious problem, *i.e.* it is insensitive to chemical details of the molecule in the junction. For example, in the coherent tunneling, almost any tunneling model that produces symmetric I–V can be approximated into a cubic function at low bias of the form:

$$I(V) = AV + BV^3 \tag{4.20}$$

where A and B are constant. Indeed, the variety of the chemical structure in coherent transport can easily get lost into the above I–V relation. Therefore, the I–V relation analysis has to be done with the great caution if the I–V curves have no much structure to make sure the physical parameters obtained are sensible.

4.2 Inelastic Tunneling Spectroscopy (IETS)

The featureless I–V relation as we have seen above cannot provide fingerprint information of the molecular structure. The molecule chemical structure in the transport current, however, can manifest themselves by the inelastic tunneling process and the IETS.

The discovery of the inelastic tunneling can be traced back to Jaklevic and Lambe from Ford company. In 1966, they found the fine structures in the I–V of metal–metal oxide–metal tunneling junctions and assigned the peaks in dI^2/d^2V *vs.* V to the organic contamination on the metal oxide surface [30]. They also developed the basic theory of IETS based on perturbation of the tunneling barrier by transporting electron [31]. The method was further applied to various organic compounds, including biochemicals, using solution spin coating on the metal oxide surface and molecular vibrations were found to agree very well with the Raman or IR spectra with an extremely high sensitivity [32]. In 1998, Stipe *et al.* performed IETS at a single-molecule level using low-temperature STM, which is capable of performing vibrational feature imaging of a single molecule lying on the surface [33]. IETS measurements were applied to molecular junctions in 2004 using nanopore and cross-wire junction techniques by Kushmerick *et al.* [34] and Wang *et al.*, respectively [35]. Later on, single-molecule junction IETS was achieved in STM break junction [36] and electromigrated nanogap junctions [37]. Nowadays, IETS has been recognized and utilized as a powerful tool in revealing the presence of molecule in the gap, the structure of the junction, and even the charge transport pathways.

4.2.1 Principle and Measurement of IETS

The principle of IETS is shown in Figure 4.10 [38]. In the tunneling process, an electron in the left-side electrode tunnels into an empty state on the right side. The coherent tunneling is generally dominant in the process, where the energy of the

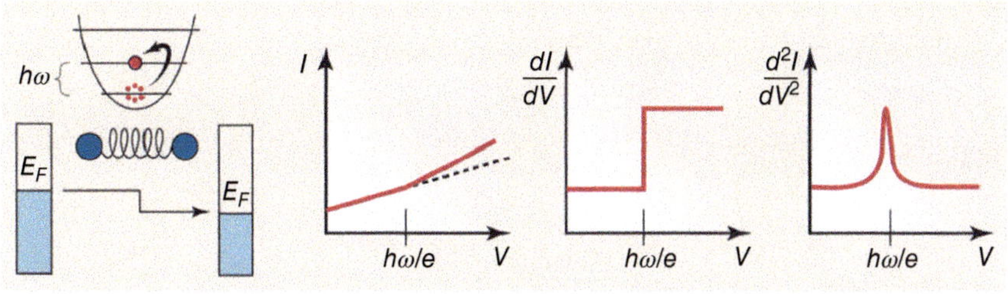

Figure 4.10 Scheme of the inelastic tunneling behavior as the voltage is increased through a vibrational resonance. Source: Galperin *et al.* [38]. 2008, American Association for the Advancement of Science.

electron is conserved, and this is also called an elastic tunneling process. However, in the finite time of the electron crosses the molecule, as shown in Eq. (4.13), there is some chance that the tunneling electron can interact with the vibration of the molecule, so that the energy of the electron and nuclei vibration will exchange. At low temperature, when all the nuclei are at the ground vibrational states, electron can only lose energy quanta $\hbar\omega_a$ to the vibrational motion of frequency ω_a. When the bias is such that $eV \geq \hbar\omega_a$, the inelastic channel for mode a will open and the conductance will take on a step increase, *i.e.* a peak in the dI^2/d^2V plot, at $V = \hbar\omega_a/e$. So if the junction is biased from low to high voltage and the dI^2/d^2V is recorded simultaneously, a series of peaks will be observed, corresponding to the opening of inelastic channels from low- to high-energy serially, and positions of the peaks tell the energy of the vibrational modes that cause the inelastic process. The dI^2/d^2V *vs.* V (or corresponding energy in unit of wavenumber) is the so-called IETS. It is the only electrical spectroscopy that can reflect the in situ chemical fingerprint in the charge transport process.

In principle, IETS signal can be obtained by a numerical differentiation of the measured I–V data. However, because the inelastic effect is very weak, this method is generally not practical. IETS, in practice, is usually measured by an AC modulation method employing a lock-in amplifier [39]. In this strategy, a small AC voltage with a certain frequency ω is superimposed on the main DC bias and the current under such a bias becomes

$$I(V) = I(V_{\rm DC} + V_\omega \cos(\omega t)) \tag{4.21}$$

By using a simple Taylor series expansion about $V_{\rm DC}$:

$$I(V) = I(V_{\rm DC}) + \left.\frac{{\rm d}I}{{\rm d}V}\right|_{V_{\rm DC}} V_\omega \cos(\omega t) + \frac{1}{2}\left.\frac{{\rm d}^2I}{dV^2}\right|_{V_{\rm DC}} [V_\omega \cos(\omega t)]^2 + \dots \tag{4.22}$$

the current can be represented as:

$$I(V) = I(V_{\rm DC}) + \left.\frac{{\rm d}I}{{\rm d}V}\right|_{V_{\rm DC}} V_\omega \cos(\omega t) + \frac{1}{4}\left.\frac{{\rm d}^2I}{dV^2}\right|_{V_{\rm DC}} [V_\omega^2 \cos(2\omega t)] + \frac{1}{4}\left.\frac{{\rm d}^2I}{dV^2}\right|_{V_{\rm DC}} V_\omega^2 + \dots \tag{4.23}$$

Therefore, by measuring the output current of the system and extracting the signal at frequency ω and 2ω, the first and second derivative, *i.e.* (dI/dV) and (d^2I/dV^2) of

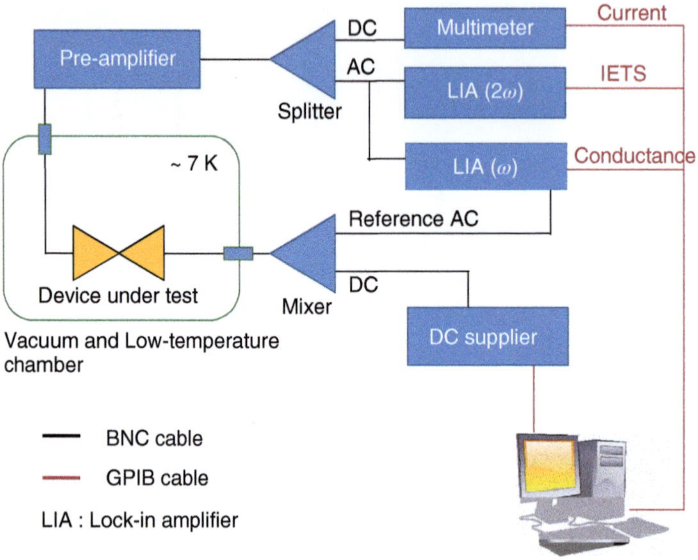

Figure 4.11 Scheme of the IETS measurement system.

the current corresponding to differential conductance and IETS signal, respectively, can be obtained. So far, the AC modulation technique using lock-in amplifier is the most popular method for IETS measurement. A typical IETS measurement system is shown schematically in Figure 4.11.

The line width broadening of the IETS, which determined the resolution of IETS, is a function of modulation voltage and temperature. Low temperature is critical to decrease the Fermi energy broadening of the injected electron and a low AC modulation voltage works like a sharper probe for local conductance variation. The full width at the half-maximum (FWHM) of the dI^2/d^2V peak can be expressed as [39]:

$$V = [(1.7V_m)^2 + (5.4k_BT/e)^2 + W^2]^2 \quad (4.24)$$

where V_m is the AC modulation voltage, k_B is the Boltzmann constant, T is the temperature, and W is the intrinsic line width.

Generally, the lower the temperature and the smaller the AC voltage, the higher the IETS resolution is. In the high-resolution STM-based IETS measurement, temperature as low as 600 mK and modulation voltage lower than 1 mV have ever been used [40]. While for a rough observation of the inelastic tunneling (IET) effect, liquid nitrogen temperature is enough.

The fully assigned IETS spectrum can provide unambiguous experimental evidence of the existence of the desired molecules in the region of the junction [41]. In a typical nanogap junction fabricated by electromigration, Song *et al.* measured IETS of 1,8-octanedithiol (ODT) and 1,4-benzenedithiol (BDT) [37] For the ODT junction, peaks were observed at 92, 119, 143, 161, 181, and 355 mV, which were assigned as ν(C–S) stretching, δ_r (CH_2) rocking, ν(C–C) stretching, γ_w(CH_2) wagging, δ_s(CH_2) scissoring, and ν(C–H) stretching modes, respectively (Figure 4.12). The absence of the ν(S–H) stretching mode at 319 mV (2575 cm^{-1})

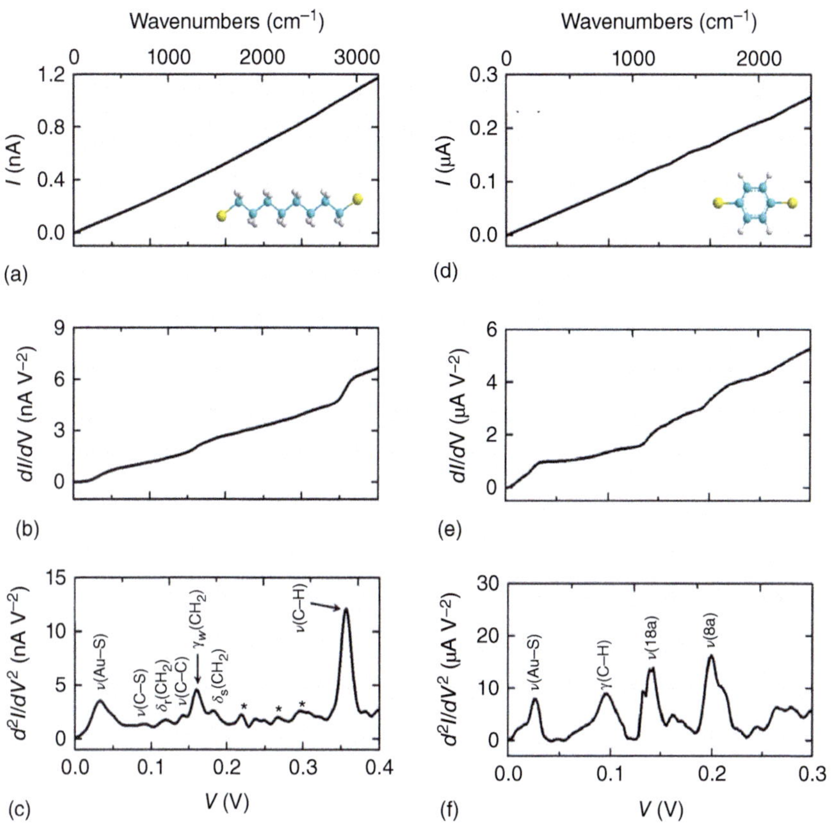

Figure 4.12 Transport properties of Au–ODT–Au (a–c) and Au–BDT–Au (d–f) molecular junctions measured at 4.2 K. (a) and (d) The *I–V* characteristics with inset chemical structure of each molecule. (b) and (e) Differential conductance obtained from lock-in first harmonic signal. (c) and (f) IETS obtained from lock-in second harmonic signal. The peaks are labeled with their assigned vibration modes. Source: Song *et al.* [37]. 2009, American Institute of Physics.

suggests that the thiol (–SH) anchoring group has reacted with the Au electrode during the assembly process. In the IET spectrum of the BDT junction, peaks at 96, 142, and 201 mV correspond to γ(C–H) aryl out-of-plane bending, ν(18a) stretching, and ν(8a) stretching modes, respectively.

4.2.2 Selection Rule and Charge Transport Pathway

Like all the spectroscopy technique, the selection rule in IETS is the key in correlating the spectrum with the junction structure information. IETS is not subject to the selection rules of infrared or Raman spectroscopy, and both IR and Raman-active vibrational modes can appear in the IETS. So IETS is a quite informative on one hand. On the other hand, however, it is difficult to tell at the first sight the vibrational modes that are active in the IETS and chemical groups that contribute to the

IETS signal. Beebe *et al.* studied IETS of hexadecanethiol molecules with varying degrees of fluorination (Figure 4.13A) [42]. They found that C–H vibrational modes did not attenuate with fluorine substitution in the IETS, suggesting that either the scattering cross section is not an additive quantity following the length of current path or the physical position of a vibrational mode within the junction influences its activity in IETS.

However, different conclusion was obtained by Okabayashi *et al.* [43] in the study of the IETS of deuterium isotope-labeled alkanethiols (Figure 4.13B). Using alkanethiols that are deuterated at different parts of the $CD_3(CH_2)_7S$, $CH_3(CH_2)_6CD_2S$, and $CD_3(CD_2)_7S$, they found that the IETS weights of the peaks of C–C stretching, CH_2 wagging, and CH_2 scissoring did not change with the deuteration of the CH_3 closest to the end of the molecule, while the C–H stretching peak decreased partially. Supported by the first principle calculations, they found the terminal groups provided almost the same contribution to the IETS signals of C–H stretching as the middle part of the molecules.

Later on, Lin *et al.* reinvestigated the IETS of fluorinated alkanethiol system by an initio calculation, which shed light on the contradiction [44]. They found, shown in Figure 4.14, that the strong C–H stretching peak, which did not change upon fluorination of the far end of the alkanethiol, originates from the CH_2 groups next to the S–Au interface and is largely dependent on the orientation of the molecule. The titling of the molecule makes the CH_2 group next to the S—Au bonding site to get close to the gold electrode and couple with it, which in turn opened a new transport route bypassing the S—Au bond. The strong C–H peak in the IETS thus is due to the strong perturbation of the C–H stretching to the coupling between C–H group and gold electrode of the new pathway.

This conclusion is in agreement with the study of the IETS by Troisi and Ratner, which shows that only a few scattering states – namely those belonging to the most transmitting eigenchannels – need to be considered for a complete description of the electron transport and the IET process [45]. They proposed that only vibrations that involve atoms lying along favorable tunneling pathways are active in IETS, and so that IETS can be used to trace the electron pathway in the junction. As shown in Figure 4.15, they studied the IETS both experimentally and by ab initio calculation of the junctions with two molecules. For molecule 1, the π band is due to the electron tunneling through the π orbital of the anthracene. The σ–π band below 1100 cm^{-1}, *i.e.* the active normal modes (*i.e.* 15, 17, 22, and 36), contains an out-of-plane displacement of hydrogen atoms, and they facilitate molecule–electrode coupling by vibronically enabling σ–π mixing for the tunneling pathway. At the other end of the molecule since the electron prefers to tunneling directly from π orbital to the electrode, bypassing the s orbital of the H atom, the C–H stretching mode is absent. For molecule 2, extra σ peaks appears due to the tunneling through the alkane chain. Their study implied the importance of both the molecule–electrode interface and internal interfaces of the molecule for IETS. In molecule 1, most of the distinctive peaks come from vibrations localized at the interface between the molecule and the electrode. π orbitals are coupled by the out-of-plane vibrations to the σ orbitals of the molecule that coupled to the gold electrode. For molecule 2, the interface is within

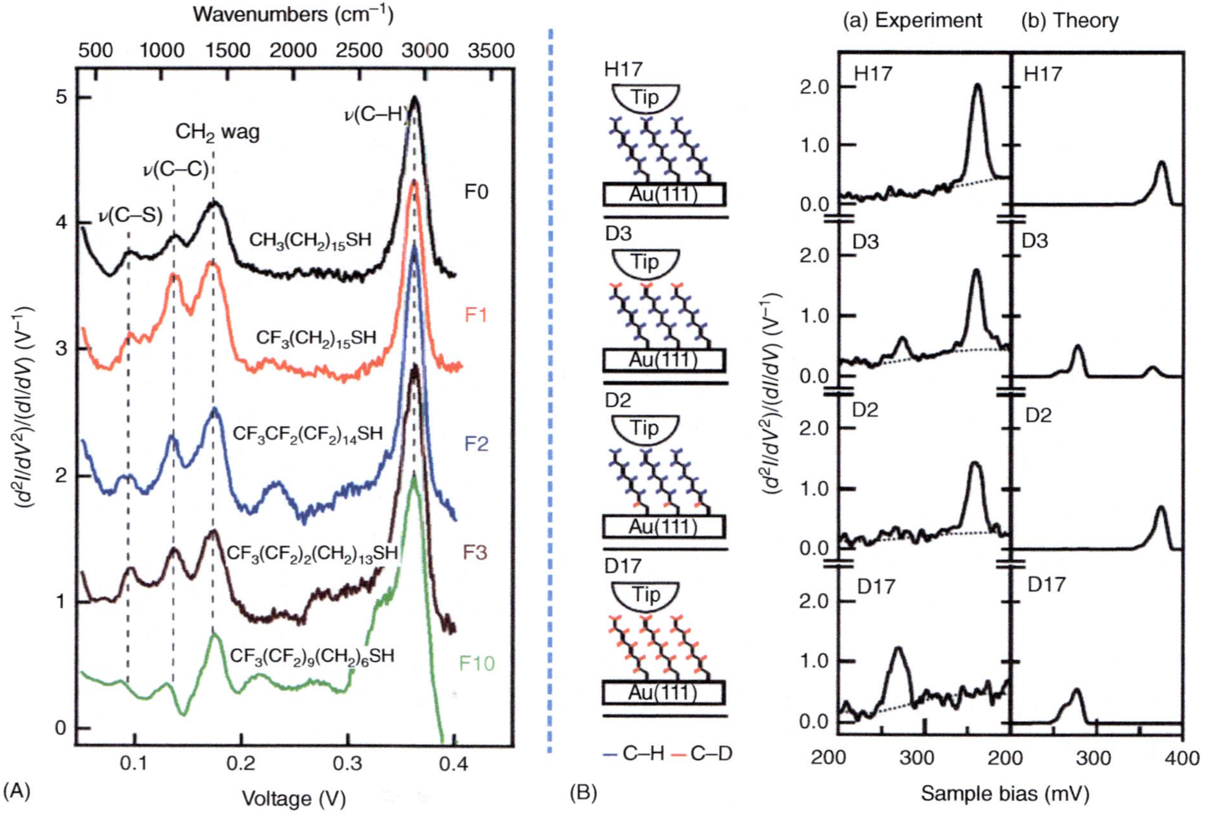

Figure 4.13 (A) IETS of the five fluorinated molecules (F0, F1, F2, F3, and F10). The dashed vertical lines represent the average position of each peak [42]. (B) theoretical and experimental IETS of the deuterated alkanethiol molecule in the ν(C–D) and ν(C–H) region, with left scheme showing the position of the deuteration. Source: Okabayashi *et al.* [43]. 2010, American Chemical Society.

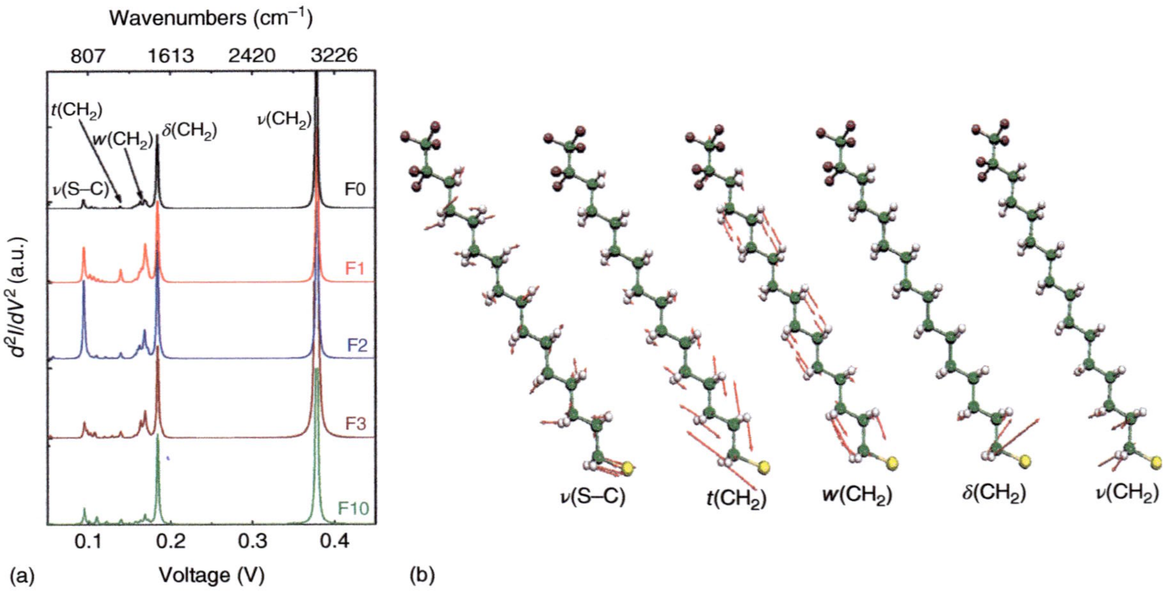

Figure 4.14 (a) Calculated IETS spectra of F0, F1, F2, F3, and F10 molecular junctions and (b) the five active vibration modes in the IETS spectrum. Source: Lin *et al.* [44]. 2011, American Chemical Society.

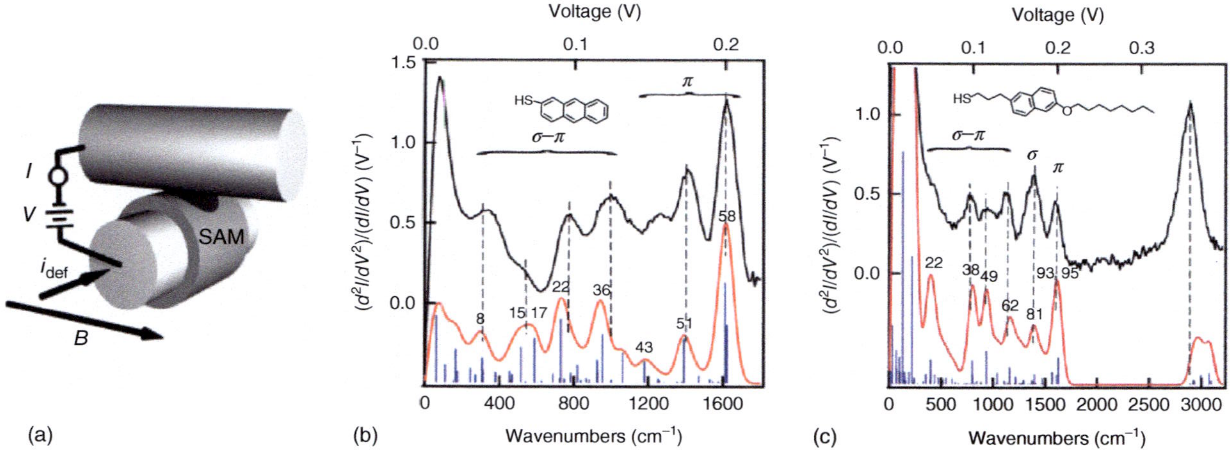

Figure 4.15 (a) Schematic of a crossed-wired tunnel junction. (b) IETS of experimental (black) and computed (red) anthracene junction. (c) IETS of experimental (black) and computed (red) naphthyl ether junction. Source: Troisi *et al.* [45]. 2007, National Academy of Sciences.

the molecule itself where vibrations that favor the σ–π mixing at the contact between the naphthalene and the alkyl tail are the most characteristic signals.

The work by Trosis and Ratner actually can reconcile the contradiction between Kushmeric and Okabayashi. Both the molecule–electrode interface and molecule internal structure can contribute to the inelastic process, and the relative contribution from interface and internal structure has to be analyzed based on transport pathway and the vibrational modes that can disturb the inter orbital coupling along the pathway.

4.2.3 Line Shape of the IETS

The above-mentioned IETS all have positive peak shapes, *i.e.* conductance increases by the electron–vibration interaction in the transport. This actually is the case only at off-resonance situation when the Fermi level of the electrode is far from the electronic states responsible for transport. In this situation, inelastic tunneling results in an enhanced probability of forward scattering. In contrary, in a high-conductance regime close to resonance, the probability of electron backscattering will increase due to a momentum transfer to the excited vibrational mode in the inelastic process, which in turn decreases the net transmission probability. Therefore, the inelastic vibronic effect is present as downside steps in conductance, and so dips in the IETS. The peak-to-dip line shape transition in IETS is shown in the phase diagram of Figure 4.16 as predicted theoretically by Paulsson *et al.* [46]. The peak or dip shape is transmission and coupling symmetry-dependent. In a full symmetry junction, the line shape of IETS follow a 1/2 rule, *i.e.* it is peak below 1/2 transmission probability and dip above 1/2.

A good example of this rule in molecular nanogap junction was shown by Kim *et al.* [47]. They studied the charge transport of Au–1,4-BDT–Au single-molecule

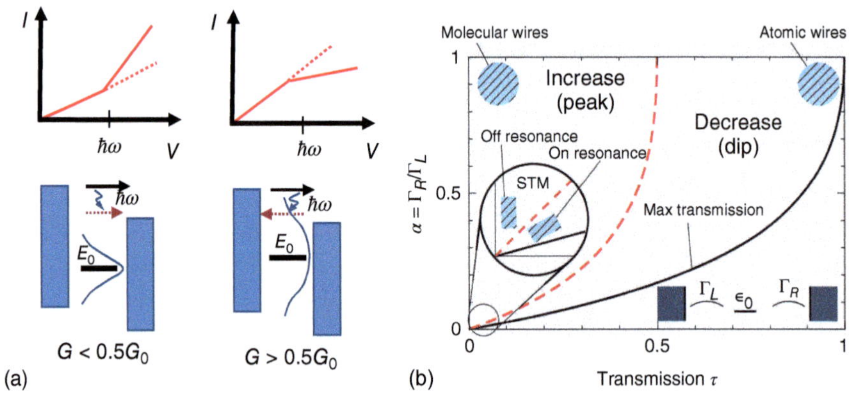

Figure 4.16 (a) Scheme of the scattering by the vibration of the molecule, which is forward (left) when the conductance (G) is lower than $0.5G_0$ and backward (right) when the conductance is higher than $0.5\,G_0$. (b) Phase diagram of a one-level model illustrating the sign of the conductance change due to electron–vibration interaction at a given asymmetry factor α and the elastic transmission. Source: Paulsson *et al.* [46]. 2008, American Physical Society.

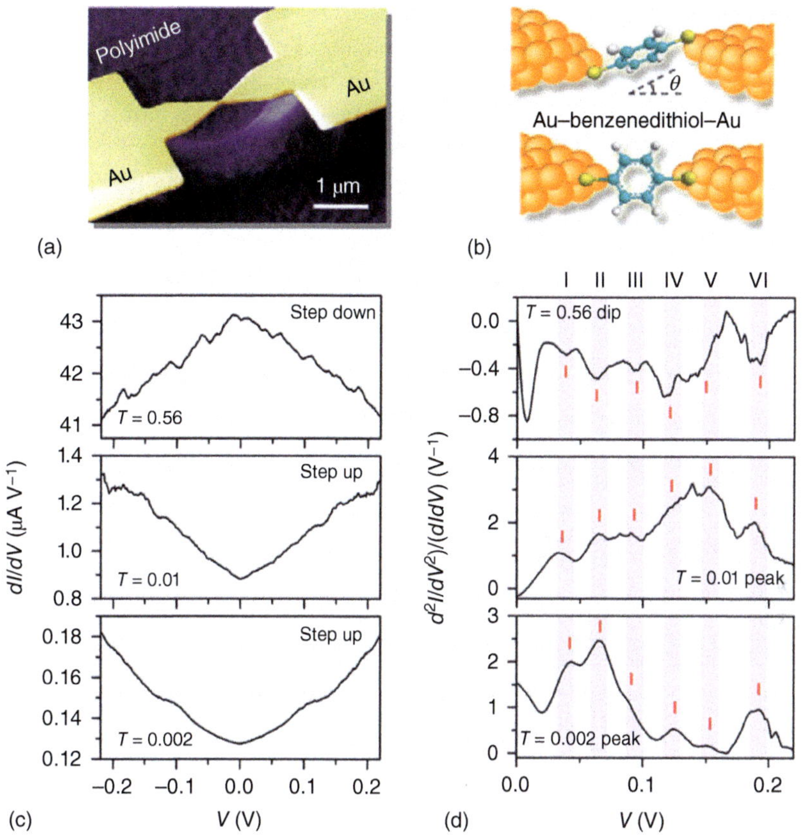

Figure 4.17 (a) Scanning electron microscopy image of the MCBJ device. (b) Schematic of an Au–BDT–Au junction with a tilt angle (θ) of the benzene with respect to the current path (top) and sulfur (S) connecting to the Au sites with hollow site. (c) Differential conductance (dI/dV) curves as a function of bias with T = 0.56 (top), 0.01 (middle), and 0.002 (bottom). (d) Normalized IETS with $T = 0.56$ (top), 0.01 (middle), and 0.002 (bottom). Vibration modes of Au–BDT–Au junctions are assigned: I, gold–sulfur stretching; II, C–S stretching; III, C–C–C bending; IV, C–H in-plain stretching; V, C–H in-plain bending; VI, C=C stretching. Source: Kim *et al.* [47]. 2011, American Chemical Society.

junctions at 4.2 K, using a MCBJ technique. By modulating the gap size through mechanically stretching or pushing the molecular junctions, therefore the molecule–electrode coupling, they were able to form stable molecular junctions in a broad range of transmission from T = 0.002 to 0.56. As shown in Figure 4.17, the symmetric conductance obtained at T = 0.56 shows a reduced conductance at the threshold voltages corresponding to molecular vibrational modes, and the IETS (Figure 4.17) shows dips. In contrast, at T = 0.01 and 0.002, the dI/dV and IETS are stepwise enhanced and show peaks. Crossover between dips and peaks of IETS was further evaluated to be in the range of T ~ 0.420–0.56 for the case of symmetric coupling, which is in well agreement with the theoretical prediction. Moreover, they found the normalized IETS intensity follows the predicted relation of 1–2 T.

In addition to the peak and dip, it was observed in experiment that the IETS can also show a zig-zag shape in certain conditions. For example, Yu *et al.* studied the IETS of multilayer molecular junctions with and without incorporated metal nanoparticles. The incorporation of metal nanoparticles into devices leads to enhanced IET intensity and a modified line shape for some vibrational modes [48]. As shown in Figure 4.18, the IETS line shape changed into zig-zag shape in present of metal ion. They attributed this change to a near resonance tunneling by an energy level close to the Fermi level originated from metal particle–molecule hybrid. In the

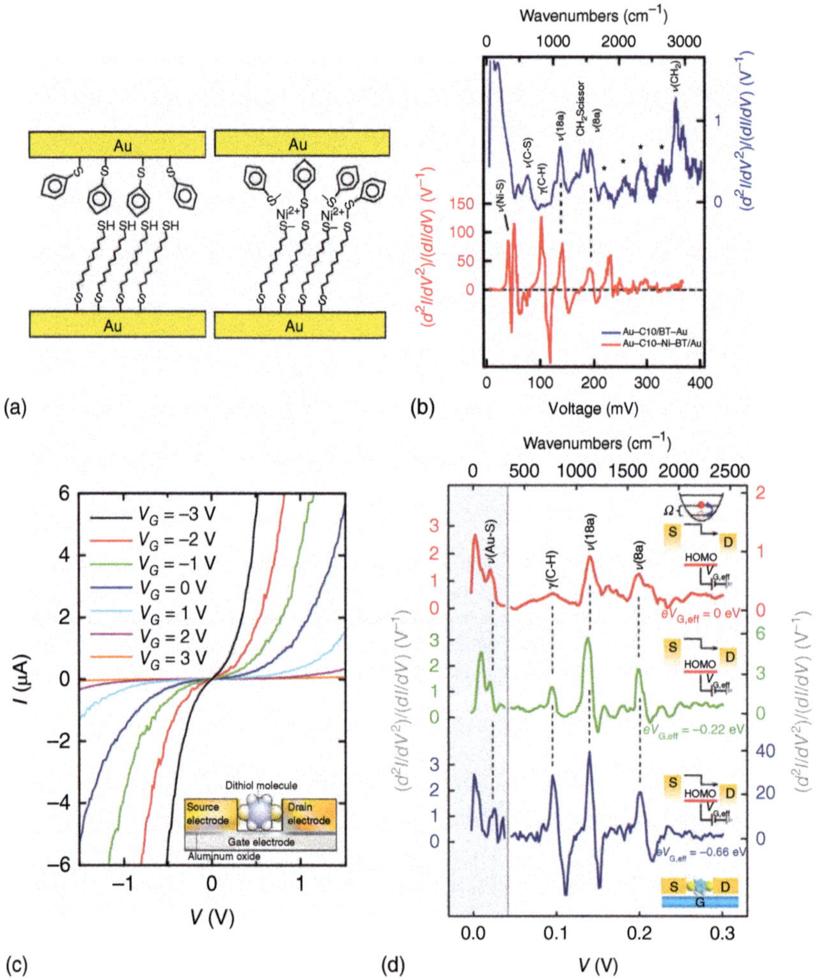

Figure 4.18 Panels (a) and (b) are the schematic of the molecular junction with metal nanoparticle and its IETS result. Note in the control experiment where no metal particle, zig-zag shape in IETS is absent. Source: Yu *et al.* [48]. 2007, American Physical Society Panels (c) and (d) are the *I*–*V* curves of benzene dithiol junction with different gate voltage at 4.2 K and its resonantly enhanced IETS of the junction by gate voltage. Source: Song *et al.* [49]. 2009, Springer Nature.

near resonance situation, higher-order elastic cotunneling processes can contribute substantially to the electronic transport near the onset of an inelastic channel.

Song *et al.* in the study of molecular transistor using gated nanogap junction with a benzene dithiol molecule also observed similar line shape transition in IETS upon applied gate voltage (Figure 4.18c) [49]. Specifically, when a negative gate voltage is applied, which brings the energy of the molecular junction's HOMO closer to that of the electrode's Fermi level, *i.e.* near resonance tunneling situation, the IETS signal increased greatly and the shapes of the vibrational peaks changed from peak into zig-zag shape. It was believed that as HOMO of the molecule was brought close to resonance, it would be strongly coupled to the internal vibrational modes (Figure 4.18d, insets), resulting in resonantly enhanced IET spectra. In addition, strong coupling between the molecular resonance and the vibrational modes is expected only when the energy states of the molecular orbitals are distributed around the bond directly related to the vibrational modes. For benzene dithiol molecule, the resonance HOMO is localized on conjugated benzene ring. Indeed, the vibrational modes associated with the benzene ring are the only modes enhanced and line shape changes by gate voltage. So the enhanced IETS signal and the line shape change presented as an unambiguous evidence for molecular orbital gating in their study.

Theoretical study later on found that the zig-zag line shape actually came from the higher-order elastic cotunneling of the charge cross the molecule that consists phonon emission and re-absorption processes [50, 51]. Unlike the tunneling and inelastic tunneling, which are the odd function of bias therefore produce symmetric I–V, the cotunneling current is even function of the bias and its second derivative is zig-zag shape. In the near resonance situation, this higher-order cotunneling becomes significant due to the asymmetric coupling of the two electrodes to the electronic level. So the experimental observation by Kushmeric and Lee is actually the results of the overlap of the enhanced IETS peak near resonance with the zig-zag of the phonon-assisted cotunneling. Theory also predicted that the zig-zag shape can be observed in off-resonance situation as long as the coupling of the two electrodes to the molecule electronic state is quite asymmetric. Xu and Ho proved this in a STM study by controlling the relative coupling of a CO molecule to the substrate and to the STM tip [52].

The last special IET process we would like to mention is the so-called resonance inelastic tunneling process (RIET). As the name says, RIET process takes place in the resonance tunneling scenario [53]. As shown in Figure 4.19 [54], when the molecular energy level alignes with one of the Fermi level electrodes under bias, resonance tunneling is initiated, and we will see a resonance peak in the conductance spectrum (*vs.* bias). As the bias increases further, the Fermi level of the electrode will surpass the resonant state. However, there is still some chance the electrons near the Fermi level can lose vibrational energy quanta and fall into resonance with the molecule energy level. In the conductance spectrum, this process gives rise to a satellite peak (sometimes a progression of satellite peaks) at a bias, which is in the order of the vibrational energy quanta, higher than the resonance peak. Distinct from aforementioned IET process, the vibrational energy in RIET is read as the energy

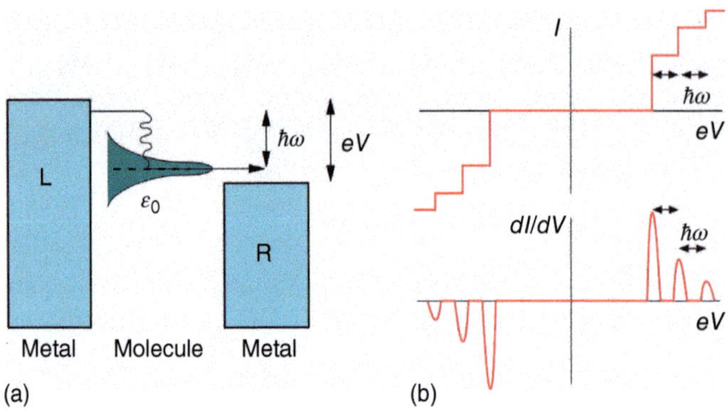

Figure 4.19 (a) Schematic of resonance inelastic tunneling of a metal–molecule–metal junction under bias eV. (b) *I vs.* bias voltage plot shows the case when molecular energy level aligned with one of the Fermi level and beyond (top) and its corresponding *dI/dV vs.* bias voltage (bottom). Source: Cuevas and Scheer [54].

difference between satellite peak and the resonance peak instead of the absolute bias value.

We can make a good comparison between IET *vs.* RIET and Raman *vs.* resonance Raman spectroscopy in that in both the former processes electronic state did not directly get involved into the process while the latter two it did. Even in the resonant enhanced IETS that we discussed above, molecular electronic state only plays an assistant role. The resonance inelastic tunneling process spectrum (RIETS) can generally be observed when coupling between the electrode and molecule electronic state is not too large, while electron–vibration interaction is relatively high [55].

It is important to point out that, due to the bias dividing effect in the junction, the bias difference between the RIETS peak and resonance peak does not always equal to vibrational energy quanta. In the case bias dividing is evenly distributed, electroactive group is almost at the center of the junction, the bias gap is two times of the vibrational energy quanta, while in contrast when the electroactive group is much closer to one of the electrodes than the other, the bias gap is close to the vibration energy.

4.2.4 Application of the IETS

We have seen that IETS, not only the peak position and amplitude but also the line shape, can provide quite rich information about the junction structure and charge transport mechanisms. Therefore, the IETS can be a powerful tool in molecular electronic study. We will give several examples in this section.

First, in molecular switch study, to date, various possible mechanisms for conductance switching behavior have been proposed. IETS has proven to be an extremely useful tool in studying the switching mechanism, especially by comparing the theoretical and experimental results. For example, Cao *et al.* conducted density functional theory calculations for the IETS of an oligoaniline dimer dithiolate junction [56]. Both the conjugation changes and the oxidation effects on the IETS have been explored. Comparison between the calculated IETS and the experimental spectra indicated that the observed conductance switching was most likely due to the conjugation change rather than the oxidation.

Experimental and theoretical analyses of IETS in azobenzene-based switch using break junctions have also been performed [57]. As seen in Figure 4.20, IETS of trans isomers showed a strong peak at ~160 meV, while cis isomers is at ~190 meV. These distinct changes in IETS revealed that the molecules were isomerized by the UV irradiation. Theoretical study showed that these variations in peaks originated from modes with N=N stretching and the atomic arrangement around the N=N bond of azobenzene. This observation demonstrated IETS as a precise tool in revealing the cis–trans structure change of molecules by different vibrational modes.

The molecule–electrode interfacial bonds were also characterized by IETS. For example, thiol-ended and amine-ended molecular junctions were systemically examined by Kim *et al.* by comparing the vibrational modes ν(Au–S) and ν(Au–N) in the IETS while increasing the electrode separation (Figure 4.21) [58]. It was found that the ν(Au–S) mode was so stable that no significant changes in the vibrational energy were observed during the stretching of the Au–ODT–Au junction. In contrast, the energy of the ν(Au–N) mode in the Au–ODA–Au junction red-shifted while increasing the electrode separation. This behavior can be explained by an increase in the Au—N bond length, therefore the reduction of the force constant due to the relative weak Au—N bond. The IETS measurements thus provided direct evidence that the Au—N bond was weaker than the Au—S bond.

4.2.5 Mapping the Charge Transport Pathway in Protein Junction by IETS

The idea of charge transport pathway initially originated from the study of long-rang charge transfer, for example in redox protein, where the electrons can tunnel a long distance across the peptide matrix from the surface to the cofactor of the protein. Charge transfer pathway was then theoretically proposed based on super-exchange tunneling mechanism, in which the peptide matrix was divided into various tunneling elements with their own tunneling efficiency, respectively [59]. Using this model, mapping the electron tunneling pathways was realized in protein by an algorithm that search for the most efficient one out of the many possible combinations of the tunneling elements across the peptide matrix [60]. It was proposed then, similar to the pathway idea in the IETS study by Troisi that the vibrational modes of the chemical bonds along the pathway in protein should manifest themselves in the IETS of protein junction. Yu and coworkers have done series of works in this line of research [61].

To explore the effect of the peptide composition and sequence on charge transport in protein, they designed model peptide system, where 7-alanine peptide was doped by a tryptophan and with varying position, and performed charge transport and IETS measurement. The junction was made by aligning gold nanowire, as a soft and stable top contact, on peptide monolayer on top of lithographically defined Au electrodes using electrophoresis and the IETS of the Au–peptide–Au junction was measured at around 15 K (Figure 4.22). The C=C stretching mode of the indole

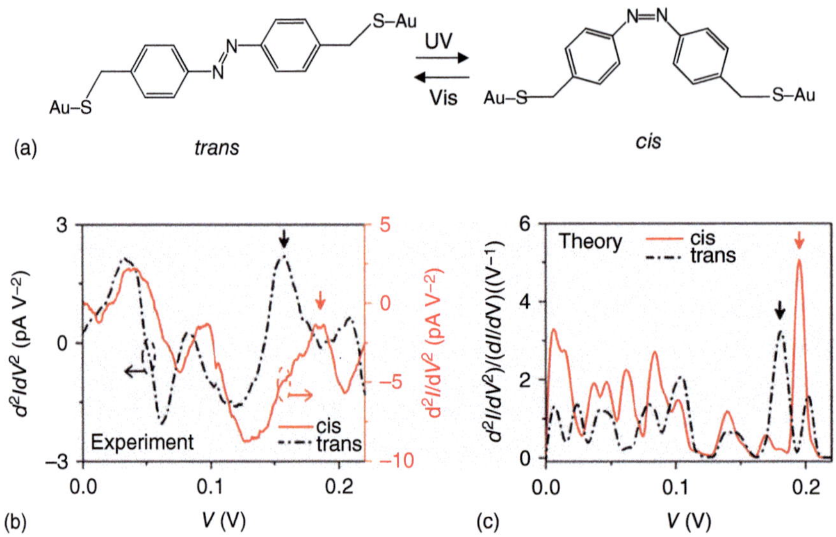

Figure 4.20 (a) Chemical structure of trans and cis isomers of AzoTM measured in single-molecule junctions (SMJs). (b) Experimental and (c) theoretical symmetrized IETS for cis and trans isomers (average of several configurations) at 4.2 K. Measured and calculated at 6 and 5 meV, respectively. Source: Kim *et al.* [57]. 2012, American Physical Society.

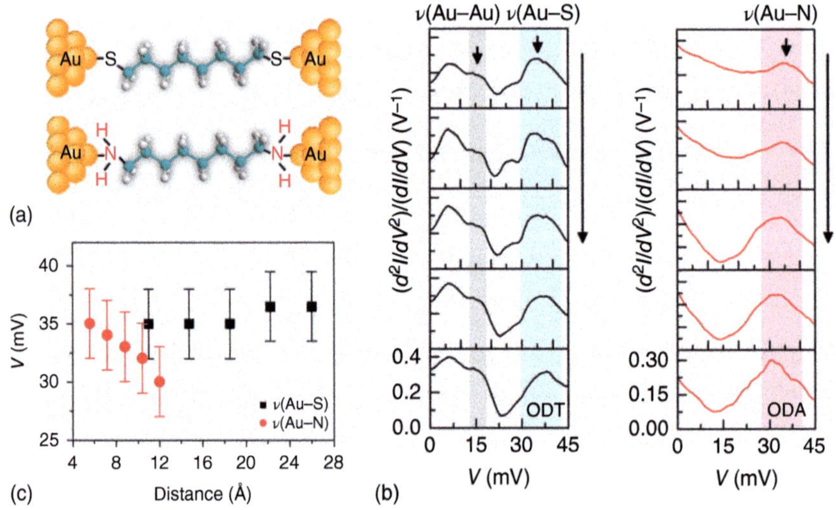

Figure 4.21 (a) Illustration of ODT and ODA single-molecule junctions. (b) The evolution of IETS spectra for the stretching of a single junction for ODT (left panel) and ODA (right panel) measured at the lowest conductance plateaus for DC bias up to 45 mV. (c) Evolution of the ν(Au–S) and ν(Au–N) modes extracted from panels in (b). Source: Kim *et al.* [58]. 2011, American Chemical Society.

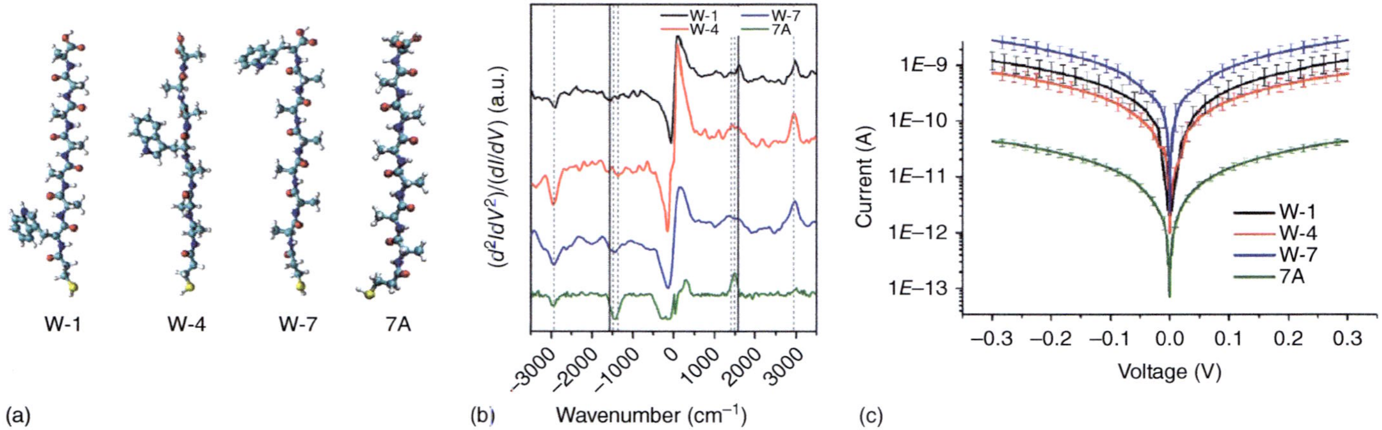

Figure 4.22 (a) Structure of peptides used: W-1: MPA-WAAAAAA; W-4: MPA-AAAWAAA; W-7: MPA-AAAAAAW; 7A: MPA-AAAAAAA (A = alanine, W = tryptophan). (b) IETS of W-1, W-4, W-7, and 7A measured at ~10 K. The characteristic C=C stretching of the indole residue of Trp (solid gray) is absent in 7A. (c) Averaged I–V plots at 300 K for Au–peptide–Au junctions of W-1, W-4, W-7, and 7A of ~20 curves. Source: Based on Betts *et al.* [60].

ring of tryptophan appears in the IETS of the doped peptide, accompanied by a much higher conductance than the pure alanine counterpart. They modeled the system using the same idea of super-exchange pathway mechanism and indicated that the conjugate indole group of tryptophan worked as an efficient tunneling footstone.

The charge transport pathway in a natural copper-containing redox protein, azurin, which functioned as an electron carrier in a bacterial energy conversion system, was also investigated using IETS by Yu *et al.* [62] As shown in Figure 4.23, in IETS of the azurin junction, Au—S and S—C stretching bonds were observed, indicating charge tunneling from gold electrode to protein through Au—S—C bond. At an even lower temperature (~6 K), fine structures at around 1600 cm^{-1} were resolved and peaks at 1670 and 1550 cm^{-1} were found, which were amide I and II, the signature of peptide chain. Between them, one extra peak could be observed at around 1580 cm^{-1}, which should be assigned to modes of the side groups of amino acids, such as the NH_2 in-plane bending, and C=C or C=N bond in conjugated side groups present in tyrosine, phenylalanine, and tryptophan. These results confirmed that both the peptide chain and the side groups, mostly the conjugate side groups, of the amino acids contributed to the charge transport in the protein.

Meanwhile, they found in the IETS of azurin junction, only peaks, never dips or zig-zag shapes, were present, indicating off-resonance tunneling transport. The observation raised a big confusion then: why no resonant transport was observed in azurin junction despite the presence of redox-active Cu site whose electronic state is near the Fermi level of the gold electrode?

In a further study, they made a small modification to the junction structure [63]. As shown in Figure 4.24, they added a spacing layer between azurin and the top nanowire electrode by mercapto propionic acid SAM. The transport property changed significantly in present of the thin molecule layer. The I–V changed from near linear shape in the absence of the spacing layer into step-like shape and a series of peaks in the conductance spectrum. They recognized the peaks in the conductance spectrum as the resonance peaks assisted by theoretical model and temperature-dependent measurement. More astonishingly, they observed the RIETS satellite peaks following resonance peaks, as shown in Figure 4.24, and the spacing between peaks were precisely in agreement with the energy of Cu–S stretching mode of the Cu site documented by resonance Raman spectroscopy. The results provided the first in situ vibrational spectroscopy fingerprint of resonant tunneling charge transport across protein redox center and represented a significant progress in protein-based molecular device and charge transport study.

4.2.6 STM Imaging by IETS

IETS can also be used to in the STM imaging to provide visible information on molecular structure and the chemical bonding more than traditional current mapping. Ho's group has done a series of amazing works in this aspect. For example,

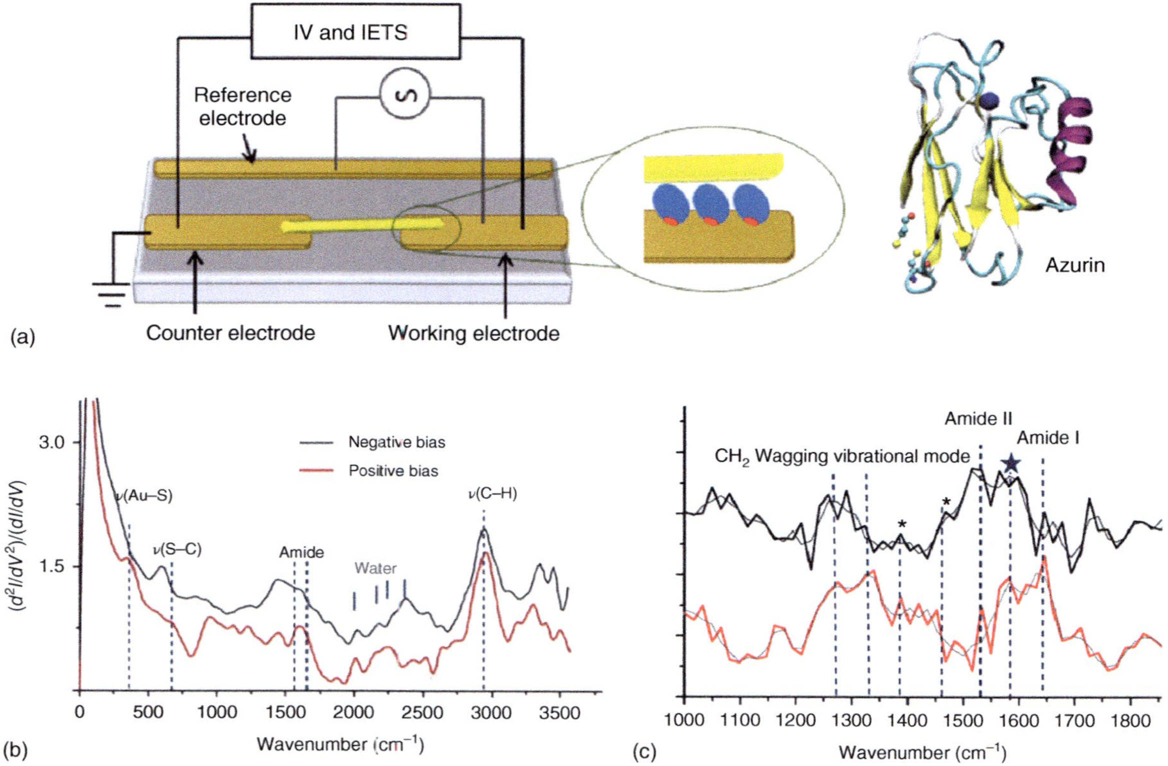

Figure 4.23 (a) Scheme of the Azurin junction fabricated by trapping nanowires by AC electric field for $I-V$ and IETS measurement. Panels (b) and (c) are IETS of an Azurin junction. The gray and red color lines are the spectra at negative and positive bias, respectively. Panel (c) is a higher resolution IET spectrum at ~5.5 K with AC modulation amplitude of 6 mV. Note the IETS (d^2I/dV^2) was normalized with (dI/dV) in order to correct the baseline in conductance. Source: Yu *et al.* [62]. 2015, American Chemical Society.

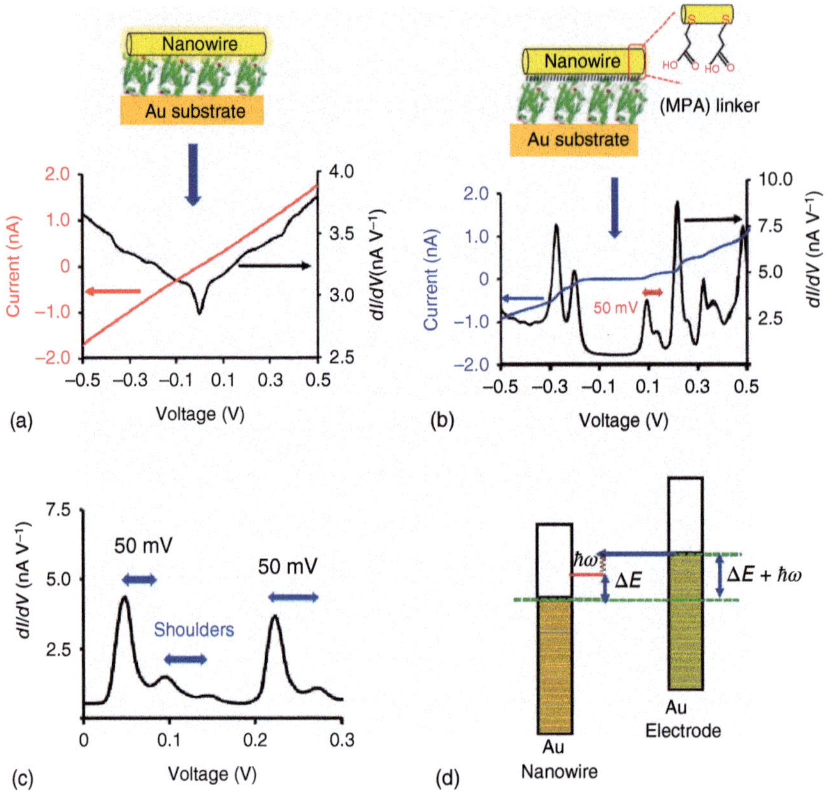

Figure 4.24 (a) *I–V* (red) and conductance–voltage *dI/dV–V* (black) plots of the Az junction. (b) *I–V* plots (blue) of Az with mercapto propanoic acid (MPA) linker and conductance–voltage *dI/dV–V* (black) plot. (c) Conductance–voltage plot of the Az junction with MPA linker showing the presence of the satellite peak ~50 mV away from the main peak. (d) The schematic of the vibrational modes of the resonant electronic state due to vibronic interaction. Source: Fereiro *et al.* [63]. 2018, National Academy of Sciences.

they performed inelastic tunneling measurement using a single carbon monoxide CO molecule on STM tip as probe to image the skeletal structure and bonding in an adsorbed molecule. As the CO-terminated tip scanned over the molecule during imaging, changes in the potential energy would hinder translational vibration of CO, which were measured by IETS as the vibrational energy change. They found this method can sense the spatially varying potential energy landscape of the molecule and its surroundings. So they can image not only the skeletal covalent bond of the molecule, but also the noncovalent bond like hydrogen bond and halogen bond for the first time (Figure 4.25) [40].

Although the IETS technique possesses many merits and becomes a powerful method for characterizing molecular junctions, there are still several obstacles that limits its further applications. For example, the IETS requires low temperatures (typically liquid He) to suppress the thermal broadening effect for high-resolution

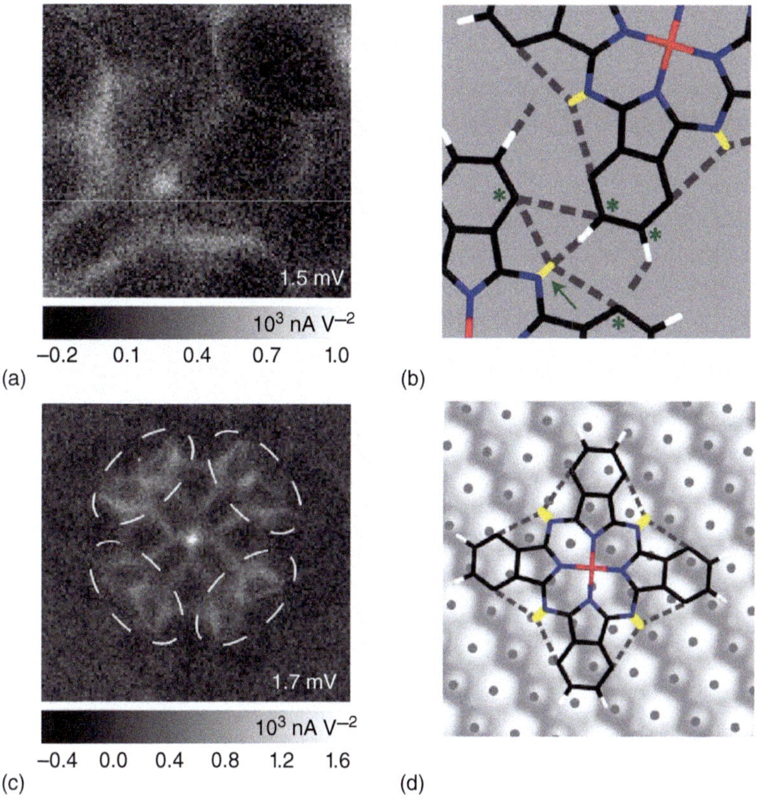

Figure 4.25 (a) and (b) Imaging the chemical bonding between a pair of CoPc on Ag (110) surface. (a) Constant current inelastic tunneling probe (itProbe) image (128 × 128 pixels or 8.1 Å × 8.1 Å) at 1.5 mV. (b) Schematic diagram of the image of (a) showing in dashed lines the intermolecular and intramolecular bonds involving hydrogen, and in yellow the lone pairs of imine nitrogen. Four hydrogen atoms are shared among five centers (four carbon atoms [indicated by asterisk] and one imine nitrogen [indicated by arrow]) in the five-center four hydrogen bonds. (c) and (d) Skeletal images of CoPc obtained by itProbe. (c) Constant current image of CoPc(+) on Ag(110). (d) Schematic diagram showing the skeletal structure of CoPc(+) and the intramolecular hydrogen bonds highlighted in dashed ellipses in (c). Source: Chiang *et al.* [40]. 2014, American Association for the Advancement of Science.

spectrum, self-made lock-in measurement system to detect the weak inelastic signal and stable, low-noise junction structure for broad temperature range. These conditions are not available in many laboratories. Additionally, the intrinsic weakness of vibrational spectroscopy applies to IETS as well, *i.e.* many fingerprint peaks of various organic functional groups are present in a relatively small energy range between 500 and 1800 cm^{-1} and they can be hard to resolve. So the analysis of the IETS is not always unambiguous, especially for systems with a large number of nuclei and numerous bonds.

4.3 Optical and Optoelectronic Spectroscopy

The size of nanogap junction device and the weak signal from a single molecule largely restricted general application of in situ optical and optoelectronic characterization to the junction. Fortunately, the development of surface-enhanced Raman spectroscopy (SERS), makes measurement of in situ optical and optoelectronic properties of nanogap junctions possible. The enhancement of the Raman signal originates from the strong localized plasmon filed between the nanogap of two metal electrodes upon light illumination, and therefore, small group of, or even single, molecule measurement is possible. Combing the high sensitivity of SERS optical measurement technique and the electrical characterization like I–V and IETS setup, researchers can study the in situ fine structure of molecular junction and the relation between structure and electrical function. We will introduce several fascinating examples in this section.

An efficient on-wire lithography method was invented by Mirkin's group to fabricate multi-segmented metal nanowires, as shown in the Figure 4.26a [64]. The metal on both sides of the nanogap acted as plasmonic antennas. The long-range SERS was acquired from different parts of nanowire with the phthalic acid (benzene-1,2-dicarboxylic acid) molecules on the nickel segment (see Figure 4.26b,c). The gold and nickel segment (point A and C in Figure 4.26a) did not show any signal while Raman spectrum of point B was clear and consistent with the peaks of phthalic acid in the solid state, which indicated the SERS enhancement was only due to the molecules between the gap between gold and nickel (point B in Figure 4.26a).

This SERS method was further utilized in the nanogap molecular junction characterization. Chen *et al.* used alkyne–azide "click reaction" between the nanogap to grow molecule conducting wire stepwise and the process was monitored by SERS (Figure 4.26d) [65]. 4-Ethynyl-1-thioacetylbenzene (1) was first assembled as a monolayer on the electrode surfaces, and then the alkyne terminal group of (1) reacted with 2,7-di-azido-fluorene (2) forming a 1,2,3-triazole unit at one terminal. The structure could be increased by further reaction with 2,7-diethynyl-fluorene (3). After several reaction cycles, the molecular wires would connect the opposing electrodes (Figure 4.26d, e). Figure 4.26f(1) is the spectrum of nanogap after assembled with 1, which showed the alkyne C≡C symmetric stretch and benzene rings C–C aromatic stretch peaks at 2108 and 1585 cm^{-1}, and the thioacetyl and S–H bending vibration (634 and 915 cm^{-1}, respectively) disappeared indicating the monolayer formed successfully. After click reaction with 2, the triazole ring stretch peaks (967 and 1010 cm^{-1}) and aromatic C–C stretch of 2 (1606 cm^{-1}) emerged, and asymmetric stretching of azide groups (2108 cm^{-1}) disappeared (Figure 4.26f(2)), suggesting that 2 has connected to 1. In addition, the relative intensity of the peak at 1606 cm^{-1} increased, which tells 3 reacted with 2 successfully.

Detection of the presence of single molecule in the nanogap junction using SERS was performed by combing the MCBJ and confocal SERS microscope by Tian's group [66]. SERS intensity was found to decrease significantly when electrodes pair changed from the direction parallel to the polarization, which was the direction of

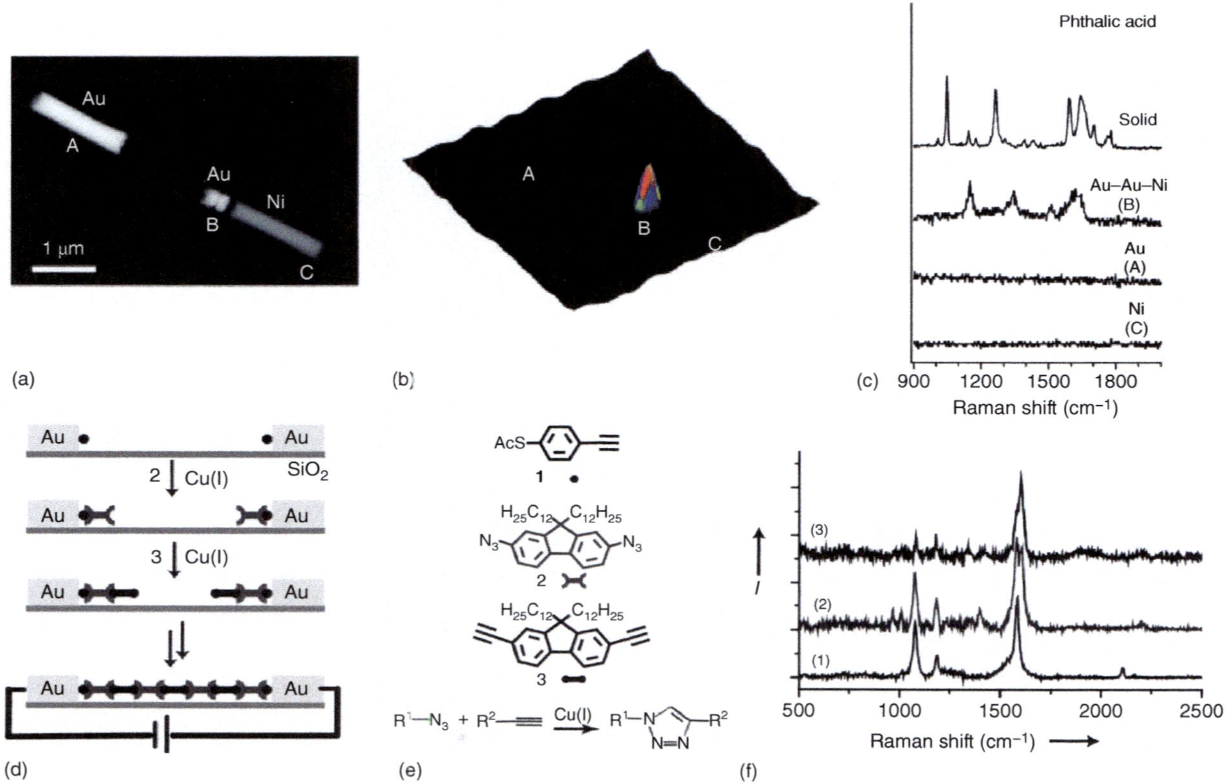

Figure 4.26 (a) The scanning electron microscopy (SEM) image of nanowires, from left to right: (1) 1.5-mm segment of gold; (2) a 1.5-μm gap; (3) a 120-nm gold nanodisks; (4) a 30-nm gap; (5) another 120-nm gold nanodisks; (6) a 120-nm gap; (7) a 1.5-μm nickel wire segment. (b) The confocal Raman microscopy images corresponding to nanowires that have functionalized with phthalic acid. (c) Raman scattering curves of solid phthalic acid and different segments of nanowires; (d) Schematic diagram of nanowire structure and click reaction within the nanogaps. (e) The structure of molecules and equation of click reaction. (f) The Raman spectrum for the different steps: (1) after 1 assemble grown in the nanogap, (2) after the reaction of 1 with 2, and (3) after addition of 3. Source: Panel (a)–(c) Wei *et al.* [64] and (d)–(f) Chen *et al.* [65]. 2009, John Wiley & Sons.

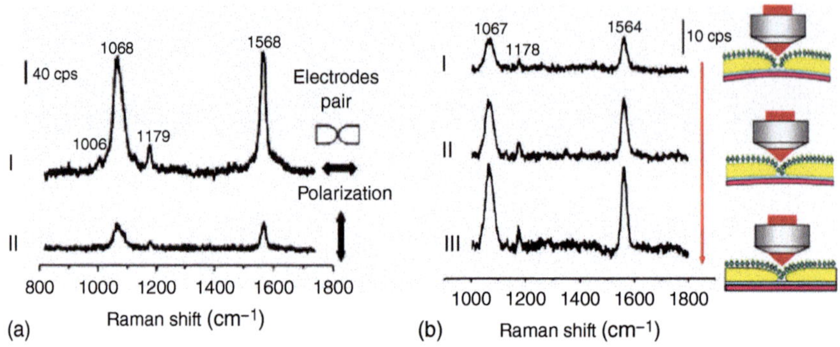

Figure 4.27 SERS of 1,4-benzenedithiol in MCBJ. (a) The incident polarization is in parallel (I) and perpendicular (II) to the electrodes pair. (b) The SERS recorded with the bending of the Au electrodes pair of nanogap width 8 Å (I), 6 Å (II), and 4 Å (III). Source: Tian *et al.* [66]. 2006, American Chemical Society.

1,4-benzenedithiol attached between two Au electrodes on a Si chip, to the direction perpendicular to the polarization, see Figure 4.27a. The origin of SERS intensity has been proven to be the molecule in the nanogap, since molecule perpendicular to the polarization exhibits no SERS signal and the background SERS intensity is very weak in Figure 4.27a(II). As is shown in Figure 4.27b, SERS intensity presented a correlation with nanogap width, *i.e.* with increasing nanogap width the SERS intensity drops, which may indicate a possible new method to examine the width exits in molecular junction linked to nanogaps.

Tian's group later used simultaneous measurement of conductance and tip-enhanced Raman spectroscopy (TERS) to show a correlation of current on-off switch and Raman peak shifts in 4-bipiridine (4-Bipy) molecule linked between nanogaps with MCBJ technique, see Figure 4.28a–d [67]. When 4-Bipy was switched to on state, TERS showed a peak broadening near 1609 cm^{-1}. In addition, the peak broadening at 1631 cm^{-1} was found to be dependent on voltage bias. In low-bias range, 1631 cm^{-1} peak intensity was negligible while in high-bias range it becomes prominent. It was believed that under high bias, the pyridine ring attached to the tip contact undergoes a deformation with stronger carbon–carbon bonds and weaker nitrogen–carbon bonds, thus a change to the frequency of the C–N/C–C at 1631 cm^{-1} (Figure 4.28e, f). Simulation of TERS spectra under different bias voltage using cluster model-based density functional theory showed good agreement with experimental results, which proved the conductance switch took place when the bonding strength of carbon–carbon bonds and nitrogen–carbon bonds of the pyridine ring anchored to the tip electrode changes. The combination of MCBJ and TERS thus offered a way to study the relation between molecule structure and its electronic transport properties and optical properties in a wide range of molecule systems.

SERS of the molecular junction was also used for the assignment of the vibration modes present in IETS and to cross-prove each other. Yoon *et al.* fabricated a crossed-nanowire molecular junction platform which was capable of both SERS

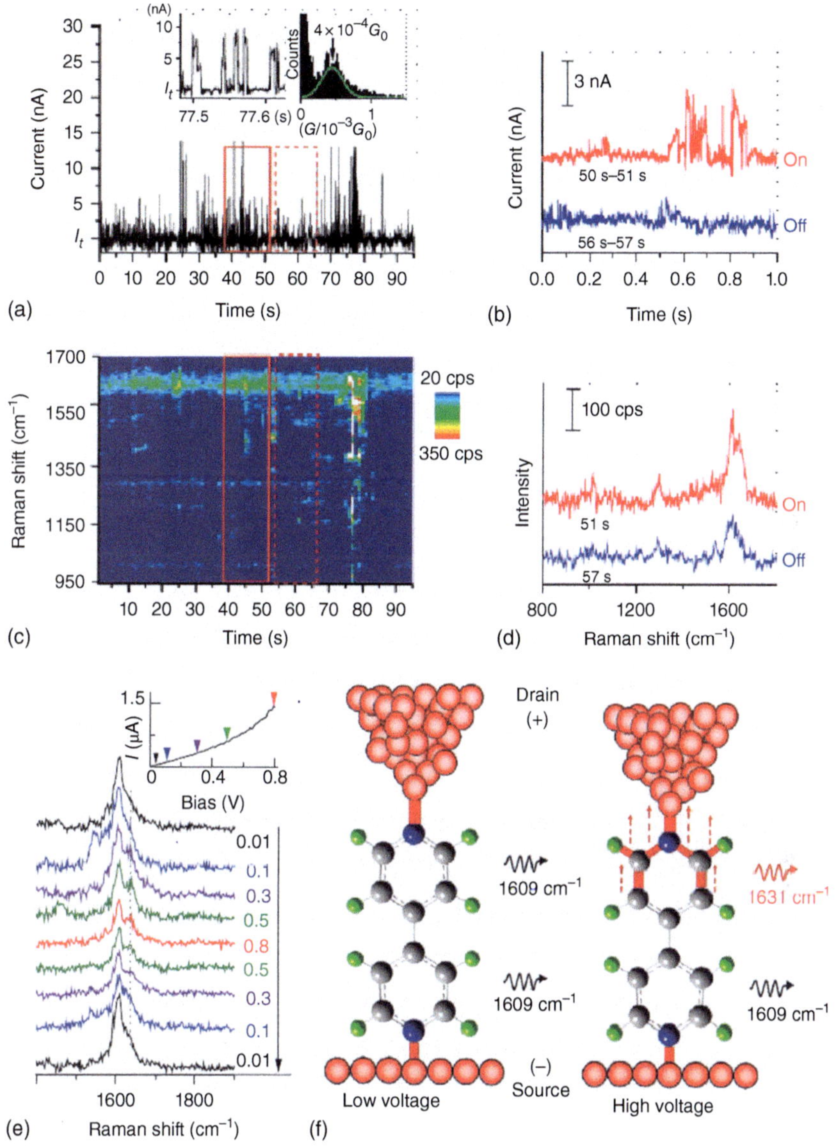

Figure 4.28 Simultaneous conductance and TERS measurement of 4-BiPy. (a) Conductance measurement of Au–4-Bipy–Au junction displayed with time. Left inset: enlarged image of (a) in 0.1 seconds; right inset: conductance histogram of Au–4-Bipy–Au junction. (b) Selected conductance curves within 1 second of both on and off state. (c) TERS spectra measured simultaneously with conductance in (a). (d) TERS spectra obtained at the same time of conductance curves in (b). (e) N–C/C–C bond changes along bias voltage. (f) Schematic image of the origin of 1609 cm^{-1} band splitting correlates with applied bias voltage. Source: Liu *et al.* [67]. 2011, Springer Nature.

and IETS on the same molecular–scale junction simultaneously [41]. Figure 4.29a,b show schematic structure of the junction and SERS, I–V, IETS integrated setting. SERS spectra (Figure 4.29c) taken at Au–molecule–Au crossed-nanowire junction with laser irradiation over different sample positions confirmed that only molecule connected with the electrodes of the nanogap generates high signal-to-noise spectrum. Figure 4.29d compared the IETS and SERS of the molecule junction. Three main IETS peaks, ν(18a) (1065 cm^{-1}, 132 mV), ν(8a) (1581 cm^{-1}, 196 mV), and ν(C–C) stretch (2226 cm^{-1}, 276 mV), and Raman signals at 1076.6 cm^{-1}(133 mV), 1135.3 cm^{-1}(141 mV), 1600.0 cm^{-1}(198 mV), and 2210.4 cm^{-1} (274 mV) showed good agreement, respectively.

Another application of SERS in molecular junction is to study the heating of the junction in charge transport [68]. Stokes (S) and anti-Stokes (AS) scattering relates to molecular steady-state nonequilibrium population in the presence of inelastic tunneling current and the ratio of AS/S of each Raman active vibrational mode can be used to determine the effective temperature of the junction by the following equation:

$$\frac{I_{\mathrm{AS}}}{I_S} = \frac{(\nu_L + \nu_V)^4}{(\nu_L - \nu_V)^4} \exp(-h\nu_v/k_B T_{\mathrm{eff}}(\nu)) \frac{\sigma_{\mathrm{AS}}}{\sigma_S} \frac{A_{\mathrm{AS}}^2}{A_S^2} \tag{4.25}$$

where $I_{\mathrm{AS}(S)}$ is the intensity of the anti-Stokes (Stokes) Raman mode, $\nu_{L(v)}$ is the frequency of the laser (Raman mode), $\sigma_{\mathrm{AS}(S)}$ is the anti-Stokes (Stokes) scattering cross section of the adsorbed molecule, and $A_{\mathrm{AS}(S)}$ is the average local field enhancement at the molecules of the anti-Stokes (Stokes) frequency.

Based on this idea, Ioffe *et al.* fabricated "on edge" junctions of the form silver–SAMBPDT–silver (SAMBPDT stands for self-assembled monolayer of 4,4′-biphenyldithiol) to carry out both I–V measurement and Raman measurements (Figure 4.30).

The presence of BPDT was confirmed from both Stokes and anti-Stokes SERS as illustrated in Figure 4.29a,b which showed only at the junction both Stokes (Figure 4.30d) and anti-Stokes (Figure 4.30e) signals were significantly enhanced. Further analysis of the Stokes and anti-Stokes SERS indicated that T_{eff} increased with bias voltage at both polarities, suggesting an increased energy dissipating at higher bias. The fact that the T_{eff} of all vibrational modes was quite similar, changing with bias in a similar way, suggested the fast intramolecular vibrational relaxation or redistribution (IVR) within the junction. The experimental results in the work revealed a rich electron–hole relaxation in the charge transport and heating–cooling behavior in the junction that was inexplicable using existing models, which deserve further investigation.

The experimental effort to simultaneously probe molecular junctions by electrical measurement and by optical spectroscopy methods such as Raman is only at its beginning. We have seen in the section that Raman spectroscopy is emerging to be a powerful tool to explore inelastic transport processes in junctions, heating of vibrational modes, and structural changes.

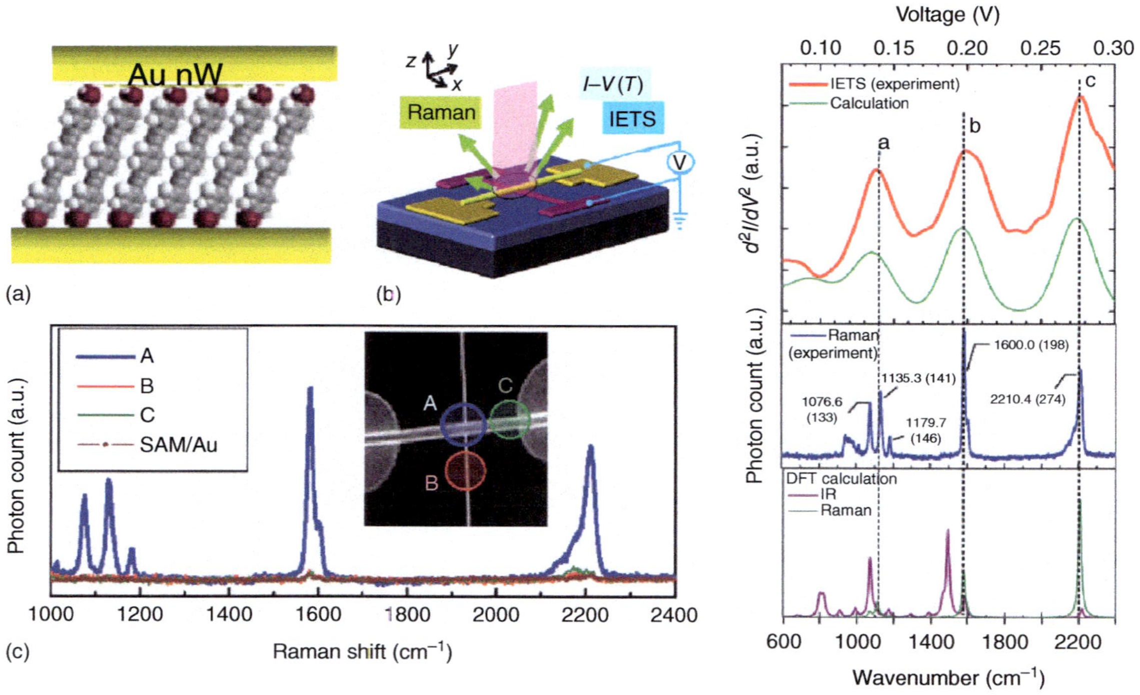

Figure 4.29 (a) Schematic of the OPE sandwiched between the crossed-area of the top and bottom Au contacts. (b) Schematic of the electrical and multi-spectroscopic measurements setup of the crossed-nanowire platform. (c) Raman spectra of laser irradiated on different positions of the crossed-nanowire junction while pointing away from the junction gives no observable spectra. (d) Comparison of measured IETS, Raman spectra, and calculated infrared (IR) and Raman spectra. Source: Yoon *et al.* [41]. 2010, American Chemical Society.

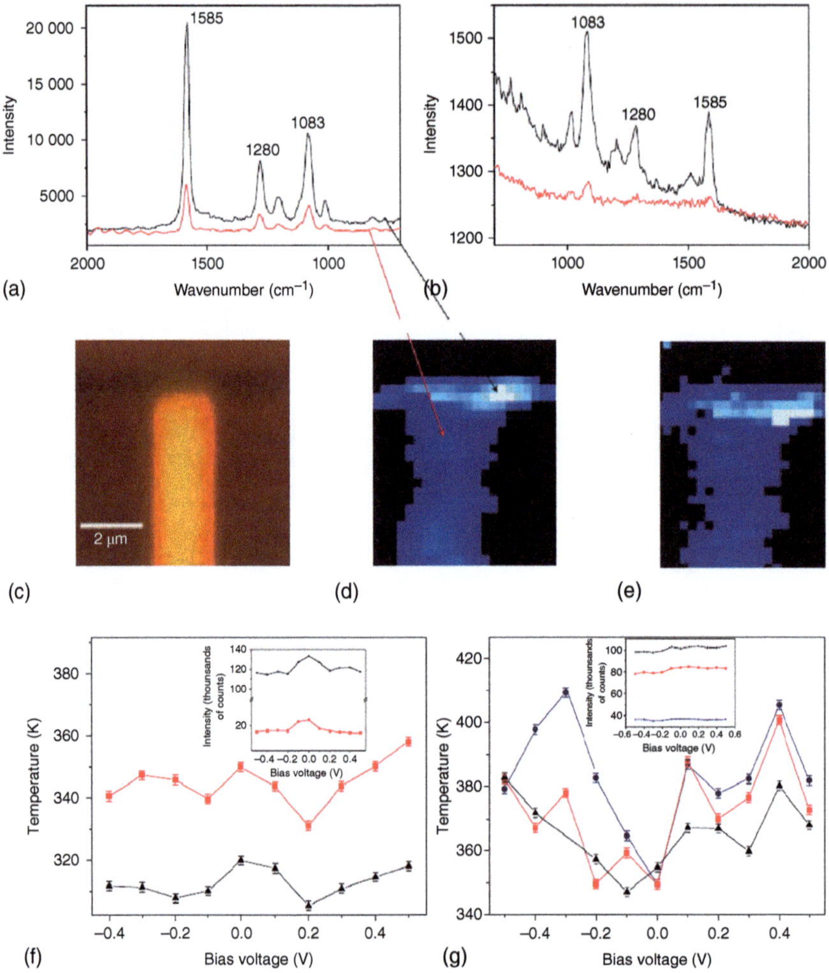

Figure 4.30 (a) Raman spectrum in Stokes region of BPDT molecules in a junction (black) and as a monolayer on the electrode. (b) The corresponding anti-Stokes spectrum. (c) Optical microscope to view of a junction. (d) Raman map of a junction in Stokes mode (1585 cm^{-1}). The brighter pixels are the SERS of the junction. (e) Raman map of the 1585 cm^{-1} anti-Stokes peaks of the same junction. (f) T_{eff} as a function of bias voltage for the 1585 cm^{-1} (black triangles) and 1083 cm^{-1} (red squares) modes (532 nm laser). Inset: the Stokes signal of this junction as a function of bias. (g) T_{eff} as a function of bias voltage for the 1585 cm^{-1} (black triangles), 1280 cm^{-1} (blue circles), and 1083 cm^{-1} (red squares) modes (671 nm laser). Inset: bias independent Stokes signal of this junction. Source: Ioffe *et al.* [68]. 2008, Springer Nature.

4.4 Concluding Remarks

The characterization techniques and analyzing methods we introduced in this chapter include only those ones most commonly used in the molecular electronic study and are apparently not complete. The characterization techniques for molecular junction are fast developing and a full introduction on the characterization of the molecular junction will require much more space than is possible in this short chapter. Other techniques like UPS, which can provide the charge injection barrier (or energy offset) of the molecular layer on metal surface, is a relatively independent technique widely used for the study of molecule–electrode interface, details of which have been well documented and can be found in Ref. [69]. Molecular junction is only small portion of its application, and so we omitted it. The noise spectroscopy studied the noise signal in the current through device, which can provide fundamental information not available from other techniques when applied to molecular junctions, such as the stability of the molecule–electrode interface [70], the number of tunneling channel [71, 72], and molecular vibration effect [73]. For gated molecular junctions (*c.f.* Lee's work introduced in this chapter), the conductance can be plotted into 2D intensity diagram *vs.* bias and gating voltage. This so-called stability diagram is a powerful tool in providing rich information about the electronic and also vibronic states in the quantum transport study by physicists. The readers are referred to several excellent review articles by Lee and coworkers [74], and monographs by Cuevas and Scheer for more detailed information [27].

Appendix

d	width of the tunneling barrier
e	electron charge
E_a	hopping activation energy
$f(\varepsilon)$	fermi distribution of the electron in electrode
G	conductance
G_0	conductance quantum
$\hbar$	reduced Planck constant
I	current
I_R	reduced current
I_{trans}	current at bias V_{trans}
k	electron transfer rate
k_B	Boltzmann constant
m	electron mass
m_e	electron effective mass
N	number of energy levels in the junction
T	temperature
V	applied bias voltage
V_{DC}	DC voltage
V_m	AC modulation voltage
V_R	reduced voltage
V_{trans}	transition voltage
V_ω	AC voltage at a certain frequency

α	unitless parameter that accounts for adjusting the effective mass of the tunneling electrons and modifying the rectangular barrier model
β	tunneling decay coefficient
Γ	coupling strength between molecule and electrodes
ε_A	energy level of the redox molecule
ε_h	energy offset between the electronic state and Fermi level of the electrode
λ	reorganization energy
μ	electrochemical potential of the electrode
τ	electron transversal time
Φ	energy barrier

References

1 Song, H., Reed, M.A., and Lee, T. (2011). Single molecule electronic devices. *Adv. Mater.* 23 (14): 1583–1608.

2 Simmons, J. (1963). Generalized formula for the electric tunnel effect between similar electrodes separated by a thin insulating film. *J. Appl. Phys.* 34 (6): 1793–1803.

3 Salomon, A., Cahen, D., Lindsay, S. *et al.* (2003). Comparison of electronic transport measurements on organic molecules. *Adv. Mater.* 15 (22): 1881–1890.

4 Wang, W., Lee, T., and Reed, M.A. (2003). Mechanism of electron conduction in self-assembled alkanethiol monolayer devices. *Phy. Rev. B* 68 (3): 245302.

5 Engelkes, V., Beebe, J., and Frisbie, D. (2004). Length-dependent transport in molecular junctions based on SAMs of alkanethiols and alkanedithiols: effect of metal work function and applied bias on tunneling efficiency and contact resistance. *J. Am. Chem. Soc.* 126 (43): 14287–14296.

6 Vilan, A. (2007). Analyzing molecular current-voltage characteristics with the Simmons tunneling model: scaling and linearization. *J. Phys. Chem. C* 111 (11): 4431–4444.

7 Garrigues, A.R., Yuan, L., Wang, L. *et al.* (2016). A single-level tunnel model to account for electrical transport through single molecule- and self-assembled monolayer-based junctions. *Sci. Rep.* 6: 26517.

8 Zotti, L., Kirchner, T., Cuevas, J. *et al.* (2010). Revealing the role of anchoring groups in the electrical conduction through single-molecule junctions. *Small* 6 (14): 1529–1535.

9 Xie, Z., Bâldea, I., Smith, C.E. *et al.* (2015). Experimental and theoretical analysis of nanotransport in oligophenylene dithiol junctions as a function of molecular length and contact work function. *ACS Nano* 9 (8): 8022–8036.

10 Baghbanzadeh, M., Bowers, C., Rappoport, D. *et al.* (2015). Charge tunneling along short oligoglycine chains. *Angew. Chem. Int. Ed.* 127 (49): 14956–14960.

11 Beebe, J., Kim, B., Gadzuk, F., and Kushmerick, J. (2006). Transition from direct tunneling to field emission in metal-molecule-metal junctions. *Phys. Rev. Lett.* 97 (2): 026801.

12 Bâldea, I. (2012). Ambipolar transition voltage spectroscopy: analytical results and experimental agreement. *Phys. Rev. B* 85 (3): 035442.

13 Jia, C., Wang, J., Yao, C. *et al.* (2013). Conductance switching and mechanisms in single-molecule junctions. *Angew. Chem. Int. Ed.* 52 (33): 8666–8670.

14 Bâldea, I. (2012). Interpretation of stochastic events in single-molecule measurements of conductance and transition voltage spectroscopy. *J. Am. Chem. Soc.* 134 (18): 7958–7962.

15 Bâldea, I., Xie, Z., and Frisbie, C.D. (2015). Uncovering a law of corresponding states for electron tunneling in molecular junctions. *Nanoscale* 7 (23): 10465–10471.

16 Guo, S., Hihath, J., Diez-Pérez, I., and Tao, N. (2011). *J. Am. Chem. Soc.* 133: 19189–19197.

17 Widawsky, J.R., Kamenetska, M., Klare, J. *et al.* (2009). *Nanotechnology* 20: 434009.

18 Lee, W. and Reddy, P. (2011). *Nanotechnology* 22: 485703.

19 Xu, B. and Tao, N.J. (2003). *Science* 301: 1221–1223.

20 Xie, Z., Bâldea, I., Oram, S. *et al.* (2017). Effect of heteroatom substitution on transport in alkanedithiol-based molecular tunnel junctions: evidence for universal behavior. *ACS Nano* 11 (1): 569–578.

21 Galperin, M., Ratner, M., and Nitzan, A. (2007). Molecular transport junctions: vibrational effects. *J. Phys. Condens. Matter* 19 (10): 103201.

22 Selzer, Y., Cabassi, M., Mayer, T., and Allara, D. (2004). Thermally activated conduction in molecular junctions. *J. Am. Chem. Soc.* 126 (13): 4052–4053.

23 Choi, S., Kim, B., and Frisbie, D. (2008). Electrical resistance of long conjugated molecular wires. *Science* 320 (5882): 1482–1486.

24 Choi, S., Risko, C., Delgado, C. *et al.* (2010). Transition from tunneling to hopping transport in long, conjugated oligo-imine wires connected to metals. *J. Am. Chem. Soc.* 132 (12): 4358–4368.

25 Zhang, J., Chi, Q., Kuznetsov, A.M. *et al.* (2002). Electronic properties of functional biomolecules at metal/aqueous solution interfaces. *J. Phys. Chem. B* 106 (6): 1131–1152.

26 Migliore, A. and Nitzan, A. (2011). Nonlinear charge transport in redox molecular junctions: a Marcus perspective. *ACS Nano* 5 (8): 6669–6685.

27 Cuevas, J.C. and Scheer, E. (2017). *Molecular Electronics: An Introduction to Theory and Experiment*, 2e. World Scientific.

28 Li, X., Hihath, J., Chen, F. *et al.* (2007). Thermally activated electron transport in single redox molecules. *J. Am. Chem. Soc.* 129 (37): 11535–11542.

29 Artés, J.M., López-Martínez, M., Giraudet, A. *et al.* (2012). Current-voltage characteristics and transition voltage spectroscopy of individual redox proteins. *J. Am. Chem. Soc.* 134 (50): 20218–20221.

30 Jaklevic, J. and Lambe, R.C. (1966). Molecular vibration spectra by electron tunneling. *Phys. Rev. Lett.* 17 (22): 1139–1140.

31 Lambe, R.C. and Jaklevic, J. (1967). Molecular vibration spectra by inelastic electron tunneling. *Physiol. Rev.* 165 (3): 821–832.

32 Hansma, P. and Coleman, R.V. (1974). Spectroscopy of biological compounds with inelastic electron tunneling. *Science* 184 (4144): 1369–1371.

33 Stipe, B.C., Rezaei, M.A., and Ho, W. (1998). Single-molecule vibrational spectroscopy and microscopy. *Science* 280 (5370): 1732–1735.

34 Kushmerick, J., Lazorcik, J., Patterson, C. *et al.* (2004). Vibronic contributions to charge transport across molecular junctions. *Nano Lett.* 4 (4): 639–642.

35 Wang, W., Lee, T., Kretzschmar, I., and Reed, M. (2004). Inelastic electron tunneling spectroscopy of an alkanedithiol self-assembled monolayer. *Nano Lett.* 4 (4): 643–646.

36 Hihath, J., Bruot, C., and Tao, N. (2010). Electron–phonon interactions in single octanedithiol molecular junctions. *ACS Nano* 4 (7): 3823–3830.

37 Song, H., Kim, Y., Ku, J. *et al.* (2009). Vibrational spectra of metal-molecule-metal junctions in electromigrated nanogap electrodes by inelastic electron tunneling. *Appl. Phys. Lett.* 94 (10): 103110.

38 Galperin, M., Ratner, M., Nitzan, A., and Troisi, A. (2008). Nuclear coupling and polarization in molecular transport junctions: beyond tunneling to function. *Science* 319 (5866): 1056–1060.

39 Okabayashi, N., Paulsson, M., and Komeda, T. (2013). Inelastic electron tunneling process for alkanethiol self-assembled monolayers. *Prog. Surf. Sci.* 88 (1): 1–38.

40 Chiang, C., Xu, C., Han, Z., and Ho (2014). Real-space imaging of molecular structure and chemical bonding by single-molecule inelastic tunneling probe. *Science* 344 (6186): 885–888.

41 Yoon, H., Maitani, M., Cabarcos, O. *et al.* (2010). Crossed-nanowire molecular junctions: a new multispectroscopy platform for conduction–structure correlations. *Nano Lett.* 10 (8): 2897–2902.

42 Beebe, J., Moore, L., and Kushmerick, J. (2007). Vibronic coupling in semifluorinated alkanethiol junctions: implications for selection rules in inelastic electron tunneling spectroscopy. *Nano Lett.* 7 (5): 1364–1368.

43 Okabayashi, N., Paulsson, M., Ueba, H. *et al.* (2010). Site selective inelastic electron tunneling spectroscopy probed by isotope labeling. *Nano Lett.* 10 (8): 2950–2955.

44 Lin, L.-L., Zou, B., Wang, C.-K., and Luo, Y. (2011). Assignments of inelastic electron tunneling spectra of semifluorinated alkanethiol molecular junctions. *J. Phys. Chem. C* 115 (41): 20301–20306.

45 Troisi, A., Beebe, J., Picraux, L. *et al.* (2007). Tracing electronic pathways in molecules by using inelastic tunneling spectroscopy. *Proc. Natl. Acad. Sci. U.S.A.* 104 (36): 14255–14259.

46 Paulsson, M., Frederiksen, T., Ueba, H. *et al.* (2008). Unified description of inelastic propensity rules for electron transport through nanoscale junctions. *Phys. Rev. Lett.* 100 (22): 226604.

47 Kim, Y., Pietsch, T., Erbe, A. *et al.* (2011). Benzenedithiol: a broad-range single-channel molecular conductor. *Nano Lett.* 11 (9): 3734–3738.

48 Yu, L.H., Zangmeister, C.D., and Kushmerick, J.G. (2007). Origin of discrepancies in inelastic electron tunneling spectra of molecular junctions. *Phys. Rev. Lett.* 98 (20): 206803.

49 Song, H., Kim, Y., Jang, Y.H. *et al.* (2009). Observation of molecular orbital gating. *Nature* 462 (7276): 1039–1043.

50 Lü, J.-T., Christensen, R., Foti, G. *et al.* (2014). Efficient calculation of inelastic vibration signals in electron transport: beyond the wide-band approximation. *Phys. Rev. B* 89 (8): 081405.

51 Guo, J., Lü, J.-T., Feng, Y. *et al.* (2016). Nuclear quantum effects of hydrogen bonds probed by tip-enhanced inelastic electron tunneling. *Science* 352 (6283): 321–325.

52 Xu, C., Jiang, J.L., Han, Z.M., and Ho, W. (2016). Nature of asymmetry in the vibrational line shape of single-molecule inelastic electron tunneling spectroscopy with the STM. *Phys. Rev. Lett.* 116 (16): 166101.

53 Galperin, M., Nitzan, A., and Ratner, M. (2006). Resonant inelastic tunneling in molecular junctions. *Phys. Rev. B* 73 (4): 166101.

54 Cuevas, J.C. and Scheer, E. (2010). *Molecular Electronics: An Introduction to Theory and Experiment*. World Scientific Publishing Co. Pte. Ltd.

55 LeRoy, B.J., Lemay, S.G., Kong, J., and Dekker, C. (2004). Electrical generation and absorption of phonons in carbon nanotubes. *Nature* 432 (7015): 371–374.

56 Cao, H., Jiang, J., Ma, J., and Luo, Y. (2008). Identification of switching mechanism in molecular junctions by inelastic electron tunneling spectroscopy. *J. Phys. Chem. C* 112 (29): 11018–11022.

57 Kim, Y., Garcia-Lekue, A., Sysoiev, D. *et al.* (2012). Charge transport in azobenzene-based single-molecule junctions. *Phys. Rev. Lett.* 109 (22): 226801.

58 Kim, Y., Hellmuth, T.J., Bürkle, M. *et al.* (2011). Characteristics of amine-ended and thiol-ended alkane single-molecule junctions revealed by inelastic electron tunneling spectroscopy. *ACS Nano* 5: 4104–4111.

59 Naleway, C., Curtiss, L., and Miller, J. (1991). Superexchange-pathway model for long-distance electronic couplings. *J. Phys. Chem.* 95 (22): 8434.

60 Betts, J., Beratan, D., and Onuchic, J. (1992). Mapping electron tunneling pathways: an algorithm that finds the "minimum length"/maximum coupling pathway between electron donors and acceptors in proteins. *J. Am. Chem. Soc.* 114 (11): 4043–4046.

61 Guo, C., Yu, X., Refaely-Abramson, S. *et al.* (2016). Tuning electronic transport via hepta-alanine peptides junction by tryptophan doping. *Proc. Natl. Acad. Sci. U.S.A.* 113 (39): 10785–10790.

62 Yu, X., Lovrincic, R., Sepunaru, L. *et al.* (2015). Insights into solid-state electron transport through proteins from inelastic tunneling spectroscopy: the case of Azurin. *ACS Nano* 9 (10): 9955–9963.

63 Fereiro, J.A., Yu, X., Pecht, I. *et al.* (2018). Tunneling explains efficient electron transport via protein junctions. *Proc. Natl. Acad. Sci. U.S.A.* 115 (20): E4577–E4583.

64 Wei, W., Li, S., Millstone, J.E. *et al.* (2009). Surprisingly long-range surface-enhanced Raman scattering (SERS) on Au–Ni multisegmented nanowires. *Angew. Chem. Int. Ed.* 48 (23): 4210–4212.

65 Chen, X., Braunschweig, A.B., Wiester, M.J. *et al.* (2009). Spectroscopic tracking of molecular transport junctions generated by using click chemistry. *Angew. Chem. Int. Ed.* 48 (28): 5178–5181.

66 Tian, J.-H., Liu, B., Li, X. *et al.* (2006). Study of molecular junctions with a combined surface-enhanced Raman and mechanically controllable break junction method. *J. Am. Chem. Soc.* 128 (46): 14748–14749.

67 Liu, Z., Ding, S.-Y., Chen, Z.-B. *et al.* (2011). Revealing the molecular structure of single-molecule junctions in different conductance states by fishing-mode tip-enhanced Raman spectroscopy. *Nat. Commun.* 2 (2): 305.

68 Ioffe, Z., Shamai, T., Ophir, A. *et al.* (2008). Detection of heating in current-carrying molecular junctions by Raman scattering. *Nat. Nanotechnol.* 3 (12): 727–732.

69 Cahen, D. and Kahn, A. (2003). Electron energetics at surfaces and interfaces: concepts and experiments. *Adv. Mater.* 15 (4): 271–277.

70 Xiang, D., Sydoruk, V., Vitusevich, S. *et al.* (2015). Noise characterization of metal-single molecule contacts. *Appl. Phys. Lett.* 106 (6): 063702.

71 Ben-Zvi, R., Vardimon, R., Yelin, T., and Tal, O. (2013). Electron–vibration interaction in multichannel single-molecule junctions. *ACS Nano* 7 (12): 11147–11155.

72 Karimi, M., Bahoosh, S., Herz, M. *et al.* (2016). Shot noise of 1,4-benzenedithiol single-molecule junctions. *Nano Lett.* 16 (3): 1803–1807.

73 Tal, O., Krieger, M., Leerink, B., and van Ruitenbeek, J.M. (2008). Electron-vibration interaction in single-molecule junctions: from contact to tunneling regimes. *Phys. Rev. Lett.* 100 (19): 196804.

74 Xiang, D., Wang, X., Jia, C. *et al.* (2016). Molecular-scale electronics: from concept to function. *Chem. Rev.* 116 (7): 4318–4440.

5

Single-Molecule Electronic Devices

Shengxiong Xiao

Department of Molecular Functional Materials, College of Chemistry and Materials Science, Shanghai Normal University, No. 100, Guilin Road, Xuhui District, 200234 Shanghai, P. R. China

5.1 Introduction

The developments of micro fabrication and field effect transistors (FETs) play a key role in today's information society. Without microprocessors which continuously become cheaper and smaller, it is impossible to acquire superfast and omnipresent electronic devices, information technology, the internet, and mobile communication technologies. From the first computer to today's mobile devices, the huge leap in performance is achieved by miniaturization of the active components to a large extent. However, there is still an enormous challenge in the field of miniaturization of electronic devices. Molecular electronics might be a potential solution, 1 nm scale chemically identical molecules which can be synthesized in molar amounts and possess various electronic characteristics including current limiting, rectification, and switching, reminiscent of their solid state device counterpart. The experimental results of single-molecule electron transport put forth the idea that the realization of single-molecule electronics (SMEs) will surpass the miniaturization limit of silicon complementary metal-oxide semiconductor (CMOS) technology, and create the next era in the field of miniaturization of electronic components [1].

The field of SME utilizes engineering, physics, and chemistry expertise to realize circuit components at the limits of miniaturization. It is a subfield of nanoelectronics in which the electronic components are single molecules. Semiconducting silicon-based devices with increasingly small dimensions, such as diodes, transistors, and memory elements have been developed over the past century. Meanwhile, the continuing study through reaction chemistry, advanced physical chemistry, and theoretical methods gives a detailed understanding of the relationship between molecule's chemical structure and electronic properties [2]. These multidisciplinary efforts have converged in the field of SMEs, with the ultimate goal of using molecules as active elements in electronic circuitry following first theoretical proposal by Aviram and Ratner in the 1970s [3]. With such a rich knowledge base of detailed molecule's electronic properties in hand, tremendous advancements

Nanogap Electrodes, First Edition. Edited by Tao Li.

might be realized if it is reimagined in the frame work of SMEs. Powerful methods have been developed to characterize and manipulate the conductance properties of single molecules [4–7]. Scanning tunneling microscopy (STM) [8], mechanically controlled break junctions [9–11], STM break junctions [12, 13], conductive atomic force microscopy [14], electromigration [15], nanoparticle arrays [16, 17], and other approaches [18–21] can be used to measure molecular conductance. The sophistication of these techniques requires chemists to collaborate with physicists and engineers, with the effort to incorporate well-understood chemical principles into the study of structure-conductivity relationships in molecular wires.

5.2 Wiring Molecules into "Gaps": Anchoring Groups and Assembly Methods

When discussing the structure–property relationships of single-molecule junctions, there are three distinct components to be considered, that is the anchor, the electrode, and the bridge (Figure 5.1) [2].

5.2.1 Anchor Groups

The anchor groups (also known as the linkers or contact groups) both mechanically and electronically connect molecular lines to the electrodes. Usually, a single anchoring group terminates each end of the molecule to form the metal–molecule–metal junction; however, additional handles for tuning conductance can be offered by including more anchoring units along the molecular bridge, determined by which two anchors form the most conductive pathway [22–27]. Direct contact between metal and the molecular kernel, the mechanical stability and also electronic transparency (weak or strong coupling) of the anchoring group need to be considered [28] when forming a molecular junction.

The most explored example of anchoring groups is thiols on gold surfaces. But as shown in Figure 5.2 [29], in addition to thiol (–SH) [30–34], some functional groups which have good affinity for gold could also be used as anchoring groups, such as selenium (–SeH) [35, 36], amino (–NH_2) [37, 38], isocyano (–NC)

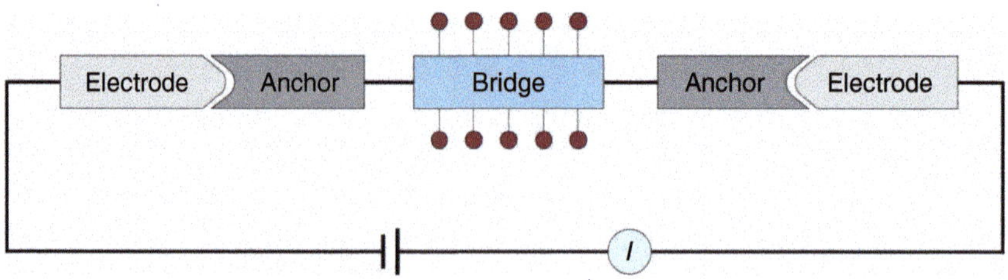

Figure 5.1 A schematic of a single-molecule junction with electrode, anchor and bridge components. The bridge unit can be further deconstructed into backbone (blue block) and substituent (red circles) subunits. *I*, current. Source: Su et al. 2016 [2]. 2016, Springer Nature.

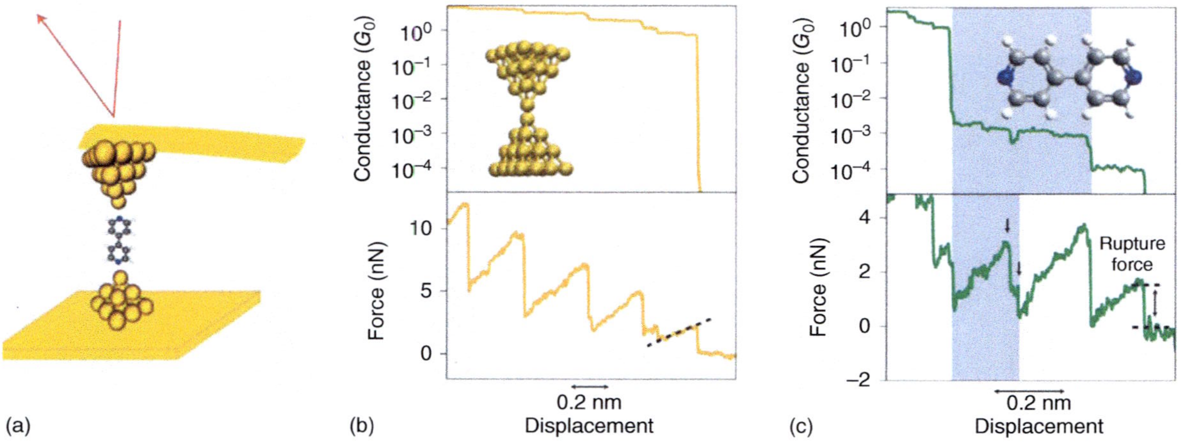

Figure 5.2 Working principle of AFM: (a) Schematic of a 4,4′-bipyridine junction formed between a gold-coated AFM tip and a gold substrate. Using a laser beam, the deflection of the cantilever can be monitored, and the corresponding mechanical force between the tip and the sample can be obtained. (b, c) Experimentally measured conductance (upper panel) and corresponding simultaneously acquired force (lower panel) traces during the junction elongation of bare gold junctions (b) and molecular junctions (c). The shaded areas in the upper panel represent high-conductance molecular regimes. The shaded areas in the low panel represent structural rearrangements within the high-conductance regime. The dashed lines indicate a complete rupture of the molecular junctions. Source: Aradhya et al. [29]. 2012, Springer Nature.

[37, 39], cyano (-CN) [40, 41], pyridyl [12, 42], carboxyl acid (-COOH) [38], nitro ($-NO_2$) [41], methylthioether ($-SCH_3$) [35], phosphino [35], and fullerenes [43–45].

The first explored anchoring groups for charge transport experiments were thiols, which are still the most widely used. The thiols–gold (as well as silver and copper) bond is very strong, even stronger than the gold–gold bond itself [10, 46]. Amines form a donor–acceptor bond to gold surfaces, preferably with under-coordinated gold [37], and carboxyl acids are considered to bind by ionic and coordinating interactions [47]. The contact resistance is sequenced as Au–S > Au–NH_2 > Au–COOH [38, 48], due to the fact that different types of anchoring groups have different electronic coupling. Because of weaker binding strength of amines and carboxyl groups, a higher bias on the electrode causes instability in the metal–molecule bond. The stability of methyl phosphine, methyl sulfide, and primary amine has also been examined in the order: PMe_2 > SMe > NH_2 [35, 49]. The probability and stability of the junction formation of pyridyl, thiol, amine, and cyano have been compared and resulted in the sequence: pyridyl > SH > NH_2 > CN [50].

The formation of efficient and reproducible molecule–metal electrode coupling is one key factor for the integration of molecules into circuits in the future [51]. To date, gold has been the electrode of choice due to its noble metal properties. Future developments could be focused on other materials, such as silicon or carbon. Nanoscale junctions of Si have been fabricated, but the Si—C bond is not as mobile as the S—Au bond, which makes it more difficult to make a well-defined self-assembled monolayer (SAM) [52]. Another approach is using carbon-based electrodes. Graphene and other 2D materials can be grown without defects at the wafer scale, this 2D structure might bring advantages in the scope of number of contact geometries compared to 3D structure. As shown in Figure 5.3, Guo and coworkers [53, 54] functionalized graphene with molecular wires in situ by covalent bonds [54]. The junction was fabricated by dash-line lithography, then the nanogaps were cut out by oxygen plasma ion etching. The carboxyl acid-terminated graphene was produced by etching, and was further functionalized with groups such as amine terminated molecular wires. van der Zant et al. created graphene electrodes by feedback-controlled electro-burning. As shown in Figure 5.4, the molecular system was introduced in a second step, and the coupling between molecule and graphene was established by π–π stacking [19].

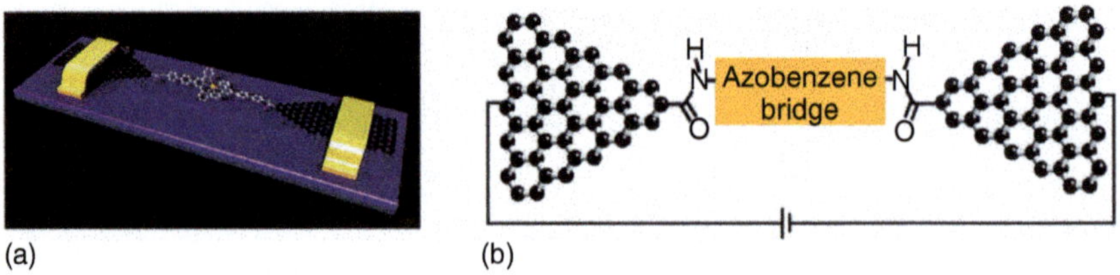

Figure 5.3 (a) Depiction of a single-molecule device based on graphene [53]. (b) Schematic of a graphene–azobenzene junction [54]. Source: Cao et al. [53] and Cao et al. [54]. 2013, John Wiley & Sons.

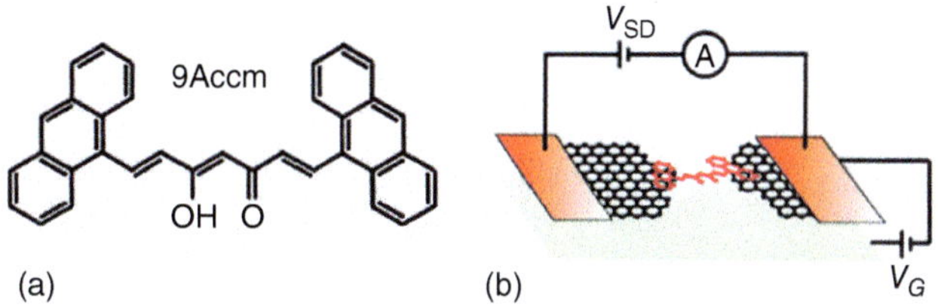

Figure 5.4 Anthracene terminated curcuminoid wire (a) brought into a graphene nanogap. (b). Source: Prins et al. [19]. 2011, American Chemical Society.

Molecular electronics experiments have also been conducted with other electrode materials in which similar anchoring groups and the same molecular wires can be used. Zhou et al. used an electrochemical jump-to-contact scanning tunneling microscope break junction (EC-STM-BJ) to create gold and copper clusters [55]. Different bipyridyl molecules were used, and the conductance between Au and Cu electrodes were compared, where the Cu electrodes showed a lower conductance. The different electronic coupling efficiencies between the metal and the wire were considered to be the cause of this effect. In another study, Danilov et al. compared devices with gold (Au) and lead (Pb) electrode material and found that similar molecular charging energies occurred in the Pb electrode case, but with different open state voltages compared to the gold electrodes. For example, the −2.6 V gate voltage separates the same open state at +0.8 V on the gold electrode [56]. The differences in the Fermi level of Au and Pb is believed to be the cause of this effect.

Anchoring groups typically bind to electrodes either through donor–acceptor (dative) interactions or through covalent bonding. Prototypical anchor groups for each type of electrode–linker interaction are shown in Figure 5.5a. Due to the fact that gold is the predominant electrode material in SME studies, the focus is on the interaction between anchor groups and gold electrodes. The electron donation from a π donor or a lone pair donor to a Lewis acidic Au atom is involved within dative interactions. Common π donors include fullerenes [43, 44] and other π-conjugated hydrocarbons [19, 26, 29, 57]. Within coordination chemistry, many lone pair anchoring groups are common σ-donor ligands [58]. Dative contacts, such as amines, can narrows the conductance distribution, because they limits the Au–linker contact geometry by selectively binding to undercoordinated adatoms on the electrode surface, which is advantageous [37]. Covalent bonding between molecular radicals and metallic electrode surfaces produce covalent contacts between the metal and the molecule, which are valuable because they are physically robust linkages that strongly couple the electronics of the molecule and the metal.

Other than the class of the anchoring group, the interaction between the anchor group and other components of the junction (the bridge and the electrode), are also decisive factors of the conductance. In Section 5.2.2, we explore how the spatial overlap between the orbitals of these three components affects the charge-transport properties of the junction.

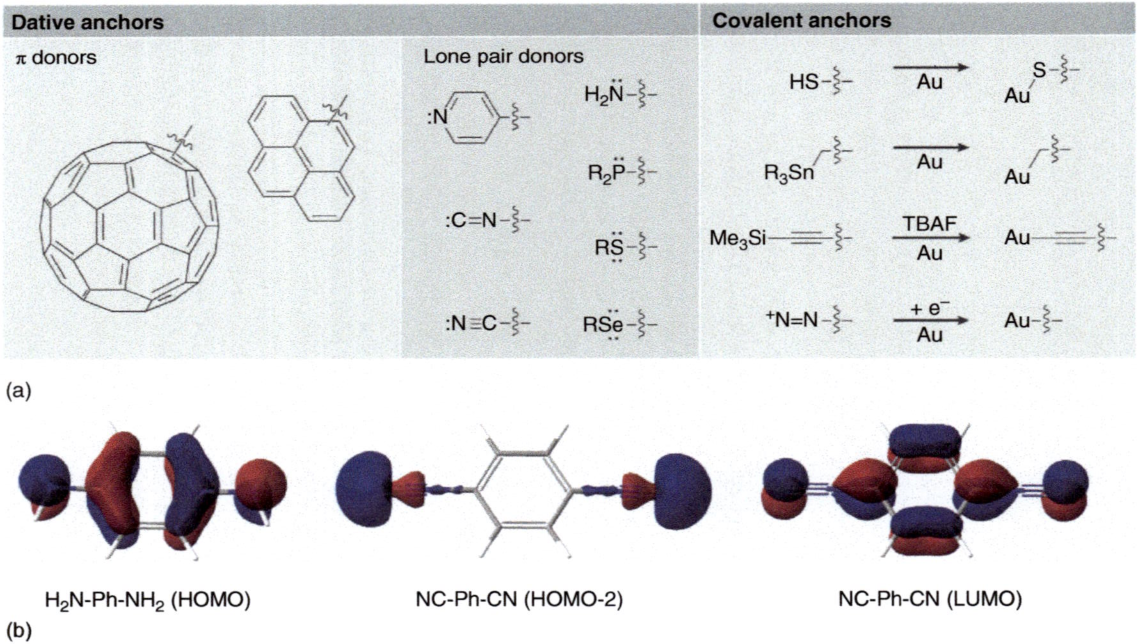

Figure 5.5 Anchor group archetypes and the nature of charge carriers for common dative anchors. Source: Su et al. [2]. 2016, Springer Nature. (a) Molecular structures of common anchors. Dative anchors can be classified as π donating or lone pair donating. For lone pair donors, the anchors shown in the left and in the right columns impart lowest unoccupied molecular orbital (LUMO)- and highest occupied molecular orbital (HOMO)-dominated conductance, respectively, in simple π-conjugated systems. Covalent anchors commonly used to generate direct Au–S and Au–C contacts are shown in the last column. These contacts can be generated from thiol oxidation, Au–Sn transmetalation, fluoride-initiated desilylation and diazonium electroreduction reactions. (b) The highest energy molecular orbital surfaces that feature strong lone pair character are depicted for 1,4-diaminobenzene (left panel) and 1,4-dicyanobenzene (middle panel) (B3LYP/6-31G**). In 1,4-diaminobenzene, the N-centered lone pairs occupy p orbitals that, like the benzene π orbitals, are perpendicular to the plane of the ring. In 1,4-dicyanobenzene, these N-centered lone pair orbitals are orthogonal to the π channel of the benzene ring; conductance in this system is dominated instead by transport through the LUMO (right panel).

There is a fact that significant measurement-to-measurement variability experiments exists in single-molecule, this is due to fluctuations in the molecular conformation, the electrode–anchor contact geometry and the electrode surface geometry. To better understand the nature of conductance in a single molecule, researchers must deal with this variability, a common approach is to compile hundreds to thousands of measurement traces into conductance histograms to obtain a distribution of all measured conductance values. The conductance peak values from such histograms reported in units of G_0 is the conductance values that we report here. G_0 is the conductance quantum and is defined as $G_0 = 2e^2/h = 77.5\,\mu\text{S}$, where e is the charge of an electron and h is Planck's constant; it is the predominant unit used to describe the conductance between metal point contacts as well as molecular conductance.

5.2.2 Effect of Anchor–Bridge Orbital Overlaps on Conductance

The anchor group usually dictates whether the molecular wire transports holes (highest occupied molecular orbital [HOMO]-dominated conductance) or electrons (lowest unoccupied molecular orbital [LUMO]-dominated conductance). Typically, the orbital that is closest in energy to the electrode Fermi level E_F, is the dominant conducting molecular orbital. Conductance depends on two variables: ΔE, the energy offset between E_F and the conducting orbital, and Γ, the strength of the electrode–molecule hybridization (Figure 5.6).

The conductance of a molecular junction is related to the probability of electron transmission between the electrodes through the Landauer formula in the limit of coherent transport [59]. Often, this transmission probability strongly depends on the degree to which a single molecular orbital hybridizes to a continuous band of energy level on the metal electrode. In these cases, transmission primarily depends on the orbital's energetic alignment (ΔE) with respect to the metal Fermi level (E_F) and on the orbital's broadening (Γ) caused by hybridization, which gives electrons on this level a finite lifetime $\sim\hbar/\Gamma$. In the figure, the red Lorentzian curves (LUMO) and blue Lorentzian curves (HOMO) depict a typical energy transmission function. The transmission function shown here describes HOMO-dominated conductance where the HOMO transmission peak is closest in energy to and intersects the E_F; in contrary, picture for LUMO-dominated conductance would show the LUMO transmission peak intersecting the E_F. The parameters Γ and ΔE mainly depend on intrinsic molecular and metallic characteristics, such as the nature of the molecular orbital, the molecule–metal bond character (covalent or donor–acceptor), the electronic density of states of the metal and the charged interactions at the metal–molecule interface. Often, a broadened transmission resonance (large Γ) occurs within orbitals that have density on the metal-binding anchor groups, whereas a very narrow resonances (small Γ) can be resulted from orbitals that localize electron density on the molecular backbone. The width of the resonance can also be affected by the length of the molecule; small molecules, such as H_2 and short π-conjugated wires, because their orbitals effectively hybridize with the metal, are strongly coupled, which results in a large Γ. For longer molecular wires, Γ is

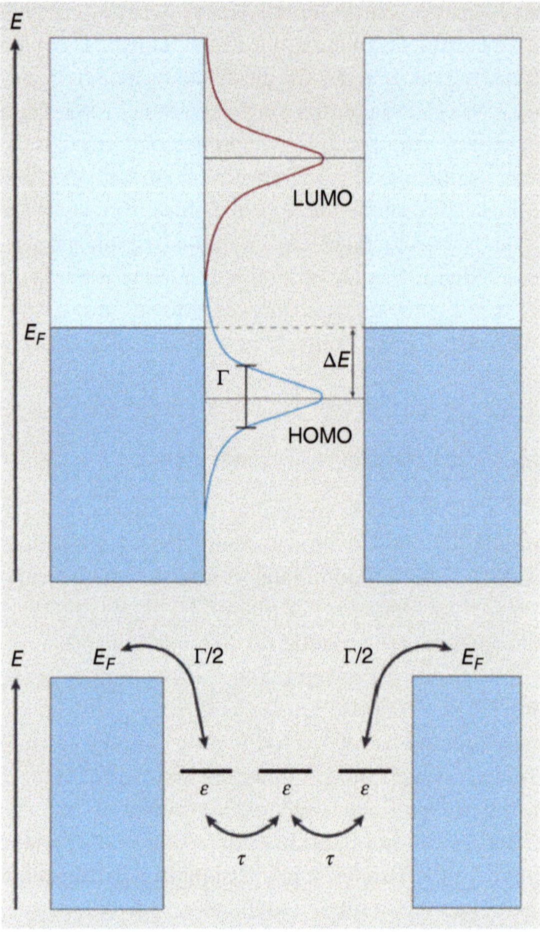

Figure 5.6 Principles of charge transport in single-molecule junctions. Source: Su et al. [2]. 2016, Springer Nature.

smaller due to the fact that a larger proportion of the hybridized orbital is localized on the molecular backbone. In oligomeric materials, with weak conjugation or coupling strength (τ) between oligomer units, Γ decreases with increasing n more quickly. This is related to the reason why the length-dependent conductance decay parameter (β) is small for conjugated repeat units but large for non-conjugated repeat units. The lower panel of the figure schematically represents a tight-binding model that assigns a single level (ε) to each repeat unit and the parameter τ which describes the electronic coupling between them. The coupling of these units with the electrodes is described by $\Gamma/2$.

The nature of the conducting orbital can be determined experimentally through thermopower measurements [60, 61] or computationally by transmission calculations [62]. However, in cases with simple structures, we can use basic chemical principles to predict the nature of charge carriers, such as considering the nature of the molecular backbone and the geometry of the lone pair relative to the conjugated orbitals of the molecule (Figure 5.5b). We use a phenyl ring terminated at the para positions by the dative linker groups from Figure 5.5a to illustrate this method.

Other basic aromatic wires can apply this analysis as well. In cases that benzene rings terminated with linkers such as –SR, $-NH_2$, $-PR_2$, or –SeR at the 1,4-positions, the lone pair orbitals are coplanar with and energetically destabilized by the filled π-conjugated bridge orbitals, thus are included in the HOMO. As a result, when such anchor groups are used, conductance occurs strongly through the HOMO owing to the alignment of the Au–lone pair bonds with the π system of the bridge.

By contrast, conductance in phenyl rings terminated by pyridine [63], isocyanide [64], and cyanide [65] linkers occurs primarily through the LUMO. For these anchor groups, the lone pair lies in the σ plane of the molecule, rigidly orthogonal to the π channel of the wire. Conduction through the lone pair orbitals is weak because the carbon sp^2 σ orbitals are poorly conjugated [66]. Moreover, the lone pair orbital is part of the σ system thus is quite low in energy, through which transport has a marginal contribution to conductance (ΔE is large). Just as importantly, these linker groups possess electron-withdrawing nature, which facilitates LUMO-dominated conductance by lowering the energy of the π^*-antibonding orbitals toward the E_F (and the HOMO-conducting π-bonding orbitals away from the E_F). The conductance is then controlled by the coupling between the electrodes and the π^*-antibonding orbitals of the LUMO. The HOMO and LUMO energies are substantially lowered when molecular bridges are very electron-deficient structures, such as thiophene dioxides [67] and porphyrins [68], conductance through the LUMO can dominate regardless of the type of linker used.

From the perspective of coordination chemistry and of the different modes of interaction between ligands and transition metals, HOMO and LUMO conduction can also be understood. When the metal–molecule bond is formed using the molecule's σ-donor orbital (which in most cases is the HOMO of the isolated molecule) and the metal's σ-accepting orbital, hole transport or HOMO-dominated conduction occurs. The metal–molecule bond, or the "gateway," can mix with the delocalized, conjugated bridge orbitals, and establish the conductivity path because of this coplanarity. Conversely, if the metal-to-molecule σ bond cannot mix with the conjugated pathway, the charge carriers cannot use the molecular HOMO; it is geometrically unavailable and energetically distant from the E_F. However, from coordination and organometallic chemistry, it is well known that ligand-to-metal σ donation was often supplemented by d_{metal} to π^*_{ligand} "back-bonding," in which occupied orbitals of the metal mix with unoccupied orbitals of the ligand, although it usually dominates metal–ligand bonding. In the case of molecular conduction, this π^* back-bonding (LUMO-transporting) pathway may be available, when the σ-donation (HOMO-transporting) pathway is unavailable if the π^* orbital into which the metal back-donates is a conjugated orbital that spans the molecule and is connected to both electrodes.

The strength of the electronic coupling between the metal and the molecule could be determined by the stereochemistry of the Au–anchor bond with respect to the conductive channel of the molecular bridge, manipulating this stereochemistry is a powerful handle for controlling conductance. If the Au–anchor bond is aligned with the conjugated orbitals of the molecule, the charge flow between the electrodes increases. A dihydrobenzothiophene (BT) thioether linker can be used to lock the

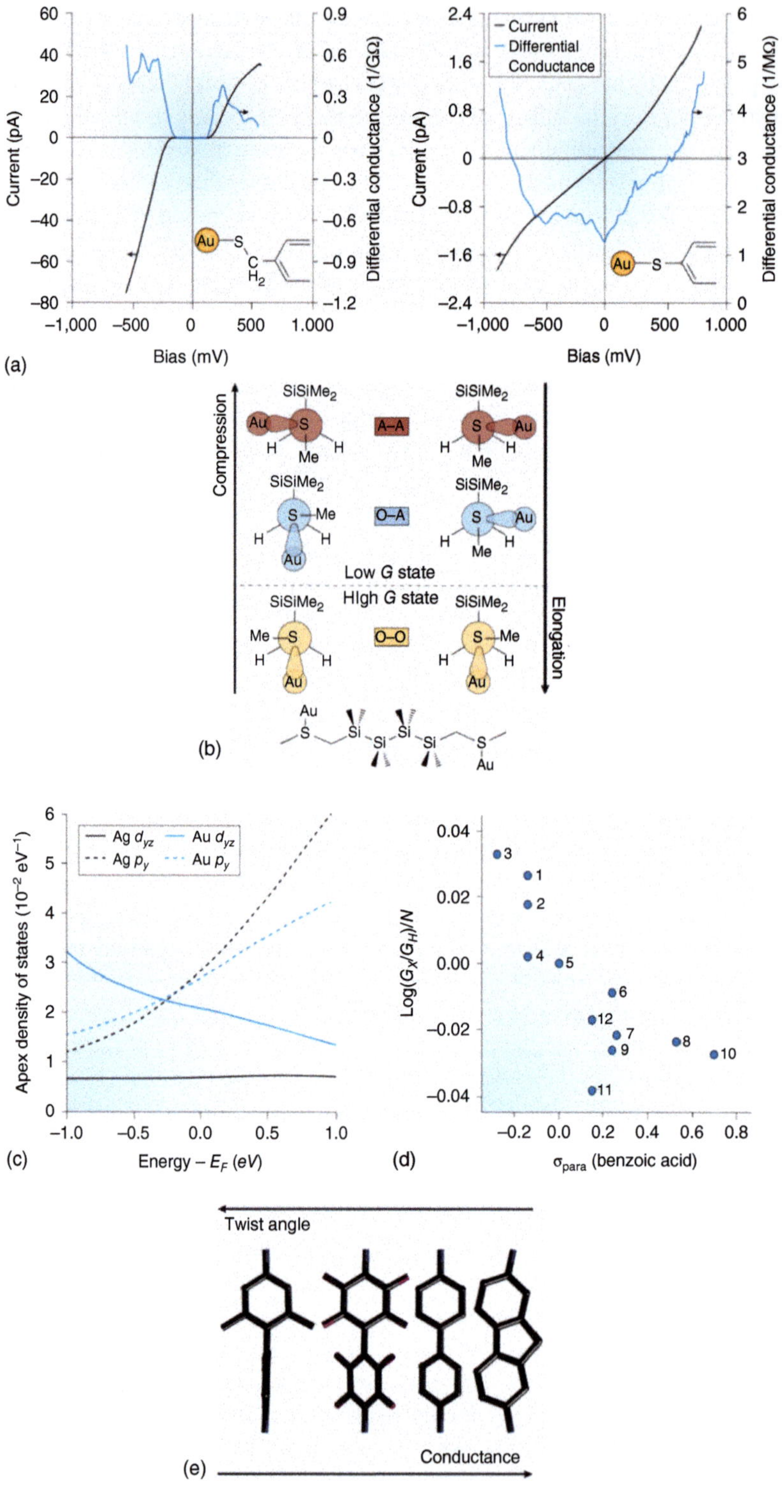

(a)
Current (pA)
Differential conductance (1/GΩ)
Bias (mV)
Au
S
C
H2
Current (pA)
Differential conductance (1/MΩ)
Current
Differential
Conductance
Bias (mV)
(b)
Compression
Elongation
SiSiMe2
A–A
O–A
O–O
Low G state
High G state
Au
S
Me
H
Si
(c)
Apex density of states (10−2 eV−1)
Energy − EF (eV)
Ag dyz
Au dyz
Ag py
Au py
(d)
Log(GX/GH)/N
σpara (benzoic acid)
(e)
Twist angle
Conductance

Figure 5.7 Tuning the structure of the anchor, electrode and bridge to modulate charge-transport properties in single-molecule junctions. Source: Su et al. [2]. 2016, Springer Nature. (a) Current and differential conductance plotted against source-drain voltage bias at a constant 1 V gate voltage for an oligo(*p*-phenylene vinylene) molecule with (left panel) and without (right panel) methylene insertions adjacent to the thiol. The tunneling barrier introduced by the methylene unit decouples the molecule from the electrodes and a Coulomb blockade is observed. Source: Danilov et al. [56]. Reproduced with permission of American Chemical Society. (b) Newman projections depicting the *anti* (A) and *ortho* (O) terminal dihedral conformations in an Au–oligosilane–Au junction. Junction elongation causes the terminal electrode–anchor dihedral geometry to shift from low conducting (low *G* state), with weakly coupled A–A and O–A configurations, to high conducting (high *G* state), with a strongly coupled O–O configuration. Source: Su et al. [73], Nature Publishing Group. (c) Density of states around the Fermi level (E_F) for p_y and d_{yz} orbitals in Ag and Au. The stronger d-orbital character of Au compared with Ag near the E_F results in a stronger d–π^* coupling for pyridine anchors and, therefore, an increased conductance. Source: Adak et al. [74], American Chemical Society. (d) The Hammett plot shows the relationship between conductance and each substituent's Hammett coefficient σ_{para} in substituted benzene rings. Source: Venkataraman et al. [75], American Chemical Society. (e) Conductance decreases with an increasing twist angle between biphenyl rings. G_H, conductance of unsubstituted molecule; G_X, conductance of substituted molecule; *N*, number of substituents. Source: Venkataraman et al. [13], Nature Publishing Group.

position of sulfur lone pairs into alignment with the molecular π backbone [69]. Compared with analogous aromatic wires with thiomethyl linkers that can freely rotate, an increased conductance and a sharper conductance peak can be observed in the frustrated rotation of the S lone pair. The BT linker has been incorporated into several different molecular wires to strengthen the anchor–bridge coupling [27, 70–72].

Poor coupling between the electrode and the molecule can also be a desirable quality. Inserting methylene (CH_2) units between the S atom and the phenyl ring can disrupt the electrode–molecule coupling. The methylene spacers decouple the molecule from the source and the drain electrodes, thus allow gating effects, such as Coulomb blockade, to occur in three-electrode systems [56] (Figure 5.7a). However, the junction conductance has decreased by 3 orders of magnitude in π-conjugated molecules due to the reduction in molecule–electrode coupling [56]. Engineering anchor groups with both strong and weak electrode–molecule coupling character can implement interesting functions into molecular electronic devices. A bulky –SPh anchor group can be used to misalign the S—Au bond with the molecular bridge, decreasing the molecule–electrode coupling [71]. A single-molecule rectifier can be created through this, with molecule strongly coupled to the electrode by a covalent Au—C bond at one end and weakly coupled by an Au—SPhR bond at the other end. A class of oligosilanes and oligogermanes that switch between different conductance values has recently been described, the conductance values of which depend on the strength of the coupling between the electrode and the molecule [73, 76]. The terminal ends of the molecule is stretched into a geometry (ortho–ortho [O–O]) that is optimal for conductance and electrode–molecule coupling by junction elongation; junction compression relaxes the molecule into dihedral geometries (anti–anti [A–A] or ortho–anti [O–A]) with diminished electrode–molecule coupling and lower conductance (Figure 5.7b). This difference in conductance arises from the weaker d_{yz}-orbital character of the Ag density of states at the E_F that results in reduced d–π* hybridization and metal–molecule coupling (Figure. 5.7c).

5.2.3 In Situ Chemical Reactions to Produce Covalent Contacts

Au–S linkages formed from the reduction of thiol (–SH) linkers on Au surfaces are the most widely studied form of covalent contact. With this strong Au–S linkages, the molecular junction can withstand harsh external conditions such as mechanical stress [77] and high-bias voltages [78]. It is problematic that thiols can oxidize to disulfides rather easily under ambient conditions; whereby dithiolated molecular wires can polymerize and form insoluble polydisulfides [79]. Although S—S bonds can be reduced on gold surfaces [80], there is no guarantee that the reduction will be exhaustive and that only the monomers will contribute to conduction measurements. Using thioacetate-protecting groups to functionalize the thiols is a common approach to enhance the ambient stability of thiol-based wires, because it can cleave on the electrode surface to form covalent Au–S contacts [81, 82]. Due to the large variability in the anchor–electrode contact geometry, thiol-based junctions tend to show broad conductance features. Several efforts have been made to study the effect of binding geometry on conductance [38, 81, 83–85], but understanding and thus gaining control over the variability of contact geometry is an ongoing challenge in the field [68, 82, 86, 87].

Au–C contacts are particularly promising covalent anchors because they give well-defined conductance peaks owing to the selective binding to undercoordinated gold [88]. Furthermore, higher conductance peak values are commonly observed in molecular wires with Au–C contacts, compared with structurally analogous wires with Au–S contacts. For example, the Au–benzenedithiol–Au junction shows a broad conductance of conductance peak values ranging from 10^{-2} to 10^{-4} G_0 [10, 89–91]. By contrast, the Au–xylyl–Au junction with similar structural conducts at 0.9 G_0 [92]. The difference in conductance arises from two factors. First, the C—Au bond is more strongly coupled to the π system than the S—Au bond due to its much shorter bond length. Second, for the Au–C gate-way states which describe the covalent metal–molecule hybridization, ΔE is much smaller. For example, the Au–S gateway state is centered at $E - E_F = -1.4$ eV for alkane dithiols [85], whereas the Au–C gateway state is centered at $E - E_F = -0.8$ eV for bis(trimethylstannyl)alkanes [92]. This difference in energy alignment with the E_F contributes significantly to the difference in conductance.

Three methods have been developed to achieve the in-situ generation of direct Au–C covalent contacts. The first method involves the generation of C—Au bonds through transmetalation of C—SnR_3 bonds on gold surfaces, which can also produce tin oxide by-products under ambient conditions [93]. Firstly, this method was used to obtain self-assembled alkane monolayers on gold surfaces from organotin species [94], then, to obtain covalent Au-arene and Au–alkane contacts in single-molecule junctions, it was used for the in situ cleavage of terminal C—$SnMe_3$

bonds [92]. A 10- to 100-fold increase in conductance is observed in alkane [92] and *para*-phenylene [46] wires terminated with Au–CH_2R contacts compared with the analogous bridges terminated with dative Au–NH_2R contacts. The applicability of this method was recently expanded to include Au–acetylene contacts [95]. The fact that this approach both uses and produces toxic and volatile trimethyltin species might be a potential shortcoming. In addition, this reaction does not occur universally for all organotin molecules; for example, this manner of cleavage does not occur in perfluorinated benzene backbones [46]. The bond between the molecular bridge and the tin atom must be the most reactive of the four organotin bonds is the most important molecular design rule for creating C–Au contacts via C–SnR_3 transmetalation on gold. For example, *n*-alkane backbones with $SnBu_3$ end groups do not show clean conductance features in the STM break junction [92], presumably because the cleavage of the four C—Sn bonds is not selective. By contrast, Au–$(CH_2)_n$–Au junctions form cleanly for trimethyltin-terminated alkane wires because of the greater stability of $RH_2C\cdot$ radicals compared with $H_3C\cdot$, which enables the preferential cleavage of the RH_2C—Sn bond. Similarly, benzyltrimethyltin molecules cleave instantaneously at the Sn—benzyl bond, even at −110 °C [88], whereas Sn—aryl bonds cleave slowly, with Au–aryl–Au junctions appearing only after 2.5 hours at room temperature [92]. This design strategy allows the programming of the junction that will form on Sn–C cleavage.

The second method for obtaining covalent Au–C contacts involves a fluoride-initiated desilylation of oligo(phenylene ethynylene) wires terminated with trimethylsilyl (TMS) end groups [96]. The terminal ethynyl—Si bonds is selectively cleaved by adding tetrabutylammonium fluoride to a solution of the TMS-protected target molecules. This approach is inspired from a classic synthetic chemistry method that exploits the strong affinity between silicon and fluorine to unmask acetylene groups [97]. The NBu^{4+} electrolytes participate in ionic conductance between the electrodes within this approach, whereby the electrolytes give rise to significant conductance noise that may cover the signal of low-conductance molecules, thus hinder the applicability of this method. However, ionic conductance can be reduced by coating the electrodes with an insulating layer [98].

The third method to produce covalent electrode anchors involves the electroreduction of diazonium salts on gold surfaces [99]. It was first used in the context of break junction experiments through electrochemically reducing the terminal diazonium end groups on a biphenyl ring to generate covalent Au–biphenyl–Au junctions [100]. This is a desirable approach, because by increasing the reduction potential via a gate electrode to irreversibly cleave the aryl—N bond, covalent Au—C contacts can be generated on demand. However, diazonium salts used in this approach are known to be thermally unstable and, in many cases, explosive [101]. In particular, alkyl diazonium salts are especially unstable, which limits the range of diazonium-functionalized structures that can be easily measured in single-molecule junctions.

There are still many unsolved issues for implementing single-molecule devices with covalent contacts. The choice of precursors for the desilylation and diazonium reduction methods has been limited thus far to those that place the Au—C bonds in the σ plane of the molecular bridge. This is an important consideration because maintaining coplanarity between the metal–carbon bond and the bridge π system is essential for optimizing the coupling between molecule and electrode. Moreover, molecular wires with Au–C contacts tend to oligomerize in situ during break junction experiments, thus are not particularly robust. This may be unavoidable under ambient conditions, as Au—C bonds are inherently sensitive to oxidation and dimerization pathways. One possible solution to avoid device failure is tuning the structure of the wire to make the metal–carbon bond more stable. Another possible route to enhance stability is using electrode materials that form more stable electrode–carbon bond; as we discuss in Section 5.3.4.1.

5.3 Electrical Rectifier

Molecular rectification, consists of a p donor system and a p acceptor with both connected to a r insulator bridge [102], was first theoretically proposed by Aviram and Ratner [3]. Since then, the molecular diode as an example of a functional molecular electronic device has attracted substantial interest from experimental demonstration. Experimental realizations have been performed in both single-molecule [71, 103–106] or SAMs [107–118]. Recently, considerable progress was made based on a better understanding of the molecule/electrodes coupling or/and intermolecules [71, 103–106, 112–118]. Although the experimental evidence demonstrates a molecule can rectify [119, 120], the molecular rectifier is still much less efficient than that in its solid state. Molecular rectifiers face two important issues: the performance of the devices and the robustness.

The performance of rectifiers is measured by the rectification ratio (RR), which is the ratio between electrical currents through the junction in the conducting and insulating directions. Although solid state diodes allow for RR ratios as large as 5 orders of magnitude, the RR ratios of molecular junctions is limited to 1 (or less than one) order of magnitude in many cases [110, 119, 121–123]. To restrict the situation to single molecules contacted to identical metallic electrodes for which a ratio above 10 has never been reported [71, 103, 104, 124] until a recent research on a DNA-based molecular rectifier with a RR of about 15 [125]. Ratios rise to 2 or 3 orders of magnitude have been measured when using SAMs contacted by chemisorption to a bottom electrode and physisorption to a top electrode [112, 126–129]. As far as we know, these systems are to date the best molecular rectifiers. However, monolayers may not be appropriate for fabricating molecular-scale digital circuitry. We only focus on single-molecule rectification here.

Basically, a molecular rectifier must have an asymmetric feature along the transport direction [130]. There are several different ways to bring in such asymmetry, affording a measurable RR, in a molecular junction. Recent research [111, 119, 121] shows that rectification can be resulted from asymmetric placement of a chromophore (A mechanism), asymmetric Schottky barriers (S mechanism), or an asymmetric electron transfer from the frontier molecular orbitals (U mechanism). The last mechanism has been named the "true" unimolecular rectification

[111, 119, 121]. The standard method to create such a single-molecule rectifier is the coupling of an electron donating segment (donor) and an electron with-drawing segment (acceptor) through a linking bridge which can be either saturated (D–σ–A) or conjugated (D–π–A), following the original rectification mechanism proposed in 1974 [3]. These segments are then coupled to external electrodes using different techniques and principles in rectification [111, 121].

The main principle of rectification is the dependence of the asymmetric current response to the external voltage [131], with extreme responses corresponding to the diode effect. The molecular LUMO and HOMO levels are slightly higher or lower than the Fermi level of two electrodes at zero bias voltage, while the energy gap between the LUMO and HOMO tends to be smaller or larger at certain voltages or at reverse voltages, which results in significant current differences under both conditions [132]. Then, Bratkovsky, Kornilovitch, and Williams also reported a resonant tunneling mechanism, but this included one conducting molecular level which was asymmetrically distributed between Fermi levels of two sides [133], which led to asymmetric tunneling barriers that resulted in rectification. Datta and coworkers discovered an asymmetric current–voltage feature with intrinsic symmetric system [134], which is ascribed to unequal coupling and charging effects (the Datta–Paulsson [DP] model). The electron-transport process in the first model contains three steps, that first from the metal electrode to the donor, then to the acceptor, and ultimately to the other electrode; the other two models are only divided into two steps. Meanwhile, the main difference in the latter two models is that the asymmetric energy level is altered by the electric field and through the charging energy, respectively [106].

Additionally, Lortscher et al. also studied rectification with single di-segment molecules where the donor and acceptor are covalently connected without the r bridge compared with the Aviram–Ratner (AR) model [106]. Due to the electronic overlap of the two moieties, the whole transport process has two steps instead of three for zero, forward and reverse voltages, as described in Figure 5.8a. Notably, the *I*–*V* curves at various temperatures demonstrated an obvious asymmetric current, and afford the largest rectification with a ratio of up to 11 [106] (Figure 5.8b). Adopting a gate electrode to a single-molecule junction is another strategy to achieve a high RR [135]. An optimized asymmetric molecule (DPE-2F) was applied to accomplish this research by Perrin et al. [135]. Figure 5.8c illustrates the electron transport mechanism, by the gate electrode controlling the level alignment and misalignment at forward and reverse bias voltages. As depicted in Figure 5.8d, the largest RR can even reach above 600 by tuning the size and direction of the gate voltage. Therefore, the rectification mechanism relying on the orbital structure can be measured in the molecule. The RR value can be improved by narrowing the energy gap between the HOMO and LUMO in molecular-level architecture [136]. In summary, the rectification is not only affected by intrinsic molecular structures and energy levels, but also factors such as the interface between single molecules and electrodes. Functional diodes have been realized based on rectification effect. Molecular diodes are considered to be the basis for more complex device structures in the nanoscale electronics field and they promote the development of molecular

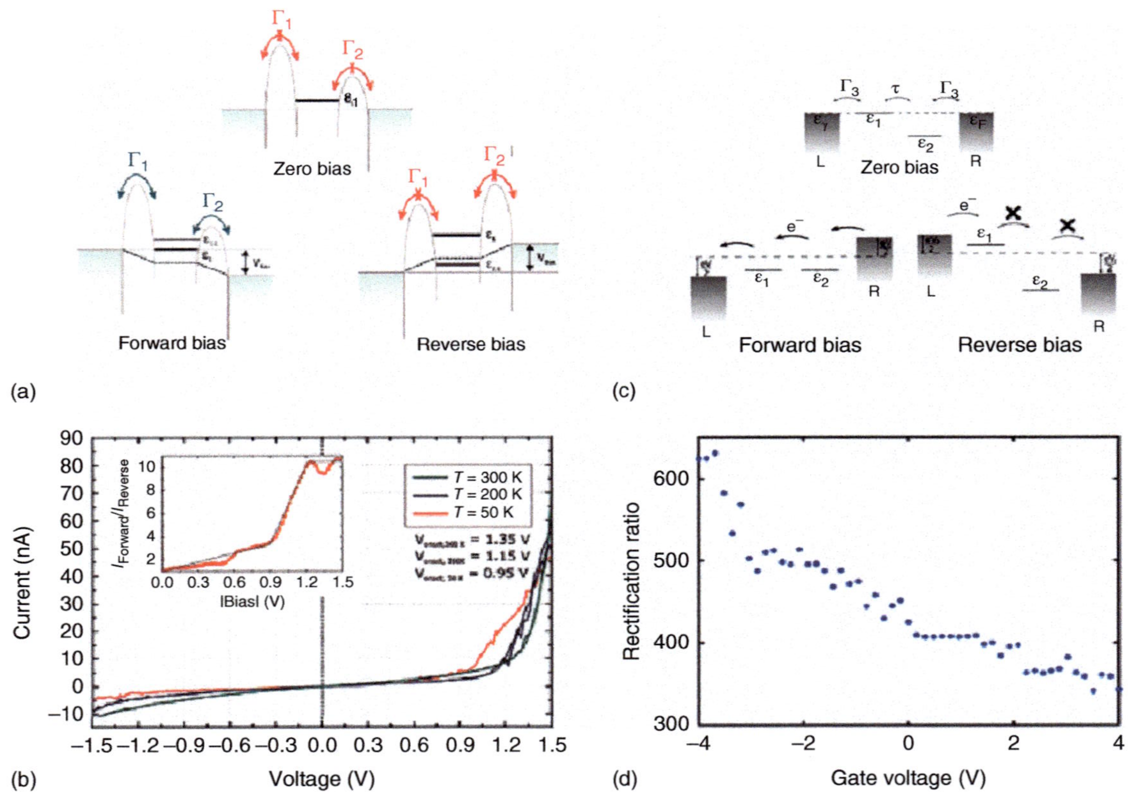

Figure 5.8 Rectification behaviors. Source: Wang et al. [132]. 2017, Springer Nature. (a) Rectification mechanism for diblock dipyrimidinyl diphenyl SB molecule. (b) $I-V$ traces for the Au–SB diode–Au junction recorded at temperatures ranging from 300 to 50 K. The inset shows the rectification ratio as a function of bias for 50 K. Source: Lortscher et al. [106]. Reproduced with permission of American Chemical Society. (c) Rectification mechanism for DPE-2F molecule. (d) Schematic diagram of the rectification ratio as a function of gate voltage. Source: Perrin et al. [135]. Reproduced with permission of the Royal Society of Chemistry.

electronics [137]. Weber and coworkers fabricated the first single-molecule diode with the mechanically controllable break junction (MCBJ) technique based on lithography [103]. Besides, various methods and experiments have been utilized to design molecular diodes. Although molecular diodes have been investigated for at least four decades and various problems impeding the progress of commercial applications, it is an inevitable trend to achieve the commercialization of molecular diodes.

5.3.1 Rectification Toward Diodes

Aviram and Ratner proposed theoretically the idea of using single molecules to fabricate a new generation of rectifiers [3] and it received considerable attention since the 1970s. In this initial prototype, they put forward an organic molecule system containing an electron-withdrawing moiety (acceptor) and an electron-donating moiety (donor) connected by the σ linkage that acts as an insulating barrier for electrically separating the two segments. It would be ideal to energetically use a simple acceptor-to-donor electron transfer to build the current rectification by bridging a pair of electrodes to construct a D–σ–A junction. Figure 5.9 indicates "D" corresponds to "n" and "A" corresponds to "p" and it's similar to the conventional rectifying systems [136]. Single-molecule diodes are of crucial importance because, for example, they enable the electrical addressing of an individual molecular node in an finally scaled crossbar-type architecture, where the node composes of an ensemble or of a single functional molecule with both current-rectifying and switching/storing moieties [106].

A tremendous improvement has been achieved for the past 40 years on molecular electronics in both theoretical and experimental aspects, especially the progress in the development of molecular rectifiers. Actually, several other models have been proposed except the AR model [106] and they not only give the explanation for molecular rectification but also provide several strategies to design and prepare novel molecular diodes. Here, we recapitulate the research on various types of diodes by using a single molecule as a core component for molecular rectification

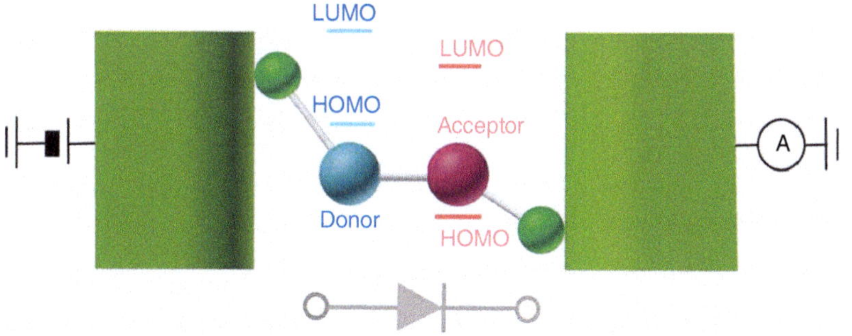

Figure 5.9 Schematic description of rectifiers based on single molecules and the corresponding simplified potential profile for rectifying mechanisms. Source: Tsutsui and Taniguchi [136]. Licensed under CC BY 3.0.

behaviors based on three fundamental mechanisms responsible. Currently, the major single-molecule rectification still relies on monolayer-based techniques that are beyond the scope of this discussion. However, we provide a brief but systematic introduction to the contributions of a thorough development of the present molecular rectifiers. For further details, interested readers are advised to read other studies [111, 121, 138, 139].

5.3.2 General Mechanisms for Molecular Rectification

As discussed above, the D moiety (electron-rich) and the A moiety (electron-poor) of the molecular rectifiers resembles the n-doped and p-doped region of the inorganic counterparts, respectively. However, the potential mechanisms of rectification are reversed for the two types of diodes. For the D–σ–A molecular rectifiers, the electron transfer under a forward bias is preferred from the metal electrode to A to D to the other metal electrode, whereas in a p–n junction rectifier, the preferred electron transfer is from n to p. Furthermore, the explanation used for understanding the conventional p–n junction diodes is inadequate for the molecular diodes, due to the complexity resulting from the considerable influence of the electrode–molecule interface on the charge transport.

On the basis of the theological development as well as the experimental discovery, three mechanisms were proposed to explain the rectification behaviors based on single molecules [106], which are depicted schematically in Figure 5.10. They are AR [3] model, Kornilovitch–Bratkovsky–Williams [133] model (KBW model), and DP [134] model.

5.3.2.1 Aviram–Ratner Model

In the original research accomplished by Aviram and Ratner, the molecule with an electron-rich moiety of tetrathiofulvalene (TTF) and an electron-poor moiety of tetracyanoquinodimethane (TCNQ) linked by a covalent triple-methylene bond was connected to a pair of electrodes (Figure 5.11a), thus fabricating a molecular diode in which the molecule acts as a core component for electrical current rectification. Afterwards, this molecule was named as a D–σ–A model consisting of a donor moiety, insulator moiety, and acceptor moiety. Actually, the corresponding mechanism supporting this proposal was the earliest strategy that has gained wide agreement. The two electrodes in the AR model are connected through three tunneling barriers, linking one electrode to the acceptor moiety, the acceptor moiety to the donor moiety, and the donor moiety to the other electrode. At zero voltage bias, the HOMO of the donor is slightly lower than the electrode's Fermi energy and the LUMO of the acceptor is slightly higher, whereas applying a voltage bias across the junction, the LUMO (or the HOMO) can be aligned to the Fermi energy of the leads or separated relying on the bias direction. Thus, the rectification behavior that was expected to occur was dominated by the electric field between the two electrodes (depicted schematically in Figure 5.10a), due to the different thresholds at the positive and negative bias voltages and shifts of the molecular orbital energy levels according to their energetic positions in the gap. The two separated energy

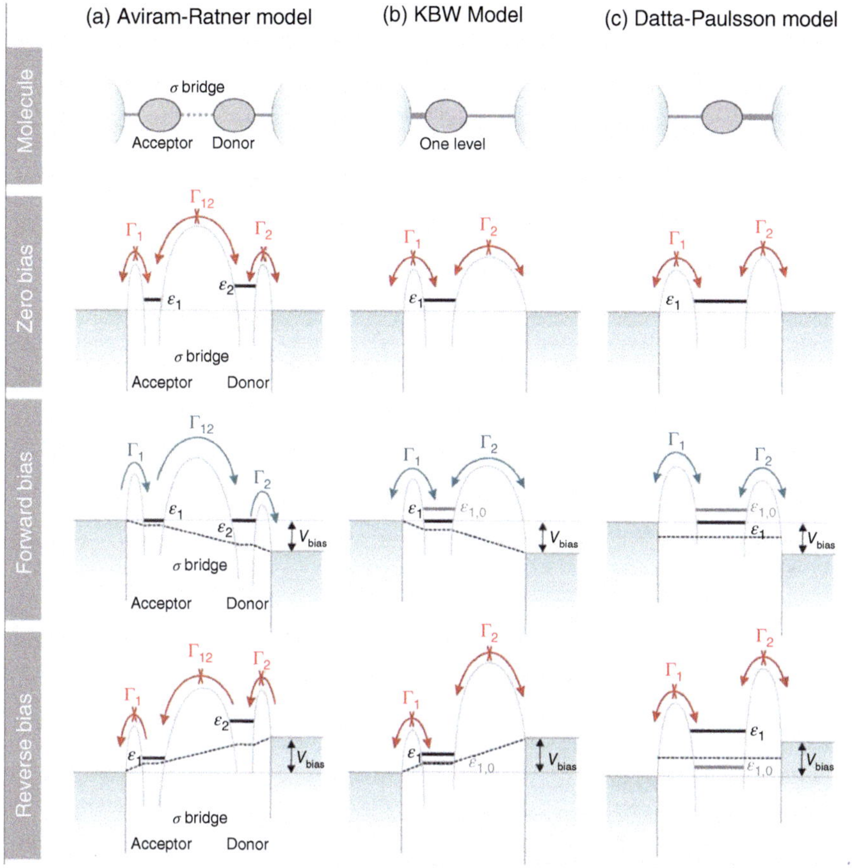

Figure 5.10 Comparison of three models for molecular diodes (Γ_i represents the rates of tunneling steps). (a) Aviram–Ratner proposal with molecule levels shifting through the applied electric field. The subtle energy-level alignment for the donor and acceptor moieties, in combination with the large electrical spacer separating the two moieties, eventually leads to a difference in the current onset. (b) Simplified model, as proposed by Kornilovitch, Bratkovsky, and Williams (KBW), with one level. The level is shifted with respect to the applied electric field. (c) Considering the charging of the energy levels (proposed by Datta and Paulsson) can lead to diode behavior, even without level shifting by an electric field. Source: Lortscher et al. [106]. 2012, American Chemical Society.

levels, ε_1 and ε_2 (corresponding to the Fermi energy of the electrodes, E_F), which arise from the donor and acceptor, respectively. They respond differently to the electric field between the electrodes when adding a bias voltage, thus resulting the alignment or misalignment in energy levels due to the asymmetric charge transport. On the whole, the rectification in an AR model could be regarded as a different onset of resonant tunneling for the two bias directions [106].

5.3.2.2 Kornilovitch–Bratkovsky–Williams Model

Later, Kornilovitch, Bratkovsky, and Williams simplified the AR mechanism and proposed a simpler system (viewed as the KBW model) with only one molecular

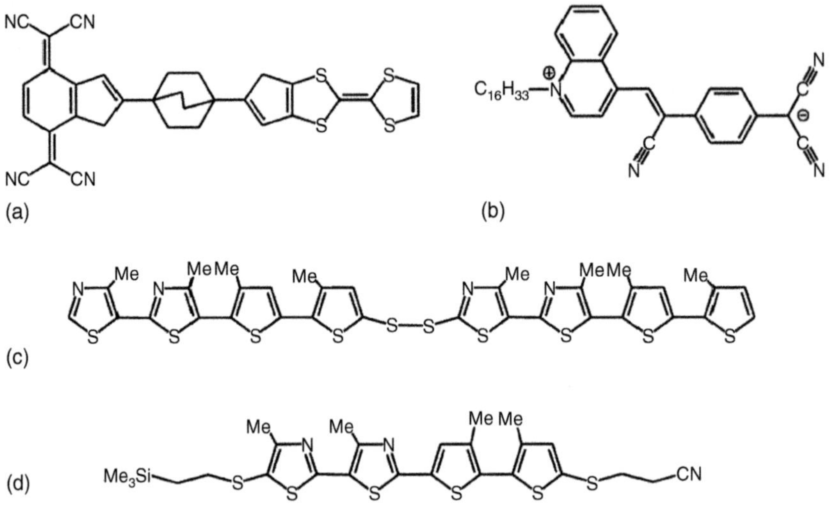

Figure 5.11 Molecular structures proposed for current rectification. (a) Initial model D–σ–A proposed by Aviram and Ratner. (b) D–π–A molecules used for rectification based on the Langmuir–Blodgett monolayer. (c) Conjugated diblock oligomers used for rectification through the formation of the SAMs. (d) p–n junction types of molecules used for rectification through the directed formation of SAMs with a specific orientation. Source: Xiang et al. [140]. 2016, American Chemical Society.

orbital, in which asymmetric tunneling barriers resulted in the current rectification (as shown in Figure 5.10b) [133]. Specifically, in this model, the orbital is not located symmetrically between the electrodes, but is closer to one of them owing to the terminated tails with different lengths. Due to most of the applied voltage decreasing on the longer insulating barrier, the required conditions for resonant tunneling through the level are realized for the two opposite polarities at different voltages. Actually, the molecular orbital, aligned, or separated with the Fermi energy of the electrodes depending on the bias direction, will lead to the rectification behavior. A similar study accomplished by Mahmoud and Lugli demonstrated that the rectification of a molecular device could be modified by imbedding a resistive molecular moiety near the metallic electrodes, thus enforcing the potential drop profile along the molecule by calculations [141]. Furthermore, the KBW model accelerated the development of molecular diodes because tunneling barriers arising from molecules with different-length tails were more accessible than that from complex structures in the AR model. At last, the RR can be systematically changed by tuning the lengths of these insulating tails.

5.3.2.3 Datta–Paulsson Model

For sterically symmetric molecules, Datta and his colleague proposed a new model based on experimental and theoretical studies, the current rectification could be also obtained by unequal coupling with the electrodes at the two ends [134]. The I–V curves were initially symmetric but appeared a weak and reversible asymmetry as the contacts were manipulated, as shown in the charge transport measurements of these molecules (Figure 5.10c). Specifically, conductance peaks of different heights occurring at symmetrically disposed voltage values demonstrated this asymmetry in resonant conduction. Herein, the conduction of these molecules at opposite voltages occurs essentially through the same molecular levels with similar wave functions, which differs from the two models discussed above. This research focus on the nontrivial contact-induced asymmetry and self-consistent shifts in the energy levels owing to the charging effects that usually exist in single-molecular devices with tunneling barriers to the electrodes. The different tunneling rates between the orbital and the two electrodes, owing to the asymmetric charging energies, demonstrate that the average population of the electron orbitals depends on the bias direction in the resonant tunneling condition, which ultimately realizes the rectification. Although this mechanism cannot be easily separated from the asymmetric coupling to the electric field described above, it can be considered as a distinct mechanism of the rectification. In principle, the molecular design strategy is that molecules have better coupling to either of the electrodes and keep the position of the resonant orbital in the center.

However, in actual experimental verification, it is difficult to discriminate between the three models discussed above for these reasons, as follows. (i) A perfect AR diode is difficult to be obtained by chemical synthesis, and a full decoupling between the donor and acceptor orbitals is not always successful [130]. (ii) The uncontrolled coupling between electrodes and the molecular terminated segments due to various interfacial interactions and random contact geometries previously mentioned above led to considerable difficulties in the experimental transport measurements [142]. In order to identify whether molecular rectification is caused by asymmetric electrode couplings or a molecular bias drop, a new study by Zhang et al. [143] demonstrates that the asymmetric couplings do not directly lead to rectification whereas the bias drop is responsible for rectification. This result is not in accordance with other reports, which indicates the complexity of differentiating specific mechanisms. In fact, the phenomenon of rectification in most studiers is always a combined effect of different mechanisms and not a pure diode prototype. Herein, molecular rectifiers based on pure operating models are relatively rare [121].

5.3.3 Rectification Originated from Molecules

5.3.3.1 D–σ–A and D–π–A Systems

Several groups pursued AR rectification by building diodes with single molecules. A standard molecular diode with D–σ–A architecture has not been reported, due to the limitations of current organic synthesis methodologies and the difficulties of controlling inter-/intramolecular energy levels in alignment. However, the D–σ–A model was thoroughly confirmed to be successful [109, 144–148] in some research on multi-/monolayer Langmuir–Blodgett (LB) films or block copolymers [149]. In addition, molecules with D–π–A structures have received a lot of attention due to the feasibility to tune the electrical current at the molecule-film level by organic synthesis [146, 150, 151]. In the rectifying process of this molecular system, a few ambiguous problems still existed in the rectification mechanism. For example, more efforts should be paid on whether the ground state is precisely $D^{+1.0}$–π–$A^{-1.0}$ and the excited state $D^{0.0}$–π–$A^{0.0}$ or their corresponding intermediate ionicities $D^{+\rho}$–π–$A^{-\rho}$ ($\rho \geq 0.8$ for the ground state and $\rho \leq 0.2$ for the excited state) [152]. However, the D–π–A structure, as a derivative of the D–σ–A structure, attracts a lot of investigation for in the field of molecular rectifiers and provides new access for realizing actual single-molecule diodes.

Several special cases of the SAM model have realized single-molecule rectifiers. In 1997, Metzger reported a single-molecular rectification behavior with a RR (the ratio of the highest conductance at the forward bias and that at the reverse bias) of up to ≈13 by using a Langmuir–Blodgett monolayer of zwitterionic hexadecylquinolinium tricyanoquinodimethanide ($C_{16}H_{33}Q$-3CNQ) embedded between two Au electrodes (Figure 5.11b) [111, 121]. Ng et al. and Jiang et al. achieved single-molecule rectifiers based on diblock molecules with a rectification of ~7 by utilizing the SAMs with the STM techniques [153, 154] (Figure 5.11c). No matter in the SAM system or in the single-molecule system, obviously, the molecular orientation in the junction will have a significant influence on the rectifying direction, which has been well reported elsewhere [155]. Therefore, tuning the molecular orientation by rational organic reactions is of huge significance and has been fundamentally realized. The molecules terminated with a thiol protected by different groups at each end, could be localized in a direct way by deprotection off thiol end sequentially, which has been adopted successfully in both the single-molecule [104] systems and the SAM [154, 156] (Figure 5.11d).

5.3.3.2 D–A Diblock Molecular System

As the issues discussed in Section 5.3.3.1, D–σ–A structures suffer from the lack of synthetic methods and controllability for the energy-level alignment, and D–π–A molecules struggle with the inferior performance of the rectifiers, which both draw researchers' attention back to the semiconductor counterparts aiming at a breakthrough in single-molecule diodes. Whereas, p–n junction molecules (viewed as diblock molecules) with direct-connected donor and acceptor moieties by C—C covalent bonds rather than a σ- or π-bridge, were obtained following the mature polysynthetic method. In 2002, Ng and Yu [149] first synthesized a regioregular, head-to-tail coupled diblock cooligomer containing an electron-rich oligo(3-alkylthiophene) as the donor moiety and an electron-deficient oligo(4-alkylthiazole) as the acceptor moiety, and its measurement demonstrated its rectification effect through the STM technique. This synthetic strategy was generally applied for coupling heterocycles and thus provided an important access to introduce other p–n junction diblock molecules, which has been regarded as the primary motive force for the progress in this field.

For example, Elbing et al. [103] designed and synthesized a series of diblock molecules with donor–acceptor structures and introduced them into electrical nanocircuits through the MCBJ technique for the studies of their rectification behaviors (Figure 5.12a). These molecules are composed of two phenyl–ethynyl–phenyl *π-blocks* which were connected by a biphenylic C—C bond. To realize electronic asymmetry, the hydrogens of one parent moiety was replaced by four electron-poor

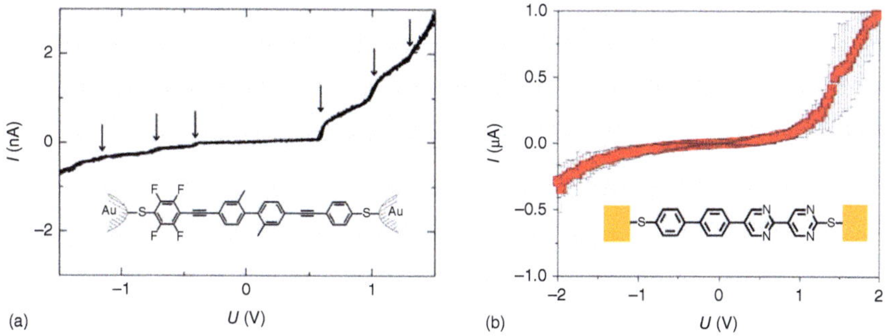

Figure 5.12 *I*–*V* characteristics of rectifiers based on single-diblock molecules. (a) Rectification of phenyl–ethynyl–phenyl molecules immobilized between two Au electrodes in an MCBJ setup. The arrows point to step-like features of the *I*–*V* curves. Source: Modified from Elbing et al. [103]. (b) Rectification of the dipyrimidinyldiphenyl molecules integrated in an STM setup at $T = 30$ K. Source: Modified from Díez-Pérez et al. [104].

fluorine atoms, while that of the other was kept primordial state with hydrogen atoms. Furthermore, the steric hindrance between biphenylic C–C link and two *ortho*-methyl groups induces a twist angle between the moieties, which reduces their π-orbital overlap and hence their electronic coupling. The single-molecule rectification exhibited a current ratio of $1:4.5 \pm 1.5$ V through systematic and rigorous control experiments, which is explained qualitatively as follows. The structurally asymmetric molecule could be regarded as two quantum dots (QDs) connected in series, named F-dot (fluorized) and H-dot (not fluorized), which localized the electronic orbitals and hence avoided overlapping. The electronic levels of both dots are relatively shifted to each other when sweeping the bias voltage, and the two levels will cross at certain voltages. An additional transport channel opens for inelastic transmission from the H-dot into the F-dot when an unoccupied level passes an occupied one. For each additional transport channel, the current grows by a certain amount, corresponding to a stepwise increase in the *I*–*V* curves (or peaks in the differential conductance (dI/dV)).

In 2009, Tao and coworkers [104] adopted STM break junctions to test a structurally similar but parent-unit-different molecule in particular bound orientation through selective deprotection strategy for its diode-like property. These asymmetric diblock dipyrimidinyldiphenyl molecules, with current flowing from the dipyrimidinyl to the diphenyl moieties, exhibited a significant rectification behavior (rectifying ratio up to 5) in comparison with its homologous symmetric counterpart (Figure 5.12b). In a further study, they used the same system and the same molecules as well as an inelastic electron tunneling spectroscopic (IETS) analysis to seek the mechanism responsible for the rectification behavior at low temperatures [157]. The rectification onset initiated near the zero bias and clear asymmetric currents appeared at biases below 200 mV, which both indicated that tunneling dominated the transport mechanism. Furthermore, this research also suggested that the bias-dependent asymmetry in the electronic coupling of diblock diode molecules was significant for both the inelastic and the elastic processes. This discovery was derived from the result that the inelastic electron tunnelling (IET) spectra with similar phonon intensities and energies were antisymmetric in both reverse and forward bias directions, regardless of the asymmetry in the *G–V* and *I–V* features at low biases. In order to further explore the mechanism of rectification behavior based on the diblock dipyrimidinyldiphenyl molecule, Lortscher et al. [106] utilized the MCBJ technique to measure its charge transport as a function of bias at different temperatures in the low-bias region. In this study, the *I–V* features could be semi-empirically interpreted by combining the DP and KBW models, which demonstrates that the rectification behavior derives from an asymmetric Coulomb blockade (charging effect) together with an electric-field-induced level shift.

It is noteworthy that the rectification in these types of D–A diblock molecules requires the electronic decoupling between the donor moiety and the acceptor moiety, which could be realized by tuning the torsion of one moiety of the molecule relative to another. In fact, in one current study conducted by Cui et al. [158], further insight was draw into the correlation between the molecular twist angle and the rectification effect (the ratio) through the density functional theory (DFT) along with the nonequilibrium Green's function (NEGF) method (Figure 5.13a). The studies demonstrate that the rectification grows along with the geometrical torsion-induced reduction of the conjugation lengths in these organic molecules. In addition, Luo et al. [160] stressed the significance of reducing the overlap between the orbitals of donor segment and acceptor segment in experiment. They designed and synthesized a series of long wire-like D–A diblock molecules with lengths ranging from 3 to 10 nm and measured their charge transport in the hopping regime through the conducting probe-atomic force microscopy (CP-AFM) technique. It was found that an electrical rectification was realized with ratios as high as 30 ± 1.0 V when an insulating cyclohexane bridge was inserted between the *N*,*N*′-di(4-anilino)-1,2,4,5-benzenebis(dicarboximide) electron acceptors (A) and the 4,4′(5′)-diformyltetrathiafulvalene electron donors (D). While the absence of an insulating bridge resulted in a current flow without rectification owing to a large conjugation for the D and A moieties. In the initial D–σ–A model, both the insulating barrier insertion and the intramolecular torsion could decrease the electronic coupling between the donor segment and the acceptor segment, which is necessary for the design of molecular diodes with high performance.

Apart from considering various methods to decouple the electronic overlap between the donor and acceptor moieties in D–A molecular systems, Hu et al. confirmed that the proportion effect on the rectification also played an important role in diblock cooligomer diodes by using theoretical calculations (Figure 5.13b) [159]. For rectifiers based on phenyl-pyrimidinyl diblock cooligomers, it was found that a short cooligomer diode with the two moieties in 1 : 1 proportion afforded a large RR. While the optimal proportion for the largest RR was not 1 : 1 in a long cooligomer diode; however, the variation between the rectifying direction and proportion was observed. In addition, Zhang et al. [161] studied the protonation effects on the electron transport by using a conjugated diphenyl-dipyrimidinyl diblock oligomer diode, and they came to the conclusion that the interference from the

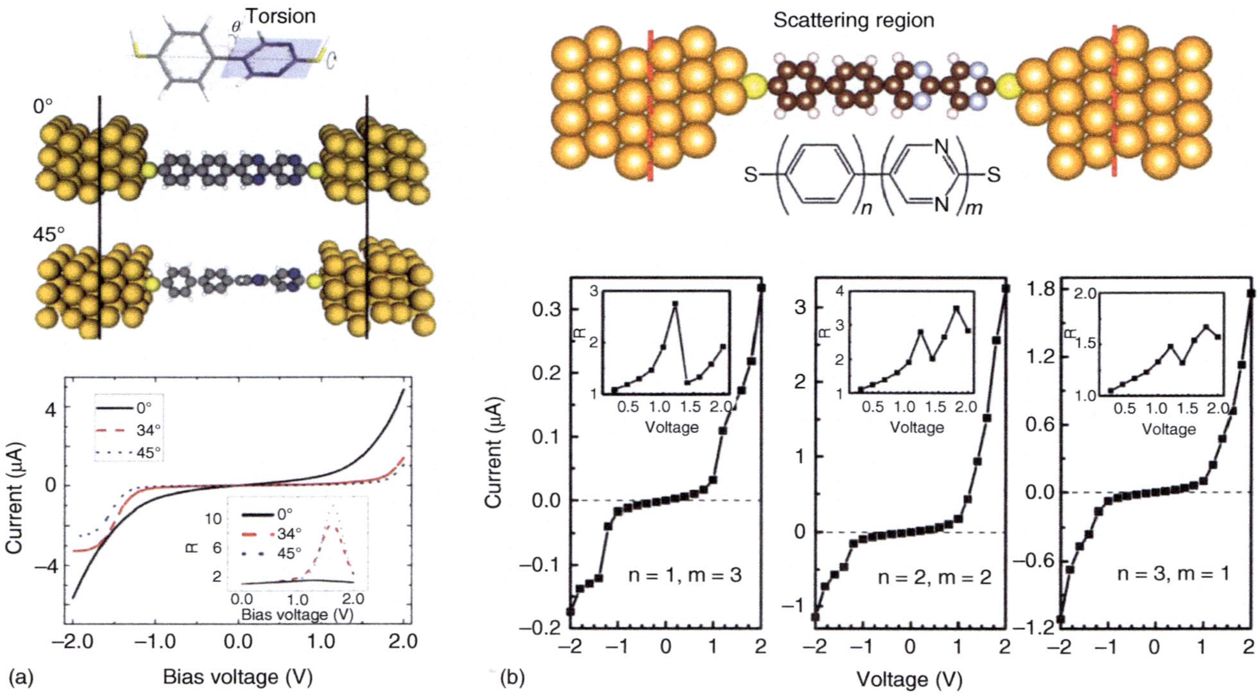

Figure 5.13 Torsion/proportion dependence of the rectification effect. (a) The influence of torsion between the donor part (phenyl segment) and the acceptor part (pyrimidinyl part) on the rectification. The corresponding rectification ratios are depicted in the inset. Source: Modified from Cui et al. [158]. Reproduced with permission of AIP Publishing. (b) The influence of proportion of the donor part (phenyl segment) and the acceptor part (pyrimidinyl part) on the rectification. Source: Hu et al. [159]. 2014, Elsevier.

external environment could impose an perturbation on the rectification behavior. The calculation demonstrated that the protonation in the outer pyrimidinyl helped to increase the RR, whereas the protonation in the inner pyrimidinyl dramatically inverted the rectifying direction. It is the first time the experimental results of the diblock molecular diodes have been clearly explained in the case of protonation in 2005 [156].

5.3.4 Rectification Stemming from Different Interfacial Coupling

D–A diblock molecular system introduces asymmetry into the molecular junctions. However, researchers may expect rectification to work based on asymmetric molecular junctions that are not necessarily consisted of donor and acceptor moieties, or symmetric systems but they have an unequal physical process under the forward and reverse bias. Actually, DP model has predicted that different couplings between the molecules and the electrodes on the two ends could afford rectification, which turned out common in nearly all published single-molecule systems. Although this type of rectification is insufficient to realize a larger diode effect especially when the bias voltage drop at the electrode–molecule interfaces cannot be tuned with subtlety, it still gives new strategies for the design of molecular diodes. In fact, molecules with asymmetric segments at two ends, a pair of different electrodes with distinct work functions, or both, which could result in unequal interfacial coupling, have been utilized to achieve molecular diodes, among which several rectifiers were prepared from films or monolayers instead of single-molecule junctions.

5.3.4.1 Different Electrodes

In the early 1990s, [102, 144, 162, 163] suitable techniques were developed by Ashwell et al. and they sandwiched LB monolayers or multilayers between two electrodes with different work functions to measure molecular rectification. Generally speaking, metals with different work functions introduce different charge injection barriers, which may lead to molecular rectification. For example, for HOMO-dominated molecules, metals with a high work function introduce a lower charge (hole) injection barrier compared with metals with a low work function, due to their HOMO is closer to the Fermi level than that of the LUMO. Although the molecules in these asymmetric systems are consisted of both donor and acceptor moieties, the presence of a low-work function metal on one side and a high-work function noble metal (Ag or Pt) on the other side [164] lays the base of the molecular diodes. In these studies, metal electrodes, especially noble ones (such as Au, Pt, or Ag), provides a nearly constant density of electronic states within a narrow energy window corresponding to the charge transport, which leads to molecular junctions with relatively featureless and smooth transmission probabilities near the Fermi energy, which thus limited their applications including rectification. Fortunately, in view of the developments in nanofabrication technology as well as material science, various types of electrodes have been developed and introduced into molecular electronics. Vuillaume et al. fabricated a molecular rectifying junction based on a sequential self-assembly on silicon by using Al on the other side of the electrode, which enriched the library of electrodes for molecular rectification [110]. Recently, Kim et al. developed a hybrid device structure with a molecule injected between the gold/graphite gap in the STM break junction system [165]. For this molecular system, they obtained a relatively high RR despite the inherent symmetry in the molecules, which could be attributed to the asymmetry in the voltage drop caused by the different interfacial coupling across the junction.

5.3.4.2 Anchoring Groups

As discussed in Section 5.2.1, molecules with different anchoring groups or different tail lengths can lead to asymmetric interfacial coupling, which implies the considerable possibility to achieve molecular rectification. In addition, single-molecule systems with movable electrodes, such as STM, CP-AFM, and MCBJ, can tune the molecule–electrode distance with an atomic resolution and therefore modulate the interfacial coupling strength. Based on this system, in 2004, Kushmerick et al. [166] experimentally achieved and tuned the current rectification in metal–molecule–metal junctions by controlling the interaction strength of one metal–molecule contact through changing the anchoring groups. In 2014, Wang et al. further confirmed this effect both theoretically and experimentally [167]. A molecule terminated by amine group and thiol group was attached at two ends with a STM tip and an Au surface. Similarly, Batra et al. [71] fabricated a molecular rectifier constructed from a symmetric conjugated molecule with a single methylsulfide group connecting one end to a gold electrode and the other end to a covalent gold–carbon bond (Figure 5.14). The nonequilibrium transport calculations demonstrated that a hybrid gold-molecule "gateway" state was localized close to the Fermi level of one electrode due to the formation of the carbon–gold bond, which could shift significantly along with an applied bias, thus achieving rectification at surprisingly low voltages. This study developed a series of tunable single-molecule rectifiers through rationally designing the anchoring groups. Recently, a theoretical study confirmed that the rectifying behavior of the diphenyl–dipyrimidinyl cooligomer diode could even be largely enhanced or reversed by adjusting the asymmetric anchoring groups [161]. These studies will favor the design for a new generation of molecular rectifiers with high tunability and high rectifying ratios through interface engineering [168].

5.3.4.3 Contact Geometry

As previously talked about in this report, a significant influence on the charge transport properties of the molecular junctions were caused by the contact geometry in the electrode–molecule interface. Clearly, this effect also happens in the molecular rectification. In order to fabricate a diblock cooligomer rectifier by differentiating the anchoring groups [161], the authors determined that the asymmetric contact geometries playing a negative or positive role in improving the rectifying behavior is closely related to each molecular diode. Recently, one theoretical study systematically and accurately calculated the effects of different interfacial contact geometries on the rectifying behaviors of the molecular diodes, especially the absorption sites of an S atom in the headgroups of molecules on the surface of Au electrodes [142].

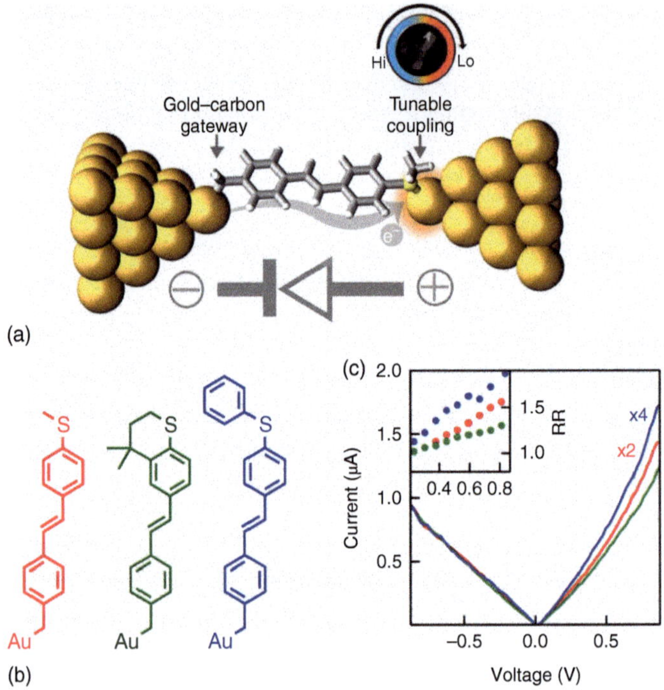

Figure 5.14 Molecular rectifiers with different anchoring groups. (a) Schematic for molecular circuits with a carbon–gold bond and a methylsulfide–gold bond. (b) Chemical structures for the three molecular rectifiers used. (c) Corresponding *I*–*V* curves for the three molecules. Inset: Rectification ratio as a function of the bias. Source: Batra et al. [71]. 2013, American Chemical Society.

These studies highlight the significance of the contact-geometry asymmetry to the rectification behavior, which serves as an aid to explain future SME behavior.

5.3.4.4 Interfacial Distance

In the same time, the length of the insulating tails connected to the electrodes could be improved to tune the coupling strength. Bratkovsky and coworkers determined that different tunneling barriers could be assigned to different tails on each side, and achieved molecular rectification in their KBW model [133]. In one case, a single electroactive group placed asymmetrically between the electrodes, and the corresponding HOMO and LUMO distributed asymmetrically with respect to the Fermi level, could rectify electrical current. Dramatically, the construction of this model was an interesting accident. The long insulating tail of $C_{16}H_{33}$ connected to the molecule core not only result in an asymmetry geometry but also helps assemble into suitable LB films [169]. Based on these studies, additional research, combined with other considerations for designing molecular diodes, were carried out on the basis of this general mechanism. Recently, the rectification of current through a single molecule with an intrinsic spatial asymmetry was reported by Liu et al. [170]. In their study, a rectifier with a high current, large voltage range and low threshold was investigated by adopting the relatively localized and high-energy d states of the cobaltocene segment within the single molecule. Based on the calculations with a self-consistent NEGF, they demonstrated the plausibility of forming excellent molecular diodes using metallocenes, which resulted in a productive class of molecules.

In addition to tune terminated tails with different lengths, the geometrically asymmetric molecule–electrode distance could also afford different interfacial coupling. In 2002, Taylor et al. [171] calculated the variation in the current–voltage characteristics by fixing one of the molecule–electrode interfaces while increasing the distance of the other, ultimately weakening the molecule–tip coupling. They confirmed that the rectification and operation range in the strongly chemisorbed molecules were limited by their proximity to the Fermi level and the width of the transmission resonances, both of which were strongly corresponded to the interfacial coupling strength. Recently, based on the theoretical calculations, the interfacial stretch or contraction effects on the rectification in molecular diodes with diblock cooligomer were studied by Zhang et al. [172], who confirmed that different mechanical treatments could inverse the rectification direction. Moreover, two competitive mechanisms, deduced from the asymmetric evolution of orbital wave functions under biases and the asymmetric molecular level shift, were proposed to be responsible for the rectification inversion due to their different sensitivities to the electrode–molecule distance. Additionally, these results demonstrated that the rectification performance in these molecular diodes could be controlled through the mechanically controllable method.

5.3.5 Additional Molecular Rectifiers

The improvements in the electrical conduction theory of single molecules provided a relatively good understanding of the rectification mentioned in the above cases. The principle of molecular rectifiers can be summarized briefly as being that the electronic resonance between the molecular energy level and the Fermi level related to the charge transport happens at different voltages in the forward and reverse bias directions, and that the barriers prohibiting the electron conduction are asymmetric at the two molecule–electrode interfaces. As further insight is obtained into the essence of rectification based on experimental and theoretical progress, as discussed briefly in this section, novel operation principles are being proposed to direct additional molecular rectifiers.

Although the development of the molecular diodes with rigid-structure molecules flourishes, investigations based on molecules with flexible nuclear degrees of freedom, such as conformational changes, used to achieve current rectification increase. Unimolecular rectification based on voltage-controlled intramolecular stereo chemical modification was reported by Ratner and coworker [173, 174] (Figure 5.15a). In brief, it was predicted that different conformations driven by an external electrical field could affect the charge transport and cause rectification owing to the inherently unequal capability to conduct electrons. As well as we know, this has not been demonstrated experimentally. To achieve a transition between configurations

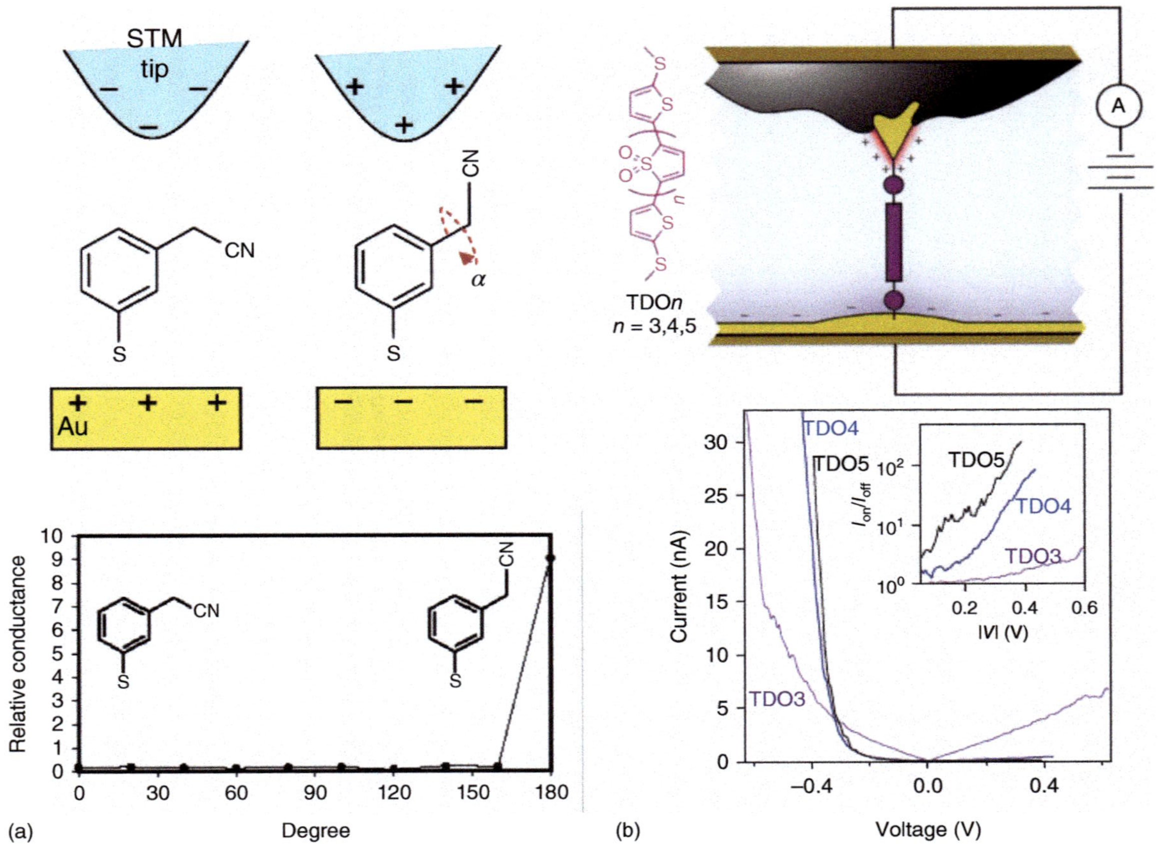

Figure 5.15 Other types of molecular rectifiers. (a) Top: Schematic representation of current rectification based on molecular conformational change. Bottom: Corresponding conductance as a function of a dihedral angle. Source: Modified Troisi and Ratner [173]. (b) Top: TDO molecular structure used and schematic of molecular junctions created using asymmetric area electrodes through environmental control. Bottom: Average I–V curves for TDOs (with $n = 3–5$). Inset: Rectification ratio (I_{on}/I_{off}) vs. magnitude of applied voltages, which depict ratios of ~200 at 0.37 V for TDO5. Source: Capozzi et al. [78]. 2015, Springer Nature.

induced by the external field, a strong dipole potentially provided by one or more C—F or C—CN bonds is necessary, and the simplest model is the molecule possessing a single flexible degree of freedom, in which the junction asymmetry can be kept through a chemical bond between at least one electrode and the molecule.

Zhang reported a theoretical gate-controlled single-molecule diode in 2004, and they proposed a clear modulation by the gate. Additionally, when the positive voltage was high enough, the rectification direction could be reversed [175]. This prediction is valuable for designing and fabricating future integrated circuits and molecular logic devices. Recently, another remarkable study on single-molecule diodes was proposed by Capozzi et al. where environmental control was used as an approach to induce current rectification [74, 78]. Based on exposing considerably different electrode areas to an ionic solution, the symmetry of a symmetric single-molecule junction with two electrodes of the same metal was broken (Figure 5.15b). This technique allowed the junction's electrostatic environment in an asymmetric manner to be controlled by simply changing the bias polarity. Using this approach, RRs in excess of 200 were achieved at voltages as low as 370 mV based on a symmetric oligomer of thiophene-1,1-dioxide (TDO). This result is reproducible as a similar rectification using other molecules/solvent systems was observed. This environmental control method offers a general route for modulating nonlinear nanoscale device phenomena that is beyond single-molecule junctions.

In summary, as the origin of molecular electronics, molecular rectifiers have received academic and commercial attention for more than 40 years. Benefiting from both the theoretical and the experimental efforts engaged in this field, this digital component pivotal for logic devices has evolved from an initial D–σ–A model to a derived D–π–A model and a simplified D–A model, in which the precise energy level alignment as well as the subtle electronic coupling between the donor region and the acceptor region in the molecule are significant for the rectification behavior. By augmenting the feasibility and the ease of constructing the single-molecule system, current–voltage curves with diode-like characteristics are common, even for molecules without donor or acceptor units, which could be ascribed to asymmetric interfacial coupling, as discussed above. Driven by this phenomenon, various methodologies have been developed into the design of single-molecule rectifiers. Furthermore, due to a thorough understanding of the charge transport through single molecules, additional models are being proposed for current rectification with novel operating principles. Clearly, a long process remains to realize commercially feasible rectifiers related to single molecules; however, a positive scenario can be imagined because of the continuous attention and effort.

5.4 Conductance Switches

The next important step in molecular electronics toward actual applications is the construction of a reliable molecular switch, a particularly attractive molecular electronic device. It is improbable to build a circuit without a switch. As a core component of current computer circuits, transistors, are ultimately operated as switches, where the conductance is efficiently and controllably turned on and off by a gate electrode. Therefore, single-molecule switches could be the prototypical device used to complement silicon in the logic and memory elements because of its obvious advantage for ultra large density integration.

In principle, a molecular switch can alter the conductance, which occurs from changes in its physical properties (such as the electronic structure of the molecule and correspondingly its density of states) upon exposed to external stimuli (e.g. light, heat, ion, magnetic, or electric field). This requires the molecules to have several (at least two) stable isomers and transition between the isomers. From an engineering's viewpoint, the challenge is to build single-molecule switches with detectable conductance changes, ideally from zero (OFF state) to infinity (ON state). Early studies on molecular switches were carried out on the basis of SAMs [176–179], particularly using the STM or AFM setups, which have been surveyed in a recent review [180]. In Section 5.4, we focus on summarizing the design, construction, and electrical properties of recent single-molecule switches, voltage pulse induced switches, light-induced switches, electrochemically triggered switches, and spintronics-based switches.

5.4.1 Voltage Pulse Induced Switches

A possibility for controlling electronics at the single-molecule level by directing molecular conformation is stereoelectronic effects, which determine how the properties and reactivities of the molecules depend on the relative spatial orientations of their electron orbitals.

Recently, this idea was first demonstrated by Su et al. [73], who developed a single-molecule switch that operated through the stereoelectronic effect. Using permethyloligosilanes with methylthiomethyl as electrode linkers, the strong σ conjugation in the oligosilane backbone were determined to be coupled with the stereoelectronic properties of the sulfur–methylene σ bonds at the contact interface. As predicted by theory, this strong coupling led to three distinct dihedral conformations that had extremely different electronic characteristics (Figure 5.16a). Remarkably, simply lengthening or compressing the molecular junction could control the transition between the three species, thus realizing conductance switching between the two states (Figure 5.16b). This study brings about a unique technique for incorporating reaction chemistry into the design of a new generation of functional molecular electronic devices. With MCBJ-based two-terminal junctions, Schirm et al. [181] demonstrated that a metallic atomic-scale contact could be operated as a reliable switch driven by the controlled current. They also provided evidence that the switching process was resulted from the reversible rearrangement of single atoms in the MCBJs. By employing MCBJ junctions, Lortscher et al. demonstrated that switching between two stable states of a single molecule in a two-terminal configuration could be reversibly and controllably realized in response to an external voltage stimulus. With the same measurement protocol but different molecules, it was concluded that the observed switching had truly a molecular

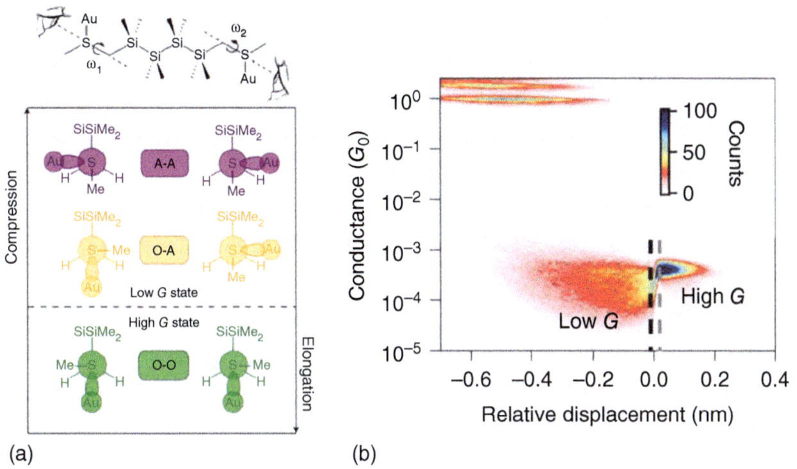

Figure 5.16 Directing molecular conformation by stereoelectronic effects. (a) Three dihedral configurations: A–A (purple), O–A (yellow), and O–O (green) from the perspective of the sulfur–methylene σ bond in the Au–Si_4–Au system. (b) 2D conductance-displacement histogram for Si_4 compiled from traces that demonstrate switching. Source: Su et al. [73]. 2015, Springer Nature.

origin and the possible reason for switching was the conformational change [182]. Because of the hysteretic behavior of conductance with two distinct states, these two-terminal switches can be potentially utilized as a nonvolatile information storage element.

5.4.2 Light-Induced Switching

Another series of examples highlighting the importance of the electrode–molecule contact interface are single-molecule photoswitches based on photochromic diarylethenes (DAEs) [183–186].

In 2003, Dulić et al. [185] first fabricated dithiolated photochromic molecular switches with a mechanically controllable break-junction technique (Figure 5.17a), and observed the molecules switching from the conducting to the insulating state when illuminated with visible light ($\lambda = 546$ nm) but failed to observe the reverse process, even when illuminated with UV light ($\lambda = 313$ nm) (Figure 5.17b). This experimental observation was attributed to quenching of the molecule's excited state in the open form due to the presence of the gold surface plasmon through the Au—S bonds. To decouple the strong interaction between the DAE molecules and the electrodes, innovative molecular design and device architectures to keep the molecular functions are required. In an attempt to solve this problem, Whalley et al. [186] synthesized different types of single-molecule switches using the single-walled nanotubes (SWNTs) as point contacts, where covalent amide linkages were adopted to form a robust contact interface (Figure 5.17c). It was observed that these SWNT-diarylethene single-molecule junctions could switch from the low conductance state (open form) to the high-conductance state (closed form) but not

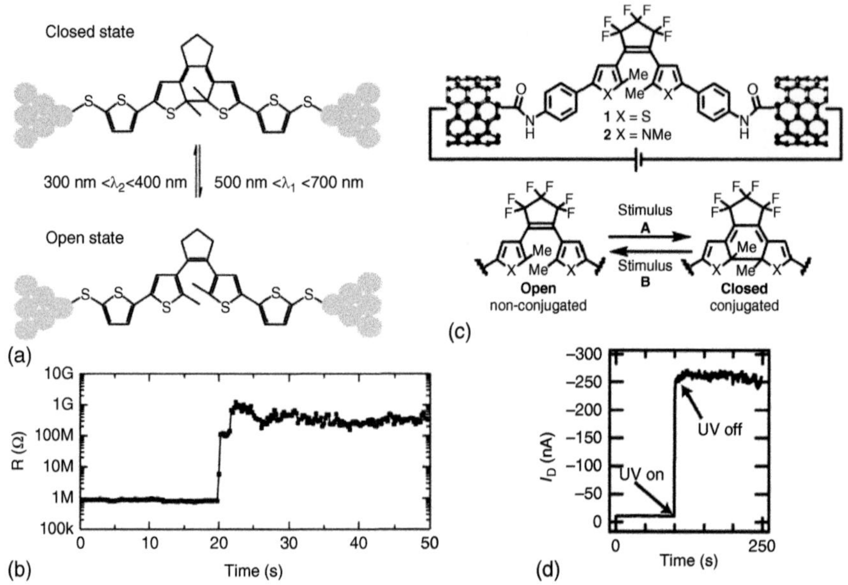

Figure 5.17 One-way photoswitching in single-molecule junctions. (a) Photochromic molecular switch between two Au contacts in the closed state (top) and the open state (bottom). (b) Resistance vs. time, which depicts the switching of the junctions from the conducting to the insulating state. Source: Dulić et al. [185]. 2003, American Physical Society. (c) Molecular bridges between the ends of an individual SWNT electrode. (d) Drain current as a function of time, which depicts the switching of the junctions from the low to the high state. Source: Whalley et al. [186]. 2007, American Chemical Society.

back again (Figure 5.17d). Interestingly, this phenomenon is in contrast to the situation occurring in devices made from gold electrode break junctions with dithiolated molecules, as discussed above. In the case of SWNT-diarylethene single-molecule junctions, the reason is most likely due to energy dissipation of the excited state of the closed form through the electrodes. This result is consistent with the fact that the lengthening of the π-conjugation on the termini of the DAE switches decreases the quantum yield for opening [187]. The nanotube is the ultimate extension of this conjugation. These results imply that the electrode-molecule contact interface is one of the most challenging issues in realizing functional molecular electronic devices.

In another attempt, Jia et al. [183] developed a series of single-molecule photoswitches with a new device platform of graphene-molecule single-molecule junctions (Figure 5.18a). In this study, to tailor the energy level alignments at the molecule–electrode interface, the diarylethene backbones were modified with rationally designed side and anchoring groups. As indicated in Figure 5.18b, the second molecule was furnished by substituting the hydrogenated cyclopentene of the first molecule with the fluorinated unit, while the third was based on the second by further introducing a methylene group (CH_2) between the terminal amine group and the functional center on each side. Demonstrated by the theoretical calculations, molecular engineering can be an efficient tool for tuning the

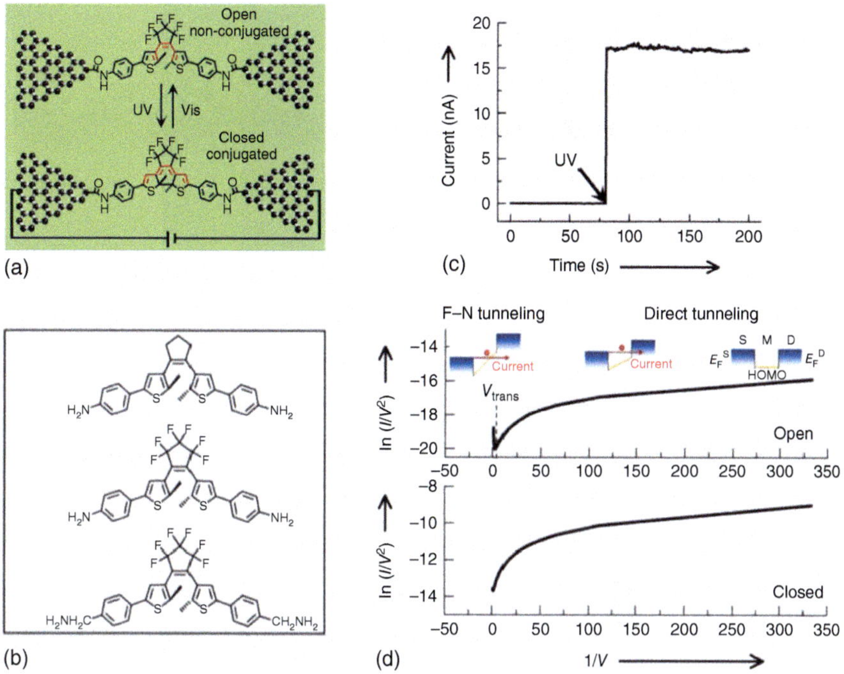

Figure 5.18 Switching properties of graphene-diarylethene single-molecule junctions. (a) Switching of graphene-diarylethene junctions. (b) Molecular structures. (c) Current-vs.-time curve of a graphene-diarylethene junction. (d) F–N plots of the I–V characteristics for the same junction. The insets indicate the conventional barrier model. Source: Jia et al. [183]. 2013, John Wiley & Sons.

molecule–electrode coupling strength. A reproducible optoelectronic switching from the off state to the on state (Figure 5.18c) was achieved. Unfortunately, continuous reversible switching was not realized, which indicated that the presence of the strong coupling between the molecules and the electrodes remained, hence consistent with the case of the SWNT-based molecular junctions. In addition to the observation of one-way photoswitching, detailed spectroscopic analyses based on both barrier and resonant tunneling models were performed, which supported the experimental observation of the photogated tunneling transition in the molecular junctions from direct to Fowler–Nordheim (F–N) tunneling (Figure 5.18d).

This approach of molecular engineering was also reported by Kim et al. [184], where four sulfur-free diarylethene molecules were developed to reduce the possibility of unintended binding. Furthermore, the charge transport of the open and closed isomers of the photoswitching molecules in the MCBJ-based single-molecule junctions was studied at low temperatures (Figure 5.19a). The single-molecule conductance was examined by breaking and forming the atomic contacts repeatedly, which revealed that both the conduction properties and the conductance switching ratios were significantly influenced by the end groups as well as the side chains of the molecules (Figure 5.19b). By analyzing the I–V curves within the framework of the single-level transport model, an unexpected behavior of the current-dominating

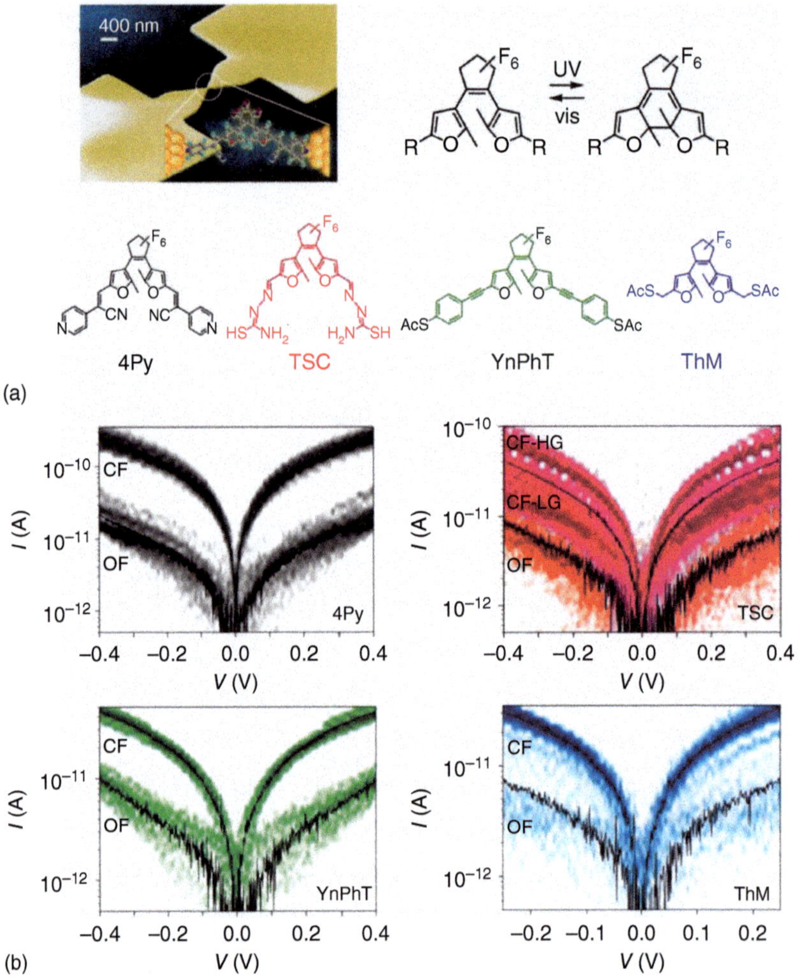

Figure 5.19 Switching properties of sulfur-free diarylethene molecules. (a) SEM image of an MCBJ device and the structure of four photochromic molecules: 4Py [188], thiosemicarbazide (TSC) (red), YnPhT (green), and ThM (blue). (b) Density plots of approximately 20 *I–V* curves of the open form (OF) and the closed form (CF). Source: (a) Ralph et al. [188]. 1997, John Wiley & Sons. (b) Kim et al. [184]. 2012, American Chemical Society.

molecular orbital was found upon isomerization. These investigations furnish new insight into designing novel types of functional molecule-based devices toward practical applications. The realization of in situ continuous reversible switching of the single-molecule photo-switches between two measurable conductance levels remains a formidable challenge, thus leaving an interesting opportunity for future studies to explore.

To create memory effects in a molecular switch, an external stimulus is required to generate independently stable states so that the information can be encoded as "0" and "1" in the same device. To achieve electrical bistability, a promising approach via molecular photochromics recently developed. Specifically, reversible photoisomerization between two isomers upon irradiation with light of different wavelengths were used [187, 189]. These conformational changes lead to remarkable differences in their electronic structures and distinct photophysical properties, which has been well observed in solution, thus forming the basis for new types of optoelectronic devices and molecular switches [190]. However, a significant challenge in practice is to maintain the photophysical properties existing in solution when sandwiched between two solid-state electrodes at the single-molecule level. Because the operational parameters, switching voltages, and bistability are sensitive to the environment. For realizing the single-molecule switches, the strong coupling with the electrodes and/or electrostatic interaction with the substrate could be unfavorable. For example, Lara–Avila et al. [191–193] demonstrated an environmental effect on the electronic structure of a bianthrone molecule connected to silver electrodes on aluminum oxide substrates. In addition to the observation of persistent switching of the electric conductance at low temperatures, it was determined that the measured activation energy of the photoisomerization process was on the order of 35–90 meV, which is an order of magnitude lower than that measured for free bianthrone ($\Delta E_{act} = \sim 0.9$ eV). The reason for this is that, when positioned in intimate contact with silver and on an aluminum oxide dielectric with high permittivity, one bianthrone isomer acquired a negative charge and hence shifted the molecular geometry (due to image forces in the substrate), which in turn leading to decreased transition barrier.

5.4.3 Switching Triggered by Chemical Process (Redox and pH)

As discussed in Sections 5.4.1 and 5.4.2, due to the consequent control over the junction conductance, wave functions and, particularly, the redox energy states of the molecular bridges can be tuned. However, it's challenging to locate three electrodes in intimate proximity of a molecule. Such result has been demonstrated in solid-state planar device architectures, where a metal gate electrode is separated from the source-drain contacts by a thin dielectric film. One disadvantage of these approaches is lacking of precise control on the degree of molecule-gate electrode coupling, which is crucial to modulating the molecular energy levels necessary for sufficient electrical coupling with both leads. Additionally, another possible challenge is to avoid the evaporation of metal grains in the electrode gap during metal evaporation in the junction fabrication, which is detrimental to the electrical characterization of the molecular junctions.

To solve these issues, electrochemical gate is an alternative approach to realizing strong gating effects. Measuring electrochemical gating of a single molecule in an electrolyte is schematically illustrated in Figure 5.20. In this setup, the bridge energies are controlled via the electrochemical potential of molecular devices immersed in an electrolyte solution relative to a reference electrode. This is typically achieved in a four-electrode electrochemical setup with counter and reference electrodes serving alongside the source and drain electrodes (both of which are the working electrodes of the electrochemical cell).

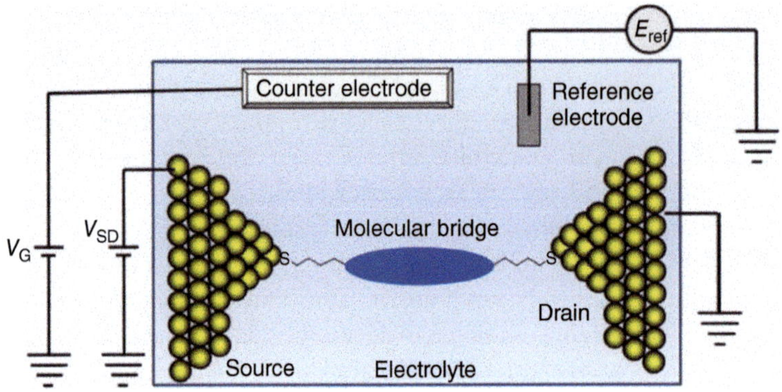

Figure 5.20 Schematic illustration of the electrochemical gating in a single molecule in an electrolyte. Source: Kay et al. [194]. 2012, American Chemical Society.

The extremely short Debye screening length (e.g. a few angstroms) is a remarkable feature of the electrochemical setup in strong electrolytes. Although the counter and reference electrodes are physically far from the molecular bridge, any charge on the molecular bridge can be monitored by the charge in the solution at a distance significantly shorter than what can be generally achieved in the solid-state planar device architectures. For example, if a 0.1 V potential difference exists across an electrochemical double layer with a thickness of 0.1 nm, a mean electric field strength of $10^9\ \mathrm{V\,m^{-1}}$ can be achieved across this double layer region. Therefore, this strong and reproducible gate field is sufficient to significantly modulate the current crossing single molecules. It should be noted that, to minimize the electrochemical leakage currents and the capacitive current components, the source-drain electrodes should be insulated by a nonconductive material, except at the very end of the electrode. However, the insulation process is relatively challenging, in the case of nanogapped electrodes in the break junction configuration.

To obtain the switching property with electrochemical gating, redox-active molecules, which can be reduced or oxidized upon an applied electrochemical gate field, represent a particularly attractive class of switches. In this context, the primary challenge is to achieve a high ratio between the ON and OFF states of the nanojunction conductance. For example, Kay et al. [194] assembled a redox active pyrrolotetrathiafulvalene (pTTF) derivative (Figure 5.21) to gold contacts using $-(CH_2)_6S-$ groups at both ends. By using an in-situ STM break junction configuration (Figure 5.21), a multiple OFF–ON–OFF–ON–OFF conductance switching behavior with a conductance ratio of ~4 were observed when the electrode potentials matched the redox peak positions (Figure 5.21). These redox peaks were ascribed to two reversible redox processes of the pTTF moiety: the formation of a radical cation and a dication. This molecular conductance vs. electrochemical potential relation could be well modeled as a sequential two-step charge transfer process with full or partial vibrational relaxation (Kuznetsov–Ulstrup [KU] model) [195]. Using this method, for both the first and the second redox transitions of the *p*TTF bridge in the 1-butyl-3-methylimidazolium trifluoromethanesulfonate ionic

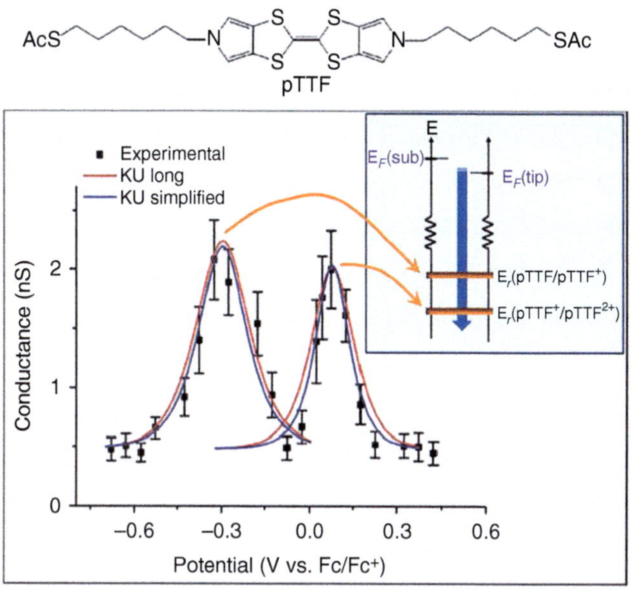

Figure 5.21 Conductance-sample potential relationship of pyrrolotetrathiafulvalene (pTTF). The red lines depict the long Kuznetsov–Ulstrup (KU) model, and the blue lines depict the simplified KU model. For both redox transitions, the KU model versions fit the experimental data well. The inset depicts a schematic energy level representation of the KU model for the two redox transitions. When the electrode potential is scanned positive, the pTTF/pTTF$^+$ redox transition is first brought into "resonance," and then at more positive potentials, the second redox transition (pTTF^{2+}/pTTF$^+$) comes into resonance. Source: Kay et al. [194]. 2012, American Chemical Society.

liquid environment, the reorganization energies of were estimated to be ~1.2 eV, which is significantly larger than that obtained for the same molecular bridge in aqueous environments (~0.4 eV). These differences should result from the large, outer-sphere reorganization energy for the charge transfer across the molecular junction in the ionic liquid environment. To improve the electrochemical switching behavior of the molecular junctions, Li et al. achieved a significant increase in both the junction conductance and the switching efficiency (ON/OFF ratio: ~8) via redox-active benzodifuran-based molecules with optimal anchoring groups ($-CS_2^-$) instead of thiolate [196].

As another example, Darwish et al. [197] developed a single molecule switch with the functional element (anthraquinone/hydroanthraquinone [AQ/H_2AQ] redox couple) merged into a rigid, structurally well-defined norbornylogous unit (Figure 5.22a). Such a design conferred additional stability to the SAMs. In-situ STM-BJ experiments shows electrochemical and reversible switches between the high-conducting H_2AQ system and the low-conducting AQ system with a conductance on/off ratio of 1 order of magnitude (Figure 5.22b,c).

Destructive quantum interference (QI) effects operating in the AQ form is accountable for this high magnitude, according to both the theoretical prediction and the bulk conductance results across the SAMs. Additionally, different pH values

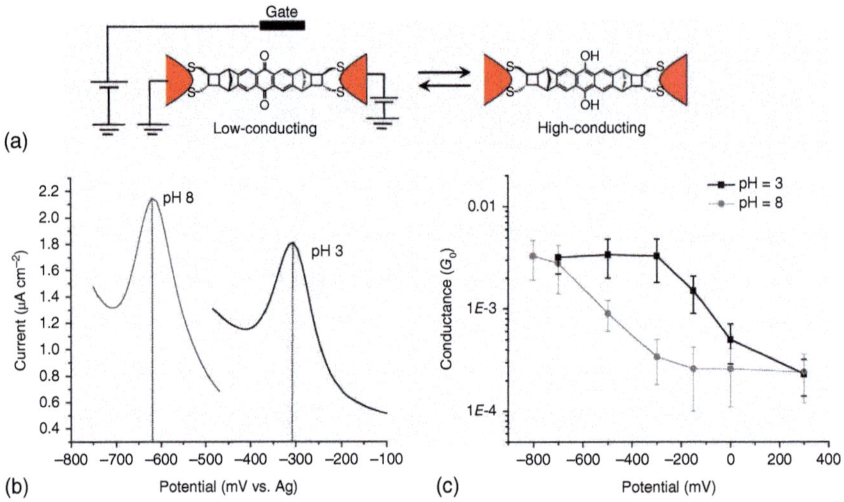

Figure 5.22 Electrochemical gating of an anthraquinone/hydroanthraquinone redox couple. (a) Redox reaction of the AQ moiety to an H_2AQ moiety. (b) Alternating current voltammograms at pH 3 (black line) and pH 8 (gray line). (c) Evolution of the conductance of the AQ molecule with a gate potential at pH 3 (black line) and pH 8 (gray line). Source: Darwish et al. [197]. 2012, John Wiley & Sons.

were also shown to modulate the potential range of the conductance enhancement. Recently, a similar result was also reported by Baghernejad et al. [198], where a conductance ratio of over >1 order of magnitude was achieved on the basis of organic molecules containing an AQ center [199]. In comparison with examples discussed above, one important distinction observed in this study was that in the redox-inactive potential region, the effect of the gating is to shift the Fermi level relative to the molecular resonances, thus leading to a modest change in the conductance.

It should be noted that, in addition to redox active molecules, nonredox active molecular systems, such as 1,2-bis(4,4-dimethylthiochroman-6-yl)ethylene and 4,4-bipyridine, also indicated a tunable charge transport, as demonstrated experimentally and theoretically by Capozzi et al. [200] and Baghernejad et al. [201]. The working principle is that the electrochemical gate potential could suitably tune the energy level alignment between the LUMO or the HOMO in the molecular orbitals with the Fermi level of the electrode, thus resulting in the modulation of the charge transport in the single-molecule junctions.

5.4.4 Spintronics-Based Switch

Molecular spintronics, combining the ideas and advantages of spintronics and molecular electronics, have emerged as a new field and gone through rapid growth in recent years [202]. Using molecule(s) responsive to magnetic field to create molecular device(s) is the key point of molecular spintronics. The so-made molecular devices are very appealing as incredibly versatile and unique platforms (playgrounds) for implementation of present spintronics concepts and/or exploration of new ones, due to their obvious(apparent) advantages, such as fine-tunability properties (i.e. spin–orbit and hyperfine interaction), high degree of purity, and ease of synthesis. Among these molecule devices, single molecule magnet (SMM) compounds appear to be particularly attractive as a result of extremely long magnetization relaxation time at low temperatures (i.e. reaching years below 2 K). In conjunction with the advantages of the molecular scale, these systems are promising for high-density information storage as well as quantum computing due to their long coherence times, compared with bulk magnetic materials [203]. Furthermore, several appealing quantum effects of the static and dynamic magnetic properties may be derived from their molecular nature [204, 205]. The rich physics behind the magnetic behaviors of the molecular spintronic devices present very interesting effects, such as negative differential conductance and complete current suppression, which are useful for molecular electronics. More importantly, specific functions (i.e. switching ability with light and electric field) can be directly integrated into such molecule devices. In this section, we are going to briefly review the primary developments in exploring the switching behaviors in single molecule spintronics. For additional fundamental information on the different aspects of molecular spintronics, interested readers are recommended to refer to several comprehensive reviews [202, 206–214].

Currently, there are two possible techniques to attach a magnetic molecule between two electrodes. One is to use an STM tip as the first electrode and the conducting substrate as the second electrode. Although STM is a powerful tool for understanding the fundamental mechanism of the magnetic interaction between a molecule and a substrate, such as Kondo effects [215] and single-atom magnetic anisotropies [216], it is not a device fabrication platform. As this method has been detailed [206, 217], and is irrelevant to the topic in this section, we omit this approach here. The other is molecular spin-transistor, which is a three-terminal transport device, based on the three- (or two-) terminal break junctions (i.e. an additional gate electrode in addition to the source and drain electrodes). This device is a single-electron transistor with nonmagnetic electrodes and a single magnetic molecule as the island. Mechanism of this device is that the current passes through the magnetic molecule from the source to the drain electrode, and the electronic transport properties are tuned using a gate voltage V_G (Figure 5.24). It is worthwhile to point out that Schwarz et al. recently demonstrated that the oxidation and reduction of Mo compounds could be solely triggered by the electric field applied to a two-terminal molecular junction presenting voltage-induced conductance switching [218].

For reliable molecular spin-transistors, a fundamental issue to be solved is the fact that, most of magnetic molecules chemically synthesized are extremely fragile away from the solution and often sensitive to a metallic surface. In a few cases, because of coupling with electrodes, certain molecules will lose their magnetic moment entirely. This disadvantage brings about a substantial difficulty in constructing robust magnetic single-molecule devices. By assuming that these three-terminal devices incorporating magnetic molecules, which preserve their SMM properties, can be achieved, the next important question is how should these devices operate? Specifically, how do these devices function as a switch or sensor to read, write, and process information upon exposure to external stimuli? In the case of the SMMs, two key ingredients of the electrodes useful for their functionalization are the ability to produce a spin-polarized current and the possibility of altering the mutual orientation of magnetization vectors. The first property depends on the ability of manipulating the exchange interaction controllably, thus converting spin information to molecular orbital information. The second property counts on the magneto-crystalline anisotropy, i.e. spin–orbit coupling.

In fact, recent experimental and theoretical developments demonstrated the possibilities for switching the magnetic molecules and the produced effects on electron transport [206, 217] using two classes of magnetic molecules: spin-crossover compounds and molecules presenting valence tautomerism [219]. These two broad classes of molecules are particularly promising for several reasons. First, their extremely long-living magnetization relaxation time signifies intrinsic nonvolatile nature of these molecules. Second, as the crossover generally depicts an internal charge rearrangement and an atomic relaxation, the two different molecular states display distinct electronic structures, which are expected to indicate different I–V curves for different spin-states, indicating that the spin-state is likely to be electrically readable. Finally, because the crossover transition is sensitive to the local electrostatic environment, an electric field may be utilized to induce the transition, suggesting that a potential device may be electrically switchable.

For example, in 2005 Rocha et al. [220] theoretically predicted a type of organic spin valve obtained by sandwiching an organic molecule between magnetic contacts. With a combination of state-of-the-art nonequilibrium transport methods and the DFT, they demonstrated that these spin valves could display a large bias-dependent magnetoresistance which could be engineered by an appropriate selection of molecules and anchoring groups. Four years later [208], the same group proposed another general route, that is, switching the sign of the exchange coupling between two magnetic centers with an electric potential. This general effect, named as the electro-static spin crossover, occurs in insulating molecules with super exchange magnetic interaction and inversion symmetry breaking. To prove its efficiency, a family of dicobaltocene-based molecules demonstrated approximately $1\,V\,nm^{-1}$ critical fields for switching, according to calculations from the first principles. More importantly, the critical fields could be engineered by incorporating appropriate substituents into the basic dicobaltocene unit. Consistently, this possibility of controlling the molecular spin state by an electric field is also proved by Diefenbach and Kim [221]. $Ba_2(C_6H_6)$ was used to obtain the distinct difference in response to the molecular electronic states upon perturbation with an external electric field. These results may inspire and benefit future experimental studies toward the synthesis and spectroscopic characterization of molecular magnetic switches to prove this switching concept.

In fact, these theoretical predictions have inspired a series of remarkable experimental studies. For example, in 2011 Prins et al. [222] reported that a temperature hysteresis in conductivity for nanoparticles consisting of spin crossover molecules. Such hysteresis was attributed to the spin crossover transition. Intriguingly, although this activity used nanoparticles or crystals, not single molecules, this experiment demonstrated that the crossover could be induced by a static electric field, thus paving the way for electrically controlled spin devices at the molecular level. In fact, these functional single-molecule spin devices were realized recently by Osorio et al. [223] and Wagner et al. [224]. In the former study, electromigration was used to create a three-terminal device incorporating a single Mn^{2+} ion coordinated by two terpyridine ligands. This device demonstrated an electrically controlled high-spin ($S = 5/2$) to low-spin ($S = 1/2$) transition by adjusting the gate voltage. Using low-temperature inelastic cotunneling spectroscopy, a strongly gate-dependent singlet-triplet splitting on the low-spin side was also uncovered, which was consistent with an exact diagonalization of the Mn-complex. In the latter study, the MCBJ technique was utilized to measure the electron transport through a single-molecule junction containing two coupled spin centers that were confined to two Co^{2+} ions (Figure 5.23a). According to the authors, this method could detect a coupled spin pair in a single magnetic molecule and a bias voltage could be used to switch between two states of the molecule. This claim was proven experimentally and theoretically. In theory,

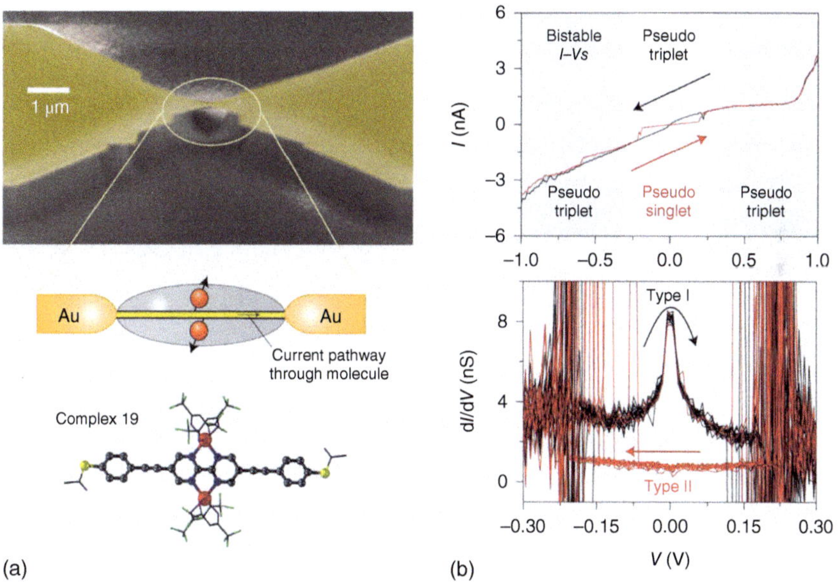

Figure 5.23 Switchable single-molecule spin devices created using the MCBJ technique. (a) Schematic of the experiment and molecular structure of complex 19. A molecule with a pair of spin centers is contacted in a single-molecule junction. The magnetic ion pair is attached orthogonally to the current pathway. (b) Bistable I–V characteristics due to the hysteresis of the coupled spin pair. Type I is a Kondo-like zero-bias anomaly; this is identified as the electrical fingerprint of the pseudotriplet state. Type II is not a Kondo-like anomaly; this is assigned as the pseudosinglet. Source: Wagner et al. [224]. 2013, Springer Nature.

spin–orbit configuration interaction methods were used to calculate the combined spin system, which indicated that the ground state was determined to be a pseudosinglet whereas the first excitation behaved as a pseudotriplet. In the experiments, these states could be assigned to the absence and the occurrence of a Kondo-like zero-bias anomaly in the low-temperature conductance data (Figure 5.23b). This result is remarkable because one can repeatedly switch the molecules between the pseudosinglet state and the pseudotriplet state by simply applying a finite bias.

5.5 Gating the Transport: Transistor-Like Single-Molecule Devices

Transistors, typically three-terminal electrical devices constituting the heart of the electrical circuits, control the current flow between two electrodes (source and drain electrodes) by adjusting the voltage applied at a third electrode (gate electrode). As the microelectronics industry is enthusiastically developing highly integrated circuits that require devices with smaller dimensions, a logical limit to device miniaturization is transistors with channels defined by a single molecule (Figure 5.24).

Therefore, the construction and understanding of such a device has been a long-standing goal of nanoelectronics. For creating nanometer-spaced electrodes, discrete approaches have been developed, among which, carbon electrode–molecule junctions and electromigration break junctions are particularly appealing because of their intrinsic three-terminal device architectures [225]. One important feature of the former is the improved device stability due to formation of molecular junctions with covalent chemistry, which is of considerable importance to build molecular devices with desired functions toward practical applications. In this section, we focus on the primary progress of single-molecule transistors with back gating, side gating, and chemical gating configurations.

5.5.1 Electrostatic Gate Control

One of the most generalized gate configurations, the back gate, is commonly found in solid-state device architecture, in which the entire nano-circuit is supported by heavily doped Si, Al, or polymer substrates and insulated from the gate electrodes

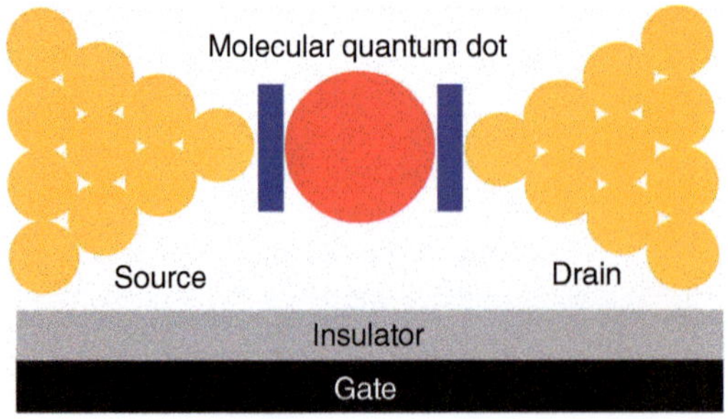

Figure 5.24 Schematic representation of single-molecule transistors with back gating architecture. Source: Xiang et al. [140]. 2016, American Chemical Society.

in the bottom by an insulating layer to ensure that only molecules in the nanogaps between the source and the drain electrodes would be electrostatically modulated. The insulating layer above could be produced by oxidizing the substrate to achieve the corresponding oxide insulator, such as SiO_2, Al_2O_3, etc., or a tens-of-nanometer vacuum gap [4], for which the thickness and dielectric constant are the two most critical parameters for determining the scale that the gate modulation could reach without incurring gate–drain leakage or gate breakdown. Therefore, it's important to select the gate and insulator materials for the total level shift, and to obtain high-performance single-molecule devices, an insulator with a high breakdown voltage, reduced thickness, and easy fabrication is required.

The electromigration technique is one of the most important techniques used to fabricate source-drain electrodes with a functional back-gating configuration. In fact, solid-state single-molecule transistors with back gating were reported by two separate groups as early as over 10 years ago [226, 227]. To achieve the strong effects of the gate voltage on the molecules, electromigration-induced break junctions were used by both groups to fabricate their device on silicon wafers with extremely thin dielectrics (30 nm silicon oxide or 3 nm aluminum oxide) (Figure 5.25a). However, this approach is difficult because with the gate dielectric scaling down, parasitic currents generally caused by various mechanisms (e.g. quantum tunneling, electron hopping) can lead to a current between the gate and the drain, thus overshadowing the molecular signature. Therefore, it's the primary challenge in device fabrication to design a junction geometry, combining with the proper material selection, which can reduce the gate leakage as much as possible and maximize the electric field reaching the molecule. The need for such precise control may partly explain why only few of the devices prepared by the authors functioned properly as transistors. In their studies, they observed novel physical phenomena based on two mechanisms: Coulomb blockade, in which the flow of electrons is controlled by the sequential charging of a molecule; and the Kondo effect, in which conducting electrons interact with a local spin (intrinsic angular momentum) in a molecular junction. Interestingly, changes in the length of the insulating tether alter the coupling between the molecules and the electrodes, enabling the fabrication of devices that exhibit either of the single-electron phenomena mentioned previously. For example, Park et al. [226] reported that by examining two related molecules containing a Co ion bonded to polypyridyl ligands and attached to insulating tethers of different lengths (Figure 5.25b), it was determined that devices formed using the longer molecules indicated a blocking of the current due to the presence of the Co ion island (Coulomb blockade) (Figure 5.25c), whereas devices created using the shorter molecules indicated the Kondo effect (Figure 5.25d). This Kondo effect is indicative of the formation of a bound state between a local spin on an island and the conduction electrons in the electrodes that enhance the conductance at low biases. In another example of single-electron-transistor measurements [228], Kubatkin et al. observed that the molecular electronic levels were strongly perturbed as compared to those of the molecules in the solution, leading to a significant reduction in the gap between the HOMO and the LUMO. This surprising observation could be caused by image charges generated in the

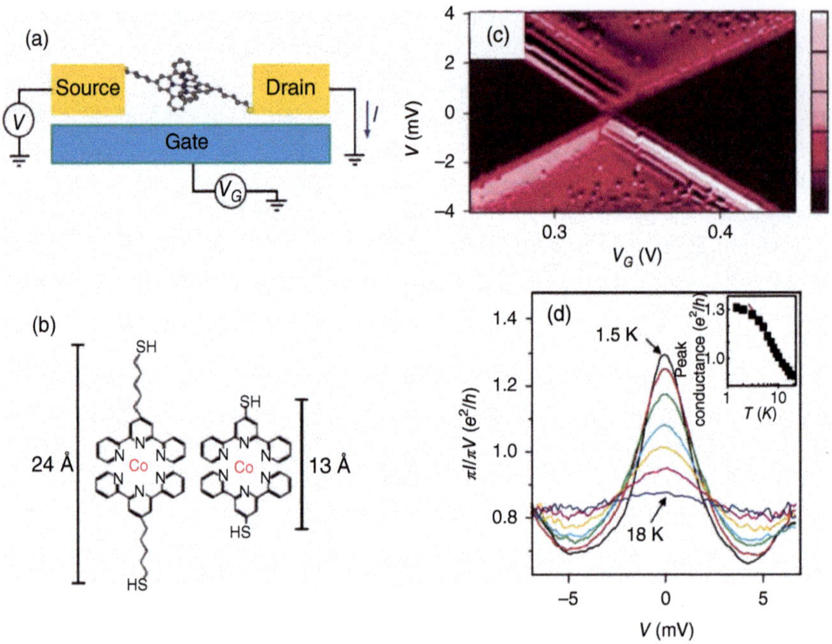

Figure 5.25 Solid-state single-molecule transistors. (a) Schematic diagram of the single-molecule transistors. (b) Molecular structures of the Co complexes. (c) Coulomb blockade observed in devices formed from longer molecules. (d) Kondo peaks as a function of the voltage. The inset depicts the conductance as a function of the temperature at $V = 0$. Source: Park et al. [226]. 2002, Springer Nature.

source and drain electrodes, thus resulting in a strong localization of the charges on the molecule.

In addition to the two single-electron phenomena discussed above, Ghosh et al. [229] predicted a third phenomenon based purely on the electrostatic modulation of the molecular-orbital energy of a single molecule. Remarkably, this result was later realized by Song et al. [230], where unprecedented insight was provided into the underlying physics of the charge transport in their molecular transistors created with the electromigration technique. In this study, an IETS measurement was performed on a gate voltage with two prototype molecules, a 1,8-octanedithiol (ODT) with an alkyl backbone and a 1,4-benzenedithiol (BDT) with a delocalized aromatic ring, to examine the interactions between the tunneling electrons and the vibrational modes of the molecules. Because each dithiol had its own vibrational “fingerprint,” this technique provided definitive proof that the measured currents actually passed through the molecules in single-molecule transistors (Figure 5.26a). More importantly, it was determined that the spectra of the ODT junction remained essentially unchanged at different values of the gate voltages, which is attributed to that the HOMO level for the ODT is energetically considerably farther away from the Fermi level (E_F). However, the IET spectra of the BDT junction were drastically modified by the gate voltages, thus exhibiting an enhanced spectral intensity. This observation was attributed to the dominant transport orbital (the HOMO) being strongly coupled

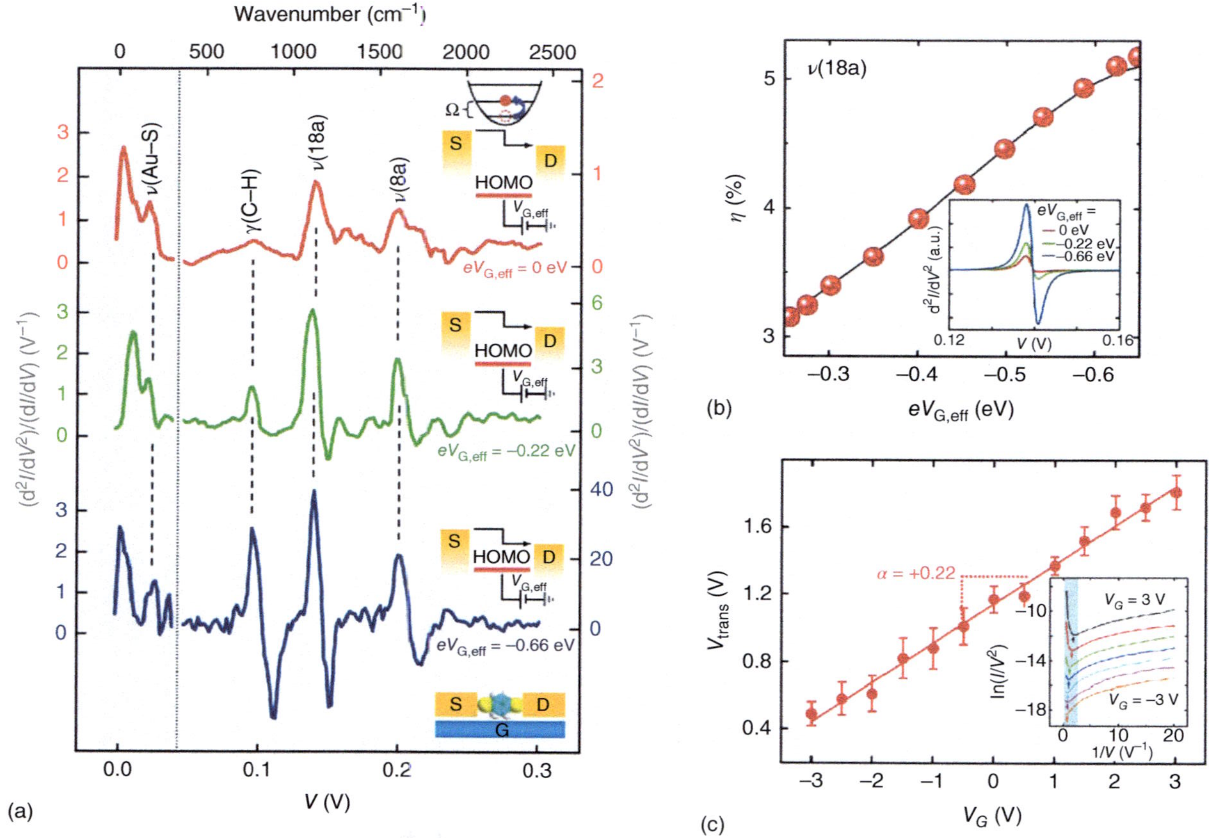

Figure 5.26 Gate modulation of molecular orbitals in single-molecule transistors. (a) inelastic electron tunneling spectroscopic (IETS) measured at 4.2 K for different values of the effective molecular orbital gating energy ($eV_{G,eff}$), with the vibration modes assigned. The left-hand y-axis corresponds to the gray shaded region of the spectra, and the various right-hand y-axis (with different scales) correspond to the related (color-coded) spectra in the nonshaded region. (b) The relative change, η, in the normalized conductance for the ν(18a) mode as a function of $eV_{G,eff}$. The circles depict the experimental data, and the solid curve represents the theoretical fit. The inset is the gate-variable IETS for the ν(18a) mode. (c) Plot of V_{trans} vs. V_G. The inset is the corresponding TVS spectra. Source: Song et al. [230]. 2009, Springer Nature.

to the internal vibrational modes of the BDT, resulting in a resonantly enhanced IETS (Figure 5.26b). Additionally, the transition voltage spectroscopy (TVS) technique was used to reveal the energy offset (V_{trans}) between the gating orbital of the molecular junction (the orbital that modulates electron tunneling) and the Fermi levels of the source and drain electrodes. By measuring V_{trans} with the TVS at different applied gate voltages, a linear relationship was determined to exist between the gate voltage and the molecular-orbital energy in their devices, as expected for single-molecule transistors (Figure 5.26c). These findings demonstrated the direct gate modulation of the molecular orbitals, thus enhancing the prospects for true molecularly engineered electronic devices.

Furthermore, the three-terminal electrostatic back gate can be achieved via a MCBJ. Champagne et al. first reported a device in which a gold wire was lithographically patterned on top of a silicon/silicon-oxide substrate [231]. The silicon was heavily doped, thus acting as a back-gate electrode. After the oxide layer was etched, a suspended gold wire was obtained above the substrate using a vacuum as an insulator. Their gated MCBJs provided the capability to simultaneous adjust the gap between the electrodes and the molecule energy levels. It was determined that the mechanical motion could change a molecule's capacitance, conductance, and gate voltage value for the Coulomb blockade degeneracy. It is anticipated that this new experimental tool will enable systematic studies of how changes in molecular conformation affect the electrical conduction and how transport varies between the weak-coupling Coulomb blockade and the strong-coupling Kondo regimes. A few years later, using the same device, further experiments were reported in which the spin states in the Spin-1 molecules could be mechanically controlled in the absence of a magnetic field by stretching the individual cobalt complexes [232]. This control enabled quantitative studies of the underscreened Kondo effect and could serve as model systems for performing precise tests on correlated-electron theories.

A similar back-gate approach was introduced by Martin et al. [233]. In contrast to Champagne et al. the greatest difference is that the flexibility of the substrate was used instead of that of the rigid silicon substrate; thus, the junctions could be easily broken by mechanical means. However, upon application of a gate voltage, the freshly broken electrodes were pulled toward the substrate due to the electric field. To prevent the drift of the source–drain electrodes under electrostatic gating, Martin et al. created an improved design in which the gate and the source–drain electrode were suspended (i.e. the source–drain electrode, insulation layer, and gate electrode were sandwiched and suspended above the substrate together) [234]. To test the gate action of this device, the C_{60} molecules were studied to determine that the total shift in the molecular levels for the devices was approximately 40 meV. Recently, the image-charge effects in single-molecule junctions were further investigated by the same group using this MCBJ with a back gate [235]. It was determined that a substantial increase in the transport gap and level shifts as high as several hundreds of millielectronvolt appeared when the electrode separation increased by a few angstroms. An analysis of this large and tunable gap renormalization based on atomic charges revealed the dominant role of image-charge effects in the single-molecule junctions.

Because carbon-based electrodes (i.e. graphene sheets and carbon nanotubes) provide a higher structural stability, even under ambient conditions, and reduce the size mismatch between the electrodes and the molecule, they are interesting candidates to replace noble metals. Another advantage of carbon-based electrodes is that they allow the molecular junctions to be brought closer to the gate as compared to bulky gold, and they typically do not have electrostatic shielding properties, thus enhancing the gating efficiency. There are several significant reports in which different strategies were used to link the molecules to the carbon electrodes, and the gating of the current was observed [225]. Here, we only focus on the novel physical phenomenon investigation using back gating with carbon electrodes. For example, Marquardt et al. first showcased an example of a single molecular junction at room temperature displaying electroluminescence [236]. The reported molecules have fluorescent cores extended with organic spacers, which not only increase the molecule length but also allow for easy electrostatical trapping of the molecules in the gap using dielectrophoresis. These results suggested a new characterization and functional possibilities as well as demonstrated a significant potential of carbon nanotubes for use in molecular electronics. One disadvantage to back gating with carbon nanotube electrodes is that, for example, an SWNT with a p-type semiconductor behavior may completely mask the gate response of the molecule. This problem could be solved using the multi-walled nanotubes (MWNTs) or thicker few-layer graphene flakes. The versatility of graphene may allow the production of electrodes with intrinsic functional devices. For example, it was predicted that graphene electrodes terminated in zigzag edges may act as Fano-like resonances and spin filters, which allow the injection of spin-polarized currents into the bridged molecules [237]. However, the fabrication of the reliable, nanometer-sized, and adjustable gaps between the source and the drain electrodes in the graphene sheets remains a considerable challenge. Particularly, improving the control of the edge chemistry is crucial.

5.5.2 Side Gating

The creation of a single-molecule FET has been considered a critical step in molecular electronics. A nanoscale gap junction formed by "electromigration" is a highly efficient method for creating a highly integrated single-molecule FET. However, this method is limited in practice by its extremely low device yield [230]. Conversely, the MCBJ technique is advantageous because it can easily form single-molecule junctions several times mechanically, and it does not suffer from low yield [131]. In the MCBJ, the primary challenge is the method used to form reliable three-terminal FET junctions from single molecules. The three-terminal MCBJ with back gating geometry may suffer from the gating electrode breaking because it bears a large strain to separate the source–drain electrodes by bending the substrate.

Xiang et al. demonstrated a completely different single-molecule FET junction with the MCBJ using a nanometer scale noncontact side gate to the molecular junctions (Figure 5.27) [238]. In this geometry, the gate electrode and source–drain electrodes were located in the same plane (thus the deformation of the gate electrode

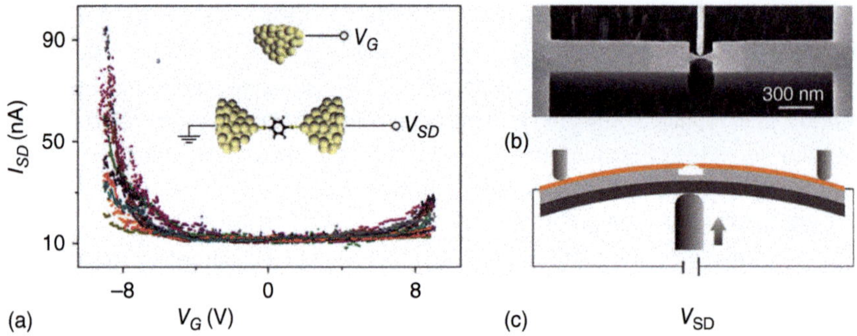

Figure 5.27 Molecular transistors with side gating. An external electric field generated by a side-gate electrode was applied to single-molecule junctions formed by the MCBJ. (a) Source drain current (I_{SD}) vs. gate voltage (V_G) curves for the Au–BDT–Au molecule. The current was modulated by the gate voltage in 12 different molecular junctions. The inset illustrates a three-terminal junction with the BDT bridged between the source and drain electrodes. (b) Top view of a scanning electron microscopy (SEM) image of a micro-fabricated MCBJ chip consisting of a freestanding metal bridge with a gate electrode. (c) Schematic of the MCBJ setup. The distance between the electrodes for opening and closing the operation modes can be tuned by bending or relaxing the substrate, respectively. Source: Xiang et al. [238]. 2013, American Chemical Society.

was extremely reduced during bending of the substrate), and the electrical field was enhanced in the thickness of the gate electrodes, especially on the gate side as compared to that of a typical back gate configuration [239]. Using this side-gating MCBJ, the ability to form a stable single-molecule junction via the MCBJ technique and shift the energy levels of the molecule by side gating was successfully demonstrated.

Generally, to directly visualize the electromigration process and discriminate trapped molecules/metal clusters from dusts inside the gap or visualize a source–drain breaking process, a transmission electron microscopy (TEM) requires the use of electron transparent substrates, which are noncompatible with the most common three-terminal devices with a back gate electrode. The side gate geometry is relatively compatible with the TEM inspection and can overcome this problem [239]. In addition to this advantage, the gate potential could be screened by the source-drain electrodes, which indicates that only those molecules facing the gate electrodes can be easily influenced by the gate electrodes. Nevertheless, the in-plane three-terminal molecular junction presents a new platform for fabricating highly stable single-molecule transistor junctions.

5.5.3 Electrochemical Gate Control

Electrochemical gating is a promising method for efficient gate coupling that provides a unique opportunity to manipulate the energy alignment and the molecular redox processes for a single-molecule junction [240]. Because the gate voltage falls across the double layers at the electrode–electrolyte interfaces, which are only a few ions thick, a field of approximately 1 V Å^{-1} can be reached. Hence, a large gate field and a high gating efficiency can be achieved using an electrochemical gate [241]. One advantage of an electrified solid/liquid interface is that the two potential differences can be controlled simultaneously: (i) the bias voltage between the source and the drain electrodes; and (ii) the potential drop between the working electrodes and the reference electrode. Generally, the electrochemical gating effect is realized via tuning the Fermi level of the junction electrodes, thus changing the redox state of the molecular functional unit. The STM break junction (EC-STM-BJ) is the most widely used technique in electrochemical gating studies thus far, and numerous significant results, such as current enhancement mechanisms, electrolyte gating models, and molecule structure changes via electrochemical gating, have been reported [241–245]. It should be noted that to reduce the electrochemical leakage currents, it is necessary to insulate the STM tip by a nonconductive material, except for the very end of the tip. The tip insulation is a simple procedure and can be performed in any laboratory in the EC-STM-BJ configuration. However, the insulation process is challenging for fixed nanospaced electrodes in the MCBJ configuration [246, 247].

It should be noted that the control of the charge transport in a single-molecule junction using electrolytic gating based on one nonredox active molecular system is also possible [198, 200, 201]. For example, Capozzi et al. reported modulation of the conductance of electrochemically inactive molecules in single-molecule junctions using an electrolytic gate [200]. This experimental study combined with the DFT calculations indicated that electrochemical gating can directly modulate the alignment of the conducting orbital relative to the metal Fermi energy, thus changing the junction transport properties. More recently, Baghernejad et al. demonstrated that the single-molecule conductance could be controlled via both the Fermi level tuning and the redox switching [198]. In the small potential region (redox-inactive), the effect of the gating is to shift the Fermi level relative to the molecular energy level, thus leading to a modest change in the conductance. A further shift in the electrode potential led to large and reversible jumps in the conductance due to the change in the molecular redox state and the molecular conjugation switching. The study indicated that the electro-chemical gating of single-molecule junctions could be modeled using computational models via conversion of the potential to the energy scale.

In summary, different approaches, including carbon electrode–molecule junctions, electromigration-induced molecular junctions, and three-terminal MCBJ-based break junctions, have been successfully developed to realize single-molecule transistors and demonstrated a direct gate modulation of molecular orbitals through the multi-spectroscopy approaches [225]. These studies set the benchmark for the validation of future studies in the charge transport for molecular systems. However, additional work must be performed before molecular electronic devices can effectively compete with their larger silicon-based equivalents. It's one of the most urgent issues to develop an efficient fabrication method that allows high yields of densely packed single-molecule devices. It is expected that the electron transport through the molecule may be controlled by a universal gate electrode, including electrical gates, electrochemical gates, magnetic gates, optical gates, terahertz wave gates, and mechanical gates. Additional new designs and device architectures are needed for transition of molecular electron transport from laboratory research to practical applications.

5.5.4 Molecular Quantum Dots

At last, another idea to be noted is to build molecular multidot devices, where a controlled number of molecules or multiple centers inside the same molecule are attached to two electrodes. This approach may lead to more interesting phenomena. For example, negative differential conductance effects were discovered in a double-metallocene molecule as a multidot system with molecular-level crossings [248]. In another study, Ouyang and Awschalom executed the chemical self-assembly approach to form multidot devices [249] containing a programmed sequence of CdSe QDs connected by *p*-benzenedithiol molecular bridges. Using a femtosecond time-resolved Faraday rotation spectroscopy, the authors probed the spin transfer process through the π-conjugated molecular bridge, which proved that the process was coherent and that the bridging molecule functioned as a quantum-information bus. The primary advantage of these multidot devices is that the intrinsic properties of the SMMs can be preserved because the coupling between the SMMs and the quantum conductor is weak, potentially enabling a nondestructive readout of the spin states.

Kondo effect: When a quantum dot (e.g. nanocrystal and nanotube) is attached to the metallic electrodes via the tunneling barriers, the electron transport is dominated by single-electron charging and energy-level quantization [227]. As the coupling to the electrodes increases, higher-order tunneling and correlated electron motion will lead to new phenomena, including the Kondo effect, which results from the exchange interaction between an unpaired localized spin and the conduction electrons in the metallic electrodes. Most of the studies on the Kondo phenomena in quantum dots were typically performed on systems where precise control over the spin degrees of freedom was difficult. Using a chemical design, the spin and the orbital degrees of freedom can be controlled for those molecules (e.g. metal ion complex), thus providing a new system for the Kondo phenomena investigations. Liang et al. first reported the observation of the Kondo effect in single-molecule junctions, where an individual divanadium molecule was trapped in the nanogap by the electromigration method and served as a spin impurity [227]. In the differential conductance maps (dI/dV as a function of the gate voltage V_G and the source–drain voltage V_{SD}), a diamond-shape pattern was observed, indicating the achievement of a single electron transistor. More significantly, the Kondo resonance (i.e. a sharp zero-bias conductance peak) was clearly observed, and it was demonstrated that this Kondo resonance could be tuned reversibly by using the gate voltage to alter the charge and the spin state of the molecules.

In another example, the Kondo effect in C_{60} single-molecule junctions was investigated [250]. Yu and Natelson used the electromigration technique to fabricate C_{60}-based single-molecule transistors, and demonstrated that each of them could act as a single-electron device in the Coulomb blockade regime at low temperatures. Interestingly, sidebands parallel to the zero-bias peak were observed, thus indicating the conduction processes that involved both multibody correlations and inelastic coupling to the vibrational modes. In the same year, Pasupathy et al. measured Kondo-assisted tunneling using C_{60} molecules in contact with ferromagnetic nickel electrodes fabricated by the electromigration method [251]. It was demonstrated that the C_{60} molecules could be strongly coupled to nickel (Ni) electrodes to exhibit the Kondo effect. More importantly, it was revealed that ferromagnetism could suppress Kondo-assisted tunneling; however, Kondo correlations were still present within the ferromagnetic electrodes. They found that a Kondo peak split was strongly affected by the magnetic moment alignment in the two electrodes under an applied magnetic field, hence proving a new window was needed to modulate the Kondo resonance.

5.6 Challenges and Outlooks

In all realms of science, great progress can be made when lessons from the past are applied to understand the challenges and opportunities of the present and future. The ideas at the base of the emerging field of SMEs are rooted in the same principles developed in chemistry over the past century. In this chapter, we have discussed the benefits and opportunities that arise from using chemical principles to guide the design and understanding of functional molecular electronic components. Nonetheless, many challenges need to be overcome before SME can be a reality out of the lab. One of the main challenges involves the realization of a technology for mass production of single molecule devices. Chemistry has been the key element in the design and synthesis of functional molecules, and it is one of the most promising ways to bridge the state-of-the-art lithography techniques (resolution ~10–50 nm) and molecular-scale dimensions (~1 nm) via techniques such as chemically driven self-assembly.

Concepts borrowed from reaction chemistry continue to inspire new SME devices. But, at the same time, discoveries in SMEs produce new knowledge in chemistry, engineering and other fields of science. Research in SMEs, particularly through scanning probe measurements on SAMs, has shed light on how molecules assemble and behave when adsorbed or bonded to metal surfaces [80, 252]. Extrapolating molecular conductance from these densely packed monolayer assemblies provides information on how molecules communicate with one another and with their surroundings in thin molecular films [79]. Single-molecule quantum interference (QI) concepts have recently been used to design new ligand dyes for dye-sensitized solar cells and new dielectric materials [253, 254]. The discoveries made in the context of SMEs are also relevant for current and future integrated circuit technologies. The insights gained from studying conductance at the molecular scale will help us to understand how the charge-transport properties of semiconducting materials might change as their dimensions decrease and they begin to resemble molecules rather than bulk solids.

The past two decades have seen a vast deepening in the knowledge of how to create, measure and understand single-molecule junctions. One of the immediate challenges is how to use this newly gained knowledge to incorporate these structures into system-level devices. These new nanoscale architectures and interconnections could revolutionize nanoscale electronics by providing access to the vast store of structures available through synthetic chemistry. The combination of synthetic chemistry and electronics is an inspiration for the creation of fields and applications yet to be imagined.

References

1 Ratner, M. (2013). *Nat. Nanotechnol.* 8: 378.
2 Su, T.A., Neupane, M., Steigerwald, M.L. et al. (2016). *Nat. Rev. Mater.* 1: 16002.
3 Aviram, A. and Ratner, M.A. (1974). *Chem. Phys. Lett.* 29: 277.
4 Tao, N.J. (2006). *Nat. Nanotechnol.* 1: 173.
5 Moth-Poulsen, K. and Bjornholm, T. (2009). *Nat. Nanotechnol.* 4: 551.
6 Aradhya, S.V. and Venkataraman, L. (2013). *Nat. Nanotechnol.* 8: 399.
7 Ward, D.R., Scott, G.D., Keane, Z.K. et al. (2008). *J. Phys. Condens. Matter* 20: 374118.
8 Bumm, L.A., Arnold, J.J., Cygan, M.T. et al. (1996). *Science* 271: 1705.
9 Moreland, J. and Ekin, J.W. (1985). *J. Appl. Phys.* 58: 3888.
10 Reed, M.A., Zhou, C., Muller, C.J. et al. (1997). *Science* 278: 252.
11 Smit, R.H.M., Noat, Y., Untiedt, C. et al. (2002). *Nature* 419: 906.
12 Xu, B. and Tao, N.J. (2003). *Science* 301: 1221.
13 Venkataraman, L., Klare, J.E., Nuckolls, C. et al. (2006). *Nature* 442: 904.
14 Wold, D.J. and Frisbie, C.D. (2000). *J. Am. Chem. Soc.* 122: 2970.
15 Park, H., Lim, A.K.L., Alivisatos, A.P. et al. (1999). *Appl. Phys. Lett.* 75: 301.
16 Andres, R.P., Bielefeld, J.D., Henderson, J.I. et al. (1996). *Science* 273: 1690.
17 Cui, X.D., Primak, A., Zarate, X. et al. (2001). *Science* 294: 571.
18 Guo, X., Small, J.P., Klare, J.E. et al. (2006). *Science* 311: 356.
19 Prins, F., Barreiro, A., Ruitenberg, J.W. et al. (2011). *Nano Lett.* 11: 4607.
20 Morpurgo, A.F., Marcus, C.M., and Robinson, D.B. (1999). *Appl. Phys. Lett.* 74: 2084.
21 Chen, X., Yeganeh, S., Qin, L. et al. (2009). *Nano Lett.* 9: 3974.
22 Quek, S.Y., Kamenetska, M., Steigerwald, M.L. et al. (2009). *Nat. Nanotechnol.* 4: 230.
23 Su, T.A., Widawsky, J.R., Li, H. et al. (2013). *J. Am. Chem. Soc.* 135: 18331.
24 Kiguchi, M., Ohto, T., Fujii, S. et al. (2014). *J. Am. Chem. Soc.* 136: 7327.
25 Moreno-Garcia, P., La Rosa, A., Kolivoska, V. et al. (2015). *J. Am. Chem. Soc.* 137: 2318.
26 Lafferentz, L., Ample, F., Yu, H. et al. (2009). *Science* 323: 1193.
27 Meisner, J.S., Kamenetska, M., Krikorian, M. et al. (2011). *Nano Lett.* 11: 1575.
28 Sun, L., Diaz-Fernandez, Y.A., Gschneidtner, T.A. et al. (2014). *Chem. Soc. Rev.* 43: 7378.
29 Aradhya, S.V., Frei, M., Hybertsen, M.S., and Venkataraman, L. (2012). *Nat. Mater.* 11: 872.

30 Love, J.C., Estroff, L.A., Kriebel, J.K. et al. (2005). *Chem. Rev.* 105: 1103.
31 Ulman, A. (1996). *Chem. Rev.* 96: 1533.
32 Yee, S.K., Sun, J., Darancet, P. et al. (2011). *ACS Nano* 5: 9256.
33 Li, C., Mishchenko, A., Li, Z. et al. (2008). *J. Phys. Condens. Matter* 20: 374122.
34 Klausen, R.S., Widawsky, J.R., Su, T.A. et al. (2014). *Chem. Sci.* 5: 1561.
35 Park, Y.S., Whalley, A.C., Kamenetska, M. et al. (2007). *J. Am. Chem. Soc.* 129: 15768.
36 Yasuda, S., Yoshida, S., Sasaki, J. et al. (2006). *J. Am. Chem. Soc.* 128: 7746.
37 Venkataraman, L., Klare, J.E., Tam, I.W. et al. (2006). *Nano Lett.* 6: 458.
38 Chen, F., Li, X., Hihath, J. et al. (2006). *J. Am. Chem. Soc.* 128: 15874.
39 Kim, B., Beebe, J.M., Jun, Y. et al. (2006). *J. Am. Chem. Soc.* 128: 4970.
40 Mishchenko, A., Zotti, L.A., Vonlanthen, D. et al. (2011). *J. Am. Chem. Soc.* 133: 184.
41 Zotti, L.A., Kirchner, T., Cuevas, J.-C. et al. (2010). *Small* 6: 1529.
42 Xu, B., Xiao, X., and Tao, N.J. (2003). *J. Am. Chem. Soc.* 125: 16164.
43 Martin, C.A., Ding, D., Sorensen, J.K. et al. (2008). *J. Am. Chem. Soc.* 130: 13198.
44 Leary, E., Gonzalez, M.T., van der Pol, C. et al. (2011). *Nano Lett.* 11: 2236.
45 Loertscher, E., Geskin, V., Gotsmann, B. et al. (2013). *Small* 9: 332.
46 Chen, W., Widawsky, J.R., Vazquez, H. et al. (2011). *J. Am. Chem. Soc.* 133: 17160.
47 Paik, W.-k., Han, S., Shin, W., and Kim, Y. (2003). *Langmuir* 19: 4211.
48 Lortscher, E., Cho, C.J., Mayor, M. et al. (2011). *ChemPhysChem* 12: 1677.
49 Kamenetska, M., Koentopp, M., Whalley, A.C. et al. (2009). *Phys. Rev. Lett.* 102: 126803.
50 Hong, W., Manrique, D.Z., Moreno-Garcia, P. et al. (2012). *J. Am. Chem. Soc.* 134: 2292.
51 Loertscher, E. (2013). *Nat. Nanotechnol.* 8: 381.
52 Guisinger, N.P., Greene, M.E., Basu, R. et al. (2004). *Nano Lett.* 4: 55.
53 Cao, Y., Dong, S., Liu, S. et al. (2012). *Angew. Chem. Int. Ed.* 51: 12228.
54 Cao, Y., Dong, S., Liu, S. et al. (2013). *Angew. Chem. Int. Ed.* 52: 3906.
55 Zhou, X.Y., Peng, Z.L., Sun, Y.Y. et al. (2013). *Nanotechnology* 24: 465204.
56 Danilov, A., Kubatkin, S., Kafanov, S. et al. (2008). *Nano Lett.* 8: 1.
57 Heimel, G., Duhm, S., Salzmann, I. et al. (2013). *Nat. Chem.* 5: 187.
58 Schlaefer, H.L. and Gliemann, G. (1969). *Basic Principles of Ligand Field Theory*. London: Wiley-Interscience.
59 Buttiker, M., Imry, Y., Landauer, R., and Pinhas, S. (1985). *Phys. Rev. B: Condens. Matter* 31: 6207.
60 Reddy, P., Jang, S.-Y., Segalman, R.A., and Majumdar, A. (2007). *Science* 315: 1568.
61 Widawsky, J.R., Darancet, P., Neaton, J.B., and Venkataraman, L. (2012). *Nano Lett.* 12: 354.
62 Derosa, P.A. and Seminario, J.M. (2001). *J. Phys. Chem. B* 105: 471.
63 Kamenetska, M., Quek, S.Y., Whalley, A.C. et al. (2010). *J. Am. Chem. Soc.* 132: 6817.

64 Lee, W., Kim, K., Jeong, W. et al. (2013). *Nature* 498: 209.
65 Koga, J., Tsuji, Y., and Yoshizawa, K. (2012). *J. Phys. Chem. C* 116: 20607.
66 Mayor, M., Weber, H.B., Reichert, J. et al. (2003). *Angew. Chem. Int. Ed.* 42: 5834.
67 Dell, E.J., Capozzi, B., Xia, J. et al. (2015). *Nat. Chem.* 7: 209.
68 Li, Z., Smeu, M., Ratner, M.A., and Borguet, E. (2013). *J. Phys. Chem. C* 117: 14890.
69 Park, Y.S., Widawsky, J.R., Kamenetska, M. et al. (2009). *J. Am. Chem. Soc.* 131: 10820.
70 Moreno-Garcia, P., Gulcur, M., Manrique, D.Z. et al. (2013). *J. Am. Chem. Soc.* 135: 12228.
71 Batra, A., Darancet, P., Chen, Q. et al. (2013). *Nano Lett.* 13: 6233.
72 Xia, J., Capozzi, B., Wei, S. et al. (2014). *Nano Lett.* 14: 2941.
73 Su, T.A., Li, H., Steigerwald, M.L. et al. (2015). *Nat. Chem.* 7: 215.
74 Adak, O., Korytar, R., Joe, A.Y. et al. (2015). *Nano Lett.* 15: 3716.
75 Venkataraman, L., Park, Y.S., Whalley, A.C. et al. (2007). *Nano Lett.* 7: 502.
76 Su, T.A., Li, H., Zhang, V. et al. (2015). *J. Am. Chem. Soc.* 137: 12400.
77 Arroyo, C.R., Leary, E., Castellanos-Gomez, A. et al. (2011). *J. Am. Chem. Soc.* 133: 14313.
78 Capozzi, B., Xia, J., Adak, O. et al. (2015). *Nat. Nanotechnol.* 10: 522.
79 Tour, J.M., Jones, L. II, Pearson, D.L. et al. (1995). *J. Am. Chem. Soc.* 117: 9529.
80 Haekkinen, H. (2012). *Nat. Chem.* 4: 443.
81 Li, X., He, J., Hihath, J. et al. (2006). *J. Am. Chem. Soc.* 128: 2135.
82 Inkpen, M.S., Lemmer, M., Fitzpatrick, N. et al. (2015). *J. Am. Chem. Soc.* 137: 9971.
83 Haiss, W., Wang, C., Grace, I. et al. (2006). *Nat. Mater.* 5: 995.
84 Haiss, W., Martin, S., Leary, E. et al. (2009). *J. Phys. Chem. C* 113: 5823.
85 Li, C., Pobelov, I., Wandlowski, T. et al. (2008). *J. Am. Chem. Soc.* 130: 318.
86 Rascon-Ramos, H., Artes, J.M., Li, Y., and Hihath, J. (2015). *Nat. Mater.* 14: 517.
87 Li, Z., Smeu, M., Park, T.-H. et al. (2014). *Nano Lett.* 14: 5493.
88 Batra, A., Kladnik, G., Gorjizadeh, N. et al. (2014). *J. Am. Chem. Soc.* 136: 12556.
89 Kim, Y.-S., Pietsch, T., Erbe, A. et al. (2011). *Nano Lett.* 11: 3734.
90 Ulrich, J., Esrail, D., Pontius, W. et al. (2006). *J. Phys. Chem. B* 110: 2462.
91 Xiao, X., Xu, B., and Tao, N.J. (2004). *Nano Lett.* 4: 267.
92 Cheng, Z.L., Skouta, R., Vazquez, H. et al. (2011). *Nat. Nanotechnol.* 6: 353.
93 Kaletova, E., Kohutova, A., Hajduch, J. et al. (2015). *J. Am. Chem. Soc.* 137: 12086.
94 Khobragade, D., Stensrud, E.S., Mucha, M. et al. (2010). *Langmuir* 26: 8483.
95 Schwarz, F., Kastlunger, G., Lissel, F. et al. (2014). *Nano Lett.* 14: 5932.
96 Hong, W., Li, H., Liu, S.-X. et al. (2012). *J. Am. Chem. Soc.* 134: 19425.
97 Wuts, P.G.M. and Greene, T.W. (2006). Chapter 8. Protection for the alkyne -CH. In: *Greene's Protective Groups in Organic Synthesis*, 4e (eds. P.G.M. Wuts and T.W. Greene), 927–933. Wiley.

98 Nagahara, L.A., Thundat, T., and Lindsay, S.M. (1989). *Rev. Sci. Instrum.* 60: 3128.

99 Saby, C., Ortiz, B., Champagne, G.Y., and Belanger, D. (1997). *Langmuir* 13: 6805.

100 Hines, T., Diez-Perez, I., Nakamura, H. et al. (2013). *J. Am. Chem. Soc.* 135: 3319.

101 Patai, S. (ed.) (1978). *The Chemistry of the Diazonium and Diazo Groups, Pt. 1.* Wiley.

102 Martin, A.S., Sambles, J.R., and Ashwell, G.J. (1993). *Phys. Rev. Lett.* 70: 218.

103 Elbing, M., Ochs, R., Koentopp, M. et al. (2005). *Proc. Natl. Acad. Sci. U.S.A.* 102: 8815.

104 Diez-Perez, I., Hihath, J., Lee, Y. et al. (2009). *Nat. Chem.* 1: 635.

105 Guedon, C.M., Valkenier, H., Markussen, T. et al. (2012). *Nat. Nanotechnol.* 7: 305.

106 Lortscher, E., Gotsmann, B., Lee, Y. et al. (2012). *ACS Nano* 6: 4931.

107 Dhirani, A., Lin, P.H., Guyot-Sionnest, P. et al. (1997). *J. Chem. Phys.* 106: 5249.

108 Yuan, L., Nerngchamnong, N., Cao, L. et al. (2015). *Nat. Commun.* 6: 6324.

109 Metzger, R.M., Chen, B., Hoepfner, U. et al. (1997). *J. Am. Chem. Soc.* 119: 10455.

110 Lenfant, S., Krzeminski, C., Delerue, C. et al. (2003). *Nano Lett.* 3: 741.

111 Metzger, R.M. (2003). *Chem. Rev.* 103: 3803.

112 Nijhuis, C.A., Reus, W.F., Barber, J.R. et al. (2010). *Nano Lett.* 10: 3611.

113 Nijhuis, C.A., Reus, W.F., Siegel, A.C., and Whitesides, G.M. (2011). *J. Am. Chem. Soc.* 133: 15397.

114 Nerngchamnong, N., Yuan, L., Qi, D.-C. et al. (2013). *Nat. Nanotechnol.* 8: 113.

115 Jeong, H., Kim, D., Wang, G. et al. (2014). *Adv. Funct. Mater.* 24: 2472.

116 Yuan, L., Breuer, R., Jiang, L. et al. (2015). *Nano Lett.* 15: 5506.

117 He, J. and Lindsay, S.M. (2005). *J. Am. Chem. Soc.* 127: 11932.

118 Mentovich, E.D., Rosenberg-Shraga, N., Kalifa, I. et al. (2013). *J. Phys. Chem. C* 117: 8468.

119 Metzger, R.M. (2006). *Chem. Phys.* 326: 176.

120 Heath, J.R. (2009). *Annu. Rev. Mater. Res.* 39: 1.

121 Metzger, R.M. (2008). *J. Mater. Chem.* 18: 4364.

122 Chabinyc, M.L., Chen, X., Holmlin, R.E. et al. (2002). *J. Am. Chem. Soc.* 124: 11730.

123 Bala, S., Aithal, R.K., Derosa, P. et al. (2010). *J. Phys. Chem. C* 114: 20877.

124 Reichert, J., Ochs, R., Beckmann, D. et al. (2002). *Phys. Rev. Lett.* 88: 176804.

125 Guo, C., Wang, K., Zerah-Harush, E. et al. (2016). *Nat. Chem.* 8: 484.

126 Ashwell, G.J., Urasinska, B., and Tyrrell, W.D. (2006). *Phys. Chem. Chem. Phys.* 8: 3314.

127 Nijhuis, C.A., Reus, W.F., and Whitesides, G.M. (2009). *J. Am. Chem. Soc.* 131: 17814.

128 Zhou, C., Deshpande, M.R., Reed, M.A. et al. (1997). *Appl. Phys. Lett.* 71: 611.

129 McCreery, R., Dieringer, J., Solak, A.O. et al. (2003). *J. Am. Chem. Soc.* 125: 10748.

130 Mujica, V., Ratner, M.A., and Nitzan, A. (2002). *Chem. Phys.* 281: 147.
131 Xiang, D., Jeong, H., Lee, T., and Mayer, D. (2013). *Adv. Mater.* 25: 4845.
132 Wang, L., Wang, L., Zhang, L., and Xiang, D. (2017). *Top. Curr. Chem.* 375: 61.
133 Kornilovitch, P.E., Bratkovsky, A.M., and Stanley Williams, R. (2002). *Phys. Rev. B: Condens. Matter Mater. Phys.* 66: 165436/1.
134 Zahid, F., Ghosh, A.W., Paulsson, M. et al. (2004). *Phys. Rev. B: Condens. Matter Mater. Phys.* 70: 245317/1.
135 Perrin, M.L., Galan, E., Eelkema, R. et al. (2016). *Nanoscale* 8: 8919.
136 Tsutsui, M. and Taniguchi, M. (2012). *Sensors* 12: 7259.
137 Batra, A., Meisner, J.S., Darancet, P. et al. (2014). *Faraday Discuss.* 174: 79.
138 Metzger, R.M. (2015). *Chem. Rev.* 115: 5056.
139 Metzger, R.M. (1999). *Acc. Chem. Res.* 32: 950.
140 Xiang, D., Wang, X., Jia, C. et al. (2016). *Chem. Rev.* 116: 4318.
141 Mahmoud, A. and Lugli, P. (2012). *J. Appl. Phys.* 112: 113720/1.
142 Ma, J., Yang, C.-L., Wang, L.-Z. et al. (2014). *Physica B* 434: 32.
143 Zhang, G., Ratner, M.A., and Reuter, M.G. (2015). *J. Phys. Chem. C* 119: 6254.
144 Ashwell, G.J., Sambles, J.R., Martin, A.S. et al. (1990). *J. Chem. Soc., Chem. Commun.*: 1374.
145 Metzger, R.M., Baldwin, J.W., Shumate, W.J. et al. (2003). *J. Phys. Chem. B* 107: 1021.
146 Honciuc, A., Jaiswal, A., Gong, A. et al. (2005). *J. Phys. Chem. B* 109: 857.
147 Shumate, W.J., Mattern, D.L., Jaiswal, A. et al. (2006). *J. Phys. Chem. B* 110: 11146.
148 Honciuc, A., Metzger, R.M., Gong, A., and Spangler, C.W. (2007). *J. Am. Chem. Soc.* 129: 8310.
149 Ng, M.-K. and Yu, L. (2002). *Angew. Chem. Int. Ed.* 41: 3598.
150 Metzger, R.M., Xu, T., and Peterson, I.R. (2001). *J. Phys. Chem. B* 105: 7280.
151 Jaiswal, A., Rajagopal, D., Lakshmikantham, M.V. et al. (2007). *Phys. Chem. Chem. Phys.* 9: 4007.
152 Honciuc, A., Otsuka, A., Wang, Y.-H. et al. (2006). *J. Phys. Chem. B* 110: 15085.
153 Ng, M.-K., Lee, D.-C., and Yu, L. (2002). *J. Am. Chem. Soc.* 124: 11862.
154 Jiang, P., Morales, G.M., You, W., and Yu, L. (2004). *Angew. Chem. Int. Ed.* 43: 4471.
155 Cuevas, J.C. and Scheer, E. (2017). *Molecular Electronics: An Introduction to Theory and Experiment*, 2e. World Scientific.
156 Morales, G.M., Jiang, P., Yuan, S. et al. (2005). *J. Am. Chem. Soc.* 127: 10456.
157 Hihath, J., Bruot, C., Nakamura, H. et al. (2011). *ACS Nano* 5: 8331.
158 Cui, B., Zhao, W., Wang, H. et al. (2014). *J. Appl. Phys.*116: 073701/1.
159 Hu, G.C., Zhang, G.P., Li, Y. et al. (2014). *Chem. Phys. Lett.* 614: 207.
160 Luo, L., Balhorn, L., Vlaisavljevich, B. et al. (2014). *J. Phys. Chem. C* 118: 26485.
161 Zhang, G.-P., Hu, G.-C., Song, Y. et al. (2012). *J. Phys. Chem. C* 116: 22009.
162 Geddes, N.J., Sambles, J.R., Jarvis, D.J. et al. (1990). *Appl. Phys. Lett.* 56: 1916.
163 Geddes, N.J., Sambles, J.R., Jarvis, D.J. et al. (1992). *J. Appl. Phys.* 71: 756.
164 Arroyo, C.R., Tarkuc, S., Frisenda, R. et al. (2013). *Angew. Chem. Int. Ed.* 52: 3152.

165 Kim, T., Liu, Z.-F., Lee, C. et al. (2014). *Proc. Natl. Acad. Sci. U.S.A.* 111: 10928.

166 Kushmerick, J.G., Whitaker, C.M., Pollack, S.K. et al. (2004). *Nanotechnology* 15: S489.

167 Wang, K., Zhou, J., Hamill, J.M., and Xu, B. (2014). *J. Chem. Phys.* 141: 054712/1.

168 Yu, H., Luo, Y., Beverly, K. et al. (2003). *Angew. Chem. Int. Ed.* 42: 5706.

169 Krzeminski, C., Delerue, C., Allan, G. et al. (2001). *Phys. Rev. B: Condens. Matter Mater. Phys.* 64: 085405/1.

170 Liu, R., Ke, S.-H., Yang, W., and Baranger, H.U. (2006). *J. Chem. Phys.* 124: 024718/1.

171 Taylor, J., Brandbyge, M., and Stokbro, K. (2002). *Phys. Rev. Lett.* 89: 138301/1.

172 Zhang, G.-P., Hu, G.-C., Song, Y. et al. (2013). *J. Chem. Phys.* 139: 094702/1.

173 Troisi, A. and Ratner, M.A. (2002). *J. Am. Chem. Soc.* 124: 14528.

174 Troisi, A. and Ratner, M.A. (2004). *Nano Lett.* 4: 591.

175 Zhang, Q. (2014). *Physica B* 450: 90.

176 Li, T., Jevric, M., Hauptmann, J.R. et al. (2013). *Adv. Mater.* 25: 4164.

177 Katsonis, N., Kudernac, T., Walko, M. et al. (2006). *Adv. Mater.* 18: 1397.

178 Kronemeijer, A.J., Akkerman, H.B., Kudernac, T. et al. (2008). *Adv. Mater.* 20: 1467.

179 Meng, F., Hervault, Y.-M., Norel, L. et al. (2012). *Chem. Sci.* 3: 3113.

180 Zhang, J.L., Zhong, J.Q., Lin, J.D. et al. (2015). *Chem. Soc. Rev.* 44: 2998.

181 Schirm, C., Matt, M., Pauly, F. et al. (2013). *Nat. Nanotechnol.* 8: 645.

182 Lortscher, E., Ciszek, J.W., Tour, J., and Riel, H. (2006). *Small* 2: 973.

183 Jia, C., Wang, J., Yao, C. et al. (2013). *Angew. Chem. Int. Ed.* 52: 8666.

184 Kim, Y., Hellmuth, T.J., Sysoiev, D. et al. (2012). *Nano Lett.* 12: 3736.

185 Dulic, D., van der Molen, S.J., Kudernac, T. et al. (2003). *Phys. Rev. Lett.* 91: 207402/1.

186 Whalley, A.C., Steigerwald, M.L., Guo, X., and Nuckolls, C. (2007). *J. Am. Chem. Soc.* 129: 12590.

187 Irie, M., Eriguchi, T., Takada, T., and Uchida, K. (1997). *Tetrahedron* 53: 12263.

188 Ralph, D.C., Black, C.T., and Tinkham, M. (1997). *Phys. Rev. Lett.* 78: 4087.

189 Feringa, B.L. (2001). *Acc. Chem. Res.* 34: 504.

190 Glebov, E.M., Pozdnyakov, I.P., Plyusnin, V.F., and Khmelinskii, I. (2015). *J. Photochem. Photobiol., C* 24: 1.

191 Lara-Avila, S., Danilov, A., Geskin, V. et al. (2010). *J. Phys. Chem. C* 114: 20686.

192 Lara-Avila, S., Danilov, A.V., Kubatkin, S.E. et al. (2011). *J. Phys. Chem. C* 115: 18372.

193 Broman, S.L., Lara-Avila, S., Thisted, C.L. et al. (2012). *Adv. Funct. Mater.* 22: 4249.

194 Kay, N.J., Higgins, S.J., Jeppesen, J.O. et al. (2012). *J. Am. Chem. Soc.* 134: 16817.

195 Zhang, J., Kuznetsov, A.M., Medvedev, I.G. et al. (2008). *Chem. Rev.* 108: 2737.

196 Li, Z., Li, H., Chen, S. et al. (2014). *J. Am. Chem. Soc.* 136: 8867.

197 Darwish, N., Diez-Perez, I., Da Silva, P. et al. (2012). *Angew. Chem. Int. Ed.* 51: 3203.

198 Baghernejad, M., Zhao, X., Baruel Oernsoe, K. et al. (2014). *J. Am. Chem. Soc.* 136: 17922.
199 Yang, F., Wang, X., Zhang, D. et al. (2014). *Nature* 510: 522.
200 Capozzi, B., Chen, Q., Darancet, P. et al. (2014). *Nano Lett.* 14: 1400.
201 Baghernejad, M., Manrique, D.Z., Li, C. et al. (2014). *Chem. Commun.* 50: 15975.
202 Sanvito, S. and Rocha, A.R. (2006). *J. Comput. Theor. Nanosci.* 3: 624.
203 Leuenberger, M.N. and Loss, D. (2001). *Nature* 410: 789.
204 Thomas, L., Lionti, F., Ballou, R. et al. (1996). *Nature* 383: 145.
205 Wernsdorfer, W. and Sessoli, R. (1999). *Science* 284: 133.
206 Sanvito, S. (2011). *Chem. Soc. Rev.* 40: 3336.
207 Sanvito, S. (2007). *J. Mater. Chem.* 17: 4455.
208 Baadji, N., Piacenza, M., Tugsuz, T. et al. (2009). *Nat. Mater.* 8: 813.
209 Bogani, L. and Wernsdorfer, W. (2008). *Nat. Mater.* 7: 179.
210 Affronte, M. (2009). *J. Mater. Chem.* 19: 1731.
211 Camarero, J. and Coronado, E. (2009). *J. Mater. Chem.* 19: 1678.
212 Lehmann, J., Gaita-Arino, A., Coronado, E., and Loss, D. (2009). *J. Mater. Chem.* 19: 1672.
213 Mas-Torrent, M., Crivillers, N., Mugnaini, V. et al. (2009). *J. Mater. Chem.* 19: 1691.
214 Stamp, P.C.E. and Gaita-Arino, A. (2009). *J. Mater. Chem.* 19: 1718.
215 Wahl, P., Simon, P., Diekhoner, L. et al. (2007). *Phys. Rev. Lett.* 98: 056601/1.
216 Hirjibehedin, C.F., Lin, C.-Y., Otte, A.F. et al. (2007). *Science* 317: 1199.
217 Wiesendanger, R. (2009). *Rev. Mod. Phys.* 81: 1495.
218 Schwarz, F., Kastlunger, G., Lissel, F. et al. (2016). *Nat. Nanotechnol.* 11: 170.
219 Guetlich, P. and Goodwin, H.A. (eds.) (2004). *Spin Crossover in Transition Metal Compounds I*, Topics in Current Chemistry, vol. 233. Berlin, Heidelberg: Springer.
220 Rocha, A.R., Garcia-suarez, V.M., Bailey, S.W. et al. (2005). *Nat. Mater.* 4: 335.
221 Diefenbach, M. and Kim, K.S. (2007). *Angew. Chem. Int. Ed.* 46: 7640.
222 Prins, F., Monrabal-Capilla, M., Osorio, E.A. et al. (2011). *Adv. Mater.* 23: 1545.
223 Osorio, E.A., Moth-Poulsen, K., van der Zant, H.S.J. et al. (2010). *Nano Lett.* 10: 105.
224 Wagner, S., Kisslinger, F., Ballmann, S. et al. (2013). *Nat. Nanotechnol.* 8: 575.
225 Perrin, M.L., Burzuri, E., and van der Zant, H.S.J. (2015). *Chem. Soc. Rev.* 44: 902.
226 Park, J., Pasupathy, A.N., Goldsmith, J.I. et al. (2002). *Nature* 417: 722.
227 Liang, W., Shores, M.P., Bockrath, M. et al. (2002). *Nature* 417: 725.
228 Kubatkin, S., Danilov, A., Hjort, M. et al. (2003). *Nature* 425: 698.
229 Ghosh, A.W., Rakshit, T., and Datta, S. (2004). *Nano Lett.* 4: 565.
230 Song, H., Kim, Y., Jang, Y.H. et al. (2009). *Nature* 462: 1039.
231 Champagne, A.R., Pasupathy, A.N., and Ralph, D.C. (2005). *Nano Lett.* 5: 305.
232 Parks, J.J., Champagne, A.R., Costi, T.A. et al. (2010). *Science* 328: 1370.
233 Martin, C.A., Smit, R.H.M., van der Zant, H.S.J., and van Ruitenbeek, J.M. (2009). *Nano Lett.* 9: 2940.

234 Martin, C.,.A., van Ruitenbeek, J.,.M., and van der Zant, H.,.S.J. (2010). *Nanotechnology* 21: 265201.

235 Perrin, M.L., Verzijl, C.J.O., Martin, C.A. et al. (2013). *Nat. Nanotechnol.* 8: 282.

236 Marquardt, C.W., Grunder, S., Blaszczyk, A. et al. (2010). *Nat. Nanotechnol.* 5: 863.

237 Carrascal, D., Garcia-Suarez, V.M., and Ferrer, J. (2012). *Phys. Rev. B: Condens. Matter Mater. Phys.* 85: 195434/1.

238 Xiang, D., Jeong, H., Kim, D. et al. (2013). *Nano Lett.* 13: 2809.

239 Mangin, A., Anthore, A., Della Rocca, M.L. et al. (2009). *J. Appl. Phys.* 105: 014313/1.

240 Huang, C., Rudnev, A.V., Hong, W., and Wandlowski, T. (2015). *Chem. Soc. Rev.* 44: 889.

241 Xu, B., Xiao, X., Yang, X. et al. (2005). *J. Am. Chem. Soc.* 127: 2386.

242 Cui, L., Liu, B., Vonlanthen, D. et al. (2011). *J. Am. Chem. Soc.* 133: 7332.

243 Chen, F., He, J., Nuckolls, C. et al. (2005). *Nano Lett.* 5: 503.

244 Pobelov, I.V., Li, Z., and Wandlowski, T. (2008). *J. Am. Chem. Soc.* 130: 16045.

245 Guo, S., Artes, J.M., and Diez-Perez, I. (2013). *Electrochim. Acta* 110: 741.

246 Yi, Z., Trellenkamp, S., Offenhaeusser, A., and Mayer, D. (2010). *Chem. Commun.* 46: 8014.

247 Yang, Y., Liu, J.-Y., Chen, Z.-B. et al. (2011). *Nanotechnology* 22: 275313/1.

248 Liu, R., Ke, S.-H., Baranger, H.U., and Yang, W. (2006). *J. Am. Chem. Soc.* 128: 6274.

249 Ouyang, M. and Awschalom, D.D. (2003). *Science* 301: 1074.

250 Yu, L.H. and Natelson, D. (2004). *Nano Lett.* 4: 79.

251 Pasupathy, A.N., Bialczak, R.C., Martinek, J. et al. (2004). *Science* 306: 86.

252 Bartels, L. (2010). *Nat. Chem.* 2: 87.

253 Maggio, E., Solomon, G.C., and Troisi, A. (2014). *ACS Nano* 8: 409.

254 Bergfield, J.P., Heitzer, H.M., Van Dyck, C. et al. (2015). *ACS Nano* 9: 6412.

6

Molecular Electronic Junctions Based on Self-Assembled Monolayers

Yuqing Liu and Zhongming Wei

Institute of Semiconductors, Chinese Academy of Sciences, State Key Laboratory of Superlattices and Microstructures, No. A35, QingHua East Road, Haidian District, 100083 Beijing, P. R. China

6.1 Introduction

Instead of using single molecules to provide the electronic functionality, self-assembled monolayers (SAMs) of molecules are another important branch of molecular electronics and often referred to as ensemble molecular junctions (also included the junctions formed from more than a few molecules). SAMs are organic assemblies formed by molecules in solution or gas phase which attached to the surface of solids or liquids via anchoring groups and have aroused great attention in interdisciplinary research areas due to their fascinating characteristics [1].

SAMs provide great opportunities to refine the fundamental understanding of self-organization, structure–property relationships, and interfacial phenomena [2]. The ability to tailor both head and tail groups of the constituent molecules in SAMs ensures necessary stability and high performance of the device. Moreover, SAMs offer the needed design flexibility. The functional moieties such as chromophores, photochromic, or electroactive groups in the molecules of SAM enable capabilities in sensing, switching, electron transfer, and molecular recognition and so on [3].

From the point of view of device fabrication, the first is to explicitly make contacts to molecular components while preserving their function [4]. Although various test beds have been reported for contacting and detecting single molecules and their ensembles [5–8], molecular junctions based on SAMs can be created and operated in a more reproducible manner, more suitable for mass production and integration [9], and it has statistical significance in which the viewed electronic behavior represents an average over many molecules [10]. However, the SAM approach has the additional problems associated with possible defects and with intermolecular interactions. For example, metal atom penetration into the molecular layer during the process of evaporating metal top contact will create metal filaments and short circuits. In this chapter, we talk about the attachment of SAMs to various substrates (metals, oxides, and semiconductors), the formation of top electrodes by

Nanogap Electrodes, First Edition. Edited by Tao Li.

direct evaporation and indirect evaporation as well as the state-of-art methods for fabricating high-quality molecular junctions such as transfer printing, liquid metal contact, and graphene as top contact.

6.2 Molecular Monolayers for Molecular Electronics Devices

SAM decoration of surfaces is of particular interest for molecular electronic applications where the monolayer quality plays a dominant role in device properties. Earlier reports on SAMs (1980s–1990s) focused primarily on the assemblies formed by the adsorption of organosulfur compounds from solution or the vapor phase onto gold or silver metal film substrates [1]. Now, the common types of substrates used to support SAMs range from metal films (Au, Ag, Pt, Cu, …) to oxides (SiO_2/Si, Al_2O_3, HfO_2, ZrO_2, …) and semiconductors (GaAs, CdSe, …). Therefore, suitable for different substrate anchoring groups came into being, such as thiol, amine, isocyanide, C_{60}, phosphonate. Table 6.1 shows chemical systems of adsorbates and substrates that form SAMs. The interface or the contact between the molecules and the electrodes can be a chemical or physical adsorption [11]; it directly affects the stability and charge transport characteristics of molecular junctions.

Table 6.1 Chemical systems of adsorbates and substrates that form SAMs.

Surface	Substrate	Adsorbate(s)
Metal	Au	R–SH, R–SS–R, R–S–R, R–NH_2, R–NC, R–Se, R–Te
	Ag	R–COOH, R–SH
	Pt	R–NC, R–SH
	Pd	R–SH
	Cu	R–SH
	Hg	R–SH
Semiconductor	GaAs(III–V)	R–SH
	InP(III–V)	R–SH
	CdSe(II–VI)	R–SH
	ZnSe(II–VI)	R–SH
Oxide	Al_2O_3	R–COOH
	TiO_2	R–COOH, R–PO_3H
	$YBa_2Cu_3O_{7-\delta}$	R–NH_2
	Tl–Ba–Ca–Cu–O	R–SH
	ITO	R–COOH, R–SH, R–$Si(x)_3$
	SiO_2	R–$Si(x)_3$

Source: Smith et al. [3]. 2004, Elsevier.

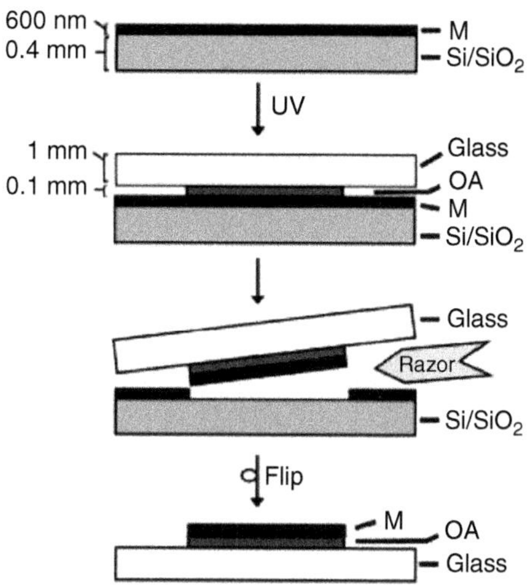

Figure 6.1 Schematic of the procedure for template stripping method. Source: Weiss et al. [12]. 2007, American Chemical Society.

6.2.1 Monolayers Covalently Bonded to Noble Metals

Metal planar substrates are used generally for covalent connection with SAMs because they are convenient (easy to prepare) and compatible with a number of techniques for surface analysis and spectroscopic/physical characterization [1]. Whitesides and coworkers reported template stripping (TS) method to produce ultraflat metal films of silver, gold, palladium, and platinum [12]. It is also the origin of the Au^{TS}, Ag^{TS}, etc. that we often see in the fabrication of molecular junctions. Figure 6.1 shows the procedure for template stripping approach; they deposited a metal onto an Si/SiO_2 substrate by evaporation with an e-beam firstly, attached a composite of a glass slide and optical adhesive (glass/OA) to the top of the metal film, and then cured the OA by exposing it to ultraviolet (UV) light and used a razor to cleave the glass/OA/metal composite manually from the template to expose the smooth face of the metal. Template stripping is a versatile method to produce ultrasmooth metal films for various applications, and the TS surfaces have unique advantages for the formation and characterization of large-area junctions.

Gold is the standard substrate for SAMs. For most purposes, its surface is satisfactorily inert, it is not susceptible of oxygen, and it does not react with most chemicals. We can obtain the gold thin films straightforward by physical vapor deposition, sputtering, or electrodeposition and further pattern them by a combination of lithographic tools and chemical etchants. In addition, gold thin films are common substrates used for many existing spectroscopies and analytical techniques, including ellipsometry, surface plasmon resonance (SPR) spectroscopy and reflectance absorption infrared spectroscopy (RAIRS) [1]. The compatibility of gold and cells provides cheerful prospect for the application of molecular junctions based on gold substrate in the biological field. For example, Long and coworkers proposed novel electrochemical probes based on gold surface-coated hybrids designed to intelligently detect specific carbohydrate–protein interactions [13].

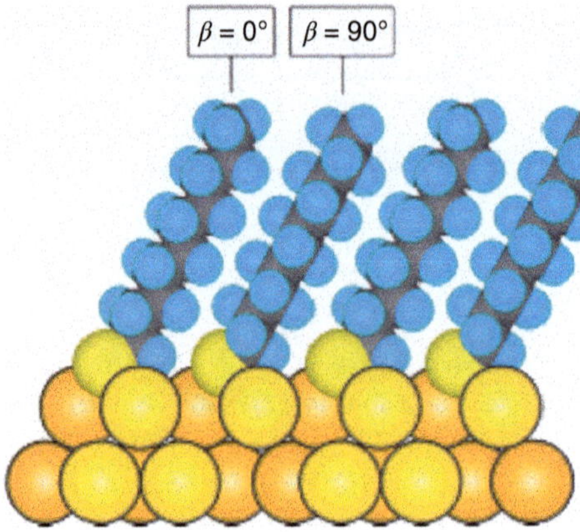

Figure 6.2 SAMs formed by *n*-alkanethiols covalently adsorbed on the gold substrate. The chains are marked with twist values (*β*) to represent the relative orientations of the neighboring chains. Source: Love et al. [1]. 2005, American Chemical Society.

The high affinity of thiols for the surfaces of noble metals makes it possible to generate well-defined organic surfaces; it is not surprising that strong and stable S—Au covalent bond is the most popular contact in the molecular junctions since 1980s. The most common protocol for preparing SAMs of alkanethiols on metal film is immersing a freshly prepared or clean substrate into a dilute (about 1–10 mM) ethanolic solution of thiols for a certain incubation time at room temperature. The SAMs formed by *n*-alkanethiols covalently adsorbed on the gold substrate via Au—S bond. Figure 6.2 shows a simple adlayer model of the Au—S bonding interactions. Wherein, bonding of the thiolates at both threefold hollows and bridge sites has been proposed according to experiment and theory [14–16].

The Au—S bond is not limited to *n*-alkanethiols and also includes interesting conjugated molecules with thiol-terminal. Oligo(phenylene ethynylene)s (OPEs) with highly conjugated and conductive backbone have been extensively used in molecular electronics. Due to the ease of self-assembly, it is viable for measuring the electrical properties and charge transport mechanism in molecular junctions [17, 18]. Additionally, a redox-active and electron-rich moiety of tetrathiafulvalene (TTF) group can act as a donor to combine with OPEs. We have developed synthetic protocols of the cruciform system based on OPE and TTF scaffolds, using acetyl-protected thiolate end caps as anchoring groups [19–23], and the molecular structures are shown in Figure 6.3. After that, we obtained high-quality SAMs based on OPE3 and OPE3-TTF molecules on ultraflat gold substrates and investigated transport properties of molecular junctions using conducting probe-atomic force microscopy (CP-AFM) [25]. The results indicated that TTF substituent changes the molecular orbital energy levels and reduces the highest occupied molecular orbital (HOMO)–lowest unoccupied molecular orbital (LUMO) energy gap and contact resistance of the molecular junction, leading to approximately a 10-fold increase in conductivity compared to OPE3. We used amine instead of thiol as anchoring group attached to the gold substrate later and found that the conductance increase resulting from the addition of the TTF moiety was much larger than the situation having thiol anchoring group [26]. Compared with the Au—S bond that has very flexible contact angle and chain rotation, amine group holds a great potential to form more uniform SAMs because of the more restrict bond and its comparatively

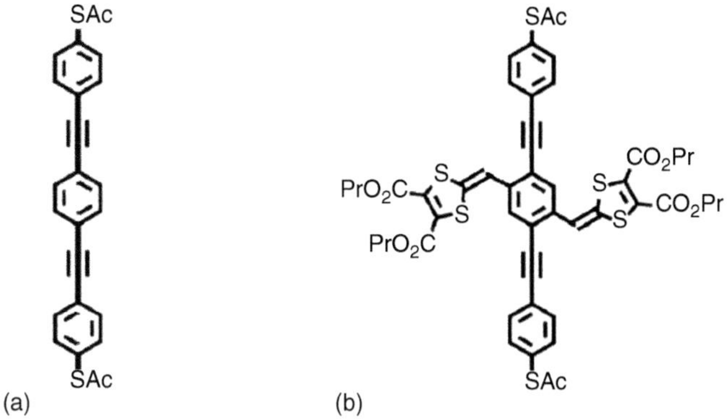

Figure 6.3 Molecular structures of OPE3 and OPE3–TTF. Source: Wei et al. [24]. 2015, John Wiley & Sons.

uniform contact geometry with a well-defined electronic coupling of the N lone pair to the Au.

Could you wonder a large system as an anchoring group to connect the gold substrate? Triazatriangulenium ($TATA^+$) ion is a large cationic π-system in which abundant side chains can be easily introduced onto the three nitrogen atoms (Figure 6.4) [28, 29]. Adding carbanions to the central carbon atom in triangulenium ions results in neutral platform molecules with the carbanion segment being perpendicular to the present curved plane of the triangulene ring system [30–32]. And, such TATA platforms are efficiently combined to the Au surface and form well-defined monolayers with the wire segment orthogonal to the substrate. We fabricated SAMs of TATA platforms on the gold with three different lengths of molecular wires and investigated electrical conductance by both CP-AFM and scanning tunneling microscopy (STM) [27]. The large footprint of the TATA platform will dispel any wire-to-wire transport and provide stable electronic contact to the Au substrate. For evaluating the potential of TATA as a binding/anchoring group in molecular junctions, we designed three phenylene wire molecules with TATA groups (TATA 1–3, Figure 6.4a) and compared the electronic transport properties of the SAMs based on these molecules with SAMs based on similar phenylene wires with thiol binding groups (S1–3, SS1–3) (Figure 6.4c). Similar β (tunneling attenuation factor) values suggest that TATA is a good anchor to detect the intrinsic properties of molecular wires compared to thiol anchoring groups. Although there is a central saturated carbon atom on the TATA bridge head and the TATA unit is noncovalently bonded, the contact resistance R_0 is unexpectedly low. TATA, as a new exciting anchoring group for molecular electronics, exhibits attractive advantages in both geometric and electronic structures.

Coating the organic SAM on the inorganic substrates such as metal oxides and metals may render them suitable for electronic or biological applications. Nakamura and coworkers reported the influence of the SAM of umbrella-shaped fullerene derivatives on the work function of indium-tin oxide (ITO) and gold surfaces [33]. (We focus on gold substrate, and ITO will be introduced in Section 6.2.2.) These

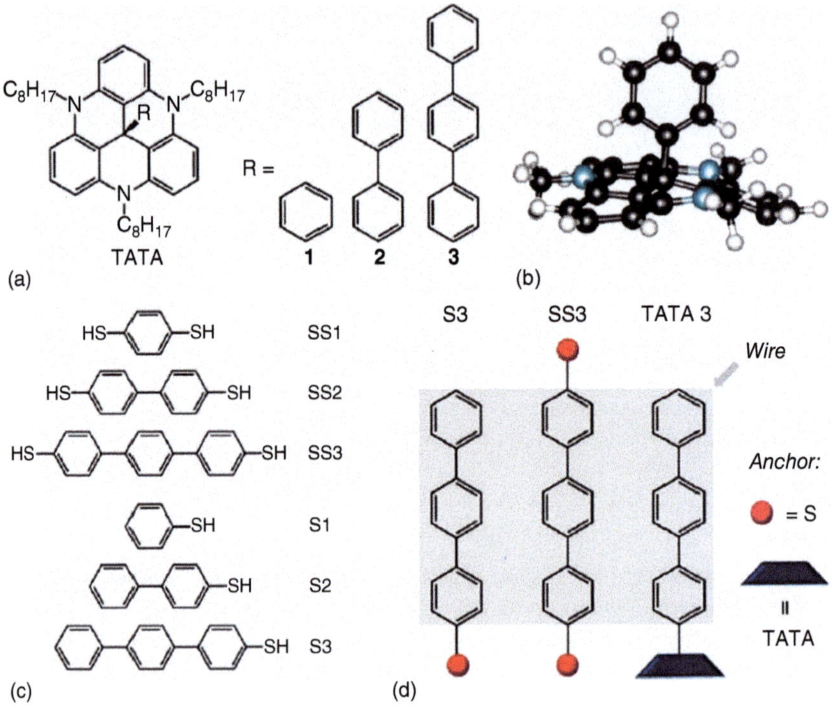

Figure 6.4 (a) Molecular structures of TATA 1–3; (b) optimized geometry of TATA 1 (changing alkyl chains into a methyl group for clarity); (c) molecular structures of bis-thiol SS1–3 and thiol S1–3; (d) schematic structure comparison between thiol, bisthiol, and TATA anchored terphenyl molecule. Source: Wei et al. [27]. 2014, American Chemical Society.

fullerene derivatives formed controlled-molecular-density SAM with a constant tilt angle value, which were standing upright through covalent bonds between the electrode surface and either the tips of the ribs (Figure 6.5a) or the handle of the umbrella tube (Figure 6.5b,c), or standing on the sides (Figure 6.5d). The SAM of a polar compound formed on an electrode surface results in an electronic field via intermolecular interaction of the molecular charges, and this field makes the total dipole of the SAM much smaller than the expectedly simple sum of the dipole of all molecules in the SAM. This effect is called depolarization, and it has been predicted by theory before. Especially when the molecules in Figure 6.5a–c form dense SAM on gold, the depolarization effect becomes so large that the work function shift becomes dependent mainly on the dipole length. And their finding showed that the work function shift for fullerene derivatives on gold is smaller than those for the structurally similar compounds on ITO, suggesting the "push-back" effect of the metal surface is crucial for understanding work function shifts.

Silver is the second most used substrate for SAMs of alkanethiols, but it is susceptible to oxidization in air and is toxic to cells. However, it can form high-quality SAMs with a simpler structure compared with gold due to smaller tilt angle. A structure characterized in low-energy electron diffraction (LEED) of elemental

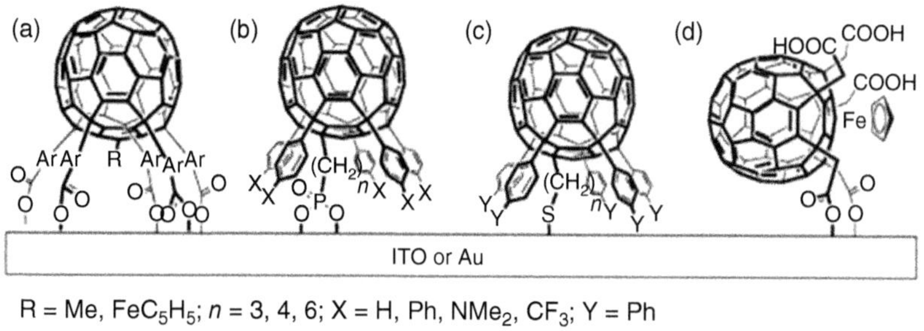

Figure 6.5 (a–c) Three classes of umbrella-shaped molecules for SAMs standing upright on ITO or gold; (d) molecules with different orientations on ITO (Ar = C_6H_4, $C_6H_4C_6H_4$). Source: Lacher et al. [33]. 2011, American Chemical Society.

sulfur on Ag(111) showed that the chain organizations observed in these SAMs conform to an organizational habit involving a $(\sqrt{7} \times \sqrt{7})$ $R10.9°$ overlayer of the thiolate sulfur atoms [34, 35]. Generally, the silver surface is present in the form of a complex oxide that can carry considerable amounts of adsorption of environmental contaminants. Fortunately, when the Ag substrates immersed promptly in a solution containing thiol, the interactions of thiols with Ag can rapidly strip these contaminating layers from the surface. In most cases, the removal of the surface oxide can be confirmed through X-ray photoelectron spectroscopy (XPS).

Whitesides and coworkers presented a mechanism for the large rectification of current viewed in molecular junctions based on SAMs of ferrocenyl alkanethiols (called $SC_{11}Fc$ in the article) on Ag^{TS} [36]. The fabrication of coating SAM on silver substrate is similar to that on gold. Swiftly immersing the freshly stripped ultraflat silver substrates in the ethanolic solutions of the corresponding thiols was to minimize contamination of the silver surface and formed the SAMs over 12 hours. In their study, eutectic indium–gallium (EGaIn) alloy with a surface layer of Ga_2O_3 are top contacts to form molecular junction and the molecules in the junction consists of two parts (Figure 6.6): an "insulating" part (i.e. alkyl chains) and a "conductive" part (i.e. ferrocene [Fc] head groups). The results showed the junctions rectified currents with rectification ratios $R > 10^2$, and the rectifying mechanism is based on a HOMO that is more strongly coupled to one electrode than to the other. At forward bias (V_f, Ga_2O_3/EGaIn is negatively biased), Fc HOMO is energetically accessible more easily than at reverse bias (V_r, Ga_2O_3/EGaIn is positively biased), the conductive part of the SAM does not significantly hinder charge transport, and the insulating part is the sole tunneling barrier. But, at reverse bias, both the conductive and insulating parts of the SAM are the tunneling barrier. Hence, the junction rectifier appears with a higher current flow at forward bias than at reverse bias.

Whitesides and coworkers used the similar tunneling junction system in the next year and changed the SAMs of $SC_{11}Fc$ into SAMs of n-alkanethiols containing odd and even numbers of methylenes to study the charge transport performance [37]. SAMs of odd-numbered n-alkanethiols are different from SAMs of even-numbered n-alkanethiols with regard to many properties, such as conformation, structure, tribology, and packing density. These are all factors that are closely related to the

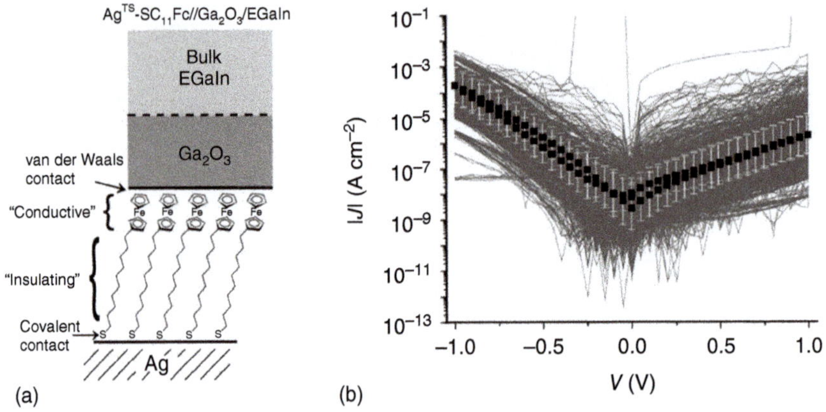

Figure 6.6 (a) Idealized schematic of the tunneling junctions consisting of Ag^{TS} substrates, SAMs of $SC_{11}Fc$, and Ga_2O_3/EGaIn top contacts. (b) Average trace of the Ag^{TS}-$SC_{11}Fc$//Ga_2O_3/EGaIn junctions superimposed on all 997 traces (gray) collected at these junctions. Source: Nijhuis et al. [36]. 2010, American Chemical Society.

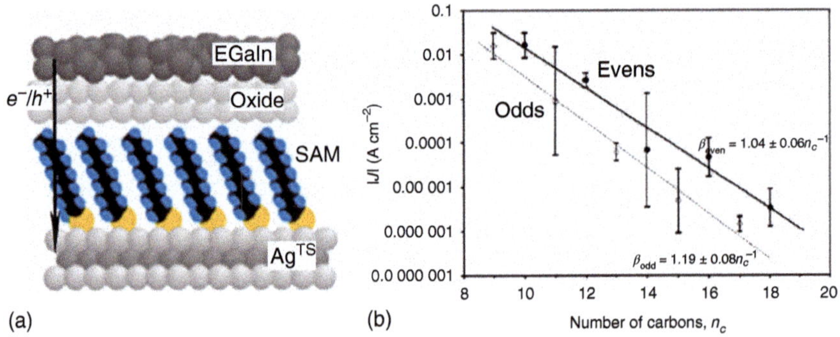

Figure 6.7 (a) Schematic diagram of Ag^{TS}-SAM//Ga_2O_3/EGaIn junctions. (b) Plot of |J| (|current density|) against the chain length of the odd- and even-numbered *n*-alkanethiols. Source: Thuo et al. [37]. 2011, American Chemical Society.

charge transport. Measurement data demonstrated a difference in charge transport between SAMs with odd- and even-numbered *n*-alkanethiols – the "odd-even effect." Although both the alkanethiols with an even number and an odd number of methylenes exhibited the exponential decrease in current density with increasing chain length, the statistical result suggested that the former typically showed higher current density than the latter (alkanethiols with an odd number of methylenes), as shown in Figure 6.7.

In addition to noble metals, copper is a relatively wide used metal to support SAMs. Copper has several outstanding properties, including high thermal and electrical conductivity and high resistance to electromigration [38]. And from a technical point of view, copper is semiconductor-compatible compared with gold because it is a common material for interconnection and seed layers for electroless deposition, but it is more easily oxidized than silver. The adsorption of an organosulfur adlayer

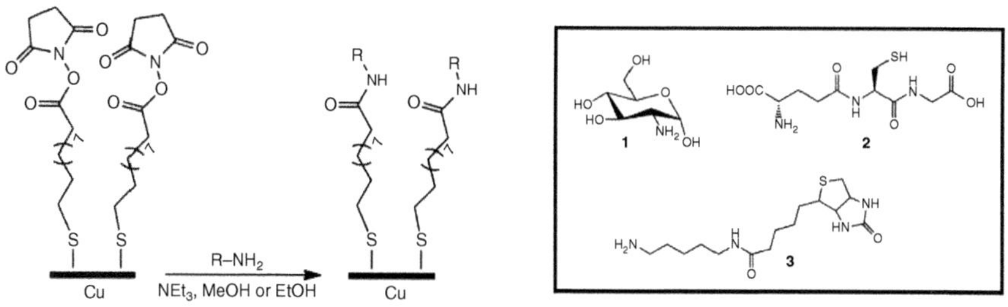

Figure 6.8 Functionalization of copper surfaces with biomolecules. Direct attachment of NHS–MUA onto the copper surface allowed for the successful attachment of biomolecules, such as β-D-glucosamine, the tripeptide glutathione, and biotin. Conjugation reactions were carried out at room temperature by immersion of the NHS-functional copper substrates in alcoholic solutions containing the respective amine, as well as triethylamine, under an argon atmosphere. Source: Caipa Campos etal. [38]. 2011, American Chemical Society.

can overcome the tendencies of copper to be oxidized and reduce the adsorption of contaminants from ambient; however, it is not as optimistic as silver. We should pay much attention to prepare high-quality SAMs on copper. And these experimental difficulties make the copper an uncertainty in the chemistry of SAM.

Many researches were devoted to the development of methods to prevent copper oxidation, for example, the electrochemical reduction of copper electrodes to remove the oxide layer before SAM formation [39]. Zuilhof and coworkers [38] proposed a simple and accessible wet-chemical route to obtain oxide-free copper surfaces (oxide layer was etched by acid), modified with alkanethiols and functional thiols, with *N*-succinimidyl mercaptoundecanoate (NHS-MUA). Detailed characterization showed that these monolayers were well-ordered and could easily be used as a platform for biofunctionalization of surface. Thus, coating these monolayers onto a copper surface significantly increases its application capacity. For example, the SAMs based on NHS-MUA were conjugated with different amine-functional biomolecules, as shown in Figure 6.8.

6.2.2 Monolayers Attached to Non-metal Substrates

Similar to the metal substrate, SAMs can modify the surface chemistry of oxide and semiconductor so that control properties including wettability, work function, and charge transfer. Loo and coworkers [40] used gallium arsenide (GaAs) as substrate to study the effects of NH_4OH concentration on the final structure of thioacetyl-protected terphenyldithiol (TPDT) assembled from ethanol (EtOH) and tetrahydrofuran (THF). The different molecular orientation, surface coverage, and structural organization of these SAMs caused by different solvent and NH_4OH concentrations suggested the quality of TPDT SAMs on GaAs is highly sensitive to the assembly conditions.

A main route for biofunctionalization of inorganic surfaces via assembling amine-containing biomolecules is using activated esters derived from NHS. We introduced the adsorption of the NHS-terminated monolayers on copper substrates in Section 6.2.1; actually, not only the copper surface but these monolayers can also covalently attach in one step onto silicon through irradiation with visible light [41]. And the resulting surface allows a further modification by substitution of the NHS–ester moieties with amines at room temperature (Figure 6.9). This light-induced attachment method can be compatible with photoinduced patterning techniques to broaden the possibility for the application of biofunctionalized silicon surfaces in biochips and biosensors.

Dong et al. [42] described a novel solvent-free chemical vapor deposition (CVD) method to deposit octadecylsilane (OTS) monolayers on silicon wafers at room temperature under vacuum. CVD method overcomes the disadvantages that have been brought by dip coating method from solution (such as the production of ill-defined particulate aggregates on the SAM) and allows precise control of the OTS surface coverage and substrate hydrophobicity. Kim and coworkers [43] measured lateral polymerization in ω-(*N*-pyrrolyl)propanol SAMs grafted on hydrogen-terminated n-type Si(111) substrate by a scanning-probe-based method. They found that the biased AFM scan could successfully induce local polymerization/oligomerization within the covalently grafted molecular layer, depending on the changes in the mechanical and electrical properties of the SAMs during the AFM scan.

In recent years, modification of oxide surfaces with organo phosphonic acid (PA)-based molecules has attracted increasing interest due to their moisture stability and lack of homocondensation [44]. Acton et al. [45] developed an efficient process to modify SiO_2 with phosphonate SAMs by spin-casting or microcontact printing. It is important that the SiO_2 substrate is activated by an in situ generated nanoscale AlO_x activation layer to increase the surface reactivity for PA molecule binding. PA molecules covalently bind to metal oxides following an acid–base condensation and coordination mechanism where the acidic PA headgroups and more basic metal hydroxyl (M–OH) groups react to produce stable P–O–M phosphonates. Surface characterization techniques indicted that the phosphonate SAMs on AlO_x/SiO_2 are covalently bound, densely packed, and highly ordered. And they fabricated SAM dielectric-based organic thin-film transistors (OTFTs) on $AlO_x/SiO_2/Si$ (Figure 6.10) demonstrating the practicality of this rapid SAM processing, of which self-organized n-channel and p-channel active areas defined by SAM/AlO_x hybrid dielectrics. Analogously, Ginger and coworkers [46] utilized microcontact printing to pattern PA SAMs on ITO to achieve the local control of the ITO work function and surface contact potential difference contrast of over 0.5 V, which is comparable to those obtained from solution-deposited SAMs. The ability to pattern PA SAMs with low-cost printing may provide valuable applications toward organic electronic devices including low-cost organic light-emitting diode (OLED) and high-throughput OTFTs. Moreover, Halik and coworkers [47] demonstrated low-voltage operating n-type SAM transistors in which densely packed fullerene SAMs are attached to AlO_x/Al gate by PA anchoring group, which is a major progress toward practical applications.

In most experiments, the preparation of SAMs is immersing a freshly prepared or clean substrate into molecular solution for an incubation time; this is also the so-called "solution dipping" method just like immersing the Oreos cookie in milk.

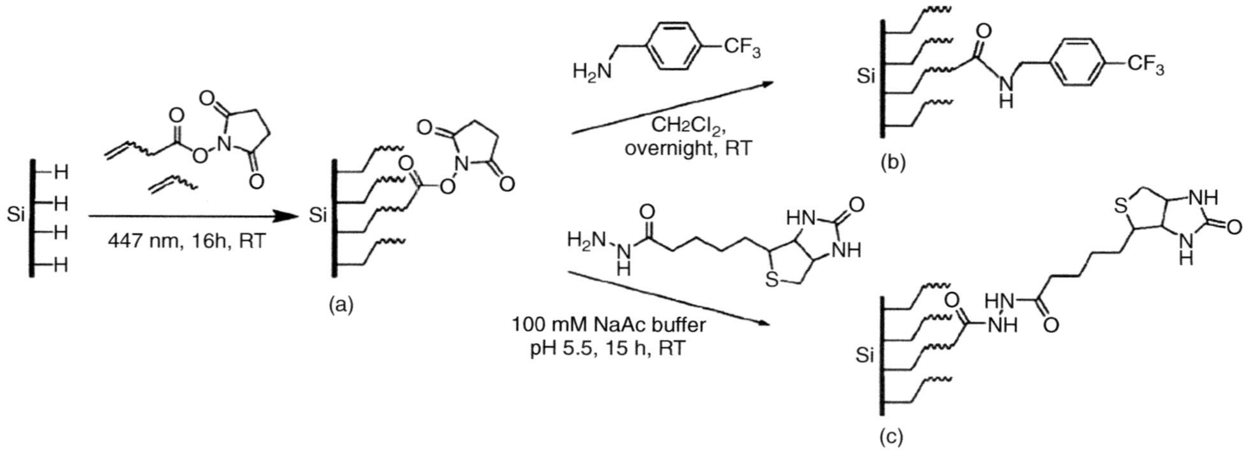

Figure 6.9 Schematic description of the formation of mixed monolayers terminated with NHS-ester moieties (a) and of the further modification by *para*-trifluoromethyl benzylamine (TFBA) (b) or biotin hydrazide (c). Source: Yang et al. [41]. 2008, American Chemical Society.

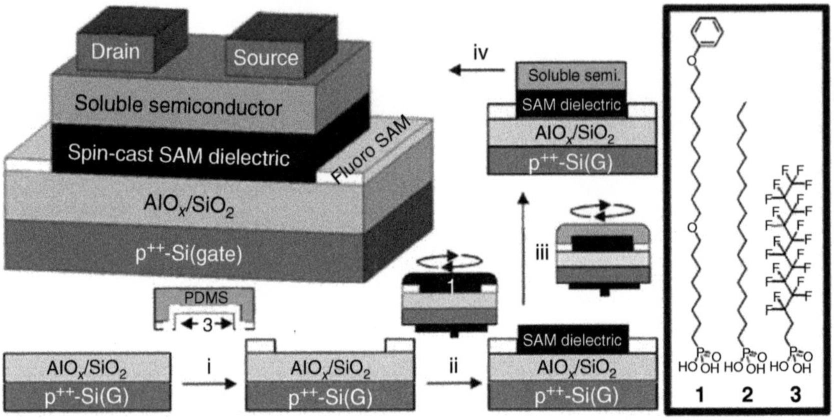

Figure 6.10 Fabrication process for spin-cast and patterned SAM dielectric-based OTFTs on $AlO_x/SiO_2/Si$: (i) microcontact printing **3** onto plasma-activated $AlO_x/SiO_2/Si$, (ii) self-organized patterned spin-cast depositing **1**, (iii) self-organized patterned solution depositing semiconductor, and (iv) thermal evaporation of metal electrodes through a shadow mask. And, structures of the PA molecules used in this study (black rectangle). Source: Acton et al. [45]. 2011, John Wiley & Sons.

The above-mentioned spin-casting is another molecular deposition method, in which the molecular solution is spin-coated on the target substrate and then rotated at a certain rotational speed. As the substrate rotates, the solution is widely diffused across the wafer to achieve very small molecular concentrations on the substrate. After spin-casting, substrates are thermal-annealed to facilitate the diffusion of unbound molecules in the as-spun film.

When it comes to the fabrication of OTFTs, linear acenes such as pentacene and tetracene are not stranger as a result of their relatively high field effect mobilities and large ON/OFF current ratios [48–50]. Tulevski et al. [51] employed a tetracene derivative to form dense, upright SAMs on the surface of aluminum oxide. These monolayers form the active channel in nanoscale OTFTs when aluminum oxide is used as the dielectric layer (Figure 6.11). This approach enables high yields of devices with a source–drain distance of <100 nm and offers a new idea for the study of OTFTs at that time. In addition, these devices have the potential to be used as highly sensitive environmental and molecular sensors because their active channel is exposed to the environment.

6.2.3 Langmuir–Blodgett Method

SAM is not the only means to obtain an ensemble molecular junction, but Langmuir–Blodgett (LB) method is also an important technique to form monolayers. LB technique [52, 53] involves surface chemistry; the LB films are formed when the amphiphilic molecules like surfactants or nanoparticles suspension deposited at the air/water interface. Amphiphilic molecules include hydrophobic "tails" and hydrophilic "heads" and arrange themselves in a pattern that exposes the hydrophobic portion to air and the hydrophilic portion to interact with water,

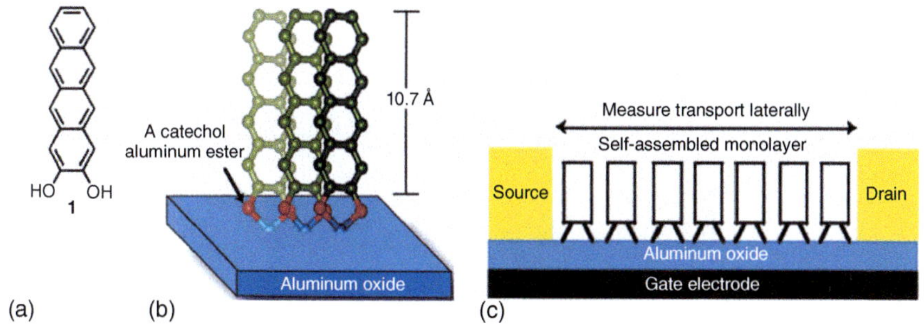

Figure 6.11 (a) Molecular structure of end functionalized tetracene **1**. (b) Schematic of bonding and orientation of **1** on aluminum oxide. (c) Schematic of the SAM transistor. Source: Tulevski et al. [51]. 2004, American Chemical Society.

so reduce the surface tension of the water. Then, the resulting film is compressed to the desired surface pressure and particle density and then transferred onto a solid substrate by immersing the solid substrate into the liquid.

LB method has the following advantages [54, 55]: (i) accurate control of the thin-film thickness and packing density, (ii) homogeneous deposition over large areas, (iii) multilayer LB film with varying layer composition, not only monolayer, and (iv) but also many types of substrates are available. Still, LB films suffer both in chemical and mechanical stabilities, and their applications are not as good as SAM. However, LB monolayers have been used for device applications early in the 1970s. Mann and Kuhn [56] reported the first direct measurement of through molecule conductance. They prepared a series of ordered fatty acid monolayers using LB method and successfully sandwiched them between metal electrodes as well as measured the electrical characteristics of the system. Results revealed that the conductivity decays exponentially with the length of the molecule.

The LB technique is a mature and experienced method for controlling the molecular orientation and packing of interfaces [57]. Moreover, it is an efficient approach toward the controllable fabrication of laterally patterned structures on solid supports [58]. Bao and coworkers [59] investigated the effect of density and degree of ordering of OTS monolayers on the performance of pentacene OTFT. In their study, LB method was applied to systematically alter the organization and density of the OTS monolayers that were used to modify the SiO_2 dielectric surface. During the process of ultrathin-film deposition, amphiphilic OTS molecules are compressed by exerting a lateral pressure to the monolayer film at the air–water interface [60]. And the film changed from two-dimensional (2D) gas to 2D liquid and finally transformed into an ordered 2D solid with the increase of the applied lateral pressure; the complete structure of pentacene OTFT is shown in Figure 6.12. The results show that phase and order of the OTS surface modification layer are important for device performance. Crystalline, dense organic dielectric surface modification layer can boost 2D semiconductor growth and achieve higher mobility. This is a key step toward morphological control of organic semiconductors and offered a new route to optimize OTFTs.

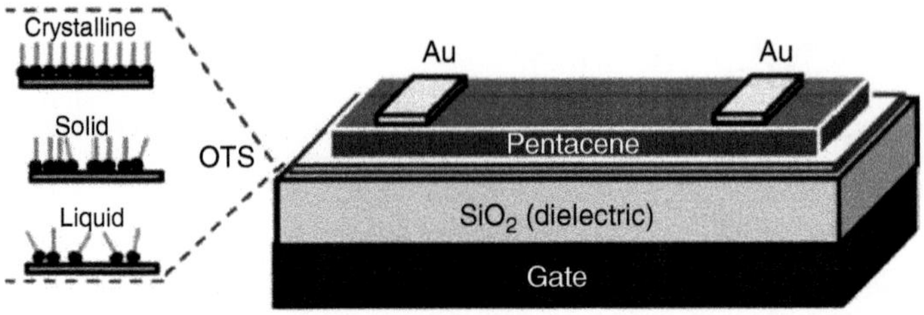

Figure 6.12 Schematic drawing of pentacene OTFT with underlying OTS surface modification layers in different 2D phases. Source: Virkar et al. [59]. 2009, John Wiley & Sons.

Furthermore, LB technique also has good applications in two-terminal devices. These two-terminal devices are useful for molecular memory and logic devices because they possess two states of conductance and can be reversibly switched with an applied voltage [61]. One of them is based on mechanically interlocked bistable complexes, such as catenanes and rotaxanes using LB films [62]. The basic structure of early molecular diodes or rectifiers was proposed by Aviram–Ratner that consisted of a donor and an acceptor separated by a σ bridge [63]. And, this D–σ–A model was demonstrated to be successful and thoroughly confirmed using LB films [64] and block copolymers [65] sandwiched between electrodes.

6.3 Top Electrodes

We have gotten a general idea of how the SAMs connect to the substrates (the bottom electrodes). We know that it is not enough for a molecular device to have only one electrode, so in this section, we will talk about the preparation of the top electrode. The formation of the top electrode is of great significance for the realization of device performance; there are three main methods: (i) direct formation of metal electrodes using either electron beam/thermal evaporation or electrochemical deposition, (ii) indirect evaporation of incorporating with buffer interlayer, and (iii) making top contact by soft methods such as liquid metal contact, transfer printing and graphene as top contact.

6.3.1 Deposited Metal

Metals with good conductivity have been the preferred electrode materials in molecular electronics. Direct evaporation is a common method for preparing metal top electrodes, but it has proved to be challenging. The major stumbling block is the formation of metal filaments that penetrates the molecular layers. They make direct contact with the bottom electrode resulting in potential damage or destruction of SAMs and short-circuit junction during top contact formation. In order to improve this situation, indirect evaporation is developed, of which very typical is the incorporation of buffer interlayer.

6.3.1.1 Direct Evaporation

Electron beam evaporation and thermal evaporation are the classic approaches of direct evaporation to form metal top contacts. The electrical contacts produced by evaporation depend on evaporating the metal by heating the metal to a sufficiently high temperature, then condensing the vapor to a cooler substrate. Often, metal atoms and clusters reach the substrate with high temperature and kinetic energy. Together with the radiation emitted by the heated source, the substrate surface can be modified [66]. If the surface is covered by SAMs, SAMs can easily be damaged and form metal filaments that penetrate through the molecules to reach the bottom electrode [67–69].

Several metals have been evaporated directly as the top electrodes, including gold, silver, aluminum, copper, and it has been found that metal filaments penetrating the monolayer to cause electrical short circuits. A few studies investigated the interaction between the evaporated metal atoms and methoxy-terminated SAMs on the Au substrate [68, 70, 71]. The results showed that evaporated aluminum, copper, and silver atoms were separated along two parallel paths: penetration through the $-OCH_3$ SAM to the monolayer/Au (S/Au) interface, and interaction with $-OCH_3$ groups resulting in overlayer metal film. Especially, for Al, penetration channel closed after about 1 : 1 Al/Au layer has been reached and an overlayer metal film formed at the vacuum interface. However, for Cu and Ag, the penetration channel does not close. Moreover, almost inert gold does not interact with methoxy end groups. We observed that the evaporated Au atoms penetrated to the Au/S interface at all coverage situations (Figure 6.13a). If the molecular end groups react with the invading metal, the damage can be controlled to some degree and the diffusion of the metal through the molecules was decreased.

The balance between metal film on top of SAM and penetration through the SAM is related to the functionality of the terminal groups and the reactivity of the intruding metal atoms. The metal atoms with low reactivity toward the end groups will penetrate through the SAM to reach the substrate and form an adlayer between the bottom electrode and SAMs. This process can build metallic "bridges" that contact top electrode and bottom electrode directly leading to short circuit

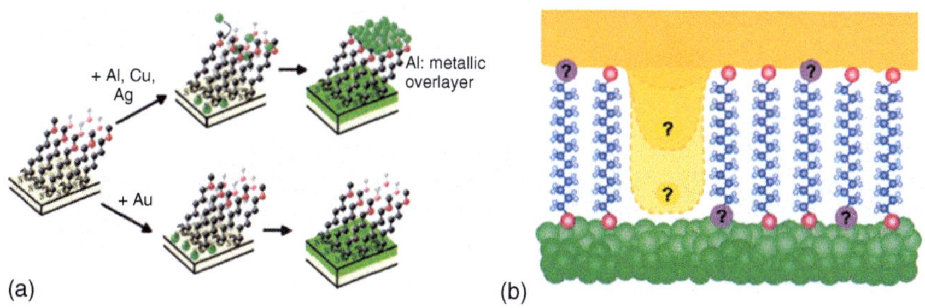

Figure 6.13 (a) Schematic of interaction between the evaporated metal atoms and SAM. Metal atom is represented in green, oxygen in red, and hydrocarbons in black. Source: Walker et al. [71]. 2004, American Chemical Society. (b) Simulation of the evaporated metal atoms penetration process. Source: Haick and Cahen [72]. 2008, American Chemical Society.

as shown by Figure 6.13b. Or forming relatively insulating pillars if the metal is easily oxidized will damage the SAM and affect the accuracy of the charge transport measurement. In contrast, the invasive metal atoms with high reactivity toward the end groups will leave the metal on the top of the SAM by interacting with its end groups. The more reactive the metal is to the end group, the more likely that the metal will remain at the top of the SAM and less likely penetrate through the SAM [72].

In summary, the fabrication of the top electrodes by direct evaporation is difficult to form good molecular electronics. The filamentary penetration and the destruction of SAM seriously affect the quality and yield of devices. We urgently need to find alternative ways to ameliorate this situation.

6.3.1.2 Indirect Evaporation

In order to reduce SAM defects and device shorts, some indirect evaporation methods have been developed, such as surface-diffusion-mediated deposition (SDMD) junctions and buffer interlayer-based junctions. The following will describe these methods in detail.

Surface-Diffusion-Mediated Deposition Junctions The novel technique for the formation of metallic contacts through SDMD is proposed by McCreery and coworker [73] to overcome the problem of metal atoms penetrating or SAM damage caused by direct evaporation, in which the metal atoms were deposited remotely and then diffused onto the molecular layer.

We can see the SDMD fabrication process from Figure 6.14a–d. In short, pyrolyzed photoresist films (PPFs) fabricated on thermally oxidized silicon substrates provide the necessary surface chemical environment to allow the formation of C—C bonds between the conducting PPF film and the SAMs. A patterned SiO_2 mask was fabricated on top of the PPF surface using optical lithography, and then the unmasked PPF was etched by O_2 reactive ion etch process to produce a nearly vertical sidewall for SAM formation and a SiO_2 "overhang" above the SAMs protects the molecules from direct impingement of metal atoms and source radiation. The SAMs were electrochemically grafted to the PPF sidewall, and gold was remotely deposited using electron beam evaporation followed by a diffusion process to finish the formation of the molecular junctions. It should be mentioned that for direct evaporation to form the metal top electrodes, metal deposition occurred in a direction perpendicular to the SAM surface and the molecular layer is exposed to radiation from the evaporation source. For SDMD process, however, the deposition angle can be changed and the overhang of the SiO_2 mask protected the SAMs from impact of both invading metal atoms and radiation from the evaporation source. This indirect approach allows for the formation of robust metal top contacts on SAMs with excellent yield (typically >90%) and reproducibility. The surface diffusion of the deposited metal atoms toward the SAMs in the SDMD process is shown in Figure 6.14e [74].

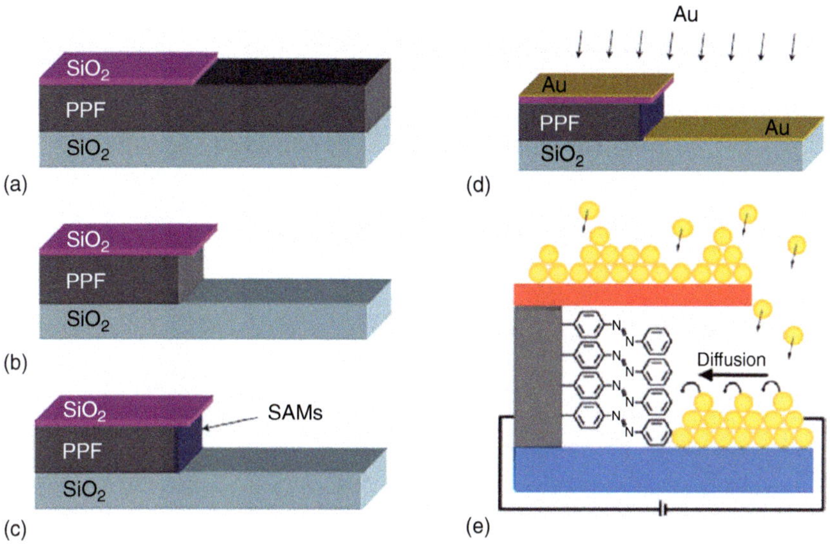

Figure 6.14 (a–d) Schematic of the SDMD process. Source: Bonifas et al. [73]. 2010, Springer Nature. (e) Schematic of the atom-diffusion process to complete the molecular junctions. Source: Bonifas et al. [74]. 2011, American Chemical Society.

Buffer Interlayer-Based Junctions SAMs are sandwiched between two metallic (top and bottom) electrodes in a nanopore which is also a common structure to constitute molecular junctions that can well define the area of the SAMs. Unfortunately, electrical shorts due to the direct evaporation of top metal thin films are also inevitable in this case. A valid solution to prevent the penetration of the metal atom into the SAMs and avoid short circuit is to insert a buffer interlayer between the top electrodes and SAMs.

Akkerman et al. introduced a method to produce large-area molecular junctions with diameters up to 100 μm with high yields (>95%) [75]. This approach involves fabricating molecular junctions in the nanopores of a lithographically patterned photoresist and introducing a conducting polymer interlayer between top electrodes and the SAMs. The preparation of molecular junctions in the nanopores is schematically illustrated in Figure 6.15. First, a thin chromium adhesion layer and a gold bottom electrode were vapor-deposited on Si/SiO_2 wafer (Figure 6.15a). Then, photoresist was spin-coated onto the Si wafer, and holes were produced by the standard photolithography method (Figure 6.15b). Next, the substrate was submerged in a freshly prepared molecular solution for at least 36 hours to grow SAMs. After the formation of SAM, a water-based suspension of poly(3,4-ethylenedioxythiophene) stabilized with poly(4-styrenesulfonic acid) (PEDOT:PSS, a commercially available, highly doped, conductive polymer) is spin-coated on top of the SAM as the buffer interlayer (Figure 6.15c). At last, the top gold electrode was vapor-deposited through a shadow mask, and the redundant PEDOT:PSS was removed by reactive ion etching (Figure 6.15d). The molecular

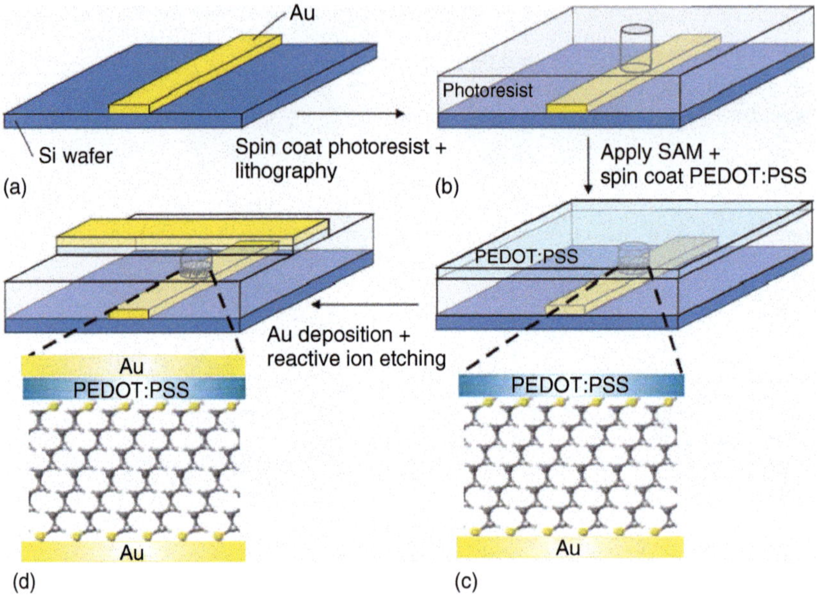

Figure 6.15 Processing steps of a large-area nanopore molecular junction. Source: Akkerman et al. [75]. 2006, Springer Nature.

junctions made by this approach showed excellent stability and reproducibility, and the conductance per unit area is comparable to that obtained through other methods.

Although the incorporation of PEDOT:PSS buffer interlayer is resultful for producing high-yield molecular devices, the use of a conductive polymer has a few limitations because the resistance of PEDOT:PSS is relatively large and PEDOT:PSS is very sensitive to temperature as well as the contact between the PEDOT:PSS and molecules is poor. The electrical properties of the junctions have a closed relationship with the type of the isolating layer (photoresist or SiO_2) and the molecular contact groups (hydrophilic or hydrophobic) [75]. Graphene, with its outstanding electronic properties, chemical stability, and mechanical properties, might be a favorable alternative to PEDOT:PSS.

Wang et al. [76] presented a new route to fabricate reliable solid-state molecular devices using multilayer graphene as the buffer interlayer, as shown in Figure 6.16. After the SAMs were formed on the gold bottom electrode, a multilayer graphene film was transferred to the substrate. And then gold top electrode was vapor-deposited on top of the graphene films using a shadow mask at a low deposition rate. The graphene buffer interlayer prevents the formation of electrical shorts and filamentary paths that would stem from penetration of the Au top metal. Graphene-based molecular devices may have potential advantages over the PEDOT:PSS-based devices in relation to proper charge transport properties in molecular systems with variable lengths and different contact groups (hydrophilic and hydrophobic). In addition, it seems that the electronic coupling of graphene to the molecules is stronger, resulting in lower contact resistance.

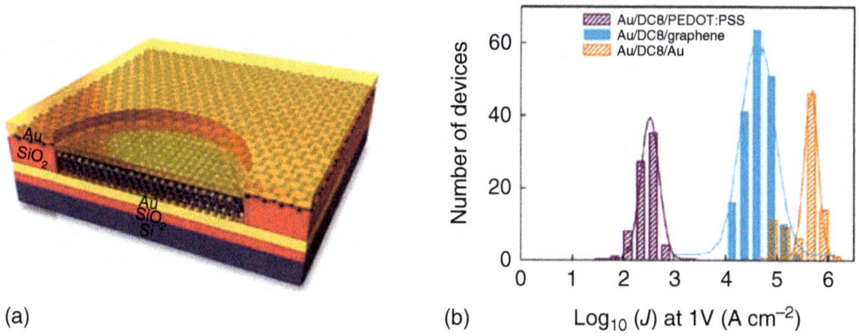

Figure 6.16 (a) The molecular junction based on graphene buffer interlayer. (b) Histogram of the logarithmic current densities at 1 V with different types of top electrodes (PEDOT:PSS, graphene, and Au). Source: Wang et al. [76]. 2010, John Wiley & Sons.

This technique is of excellent durability, thermal and operational stabilities, and improves device lifetimes, which could pave the way for practical molecular electronics.

6.3.2 Make Top Contact by Soft Methods

In this section, the fabrication of reliable and high-yield SAM-based molecular junctions using different state-of-art methods will be introduced, including lift-and-float, crosswire or crossbar, transfer printing, graphene or liquid metal as top electrode. These methods also effectively overcome the disadvantages resulting from direct evaporation of top contacts. The high-quality ensemble molecular junctions make it possible to systematically analyze the charge transport characteristics and facilitate the realization of producing functional molecular devices.

6.3.2.1 Lift-and-Float Approach

This method depends on capillary interactions at liquid–solid interfaces between two solids and a common liquid is used to transfer thin solid metal films onto substrates, which is described as lift-off float-on (LOFO) approach.

The general procedure for this method includes two major steps (Figure 6.17) [72]: (i) lift-off of evaporated metal thin film. The metal film was evaporated onto a solid support such as a glass slide and then seperated from the solid support by a certain detaching agent and finally floating on the solvent surface due to the capillary interaction. (ii) float-on of the metal film on the substrate. The target substrate was modified by SAMs and attached to the metal film in liquid. The solvent was rapidly removed to prevent the wrinkling of the metal film. Metal/SAM/substrate was formed by liquid-mediated process. This idea provides an effective way to relatively reduce the physical damage that is encountered during most vacuum deposition processes and the chemical damage that is involved in electrochemical deposition. To some extent, it is biocompatible and has the potential to apply to biological

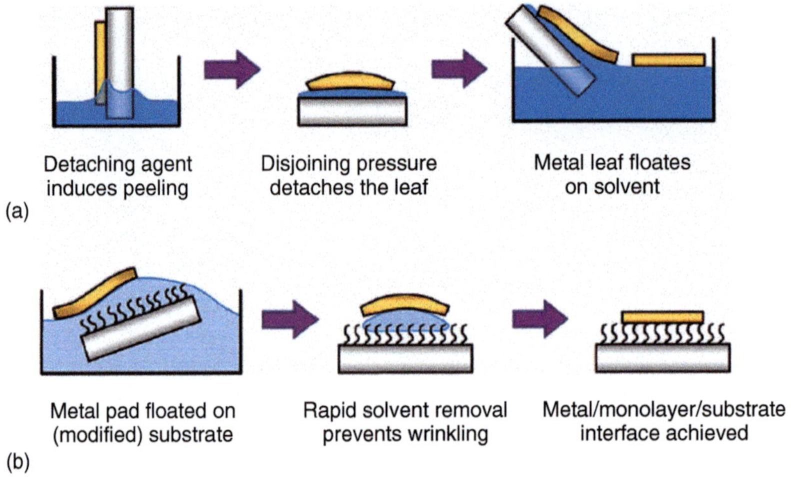

Figure 6.17 Schematic illustration of the lift-off, float-on (LOFO) procedure. (a) Lift-off of the metal leaf from a glass slide. (b) Float-on of the metal pad on the solid substrate. Source: Haick and Cahen [72]. 2008, American Chemical Society.

molecules. But there are also some challenges in the LOFO method, for instance, poor control of the force between the floating electrodes and the substrates. And air gap may appear at the interface between top electrode and the SAM. In addition, it is hard to completely prevent the metal film from wrinkling during the contact.

6.3.2.2 Crosswire Junction

A promising device architecture for the realization of the next generation of memory and logic devices based on metal–molecule–metal junctions could be the crosswire/crossbar structure [77–79]. It has the potential to reach high-throughput production at low cost due to the ability of manufacturing in parallel. In the crosswire and crossbar architectures, the electrodes cross each other and sandwich the molecular layer.

Kushmerick et al. [80] measured I–V characteristics for crossed-wire tunnel junctions where SAMs based on isonitrile OPE molecules were sandwiched between two macroscopic 10-μm-diameter gold wires controlled by the Lorentz force. Direct current in one wire deflected it in a magnetic field to bring the wires gently together to form a junction at the contact point, as shown in Figure 6.18a.

Compared with STM measurements of same individual molecules adsorbed on a flat surface [81], multiplying the original STM current data by 1000 achieved the quantitative agreement between crossed-wire and STM data surprisingly (Figure 6.18b). It demonstrated that such crosswire tunnel junctions contained ~1000 molecules and the conductance of crosswire junction scales linearly with the number of molecules in parallel. Since molecular wires act as discrete, noninteracting conductance channels, simple linear superposition law of conductance can be applicable in the design of molecular electronics. Besides, crosswire junctions have also been compared to CP-AFM devices [82, 83]. The results showed that

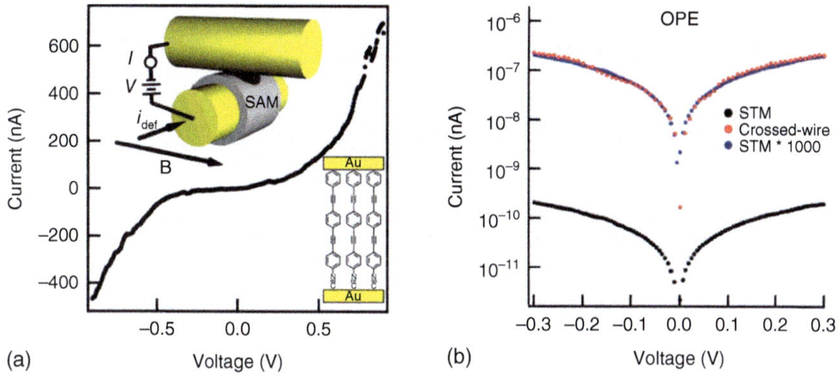

Figure 6.18 (a) Schematic of the crossed wire tunnel junction controlled by the Lorentz force. The *I*–*V* characteristics of the junction are obtained by ramping the bias voltage (*V*) while monitoring the current flow (*I*) across the junction. Source: Kushmerick et al. [80]. 2003, American Chemical Society. (b) Comparison of *I*–*V* curves for crosswire and STM-based junctions. Source: Blum et al. [81]. 2004, American Chemical Society.

the V_{trans} (which is conceptually a rough estimate of the tunneling barrier height) obtained from the crosswire junction and CP-AFM measurements was consistent for a series of molecules, and it was independent of uncertainties in junction area (the number of molecules involved in transport is uncertain for both CP-AFM and crosswire junctions). And subsequent investigation suggested that V_{trans} was molecule-specific, providing brilliant insights into electron transport mechanism.

Later, Kushmerick and coworkers further studied different materials with this Lorentz force-assisted crosswire method and made a remarkable contribution to this test bed [84]. However, due to the surface curvature of the wires in the architecture, it is of difficulty to precisely control the number of contacted molecules and their orientation. Additionally, this technique is not suitable for large-scale production of solid-state devices because of the requirement of the external magnetic field, although the yield of the device is increased compared with the metal-evaporated molecular junction.

Crossbar is similar with crosswire method; it is a set of rectangular block wires that cross each other with a functional molecular layer in the crosspoints. Nijhuis et al. [85] proposed a method to create a crossbar array based on Ga_2O_3/EGaIn stabilized in the microchannels in polydimethylsiloxane (PDMS). In their experiments, the pattern of Ag electrodes that fabricated by using a standard lithography technique embedded in a cured OA was placed on the Si wafer. A glass sheet then was affixed to the Ag electrode pattern using a cured UV-curable adhesive. And the optical adhesive interacted strongly with the Ag and the glass sheet but not the Si wafer, so that the Ag/adhesive/glass composite can be easily cleaved from the wafer to obtain bottom electrodes. Next, the microchannels in PDMS were aligned vertically to the bottom electrodes after the formation of SAMs (one microchannel in PDMS is shown in Figure 6.19). Finally, the device was fabricated after filling the microchannels with Ga_2O_3/EGaIn. This method does not require metal deposition or electron-beam

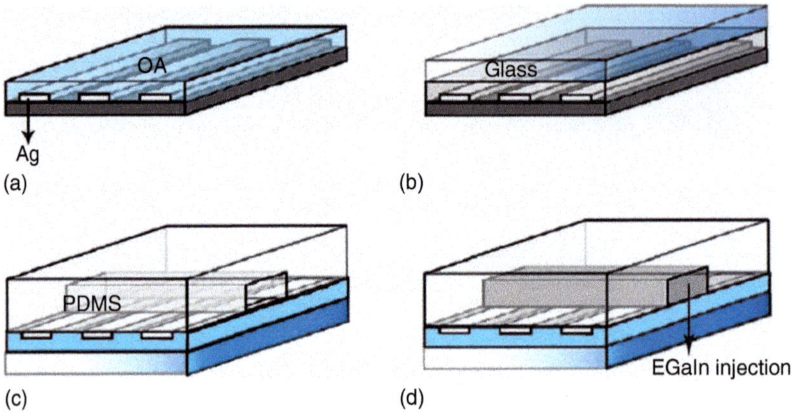

Figure 6.19 Process of the arrays of SAM-based molecular junctions formed by crossbar technique. Source: Nijhuis et al. [85]. 2010, American Chemical Society.

evaporation to form a top electrode, reducing defects in the SAMs. More information about the crossbar architecture is provided in Section 7.2.3.

6.3.2.3 Transfer Printing

Complete, off-the-shelf functionality can be transferred to the substrates without the need for energetic photons, ions, or electron beams [86, 87]. In transfer printing, a thin metal film is evaporated onto an elastomeric stamp, such as PDMS. And the electrical contacts are made by mechanical contact of the metal-coated PDMS with the SAMs. This technique is fast, simple, and stable under ambient conditions, providing a promising method to fabricate integrated molecular devices. It can be used to define nanometer-scale structures; in this case, the method is called nanotransfer printing (nTP).

The disadvantage of this method is that the metal is transferred only by mechanical contact which is lack of effective driving force. Therefore, it would be better if metal film has a higher affinity for the SAM than the stamp, where the molecules could form chemical bonds with the "to-be-transferred" metal film. Figure 6.20 shows the nTP process for fabricating metal/molecule/substrate junctions [72]. The substrate modified by SAMs with exposed reactive end groups (e.g. thiol) is in mechanical contact with the metal-clad PDMS. Next, the metal film is removed from the stamp and left on the molecular layer because metal adhesion to PDMS is weaker than that to the end group.

Loo et al. [87] used this method for fabricating top contacts to form Au/dithiol/GaAs junctions. The main procedure can be simply described by Figure 6.21. The GaAs substrate was stripped of its native oxide layer and then immediately exposed to molecular vapor to form SAMs. The majority of molecules chemisorb on the GaAs surface with a preferential orientation: one end of molecule reacts with the GaAs surface and the other is unreacted thiol terminal groups. The gold film-coated PDMS stamp is brought into contact with the GaAs substrate to form GaAs–molecule–Au junctions. The elastic and mechanical adaptability of the PDMS stamp ensure a good

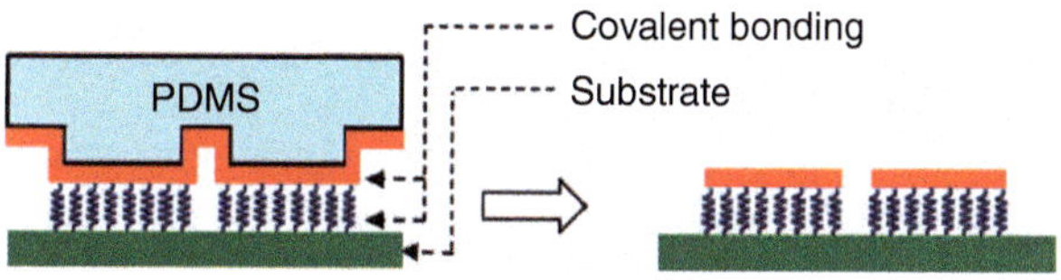

Figure 6.20 Schematic illustration of nTP process. Source: Haick and Cahen [72]. 2008, American Chemical Society.

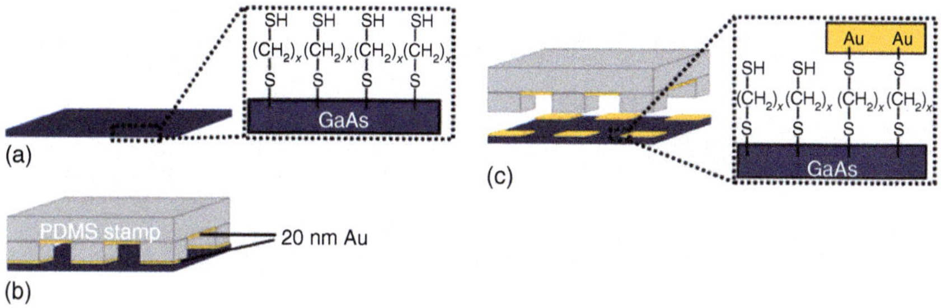

Figure 6.21 Schematic of the Au/dithiol/GaAs junctions made by nTP method. Source: Loo et al. [87]. 2003, American Chemical Society.

contact at the interface of the stamp/substrate. As compared to the Au—S bonding, Au has a poor adhesion to PDMS, so it is easy to remove the stamp from the substrate to complete the transfer printing process. I–V and photoresponse experiments were to evaluate the nature of electrical contact, demonstrating that nTP is an attractive approach for making electrical contacts in molecular devices with reproducible features.

Coll et al. [89] reported another application of nTP to form Au^{TS}–SAM–Si molecular junctions by "flip-chip lamination." As illustrated in Figure 6.22, the process can be briefly stated as follows: (i) evaporating gold onto the Si wafer with a fluorinated release layer; (ii) stacking a polyethylene terephthalate (PET) substrate on the top of the gold by using nTP and then peeling off to expose the ultra-smooth Au underside; (iii) bifunctional molecules are self-assembled onto gold film to form SAMs with the exposed functional group; (iv) inverting these films and transfering them to passivated silicon wafers. Finally, the two electrodes are laminated together with nTP to complete the junction. This work shows a rare example of the SAM contact with a silicon electrode.

6.3.2.4 Graphene as Top Electrode

Graphene possesses a natural compatibility with organic/biological molecules and outstanding electronic properties with high stability and chemical flexibility, which is not limited to the conductive interlayer but can also be directly used as a top contact. However, pristine graphene prepared by mechanical exfoliation [90], liquid phase exfoliation of graphite [91], and CVD method [92, 93] is usually accompanied by low yield or high cost. In addition, one of the key challenges in incorporating

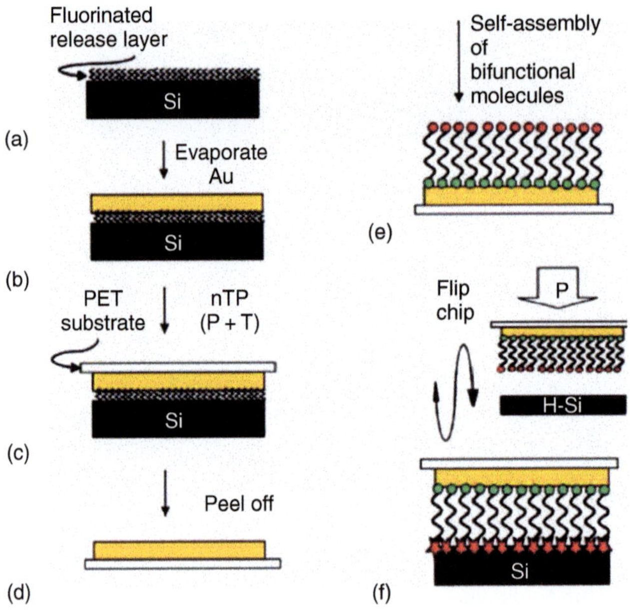

Figure 6.22 Schematic of the flip-chip lamination process to form Au^{TS}-SAM-Si molecular junctions. Source: (a–e) Zhang et al. [88]. (f) Coll et al. [89]. 2009, American Chemical Society.

graphene films into electronic devices is the manipulation of ultrathin graphene films. An extra polymer support layer is often required to keep the film from breaking down into individual sheets during transfer to the target substrate [94], but it is then difficult to remove the polymer completely. This is one of the urgent problems to be solved in the research.

Surprisingly, as another branch of graphene research, graphite oxide is chemically exfoliated into individual graphene oxide (GO) sheets and subsequently converted to chemically derived graphene (CDG) via chemical and/or thermal treatment, provides a low-cost solution-processing route, which can be integrated into large area electronics [95–97]. Moreover, GO is an attractive material for a variety of thin-film electrical and optoelectrical applications that stem from mechanical stability and tunable optoelectronic properties [98, 99]. More importantly, van der Waals interactions and hydrogen bonding between individual GO sheets are strong enough to maintain robust films, leading to a transfer without polymer support [100]. Li et al. reported the utilization of solution-processed CDG films as soft top-contacts for non-destructive fabrication of molecular junctions in 2012 [101].

In order to prepare reliable molecular junctions, in addition to non-destructive contact with the molecules, the formation of high-quality SAMs, the well-defined surface area of device, and the minimization of parasitic leakage current are all important considerations. The process steps of the graphene-based molecular electronics are schematically shown in Figure 6.23. Thermally evaporated ultra-flat

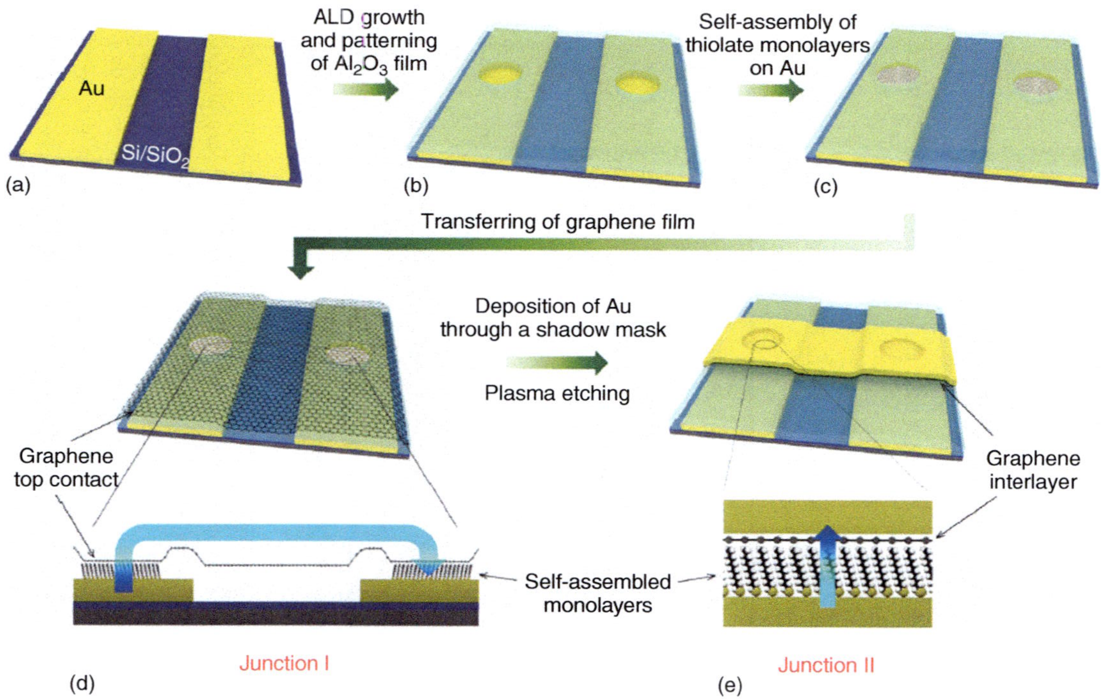

Figure 6.23 Fabrication steps of molecular junctions with CDG films as soft top contacts. Two types of molecular junctions were formed: CDG film acts as conductive interconnects between junctions in series (Junction I) and as an interlayer protecting SAMs from invasive metal damage (Junction II). Source: Li et al. [101]. 2012, John Wiley & Sons.

gold as the bottom electrode, atomic layer deposition (ALD) grown aluminum oxide as the insulating layer, and electron beam lithography (EBL) patterned micropores ensure the production of well-defined SAMs. SAMs were obtained by immersing the freshly prepared substrate in dilute ethanol solution of alkanethiols in inert atmosphere (Ar) for at least two days. After the formation of SAMs, CDG film was transferred to the device to form a top contact, which was so flexible that it can perfectly follow the surface topography of substrate. Finally, two types of molecular junctions were fabricated in this method: removing the unserviceable parts of the CDG films by oxygen plasma to obtain junction I, and evaporating Au through a shadow mask followed by a plasma etch step to form junction II. The results showed that the yield of molecular junctions exceeded 90% and these devices exhibited good operational stability and long life. This protocol would greatly enrich the test beds for molecular electronic devices because of its easy-processing, low-cost, and flexible features.

The tunable optoelectronic properties render GO an attractive material for optoelectrical applications. Li et al. further employed solution-processed ultrathin reduced graphene oxide (rGO) films as soft, transparent top contacts for light switchable solid-state devices [102]. In this study, the double junction geometry was also used where the rGO films acted as soft top contacts and interconnects between SAMs in series, without an additional top metal contact. The flexible, conductive, and transparent rGO films permitted molecular monolayers to be photoswitched in situ while enabling charge transport measurements across the molecules. The double-junction layout with well-defined electrode area improved the device yield and provided a simple reproducible structure.

Relatively less studies have been focused on the molecular junctions based on conjugated polymers (CPs) despite they may transport charge more efficiently than the widely investigated small-molecular systems [103]. Hence, Wang et al. again applied the double junction structure with graphene as the top contact to study the self-assembly manner of long CPs on Au surface and compared with vertical alignment of small molecules (Figure 6.24). Experimental results showed that, unlike the vertically oriented SAMs, CPs are self-assembled into ultrathin films (2–3 nm) in a unique "planar" manner. And they can achieve their molecular-specific features in tunneling junctions, like the less conjugated small molecules. In addition, they demonstrated the effect of redox groups that modified the molecular backbone in modulating charge transport through energy level engineering and external stimuli. This work presents the potential of using CPs instead of typical small-molecule monolayers to build molecular junctions, emphasizing the possibility of incorporating satisfactory functionality into polymeric molecules for switchable electronic devices, which are considered to be the basic elements in molecular electronics.

6.3.2.5 Liquid Metal Contact

Similar to the above techniques, the liquid metal contact is relatively stable and can effectively prevent metal atom penetration or related defects in SAMs, which forms soft top contacts by mechanically manipulating the metal or alloy with a low melting point. Reynolds and coworker [104] first introduced this method to form

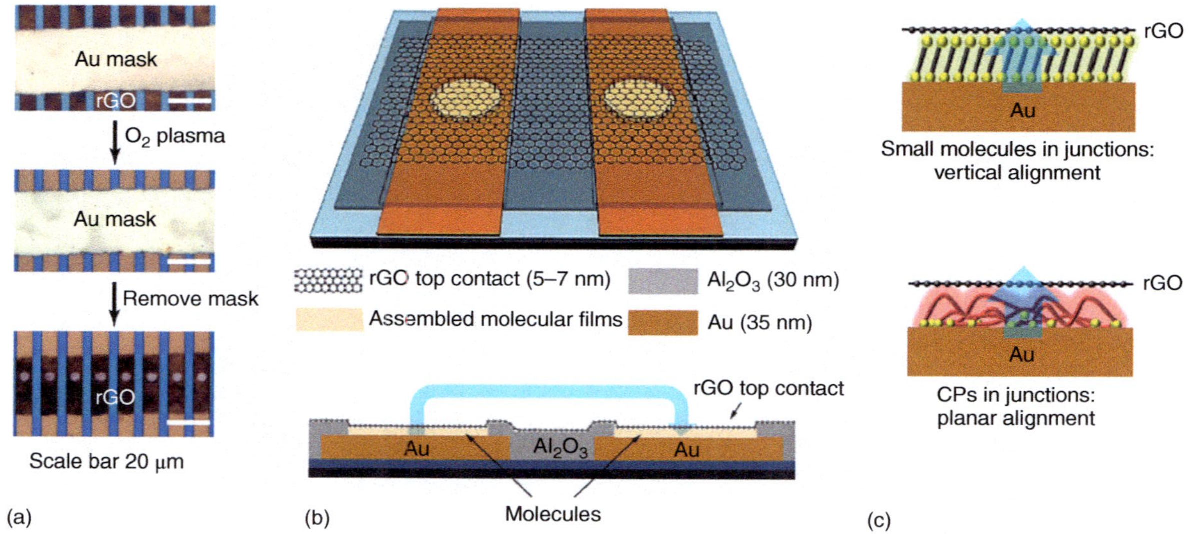

Figure 6.24 (a) Optical images describing the process of removing the unwanted portion of rGO films using thin Au films as masks. (b) Schematic of rGO top-contact test bed for molecular junctions. The blue arrow shows charge transport between adjacent junctions. (c) Schematic of self-assembled small molecules and CPs within the junctions. The blue arrows show charge transport by direct tunneling across the molecular barriers. Source: Wang et al. [103]. Licensed under CC BY 4.0.

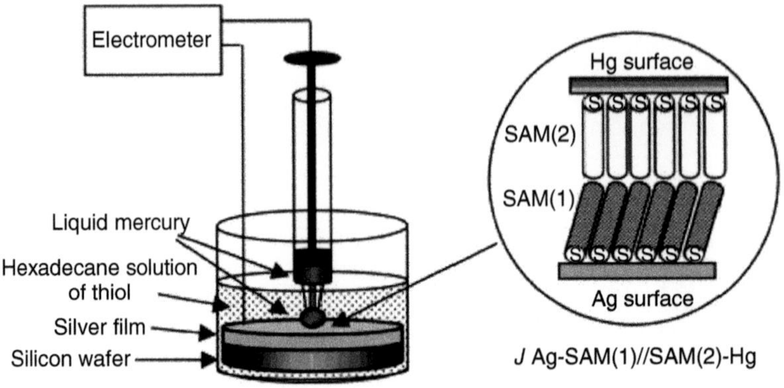

Figure 6.25 Schematic illustration of the fabrication of the Hg–SAMs–Ag junction. Source: Chabinyc et al. [105]. 2002, American Chemical Society.

the molecular junctions where a clean mercury (Hg) drop electrode acted as the top electrode and many different bottom contacts based on barium, cadmium, calcium, copper, and lead were used.

The fabrication of the Hg–molecule–Ag junction is shown in Figure 6.25 [105]. The molecular junction was formed by mechanical contact between a SAM on a mercury drop and a SAM on silver film supported by a silicon wafer. The silver film coated by SAM was placed in a beaker and covered with a solution containing alkanethiol for forming the SAM on the mercury drop. The SAM on the Hg drop was formed in situ on the drop surface suspended from a syringe. The micromanipulator allowed the mechanical contact between the two SAMs. This method can be used to fabricate the ensemble molecular junctions based on both the monolayer and the bilayer SAMs, in which the choice of bottom electrode is broad, including various metals and semiconductors [105–108]. For example, the Hg–SAMs–Si junction can be produced by placing a mercury drop on a monolayer above a silicon wafer using the hanging mercury drop electrode device [108].

There are several benefits to use Hg drop as the electrodes [109]. First, liquid Hg drop is free of structural features, such as edges, terraces, and pits; it can effectively avoid the defects in SAMs and form reproducible junctions so that we can collect statistical experimental data. Second, the Hg drop can follow the topography of the solid surfaces and facilitate a good conformal contact with SAMs on a solid surface. Finally, Hg can carry a variety of molecules and form well-ordered SAMs in a short time after contacting with the target molecules [110]. However, Hg is toxic and harmful to the health of the users. Moreover, the stability of junction is unsatisfactory. Especially, shape changes, electromigration, amalgamating with the bottom electrode, etc. may occur under the applied bias for the electrical measurements [111–113].

In addition to mercury, liquid metal alloy electrodes are another option and may solve the problems posed by mercury to some extent. Whitesides' group initially introduced an idea where EGaIn was used instead of Hg as one of the electrodes and investigated the charge transport characteristics across ensemble molecular

Figure 6.26 A series of photographs during the formation of EGaIn electrode. The pictures are continuous and show the formation of a single tip; this sequence takes less than five seconds. Source: Chiechi et al. [114].

junctions [114]. In fact, we have mentioned EGaIn in Sections 6.2.1 and 6.3.2.2 that it was just a supporting role but now it is the protagonist. The work function of EGaIn (~4.2 eV) is close to Hg (~4.5 eV); however, EGaIn does not flow until it is subjected to the critical surface stress. Figure 6.26 shows a series of photographs of the formation of a conical EGaIn electrode. In their experiment, an EGaIn drop suspended from the syringe needle was brought into contact with the bare Ag surface, and then the syringe was raised slowly until the EGaIn sperated into a conical tip. The EGaIn tip protruding from the needle did not retract into a semispherical droplet, which means the contact area can be significantly reduced. Finally, the EGaIn tip was brought into contact with the SAMs by micromanipulator. The thin Ga_2O_3 interlayer (1–2 nm), which is formed by oxidation of EGaIn on the surface, may play an important role in the system. It improved the stability of the molecular junction as a protective layer resulting in high device yields through preventing the formation of metal filaments. The effect of Ga_2O_3 layer has been extensively studied, and it was confirmed that the molecular structure, not the characteristics of the Ga_2O_3 film, dominated the electron transport bchavior [115]. Whitesides and coworkers further proposed a method to fabricate small arrays of tunneling junctions based on EGaIn electrodes by combining optical lithography. These junctions possessed liquid EGaIn top electrodes stabilized in microchannels and ultraflat Ag^{TS} bottom electrodes [85]. Furthermore, these junctions based on EGaIn with high yield (70–90%) did not need clean rooms, ultrahigh vacuum, and toxic substances, showing the potential to be better than mercury. However, this technique also has some drawbacks that hinder its viability for the practical application of molecule devices. The fabrication method for the molecular junctions still needs external equipment such as the Hg electrode-based technique. And, it is difficult to obtain a reproducible junction size that determines the electrical characteristics of the junction.

6.4 Experimental Progress with Ensemble Molecular Junctions

Molecular electronics provides fundamental research on intrinsic molecular structure–property relationship [116, 117] and possesses the great potential for

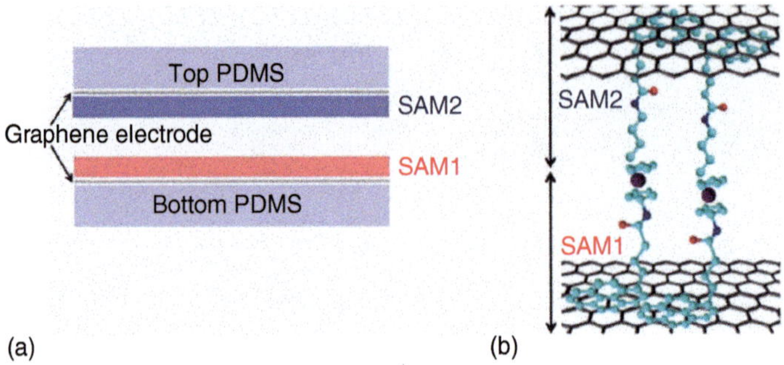

Figure 6.27 (a) An equivalent structure diagram for functional bilayer molecular junctions. (b) Schematics of molecular junctions comprising double SAMs on graphene. Source: Seo et al. [119]. 2017, John Wiley & Sons.

the miniaturization of electronics [118]. Recently, exciting research results and significant experimental advances further stimulated the enthusiasm of scientists and promoted the realization of the good prospect of molecular electronics.

Seo et al. [119] introduced functional molecular junctions derived from double SAMs intercalated between soft graphene electrodes. The bilayer molecular junctions were made by using a contact assembly technique to place a SAM/(top) electrode on another SAM/(bottom) electrode. In their experiment, a flexible PDMS substrate with patterned graphene electrodes was modified by SAM1 in good contact with another SAM2-coated PDMS substrate with patterned graphene electrodes, as shown in Figure 6.27. And they achieved crossbar molecular junctions by simple and reproducible position alignment in which the SAM-coated top and bottom electrode arrays were vertically positioned and assembled in contact with each other. Double SAMs can offer tunneling conjugation across the van der Waals gaps between the ends of each monolayer and display new electrical properties such as rectification. In addition, this contact assembly method makes it possible to form new functional molecular junctions with double SAMs and enables the formation of simple, stable, repeatable, and length-controlled molecular junctions by using either one or two types of SAMs.

McCreery group [120] also applied bilayer molecular junctions to investigate the electronic symmetry and rectification through energy level changes. The bilayers (bis-thienyl benzene [BTB]/2-anthraquinone [AQ] or BTB/NDI [a derivative of phenyl naphthalene di-imide]) were deposited on PPF bottom electrodes through successive electrochemical reduction of diazonium reagents; then, an electron beam deposited carbon (eC)/Au top contact was applied to complete a solid-state molecular junction (Figure 6.28a). The current density vs. applied voltage (J–V) curves showed assymmetry with respect to polarity for bilayer molecular junctions, while the molecular junctions containing only one molecular species were symmetric; the clearer result is shown in Figure 6.28b. When the energy levels of the two molecular layers differed, there was a pronounced rectification and the direction of the rectification will reverse with the reversal of the molecular layer. The result was an unambiguous evidence of a “molecular signature” where

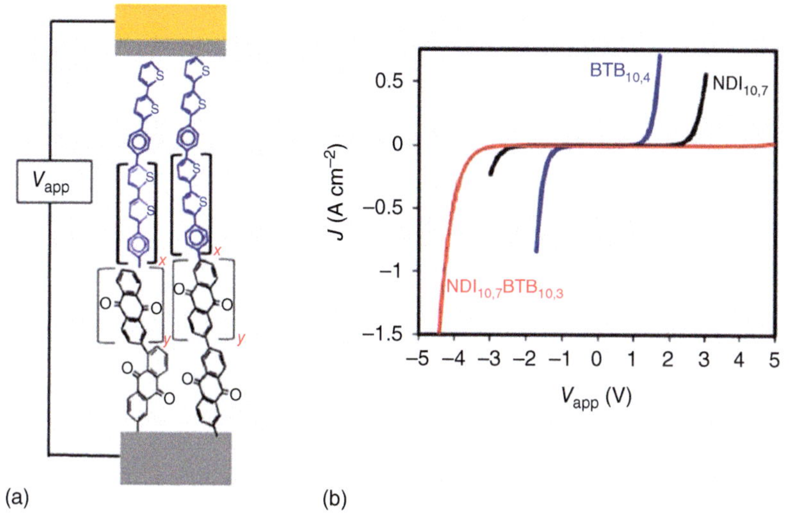

Figure 6.28 (a) Schematic of a bilayer molecular junction with an initial layer of AQ on PPF, followed by a second layer of BTB, and *x* and *y* represent different lengths of oligomers. (b) Current density vs. bias voltage (*J*–*V*) curves of NDI/BTB, NDI-only, and BTB-only junctions. Source: Bayat et al. [120]. 2016, American Chemical Society.

electronic behavior is directly influenced by molecular structure and orbital energies.

In addition to graphene, PPF and eC mentioned above are also valuable sp^2-hybridized carbon material. The SAMs with diazonium moiety can be deposited on the eC by electrochemical reduction of the corresponding diazonium ions and retaining a covalent C—C bond. McCreery group [121] still used Au/eC substrate and roughly same molecular system to observe the light emission in carbon-based molecular junctions and proposed a direct measure of energy losses occurring during charge transport. The results indicated that charge transport transitions from tunneling to hopping as the increase of device thickness. The energy loss depended strongly on molecular structure, demonstrating the "molecular signature" relating molecular structure to electronic behavior.

Trasobares et al. [122] introduced frequency response to molecular rectifier or diodes. They demonstrated radio frequency (RF) molecular diodes operated up to 17 GHz with an estimated cut-off frequency of 520 GHz. Direct current and RF performances were simultaneously measured on a large array of molecular junctions consisting of gold nanocrystal electrodes, ferrocenyl alkanethiol molecules, and the tip of an interferometric scanning microwave microscope. The results suggested that the RF-molecular diodes are very attractive for scaling and high-frequency operation and have the potential to operate in the terahertz range. Moreover, Nijhuis and coworkers [123] utilized frequency response based on an equivalent circuit to analyze impedance data and study the nature of the SAM–electrode interface by temperature-dependent and potentiodynamic impedance spectroscopy. They gave insights into how each component of two-terminal SAM-based junctions impeded charge transfer so that we can better establish the mechanism of charge transport. In addition, Nijhuis group [124] described a novel approach to build arrays of high-quality SAM-based junctions by fabricating microfluidic top electrodes, in which the liquid metal of GaO_x/EGaIn mechanically stabilized in microchannels and formed good electrical contacts with the SAMs without causing damage. The good mechanical stability and well-defined geometrical contact area of the microfluidic top electrodes ensured the relatively accurate measurement. And they explored the charge transport mechanism in the molecular junctions [125] and application in molecular diode based logic.

Farnum et al. applied self-assembled molecular p/n junctions to dye-sensitized solar energy conversion [126]. In detail, molecular chromophores (C), electron acceptors (A), and electron donors (D) were self-assembled on the surfaces of transparent, mesoporous conductive indium tin oxide nanoparticle (nanoITO) electrodes to form photoanode (nanoITO|–A–C–D) and photocathode (nanoITO|–D–C–A) assemblies. The excitation of the designed molecular assemblies is used to create controllable redox potential gradients directing the flow of light-driven intra-assembly electron transfer. And the implementation of this method is based on modular layer-by-layer structure (Figure 6.29); it enables the incorporation of light absorbers, electron-transfer mediators, and catalysts by a simple dipping process. The interfacial redox-separated (RS) lifetimes that lasted for milliseconds to seconds in their photoanode assemblies are comparable to those observed for dye-sensitized TiO_2 and core–shell SnO_2–TiO_2 photoanodes [127–129]. Even better, the photocathode assemblies exhibited interfacial RS lifetimes in the millisecond range, which is unprecedented compared with previous reports of photoinduced redox separation for dye-sensitized NiO photocathodes [130–133]. This research using high-surface-area nanoITO electrodes pointed to a new chemical strategy for dye-sensitized solar energy conversion based on molecular components rather than the traditional wide-bandgap semiconductors.

We introduced in Section 6.3.2.4 that rGO is a favorable alternative to graphene. The rGO-based molecular junctions with the characteristics of high device yield, good stability, and accessible processability have wide applications in functional molecular electronics. Lee and coworkers [134] fabricated reversible photoswitching molecular electronics with diarylethene SAMs and rGO top electrodes on flexible substrates (Figure 6.30a). It was generally accepted that it is not easy to achieve reversible switching. Low transmittance of the top electrodes, strong electronic coupling between the SAMs and bottom electrodes, lack of enough space for molecular conformational changes will cause the failure of reversible switching [135–138]. In their study, conducting and transparent rGO film with enhanced transmittance facilitated pronounced reversible switching behavior of device when exposed to UV or visible light (Figure 6.30b). Moreover, these molecular devices also showed good longtime stability and reliable electrical properties under various mechanical deformations.

Additionally, other feats achieving by unremitting efforts of scientists further open new horizons in molecular electronics, such as thorough understanding of molecular interfacial spin effects as well as the development and tailoring

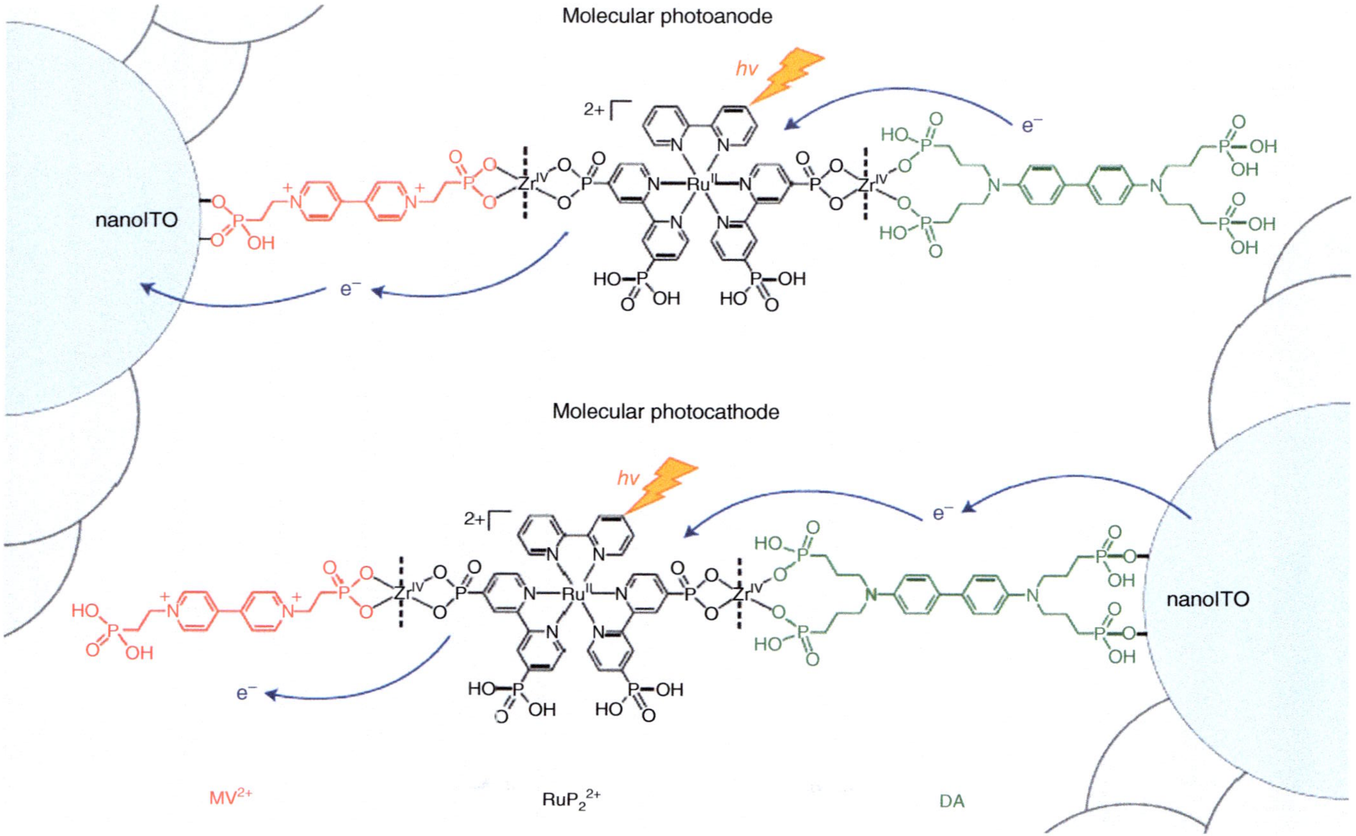

Figure 6.29 Schematic drawing of layer-by-layer molecular p/n junction assemblies on nanoITO electrodes. Molecular photoanodes and photocathodes composed of individual molecular chromophore (RuP_2^{2+}), electron acceptor (MV^{2+}), and electron donor on the surface of nanoITO. Arrows represent the direction of light-driven electron-transfer reactions. Source: Farnum et al. [126]. 2016, Springer Nature.

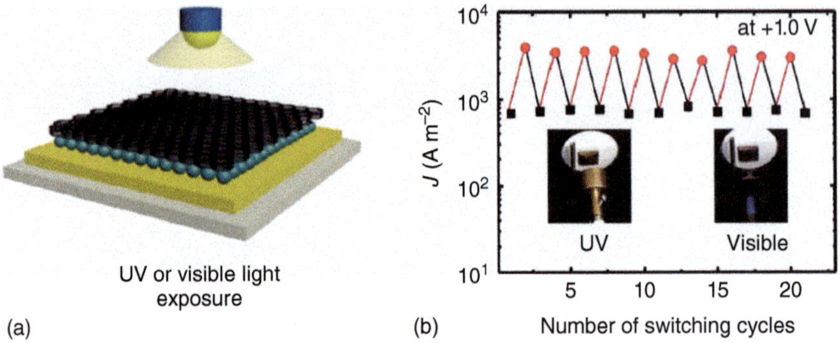

Figure 6.30 (a) Schematic illustration of photoswitching molecular devices with rGO top electrodes on a flexible substrate. (b) The reversible photoswitching phenomenon repeated 20 times by alternating UV and visible light irradiation. Source: Kim et al. [134]. 2015, John Wiley & Sons.

of molecular spinterface [139]; the progress of comprehending how heat is transported, dissipated, and converted into electricity in molecular junctions for designing high-performance thermoelectrics [140]; the appearance of "building block" approach to build robust and desirable multifunctional molecular junctions [141]; the statistical analysis for electrical characterization of molecular electronics through length/temperature variation, transition voltage spectroscopy (TVS), or inelastic electron tunneling spectroscopy (IETS) [142, 143].

6.5 Outlook

Since the advent of molecular electronics, it has attracted great attentions from scientists and engineers in different disciplines and fields and achieved tremendous advances in theory and experiment. Molecular electronics carries the future of next-generation electrical circuit units with lower power consumption and fabrication cost, higher speed, and integration density [144, 145]. Although molecular electronics has been booming, there is still a long way to achieve commercially available applications. We introduced several advanced approaches for forming top contacts to fabricate high-yield and relatively large-area molecular junctions, which is valuable but still with some drawbacks, just as we have never heard of perfect things. In addition, the electronic functionality of the molecule may be quenched due to strong coupling between molecule and electrode, while such strong molecular bonds are the guarantee for stability of molecular junctions. You cannot have your cake and eat it too, we need to modestly control the interface characteristics according to the purposes. The behaviors of molecular devices are variable and closely related to a variety of factors such as electrode fabrication process, electrode materials properties, contact chemistry, number of molecules, and the thickness of SAMs; the high degree of uncertainty in molecular devices brings challenges to real understanding of the structure–function relationships for the electronic transport through molecular junctions. And we should take the operational stability of the device into consideration. In general, the passing current of junction is sensitive to the environmental changes, and every small change that is unnoticed may cause a large change in molecular devices. Secondly, the defects of SAMs and the aging of molecules also have great effects on the performance of the device.

Philosophers told us that the development of things is the unification of progressiveness and tortuousness, and molecular electronics is no exception. Not only have we experienced the fruits of vigorous development but also constantly accepted the challenges. Molecular electronics serve as a masterpiece of the wisdom and creativity of scientists in various fields around the world, will enter our real life, and become a milestone in human history.

References

1 Love, J.C., Estroff, L.A., Kriebel, J.K. et al. (2005). Self-assembled monolayers of thiolates on metals as a form of nanotechnology. *Chem. Rev.* 105: 1103–1170.

2 Ulman, A. (1996). Formation and structure of self-assembled monolayers. *Chem. Rev.* 96: 1533–1554.

3 Smith, R.K., Lewis, P.A., and Weiss, P.S. (2004). Patterning self-assembled monolayers. *Prog. Surf. Sci.* 75: 1–68.

4 Hipps, K.W. (2001). Molecular electronics. It's all about contacts. *Science* 294: 536–537.

5 Ho Choi, S., Kim, B., and Frisbie, C.D. (2008). Electrical resistance of long conjugated molecular wires. *Science* 320: 1482–1486.

6 Kubatkin, S., Danilov, A., Hjort, M. et al. (2003). Single-electron transistor of a single organic molecule with access to several redox states. *Nature* 425: 698.

7 Li, T., Hu, W., and Zhu, D. (2010). Nanogap electrodes. *Adv. Mater.* 22: 286–300.

8 Reed, M.A. (1997). Conductance of a molecular junction. *Science* 278: 252–254.

9 Van Hal, P.A., Smits, E.C., Geuns, T.C. et al. (2008). Upscaling, integration and electrical characterization of molecular junctions. *Nat. Nanotechnol.* 3: 749–754.

10 McCreery, R.L. and Bergren, A.J. (2009). Progress with molecular electronic junctions: meeting experimental challenges in design and fabrication. *Adv. Mater.* 21: 4303–4322.

11 Akkerman, H.B. and de Boer, B. (2008). Electrical conduction through single molecules and self-assembled monolayers. *J. Phys. Condens. Matter* 20: 013001.

12 Weiss, E.A., Kaufman, G.K., Kriebel, J.K. et al. (2007). Si/SiO_2-templated formation of ultraflat metal surfaces on glass, polymer, and solder supports: their use as substrates for self-assembled monolayers. *Langmuir* 23: 9686–9694.

13 He, X.P., Wang, X.W., Jin, X.P. et al. (2011). Epimeric monosaccharide-quinone hybrids on gold electrodes toward the electrochemical probing of specific carbohydrate-protein recognitions. *J. Am. Chem. Soc.* 133: 3649–3657.

14 Nara, J., Higai, S., Morikawa, Y., and Ohno, T. (2004). Density functional theory investigation of benzenethiol adsorption on Au(111). *J. Chem. Phys.* 120: 6705–6711.

15 Sun, L. and Crooks, R.M. (1993). Indirect visualization of defect structures contained within self-assembled organomercaptan monolayers: combined use of electrochemistry and scanning tunneling microscopy. *Langmuir* 9: 1951–1954.

16 Tachibana, M., Yoshizawa, K., Ogawa, A. et al. (2002). Sulfur– gold orbital interactions which determine the structure of alkanethiolate/Au(111) self-assembled monolayer systems. *J. Phys. Chem. B* 106: 12727–12736.

17 Bumm, L., Arnold, J., Cygan, M. et al. (1996). Are single molecular wires conducting? *Science* 271: 1705–1707.

18 Lu, Q., Liu, K., Zhang, H. et al. (2009). From tunneling to hopping: a comprehensive investigation of charge transport mechanism in molecular junctions based on oligo(p-phenylene ethynylene)s. *ACS Nano* 3: 3861–3868.

19 Jennum, K., Vestergaard, M., Pedersen, A.H. et al. (2011). Synthesis of oligo(phenyleneethynylene)s with vertically disposed tetrathiafulvalene units. *Synthesis* 2011: 539.

20 Lissau, H., Frisenda, R., Olsen, S.T. et al. (2015). Tracking molecular resonance forms of donor–acceptor push–pull molecules by single-molecule conductance experiments. *Nat. Commun.* 6: 10233.

21 Parker, C.R., Leary, E., Frisenda, R. et al. (2014). A comprehensive study of extended tetrathiafulvalene cruciform molecules for molecular electronics: synthesis and electrical transport measurements. *J. Am. Chem. Soc.* 136: 16497–16507.

22 Sørensen, J.K., Vestergaard, M., Kadziola, A. et al. (2006). Synthesis of oligo(phenyleneethynylene)-tetrathiafulvalene cruciforms for molecular electronics. *Org. Lett.* 8: 1173–1176.

23 Vestergaard, M., Jennum, K., Sørensen, J.K. et al. (2008). Synthesis and characterization of cruciform-conjugated molecules based on tetrathiafulvalene. *J. Org. Chem.* 73: 3175–3183.

24 Wei, Z., Hansen, T., Santella, M. et al. (2015). Molecular heterojunctions of oligo(phenylene ethynylene)s with linear to cruciform framework. *Adv. Funct. Mater.* 25: 1700–1708.

25 Wei, Z., Li, T., Jennum, K. et al. (2012). Molecular junctions based on SAMs of cruciform oligo(phenylene ethynylene)s. *Langmuir* 28: 4016–4023.

26 Liu, Y., Santella, M., Fan, Z. et al. (2018). Diamine anchored molecular junctions of oligo(phenylene ethynylene) cruciform. *Chin. Chem. Lett.* 29: 271–275.

27 Wei, Z., Wang, X., Borges, A. et al. (2014). Triazatriangulene as binding group for molecular electronics. *Langmuir* 30: 14868–14876.

28 Laursen, B.W. and Krebs, F.C. (2000). Synthesis of a triazatriangulenium salt. *Angew. Chem.* 112: 3574–3576.

29 Laursen, B.W. and Krebs, F.C. (2001). Synthesis, structure, and properties of azatriangulenium salts. *Chem. Eur. J.* 7: 1773–1783.

30 Bosson, J., Gouin, J., and Lacour, J. (2014). Cationic triangulenes and helicenes: synthesis, chemical stability, optical properties and extended applications of these unusual dyes. *Chem. Soc. Rev.* 43: 2824–2840.

31 Laursen, B.W., Krebs, F.C., Nielsen, M.F. et al. (1998). 2,6,10-tris(dialkylamino) trioxatriangulenium ions. Synthesis, structure, and properties of exceptionally stable carbenium ions. *J. Am. Chem. Soc.* 120: 12255–12263.

32 Lofthagen, M., Chadha, R., and Siegel, J.S. (1991). Synthesis, structures, and dynamics of a macrocyclophane. *J. Am. Chem. Soc.* 113: 8785–8790.

33 Lacher, S., Matsuo, Y., and Nakamura, E. (2011). Molecular and supramolecular control of the work function of an inorganic electrode with self-assembled monolayer of umbrella-shaped fullerene derivatives. *J. Am. Chem. Soc.* 133: 16997–17004.

34 Rovida, G. and Pratesi, F. (1981). Sulfur overlayers on the low-index faces of silver. *Surf. Sci.* 104: 609–624.

35 Schwaha, K., Spencer, N., and Lambert, R. (1979). A single crystal study of the initial stages of silver sulphidation: the chemisorption and reactivity of molecular sulphur (S_2) on Ag(111). *Surf. Sci.* 81: 273–284.

36 Nijhuis, C.A., Reus, W.F., and Whitesides, G.M. (2010). Mechanism of rectification in tunneling junctions based on molecules with asymmetric potential drops. *J. Am. Chem. Soc.* 132: 18386–18401.

37 Thuo, M.M., Reus, W.F., Nijhuis, C.A. et al. (2011). Odd-even effects in charge transport across self-assembled monolayers. *J. Am. Chem. Soc.* 133: 2962–2975.

38 Campos, M.A., Trilling, A.K., Yang, M. et al. (2011). Self-assembled functional organic monolayers on oxide-free copper. *Langmuir* 27: 8126–8133.

39 Haneda, R. and Aramaki, K. (1998). Protection of copper corrosion by an ultrathin two-dimensional polymer film of alkanethiol monolayer. *J. Electrochem. Soc.* 145: 1856–1861.

40 Krapchetov, D.A., Ma, H., Jen, A.K. et al. (2008). Deprotecting thioacetyl-terminated terphenyldithiol for assembly on gallium arsenide. *Langmuir* 24: 851–856.

41 Yang, M., Teeuwen, R.L., Giesbers, M. et al. (2008). One-step photochemical attachment of NHS-terminated monolayers onto silicon surfaces and subsequent functionalization. *Langmuir* 24: 7931–7938.

42 Dong, J., Wang, A., Ng, K.Y.S., and Mao, G. (2006). Self-assembly of octadecyltrichlorosilane monolayers on silicon-based substrates by chemical vapor deposition. *Thin Solid Films* 515: 2116–2122.

43 Lee, J.S., Chi, Y.S., Choi, I.S., and Kim, J. (2012). Local scanning probe polymerization of an organic monolayer covalently grafted on silicon. *Langmuir* 28: 14496–14501.

44 Mutin, P.H., Guerrero, G., and Vioux, A. (2005). Hybrid materials from organophosphorus coupling molecules. *J. Mater. Chem.* 15: 3761–3768.

45 Acton, O., Hutchins, D., Arnadottir, L. et al. (2011). Spin-cast and patterned organophosphonate self-assembled monolayer dielectrics on metal-oxide-activated Si. *Adv. Mater.* 23: 1899–1902.

46 Knesting, K.M., Hotchkiss, P.J., Macleod, B.A. et al. (2012). Spatially modulating interfacial properties of transparent conductive oxides: patterning work function with phosphonic acid self-assembled monolayers. *Adv. Mater.* 24: 642–646.

47 Novak, M., Ebel, A., Meyer-Friedrichsen, T. et al. (2011). Low-voltage p- and n-type organic self-assembled monolayer field effect transistors. *Nano Lett.* 11: 156–159.

48 Dimitrakopoulos, C.D. and Malenfant, P.R. (2002). Organic thin film transistors for large area electronics. *Adv. Mater.* 14: 99–117.

49 Katz, H. and Bao, Z. (2000). The physical chemistry of organic field-effect transistors. *Phys. Chem. B* 104: 671–678.

50 Katz, H.E., Bao, Z., and Gilat, S.L. (2001). Synthetic chemistry for ultrapure, processable, and high-mobility organic transistor semiconductors. *Acc. Chem. Res.* 34: 359–369.

51 Tulevski, G.S., Miao, Q., Fukuto, M. et al. (2004). Attaching organic semiconductors to gate oxides: in situ assembly of monolayer field effect transistors. *J. Am. Chem. Soc.* 126: 15048–15050.

52 Roberts, G.G. (2006). An applied science perspective of Langmuir–Blodgett films. *Adv. Phys.* 34: 475–512.

53 Zasadzinski, J., Viswanathan, R., Madsen, L. et al. (1994). Langmuir–Blodgett films. *Science* 263: 1726–1733.

54 Kang, Y.J., Kim, H., Kim, E. et al. (2008). Fabrication and structural analysis of aromatic polyimide ultra-thin films with different fluorine contents. *Colloids Surf., A* 313: 585–589.

55 Petty, M. (1992). Possible applications for Langmuir–Blodgett films. *Thin Solid Films* 210: 417–426.

56 Mann, B. and Kuhn, H. (1971). Tunneling through fatty acid salt monolayers. *J. Appl. Phys.* 42: 4398–4405.

57 Roberts, G. (1990). *Langmuir–Blodgett Films*. New York: Plenum Press.

58 Chen, X., Lenhert, S., Hirtz, M. et al. (2007). Langmuir–Blodgett patterning: a bottom–up way to build mesostructures over large areas. *Acc. Chem. Res.* 40: 393–401.

59 Virkar, A., Mannsfeld, S., Oh, J.H. et al. (2009). The role of OTS density on pentacene and C_{60} nucleation, thin film growth, and transistor performance. *Adv. Funct. Mater.* 19: 1962–1970.

60 Ulman, A. (2013). *An Introduction to Ultrathin Organic Films: From Langmuir–Blodgett to Self-Assembly*. Academic Press.

61 Tao, N. (2006). Electron transport in molecular junctions. *Nat. Nanotechnol.* 1: 173.

62 Moonen, N.N.P., Flood, A.H., Fernández, J.M., and Stoddart, J.F. (2005). Towards a rational design of molecular switches and sensors from their basic building blocks. *Mol. Mach.* 262: 99–132.

63 Aviram, A. and Ratner, M.A. (1974). Molecular rectifiers. *Chem. Phys. Lett.* 29: 277–283.

64 Martin, A., Sambles, J., and Ashwell, G. (1993). Molecular rectifier. *Phys. Rev. Lett.* 70: 218.

65 Ng, M.K. and Yu, L. (2002). Synthesis of amphiphilic conjugated diblock oligomers as molecular diodes. *Angew. Chem. Int. Ed.* 41: 3598–3601.

66 Haick, H., Niitsoo, O., Ghabboun, J., and Cahen, D. (2007). Electrical contacts to organic molecular films by metal evaporation: effect of contacting details. *J. Phys. Chem. C* 111: 2318–2329.

67 De Boer, B., Frank, M.M., Chabal, Y.J. et al. (2004). Metallic contact formation for molecular electronics: interactions between vapor-deposited metals and self-assembled monolayers of conjugated mono-and dithiols. *Langmuir* 20: 1539–1542.

68 Haynie, B., Walker, A., Tighe, T. et al. (2003). Adventures in molecular electronics: how to attach wires to molecules. *Appl. Surf. Sci.* 203: 433–436.

69 Ohgi, T., Sheng, H.-Y., Dong, Z.-C., and Nejoh, H. (1999). Observation of Au deposited self-assembled monolayers of octanethiol by scanning tunneling microscopy. *Surf. Sci.* 442: 277–282.

70 Fisher, G.L., Walker, A., Tighe, T. et al. (2002). Bond insertion, complexation, and penetration pathways of vapor-deposited aluminum atoms with HO-and CH_3O-terminated organic monolayers. *J. Am. Chem. Soc.* 124: 5528–5541.

71 Walker, A.V., Tighe, T.B., Cabarcos, O.M. et al. (2004). The dynamics of noble metal atom penetration through methoxy-terminated alkanethiolate monolayers. *J. Am. Chem. Soc.* 126: 3954–3963.

72 Haick, H. and Cahen, D. (2008). Contacting organic molecules by soft methods: towards molecule-based electronic devices. *Acc. Chem. Res.* 41: 359–366.

73 Bonifas, A.P. and McCreery, R.L. (2010). 'Soft' Au, Pt and Cu contacts for molecular junctions through surface-diffusion-mediated deposition. *Nat. Nanotechnol.* 5: 612–617.

74 Bonifas, A.P. and McCreery, R.L. (2011). Assembling molecular electronic junctions one molecule at a time. *Nano Lett.* 11: 4725–4729.

75 Akkerman, H.B., Blom, P.W., de Leeuw, D.M., and de Boer, B. (2006). Towards molecular electronics with large-area molecular junctions. *Nature* 441: 69–72.

76 Wang, G., Kim, Y., Choe, M. et al. (2011). A new approach for molecular electronic junctions with a multilayer graphene electrode. *Adv. Mater.* 23: 755–760.

77 He, Z., Zhang, W., Tang, Y. et al. (2009). Crossbar heterojunction field effect transistors of CdSe: in nanowires and Si nanoribbons. *Appl. Phys. Lett.* 95: 253107.

78 Linn, E., Rosezin, R., Tappertzhofen, S. et al. (2012). Beyond von Neumann—logic operations in passive crossbar arrays alongside memory operations. *Nanotechnology* 23: 305205.

79 Sanetra, N., Karipidou, Z., Wirtz, R. et al. (2012). Printing of highly integrated crossbar junctions. *Adv. Funct. Mater.* 22: 1129–1135.

80 Kushmerick, J., Naciri, J., Yang, J., and Shashidhar, R. (2003). Conductance scaling of molecular wires in parallel. *Nano Lett.* 3: 897–900.

81 Blum, A.S., Kushmerick, J.G., Pollack, S.K. et al. (2004). Charge transport and scaling in molecular wires. *J. Phys. Chem. B* 108: 18124–18128.

82 Beebe, J.M., Kim, B., Gadzuk, J.W. et al. (2006). Transition from direct tunneling to field emission in metal-molecule-metal junctions. *Phys. Rev. Lett.* 97: 026801.

83 Kim, B., Beebe, J.M., Olivier, C. et al. (2007). Temperature and length dependence of charge transport in redox-active molecular wires incorporating ruthenium(II) bis(σ-arylacetylide) complexes. *J. Phys. Chem. C* 111: 7521–7526.

84 Lam, H.Y., Gergel-Hackett, N., Zangmeister, C.D. et al. (2008). Molecule-induced interface states dominate charge transport in Si–alkyl–metal junctions. *J. Phys. Condens. Matter* 20: 374114.

85 Nijhuis, C.A., Reus, W.F., Barber, J.R. et al. (2010). Charge transport and rectification in arrays of SAM-based tunneling junctions. *Nano Lett.* 10: 3611–3619.

86 Hsu, J.W. (2005). Soft lithography contacts to organics. *Mater. Today* 8: 42–54.

87 Loo, Y.-L., Lang, D.V., Rogers, J.A., and Hsu, J.W. (2003). Electrical contacts to molecular layers by nanotransfer printing. *Nano Lett.* 3: 913–917.

88 Zhang, Y., Zhao, Z., Fracasso, D., and Chiechi, R.C. (2014). Bottom-up molecular tunneling junctions formed by self-assembly. *Isr. J. Chem.* 54: 513–533.

89 Coll, M., Miller, L.H., Richter, L.J. et al. (2009). Formation of silicon-based molecular electronic structures using flip-chip lamination. *J. Am. Chem. Soc.* 131: 12451–12457.

90 Novoselov, K.S., Geim, A.K., Morozov, S.V. et al. (2004). Electric field effect in atomically thin carbon films. *Science* 306: 666–669.

91 Hernandez, Y., Nicolosi, V., Lotya, M. et al. (2008). High-yield production of graphene by liquid-phase exfoliation of graphite. *Nat. Nanotechnol.* 3: 563.

92 Coraux, J., N'Diaye, A.T., Busse, C., and Michely, T. (2008). Structural coherency of graphene on Ir(111). *Nano Lett.* 8: 565–570.

93 Reina, A., Jia, X., Ho, J. et al. (2008). Large area, few-layer graphene films on arbitrary substrates by chemical vapor deposition. *Nano Lett.* 9: 30–35.

94 Kim, K.S., Zhao, Y., Jang, H. et al. (2009). Large-scale pattern growth of graphene films for stretchable transparent electrodes. *Nature* 457: 706.

95 Eda, G., Fanchini, G., and Chhowalla, M. (2008). Large-area ultrathin films of reduced graphene oxide as a transparent and flexible electronic material. *Nat. Nanotechnol.* 3: 270.

96 Li, D., Müller, M.B., Gilje, S. et al. (2008). Processable aqueous dispersions of graphene nanosheets. *Nat. Nanotechnol.* 3: 101.

97 Tung, V.C., Allen, M.J., Yang, Y., and Kaner, R.B. (2009). High-throughput solution processing of large-scale graphene. *Nat. Nanotechnol.* 4: 25.

98 Eda, G. and Chhowalla, M. (2010). Chemically derived graphene oxide: towards large-area thin-film electronics and optoelectronics. *Adv. Mater.* 22: 2392–2415.

99 Yamaguchi, H., Eda, G., Mattevi, C. et al. (2010). Highly uniform 300 mm wafer-scale deposition of single and multilayered chemically derived graphene thin films. *ACS Nano* 4: 524–528.

100 Robinson, J.T., Zalalutdinov, M., Baldwin, J.W. et al. (2008). Wafer-scale reduced graphene oxide films for nanomechanical devices. *Nano Lett.* 8: 3441–3445.

101 Li, T., Hauptmann, J.R., Wei, Z. et al. (2012). Solution-processed ultrathin chemically derived graphene films as soft top contacts for solid-state molecular electronic junctions. *Adv. Mater.* 24: 1333–1339.

102 Li, T., Jevric, M., Hauptmann, J.R. et al. (2013). Ultrathin reduced graphene oxide films as transparent top-contacts for light switchable solid-state molecular junctions. *Adv. Mater.* 25: 4164–4170.

103 Wang, Z., Dong, H., Li, T. et al. (2015). Role of redox centre in charge transport investigated by novel self-assembled conjugated polymer molecular junctions. *Nat. Commun.* 6: 7478.

104 Race, H. and Reynolds, S. (1939). Electrical properties of multimolecular films. *J. Am. Chem. Soc.* 61: 1425–1432.

105 Chabinyc, M.L., Chen, X., Holmlin, R.E. et al. (2002). Molecular rectification in a metal–insulator–metal junction based on self-assembled monolayers. *J. Am. Chem. Soc.* 124: 11730–11736.

106 Duati, M., Grave, C., Tcbeborateva, N. et al. (2006). Electron transport across hexa-peri-hexabenzocoronene units in a metal–self-assembled monolayer–metal junction. *Adv. Mater.* 18: 329–333.

107 Grave, C., Tran, E., Samorì, P. et al. (2004). Correlating electrical properties and molecular structure of SAMs organized between two metal surfaces. *Synth. Met.* 147: 11–18.

108 Yaffe, O., Scheres, L., Puniredd, S.R. et al. (2009). Molecular electronics at metal/semiconductor junctions. Si inversion by sub-nanometer molecular films. *Nano Lett.* 9: 2390–2394.

109 Rampi, M.A. and Whitesides, G.M. (2002). A versatile experimental approach for understanding electron transport through organic materials. *Chem. Phys.* 281: 373–391.

110 Von Wrochem, F., Gao, D., Scholz, F. et al. (2010). Efficient electronic coupling and improved stability with dithiocarbamate-based molecular junctions. *Nat. Nanotechnol.* 5: 618.

111 Ghosh, S., Halimun, H., Mahapatro, A.K. et al. (2005). Device structure for electronic transport through individual molecules using nanoelectrodes. *Appl. Phys. Lett.* 87: 233509.

112 Heersche, H.B., Lientschnig, G., O'Neill, K. et al. (2007). In situ imaging of electromigration-induced nanogap formation by transmission electron microscopy. *Appl. Phys. Lett.* 91: 072107.

113 Trouwborst, M., Van Der Molen, S., and Van Wees, B. (2006). The role of Joule heating in the formation of nanogaps by electromigration. *J. Appl. Phys.* 99: 114316.

114 Chiechi, R.C., Weiss, E.A., Dickey, M.D., and Whitesides, G.M. (2008). Eutectic gallium–indium (EGaIn): a moldable liquid metal for electrical characterization of self-assembled monolayers. *Angew. Chem.* 120: 148–150.

115 Reus, W.F., Thuo, M.M., Shapiro, N.D. et al. (2012). The SAM, not the electrodes, dominates charge transport in metal-monolayer//Ga_2O_3/gallium–indium eutectic junctions. *ACS Nano* 6: 4806–4822.

116 Moth-Poulsen, K. and Bjørnholm, T. (2009). Molecular electronics with single molecules in solid-state devices. *Nat. Nanotechnol.* 4: 551.

117 Song, H., Reed, M.A., and Lee, T. (2011). Single molecule electronic devices. *Adv. Mater.* 23: 1583–1608.

118 Cui, A., Dong, H., and Hu, W. (2015). Nanogap electrodes towards solid state single-molecule transistors. *Small* 11: 6115–6141.

119 Seo, S., Hwang, E., Cho, Y. et al. (2017). Functional molecular junctions derived from double self-assembled monolayers. *Angew. Chem. Int. Ed.* 56: 12122–12126.

120 Bayat, A., Lacroix, J.C., and McCreery, R.L. (2016). Control of electronic symmetry and rectification through energy level variations in bilayer molecular junctions. *J. Am. Chem. Soc.* 138: 12287–12296.

121 Ivashenko, O., Bergren, A.J., and McCreery, R.L. (2016). Light emission as a probe of energy losses in molecular junctions. *J. Am. Chem. Soc.* 138: 722–725.

122 Trasobares, J., Vuillaume, D., Theron, D., and Clement, N. (2016). A 17 GHz molecular rectifier. *Nat. Commun.* 7: 12850.

123 Sangeeth, C.S., Wan, A., and Nijhuis, C.A. (2015). Probing the nature and resistance of the molecule-electrode contact in SAM-based junctions. *Nanoscale* 7: 12061–12067.

124 Wan, A., Sangeeth, C.S.S., Wang, L. et al. (2015). Arrays of high quality SAM-based junctions and their application in molecular diode based logic. *Nanoscale* 7: 19547–19556.

125 Song, P., Sangeeth, C.S., Thompson, D. et al. (2016). Noncovalent self-assembled monolayers on graphene as a highly stable platform for molecular tunnel junctions. *Adv. Mater.* 28: 631–639.

126 Farnum, B.H., Wee, K.R., and Meyer, T.J. (2016). Self-assembled molecular p/n junctions for applications in dye-sensitized solar energy conversion. *Nat. Chem.* 8: 845–852.

127 Ardo, S. and Meyer, G.J. (2009). Photodriven heterogeneous charge transfer with transition-metal compounds anchored to TiO_2 semiconductor surfaces. *Chem. Soc. Rev.* 38: 115–164.

128 Hagfeldt, A., Boschloo, G., Sun, L. et al. (2010). Dye-sensitized solar cells. *Chem. Rev.* 110: 6595–6663.

129 Knauf, R.R., Kalanyan, B., Parsons, G.N., and Dempsey, J.L. (2015). Charge recombination dynamics in sensitized SnO_2/TiO_2 core/shell photoanodes. *J. Phys. Chem. C* 119: 28353–28360.

130 Gibson, E.A., Smeigh, A.L., Le Pleux, L. et al. (2009). A p-type NiO-based dye-sensitized solar cell with an open-circuit voltage of 0.35 V. *Angew. Chem.* 121: 4466–4469.

131 Nattestad, A., Mozer, A.J., Fischer, M.K. et al. (2010). Highly efficient photocathodes for dye-sensitized tandem solar cells. *Nat. Mater.* 9: 31.

132 Warnan, J., Gardner, J., Le Pleux, L. et al. (2013). Multichromophoric sensitizers based on squaraine for NiO based dye-sensitized solar cells. *J. Phys. Chem. C* 118: 103–113.

133 Weidelener, M., Mishra, A., Nattestad, A. et al. (2012). Synthesis and characterization of perylene–bithiophene–triphenylamine triads: studies on the effect of alkyl-substitution in p-type NiO based photocathodes. *J. Mater. Chem.* 22: 7366–7379.

134 Kim, D., Jeong, H., Hwang, W.-T. et al. (2015). Reversible switching phenomenon in diarylethene molecular devices with reduced graphene oxide electrodes on flexible substrates. *Adv. Funct. Mater.* 25: 5918–5923.

135 Dulić, D., van der Molen, S.J., Kudernac, T. et al. (2003). One-way optoelectronic switching of photochromic molecules on gold. *Phys. Rev. Lett.* 91: 207402.

136 Flood, A.H., Stoddart, J.F., Steuerman, D.W., and Heath, J.R. (2004). Whence molecular electronics? *Science* 306: 2055–2056.

137 Kim, D., Jeong, H., Lee, H. et al. (2014). Flexible molecular-scale electronic devices composed of diarylethene photoswitching molecules. *Adv. Mater.* 26: 3968–3973.

138 Kim, Y., Wang, G., Choe, M. et al. (2011). Electronic properties associated with conformational changes in azobenzene-derivative molecular junctions. *Org. Electron.* 12: 2144–2150.

139 Cinchetti, M., Dediu, V.A., and Hueso, L.E. (2017). Activating the molecular spinterface. *Nat. Mater.* 16: 507.

140 Cui, L., Miao, R., Jiang, C. et al. (2017). Perspective: thermal and thermoelectric transport in molecular junctions. *J. Chem. Phys.* 146: 092201.

141 James, D.D., Bayat, A., Smith, S.R. et al. (2018). Nanometric building blocks for robust multifunctional molecular junctions. *Nanoscale Horiz.* 3: 45–52.

142 Jeong, H., Hwang, W.-T., Kim, P. et al. (2015). Investigation of inelastic electron tunneling spectra of metal-molecule-metal junctions fabricated using direct metal transfer method. *Appl. Phys. Lett.* 106: 063110.

143 Jeong, H., Jang, Y., Kim, D. et al. (2016). An in-depth study of redox-induced conformational changes in charge transport characteristics of a ferrocene-alkanethiolate molecular electronic junction: temperature-dependent transition voltage spectroscopy analysis. *J. Phys. Chem. C* 120: 3564–3572.

144 Jeong, H., Kim, D., Xiang, D., and Lee, T. (2017). High-yield functional molecular electronic devices. *ACS Nano* 11: 6511–6548.

145 Xiang, D., Wang, X., Jia, C. et al. (2016). Molecular-scale electronics: from concept to function. *Chem. Rev.* 116: 4318–4440.

7

Toward Devices and Applications

Ajuan Cui and Kasper Nørgaard

Nano-Science Center and Department of Chemistry, University of Copenhagen, Universitetsparken 5, 2100 Copenhagen, Denmark

7.1 Introduction

With the development of complementary metal oxide semiconductor (CMOS) based technologies, the size of transistors decreases continuously with their integration density doubled every two years, which is known as Moore's Law. However, it becomes much more difficult for us to follow Moore's Law in the near future because of various reasons. Economically, the significant increase in cost needed to develop sophisticated tools with reduced feature size is becoming hard to accept. Technically, with the decrease in device size, the raised current density, energy dissipation as heat, noise, etc. will put higher requirements on materials and design. Moreover, with the decrease in device size, the fundamental mechanism of the device is not the same anymore because quantum mechanical begins to play a role. In this sense, new strategies are in great need to realize further miniaturization of electronics.

Among various candidates for further miniaturization of electronics, a bottom-up approach through organic molecules is one of the most promising strategies. The first attractive feature is the nanoscale size of molecules, which promises a smaller device size and thus a higher integration density. Second, through advanced molecular synthesis tools and delicately designing of the composition and geometry, molecules with various attractive properties could be mass-produced in a cost-effective way. The synthetic tailorability of organic molecules gives us the capability not only to tune the properties of molecules to realize various advanced functionalities, but also to control their behavior such as assembly and recognition. In this way, not only will the devices with various properties be obtained by using different types of functional molecules, but they could also be assembled into integrated circuits through recognition and assembly process.

Nanogap Electrodes, First Edition. Edited by Tao Li.

In the past decades, significant progress has been made that molecular devices with functions of memories [1, 2], rectifiers [3, 4], transistors [5–7], switches [8, 9], etc. were demonstrated, yet we are still far away from the ultimate aim to use only molecules to construct ultra-high density integration circuits in a bottom-up approach. At this stage, most molecular devices are hybrid ones that molecules are used only as the active materials in the devices while other parts of the devices are still composed of inorganic materials. By using this hybrid system, remarkable progress has been made, yet challenge remains in using them for high-density integration. At molecule level, most organic molecules have poor stability that new kinds of functional molecules with improved stability needs to be developed. At device level, the device yield, reproducibility, and performance provided by various fabrication methods is not satisfying enough, and new methods to construct molecular device in a more reliable way are in great need. At integration circuits level, memory and logic functions have been demonstrated [1, 2, 10], yet the performance, functionality and integration level of the circuits are still at the early stage of development. Therefore, there are a lot of issues that need to be addressed before the actual application of molecular devices.

Here in this chapter, we mainly focus on the issues that needs to be addressed before the practical application of molecular devices. In Section 7.2 we will discuss all the relevant issues in device constructions as well as the progress made recently. In Section 7.3, we will focus on the potential integration solutions of the molecular devices. After a brief introduction of the most promising device construction method, three specific integration schemes will be discussed. It needs to be pointed out that in Chapters 3 and 6, a comprehensive description on the construction of various molecular devices were presented, and here we will put more focus on the issues related to device integration and application. Apart from applications for ultra-high intensity integration, molecular device and nanogap electrodes have demonstrated their potential applications in other fields, which will be briefly introduced in Sections 7.4 and 7.5.

7.2 Major Issues: Reliability and Robustness

With regard to the reliability and robustness, there are some common issues for molecular devices. The first one is the stability of the molecules. Different from inorganic materials, organic molecules are more fragile that the condition range (such as temperature, current density, gas, solvent, humidity, irradiation from energetic particles including electrons, ions, photons, etc.) they can withstand is much smaller. The poor stability of organic molecules makes the process window very narrow that special fabrication processes are always needed. Apart from that, each kind of molecular device such as single molecular device, or molecular junctions of self-assembled monolayer (SAM), has its own distinguishing feature, thus could lead to very different fabrication schemes and performances. In the following, we will focus on each kind of molecular devices and discuss all the relevant issues in realizing high performance devices, regarding to the reliability and robustness of the fabrication process and the device performance, toward practical application.

7.2.1 Single Molecular Device

Up to now, the most promising way to realize electronic devices with reduced size is the utilization of single molecule. In principle, by tailoring the molecular structure though chemical design and synthesis, various electronic functionalities can be realized in the form of single molecular device. Great advances have been made that single molecular device with various functionalities has been reported, making them a promising candidate for the next generation of electronics. However, there remain some formidable challenges, which must be resolved before the practical application of single molecular device. One of the biggest challenges in the construction of single molecular device is the connection of single molecule to the outer world. Because the sizes of most molecules are below 5 nm, they cannot be connected to the macroscopic electronic circuits using the same way as traditional materials, thus new approaches are needed. In the past decades, several effective ways were successfully demonstrated which can be divided into several categorizes, including top-contact junctions, planar metallic nanogap electrodes, and systems combining graphene and single-walled carbon nanotubes (SWCNTs). In the following, we will follow this classification and discuss all the relevant issues in obtaining single molecular devices in a reliable and robust way.

7.2.1.1 Top-Contact Junctions

Using the existing scanning probe microscopy (SPM) techniques such as scanning tunneling microscopy (STM) and conducting atomic force microscopy (C-AFM), top-contact junction is one of the most convenient ways to construct single molecular devices. In this approach, a conducting substrate where the molecules sit is used as bottom electrode and a metallic scanning probe is used as the top electrode to contact the other end of molecule. By controlling the size of the probe or the density of target molecules, single molecular device can be constructed. Although great progress has been made with this approach, devices constructed by this approach cannot be used for practical applications because of the constraint of the large SPM system.

7.2.1.2 Planar Metallic Nanogap Electrodes

Compared to top-contact junctions, single molecular devices constructed from planar nanogap electrodes present a flat configuration, thus is more likely to be used in practical applications for future integrated circuits. Although traditional microfabrication technologies have scaled the feature size down to tens of nanometers, fabrication of sub-20 nm features in a reproducible and reliable way is still beyond its capability, thus cannot be readily used for the construction of single molecular devices. Regarding this challenge, several creative experimental schemes for nanogap fabrications are developed, among which the most frequently used ones for the construction of single molecular devices is mechanically break junctions

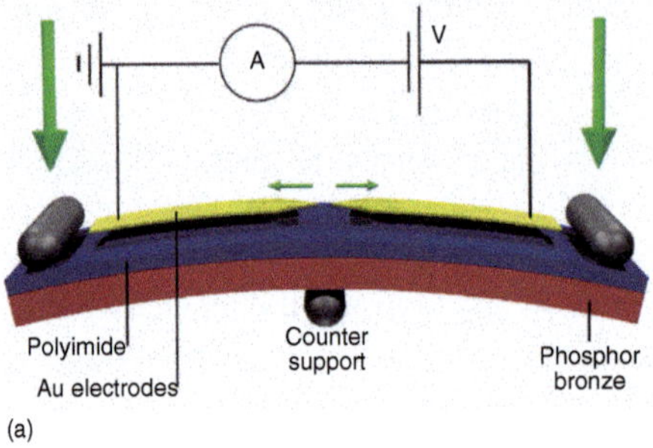

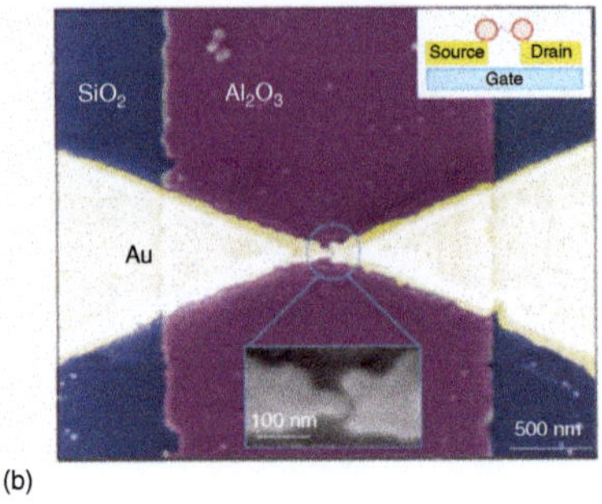

Figure 7.1 (a) Layout of a mechanically break junctions set-up. Large vertical green arrows represent the force applied to bend the sample. Small green arrows illustrate the attenuated electrode displacement as a result of bending. Source: Perrin et al. [12]. 2013, Springer Nature. (b) Scanning electron microscope image (false color) of the metallic electrodes fabricated by electromigrated break junctions. Source: Liang et al. [7]. Copyright 2002, Nature Publishing Group.

and electromigrated break junctions. With these junctions, significant progress has been made [5, 6, 11]. However, there is still a long way to go before the practical application of single molecular devices using this approach.

Taking mechanically break junctions as an example, through the bending of the supporting substrate by a three-point configuration as showed in Figure 7.1a [12], metal nanowire can be broken into nanogap electrodes. By controlling the bending degree of the supporting substrate, the gap size can be precisely controlled in a reversible way. Remarkable progresses have been made using this method in the study of various single molecules. However, due to the constraint of the existing mechanical system and the difficulty in controlling the gap size of multiple devices on one chip, the practical application of such device is severely limited. For electromigrated break junctions, single molecular devices such as single molecular transistors are successfully demonstrated as showed in Figure 7.1b [7], yet we are still far away from producing them in a reproducible way. The first reason comes from the fabrication of the nanogap electrodes. The active breaking scheme enables the fabrication of nanogaps with desired tunneling resistance, yet suffers from gap widening and the formation of metal clusters between the electrodes [13, 14], which makes it harder to distinguish the signal of molecules. The self-breaking scheme eliminated the formation of metal clusters [15], yet the controllability over the gap size is reduced. Although multiple nanogap electrodes can be produced in a parallel way now by this method [16], it is still a great challenge in controlling the shape/size of the gap. The second reason comes from the way to connect molecule to nanogap electrodes. One common way to insert molecules into the nanogap electrodes is through statistics [5, 7]. In a typical fabrication procedure, a dilute solution of molecules is used to distribute molecules onto the substrate before the formation of the electrodes. After the formation of the nanogap electrode by electromigration, there is a chance that a molecule nearby will move into the gap by thermal activated motion and a conductance enhancement will be induced to show the formation of molecular junction. In this method, the coverage of molecules on the substrate surface should be carefully controlled to increase the production probability of single molecular device on one hand and minimize the probability of trapping multiple molecules in the gap on the other hand. In this way, the yield of the device is limited. The third reason comes from the lack of control over the interface at atomic level. The properties of single molecular devices are very sensitive to the atomic structure of the metal–molecule interface and the conformation of the molecules in the gap. Although molecule can be connected to the electrodes through chemisorption by strong covalent bonds or physisorption by weaker Van der Waals (VdW) forces, the atomic structure of the metal–molecule interface is still out of control. Therefore, single-molecule devices with reproducible performance are hard to obtain at present stage. Moreover, voltage as high as several volts could be required to align the discrete molecular energy levels with the fermi level of the metal electrodes. For metallic nanogap electrodes with gap size around several nanometers, an electric filed as high as $10^9\,\mathrm{V\,m^{-1}}$ will be generated in the gap, which will cause the diffusion of metal atoms into the gap. Single molecular devices constructed from this method are usually measured at low temperature. After the measurement, the devices are taken out of the low temperature environment for economic reasons, thus no experiments are indicating the lifetime of these devices. Therefore, single molecular device constructed from this method is far from application due to the low device yield, reproducibility and stability.

7.2.1.3 Planar Nanogap Electrodes Based on Single Walled Carbon Nanotubes (SWCNTs) or Graphene

Unlike metallic nanogap electrodes with sharp tips, which are unstable under room temperature or high electric field, nanogap electrodes based on materials like SWCNTs and graphene are much more stable. Combined with their mechanical/chemical stability and extraordinary electronic properties, such materials are becoming more and more popular in single molecular devices construction and remarkable progresses have been demonstrated [17–24].

With nanoscale width, microscale length, and extraordinary electronic properties, SWCNTs offers a perfect test bed for the study of single molecular devices. A molecular-scale gap can be introduced to the SWCNTs through various methods, among which, one is of great importance that the edge of the gap can be functionalized with carboxylic acids [23]. With these functionalized nanogaps, molecules with conjugated diamines group can be connected to the nanotubes (as showed in Figure 7.2a) and the formed contacts provided by amide-linked molecular bridges are robust that they can withstand various environmental changes such as pH, metal ions and light, thus could lead to a variety of functional nanoscale devices [18]. However, ongoing challenges remain. For example, electrical properties of the SWCNTs are determined by their structures such as chirality and diameter, and the variation in the structure of the SWCNTs will definitely make the production of the molecular devices in a reproducible way more difficult. Moreover, the connection yield is relatively low (generally 3–5%) [22].

Compared to SWCNTs, graphene does not have the inherent structure variability, thus could lead to better controllability in device constructions. Remarkable

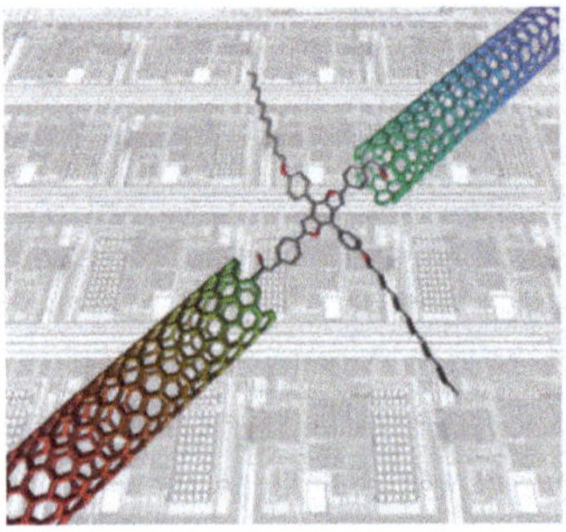
(a)

(b)

Figure 7.2 (a) A depiction of single-molecule devices based on single walled carbon nanogap electrodes. Source: Feldman et al. [18]. 2008, American Chemical Society. (b) A depiction of single-molecule devices based on graphene nanogap electrodes. Source: Cao et al. [22]. 2012, John Wiley & Sons.

progress has been made by Professor Guo' research group by using graphene nanogap electrodes in single molecular device construction [22]. They have developed a process for mass production of graphene nanogap electrodes with gap size less than or equal to 10 nm with terminal base of carboxylic acid. Through amide bond, the bond between the electrodes and the molecules are very strong that stable molecular devices can be constructed with high yield (as high as 50%). Same as SWCNTs based devices, single molecular devices based on graphene nanogaps (as showed in Figure 7.2b) can also tolerate large changes in the environment, showing excellent reproducibility and stability. Using this approach single-molecule electrical switch with outstanding performance was successfully demonstrated [24], showing unprecedented levels of accuracy (on/off ratio of ~100), stability (over a year), and reproducibility (46 devices with more than 100 cycles for photoswitching and $\sim 10^5$–10^6 cycles for stochastic switching). Compared to single molecular devices based on metallic nanogap electrodes, two main long lasting challenges are circumvented by using graphene, the low-yield device fabrication and ill-defined contacts between molecules and electrodes, pushing single molecular device one-step further on the road to practical applications.

7.2.1.4 The Absorption of Molecule on the Surface of SWCNTs or Graphene

Apart from nanogap and the SPM scheme, another scheme to construct single molecular device, mainly for single-molecule magnet (SMM) based devices are through a hybrid system [20, 25, 26]. In this hybrid system, a SWCNT or graphene bridge was connected to two electrodes, and the SMM molecules are adsorbed on the surface of them, as showed in Figure 7.3. The current between the two electrodes are determined not only by the SWCNT or graphene, but also on the spin state of the adsorbed SMM molecules through various coupling mechanisms. Using this device scheme, great progress has been made, for example, a magnetoresistance ratio up to 300% at low temperature were reported in a supramolecular spin valve (SV) device [26], and the performance of the device is relatively stable at low temperature showing repeatable conductance changes when sweeping the magnetic field.

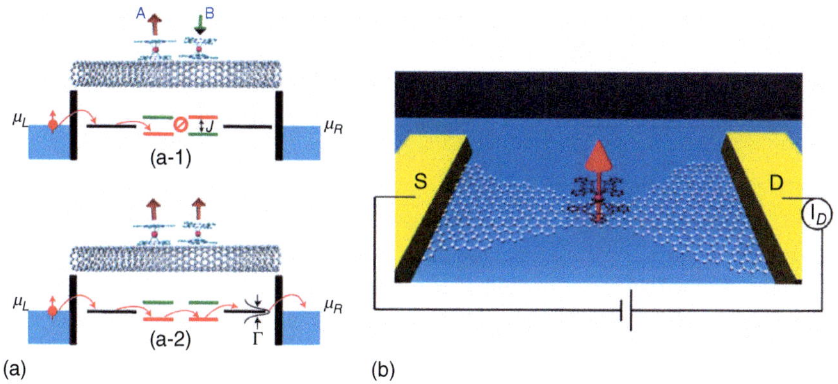

Figure 7.3 Hybrid device by graft molecules to SWCNT or graphene nanoconstriction. (a) Schematic representation of spin valves mechanism involving two $TbPc_2$ SMM molecules (A and B) grafted on an SWCNT. With increasing magnetic field, molecule A switches first (a-1), thus leading to an antiparallel configuration of the spin valve with lowest conductance. When molecule B switches (a-2), a parallel configuration is recovered leads to high conductance. Source: Urdampilleta et al. [26]. 2011, Springer Nature. (b) Schematic view of a hybrid device based on graphene, a $TbPc_2$ molecule grafted on the surface of a graphene nanoconstriction contacted by electrodes. The magnetic moments of the $TbPc_2$ SMMs on top of the graphene constriction add another degree of freedom to tune the conductivity of the device. Source: Candini et al. [20]. 2011, American Chemical Society.

However, the yield and reproducibility of this kind of devices are far from the value that could be used for practical applications. Apart from the fabrication of the base structures of SWCNT and graphene, the grafting of the molecules is also a big issue. The typical procedure used to graft the SMM molecules is by statistics, that a dilute solution of the SMM are usually drop-casted onto the sample, so there is a chance that one or two SMM molecules will graft to the surface of the SWCNT or graphene, which makes the reproductively of the device a real challenge.

7.2.2 Molecular Device Based on Molecule Monolayer

From the prospective of fundamental research, molecular device based on molecule monolayer or single molecule are two complementary approaches to provide information for us to understand the charge transport in molecule based systems. However, from the prospective of practical application, molecular device based on molecule monolayer shows several advantages. First, the device fabrication process is much less challenging, since only in one direction nanoscale size is needed. Second, the properties of molecular junction based on monolayers are far more reproducible since the transport is a sum of many electron transfer events. Last but not the least, the stability of molecular junction based on monolayers is another advantage over single molecular devices. In spite of this, to obtain devices based on molecule monolayer with robust performance in a reliable way is still a challenge considering that the device properties are very susceptible to defects and irregularities and conductance contribution from these defects and irregularities could be very high compared to that from molecules. Therefore, careful attentions are needed to realize reliable and robust performance.

Molecular device based on monolayer can be divided into two categories based on the fabrication process. In the first category, the distance between the two electrodes are pre-defined and the formation of the monolayers takes place after the fabrication of nanogap electrodes. In the second category, the devices are constructed in the process of bottom electrodes, monolayer, and top electrodes, in which the distance between the bottom electrodes and top electrodes are defined by the thickness of the monolayer. In the following, we will discuss the relevant factors that could introduce uncertainty and instability in the devices as well as typical device fabrication process for each category.

In the first category, the main step is the fabrication of nanogap electrodes and the formation of monolayer in the gap. In this category, the pre-defined gap should have similar size to match the length of the target molecules. Compared to nanogap electrodes used to construct single molecular devices, the nanogap electrodes used for monolayers have much bigger cross-sectional area, and the method to insert the molecules are carried out mostly by self-assembly. A lot of methods were reported for the fabrication of such kind of nanogap electrodes [27–30]. Apart from the state-of-the-art lithography technologies, one typical way is to use a sacrificial layer to define the size of the gap. Various materials are used as the sacrificial layer, such as metal oxide [31–33], organic molecules [29, 34], metals [35, 36], III–V group materials [37], etc. Through chemical etching, the sacrificial layer can be removed, leaving nanogap between electrodes. By this approach, a great amount of nanogap electrodes can be fabricated at the same time, some of which even showed the ability to control the size of the nanogaps precisely, such as on-wire lithography [36].

Regarding the direction of the nanogaps to the substrate, there are vertical ones and horizontal ones. For the vertical one, such as the one used for on-edge molecular junctions [38, 39], the cross section of the nanogap electrodes is parallel to the substrate, thus in this kind of nanogap electrodes there will be a space layer between the electrodes. Such vertical nanogaps can be produced in large amount using conventional microfabrication techniques, and a high integration density could be easily realized. For the horizontal electrodes, where the two electrodes lies both on the surface of the substrate without overlap, they can be used to study the molecules both electrically and spectroscopically and impressive result has been reported [40]. However, some challenges are hard to overcome at the present stage for both vertical and horizontal nanogap electrodes. For example, for vertical nanogap electrodes, the leak current between the electrodes is relatively higher, because the space layer used to define the size of the nanogap electrodes is around nanometers in thickness. Another challenge for both of them is the size mismatch between the nanogap and the molecules. Although molecular junction has been constructed through the assembly of molecules across a pre-defined gap, the assembly patterns of the monolayer in the junction, the number of molecules that actually span the nanogap is still hard to control to get a good reproducibility.

In the second category, the main step is the fabrication of bottom electrodes, the formation of monolayer, and the fabrication of top electrodes. It is much easier to construct molecular junction using this strategy because there is no need of complex process for the fabrication of nanogap electrodes. The distance between the electrodes is naturally defined by the thickness of the monolayer. Yet challenges still remain. The issues related to each process is as follows:

7.2.2.1 Bottom Electrodes

The first step to avoid defects and irregularities is the fabrication of bottom electrodes with the contacting area as uniform as possible. Apart from the conductance of the electrode materials, the roughness of the bottom electrodes should be low enough that ideally there is no protrudes that could lead to a shorter transport length between the bottom and top electrodes or cause an asymmetry in current–voltage characteristics. A low roughness also helps in getting a better reproducibility. In this respect, semiconducting materials with ultra-smooth surface, such as commercially available polished single-crystal wafers of especially Si, GaAs, GaP, etc. are naturally suitable candidates. However, from the perspective of mass production and integration, independent bottom electrodes are more preferred which makes single-crystal wafers not a good candidate. Amorphous carbon with ultra-smooth surfaces were also reported in molecular device construction [41, 42].

With good chemical stability and well-known electrical properties, noble metal such as Au is frequently used in the study of molecules. Single crystal Au beads with well-defined crystalline plane is commonly used in laboratory on molecular electronics, yet they cannot be used for practical device construction because of its big volume and incompatibility with traditional microfabrication technologies. Single crystal Au film is now commercially available, yet it is not cost-effective. A variety of effective methods to improve the roughness of Au film is developed including using different primer layers, smaller thickness, ultrafast evaporation, stripping method, etc. With respect to the uniformity of the Au films, stripping method is better in a way that the roughness of the Au film can be pre-determined by the supporting surface, so that less protrudes will be produced as compared to direct metal deposition. The advantage of direct metal deposition is its compatibility with traditional microfabrication technologies, and thus can be mass-produced. For direct metal deposition to fabricate bottom electrodes, apart from a low roughness, consideration should also be paid to the combination of the metals. For example, an adhesive layer of Cr or Ti is usually used to enhance the adhesion between Au film and the substrate. Recent research has showed that instead of forming a uniform layer under the Au film, Cr could interdiffuses with the Au layer and form a Cr–Au alloy, which can be detrimental to device since the assembly style is changed as compared to pure Au bottom electrodes [43].

7.2.2.2 Insulating Layer with Holes to Define the Size of the Bottom Electrodes

In most devices, an insulating layer with holes is used to define the contact area. Without an insulating layer, the small curvature of the electrode edge could result in loosely packed monolayers instead of closely arranged monolayers and the protrudes along the electrode edge could lead to shorter current path between the bottom and top electrodes. In this sense, the usage of an insulting layer is of great importance. In most cases, the insulating layer are fabricated on the bottom electrodes with holes to expose the bottom electrodes, so the formation of the monolayer only take place in the holes. In this way, the current path between the bottom electrodes and the top electrodes is only through the monolayer. Various kinds of materials can be used as insulating layer as long as the leak current through them is low enough and they are strong enough for the afterwards fabrication procedure.

7.2.2.3 Molecule Monolayer Formation

Compared to single molecular device, where molecules are connected to electrodes in a stochastic way, the formation of molecular monolayers are much more controllable. Yet, it is still hard to get molecular monolayer with uniform structures. For example, SAM is the most frequently used monolayers in molecular devices, yet methods to distinguish defects in the SAM are still needed. Apart from the influence from substrate, preparation method, etc., there are some intrinsic defects that cannot be distinguished. For example, grain boundaries between highly ordered domains could lead to high conductance path. Thus, new method is still needed to form molecular monolayers without defects.

7.2.2.4 Top Electrodes

One of the most challenging part in monolayer molecular device construction lies in the fabrication of the top electrodes in a reliable and permanent way without damaging the monolayer. Organic monolayers are much more fragile than its inorganic counterpart that typical technologies (such as lithography, metal deposition and lift-off process) used for the fabrication of electrodes on inorganic nanomaterials cannot be used directly. The reason is that the organic solvent, heat, energy, etc. involved in the process can damage the structure of the monolayers and cause a shortcut between the bottom and top electrodes. For example, when thermal evaporation or electron beam evaporation are used to deposit metal directly onto the SAM for top electrodes fabrication, the energetic metal atoms could penetration into the SAM and cause short circuit of the device [44–46]. The heat generated in the process could also damage the structure of the SAM and contribute to a high failure rate of the device. For this issue, four typical strategies were developed for the fabrication of top electrodes effectively including metallic liquid top electrodes, revised metal deposition, transferring techniques, and non-metallic electrodes. In the following, we will introduce more details on them.

Metallic Liquid Top Electrodes In the strategy of metallic liquid top electrodes, liquid metal is used as top electrodes to form the molecular junctions [47, 48]. Two typical liquid metals used for molecular junction construction are mercury (Hg) and liquid InGa alloy (eutectic gallium–indium [EGaIn]). There are several advantages in using liquid metal as top electrodes. First, the fluidity assured good conformal contact with the molecular layers on a solid surface. Second, the high surface tension makes them hard to penetrate small holes in the SAM. Third, the high conductivity

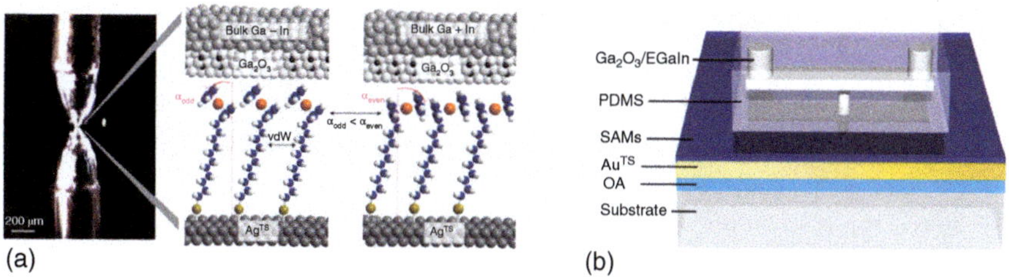

Figure 7.4 Two procedures for using EGaIn as top electrodes: (a) a cone-shaped tip of Ga_2O_3/EGaIn suspended from a microneedle in contact with a SAM on an Ag surface. Source: Nerngchamnong et al. [49]. 2013, Springer Nature. (b) Stabilizing the EGaIn electrode in PDMS microchannels with ultraflat Au bottom electrode supporting SAM. Source: Du et al. [50]. 2016, Springer Nature.

makes them ideal as electrode materials. Good contact and high device yields has been achieved by using them as top electrodes. Yet, there are still some issues that need to be solved before its practical applications. For Hg, apart from its harmful effect on human health, it is not good for the device stability because amalgamation could happen between Hg vapor and common substrate materials (Au, Ag, II–VI semiconductors). For EGaIn, the reactivity toward most of the substrates are much lower, thus becoming more popular than Hg. There are two ways to construct molecular junction with EGaIn. The first one is showed in Figure 7.4a that a suspended EGaIn drop from a microneedle is used to approach the SAM to form the molecular junctions [49]. This kind of molecular junction cannot be used for practical applications due to the existence of the mechanical system and its relatively big size. The other way is by stabilizing the EGaIn electrode in polydimethylsiloxane (PDMS) micro channels, as showed in Figure 7.4b [50]. In this way, molecular junction in a device configuration can be constructed. However, the fluidity of EGaIn makes it hard to further reduce the device size, thus they can hardly be used for high-density integration. One thing that needed to be pointed out is that due to the existence of the Ga-oxide film on the surface, the interpretation of the device properties needs more complicated modeling.

Revised Metal Deposition In the strategy of revised metal deposition, various means are taken to avoid the damage generated in the process of direct metal deposition, which mainly come from the kinetic energy of the incident metal atoms, their interaction with the monolayer, heat generated by the incident atoms or heat irradiation from the evaporation source. Besides some common procedures such as cooling down the sample, reducing heat generation by slower evaporation rate or longer distance between sample and evaporation source, several methods were successfully developed and proven to be effective, such as scattering reduced deposition [51, 52], surface-diffusion-mediated deposition [53], temporary buffer layer during deposition [54], etc.

In the method of scattering reduced deposition, as showed in Figure 7.5a [52], the samples are fixed on a low temperature stage and faced away from the evaporation source. Instead of a very low vacuum atmosphere used for normal metal deposition,

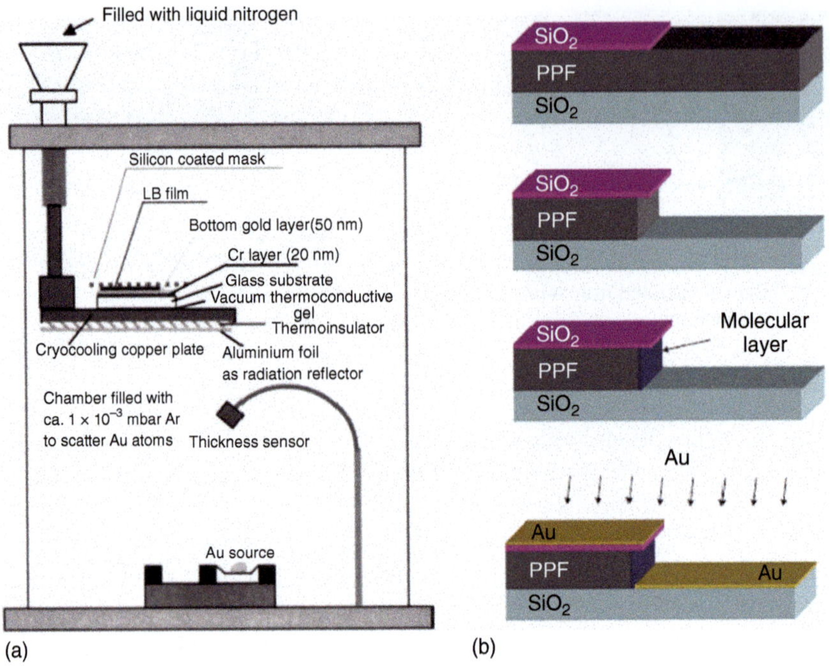

Figure 7.5 (a) Geometry of evaporation by scattering reduced deposition. Source: Metzger et al. [52]. 2001, American Chemical Society. (b) Schematic of the process to construct molecular junction, in which the top electrodes were made by surface-diffusion-mediated deposition. Source: Bonifas et al. [53]. 2010, Springer Nature.

a low-pressure inert gas such as Ar was introduced into the chamber. In this way, metal atoms from the source cannot reach the sample directly, instead, only the ones undergo multiple scatterings by the Ar have a chance to reach the surface of the sample, resulting in less damage. By this way, the device yield increases from 5% (direct evaporation) to 45% and 25% for Pd and Au top electrodes respectively. However, in this method, besides the increased cost on cooling, the cost on the electrode materials are much higher because only a very small proportion of the evaporated Au atoms can reach the sample. For example, in an evaporation process, an Au film with thickness of 600 nm was formed by direct deposition while only 17 nm was obtained on the sample.

In the method of surface-diffusion-mediated deposition [53], as showed in Figure 7.5b, the metal atoms are not deposit directly on the SAM but at a distance away from it. In this way, the kinetic energy of the incident atoms is dissipating remotely by conversion into heat or a smaller kinetic energy parallel to the substrate. With a kinetic energy parallel to the substrate, the metal atoms can diffuse to the surface of the SAM and form electrode there. Because the kinetic energy of incident metal atoms on the monolayer is much smaller, the tendency for the metal atoms to penetration into the SAM is much smaller as well as the generated heat, resulting in better device yield (typically >90%) and reproducibility. However, this method put higher requirement on the design of the device and device complexity is increased.

For example, a shadow mask is required to block the direct deposition. The direction of the shadow mask, the distance between its edge and the SAM is strictly limited to make sure that the metal atom can diffuse to the surface of the SAM with minimum damage.

Transferring Techniques In the strategy of transferring techniques, including lift-and-float approach [55–57] and transfer printing [58–60], the top electrodes are not fabricated directly on the monolayer, instead they are pre-fabricated elsewhere on a supporting substrate. After detached from the supporting substrate, the top electrodes are transferred onto the monolayer to form the contact by the force from capillary interactions or chemical affinity. Through the transferring techniques, most of the damage the monolayer went through in direct metal deposition can be avoided, thus high yields can be achieved. For example, in one lift-and-float approach as showed in Figure 7.6a [61], isopropanol (IPA) was used between the top electrode and the monolayer to assistant the formation of a fine contact through capillary action during the IPA vaporization. In this method, a device yield of 70% was reported. For lift-and-float approach, capillary interaction was demonstrated to be effective in forming the contact, making this method suitable for many kinds of molecules, especially the one with no chemical binding to the top electrodes. However, such method is not compatible with traditional micro fabrication technologies at this stage and have difficulties in high-density production. Metal transferring by means of chemical affinity such as nanocontact printing or lamination have better compatibility with traditional technologies, yet it puts higher requirement on the

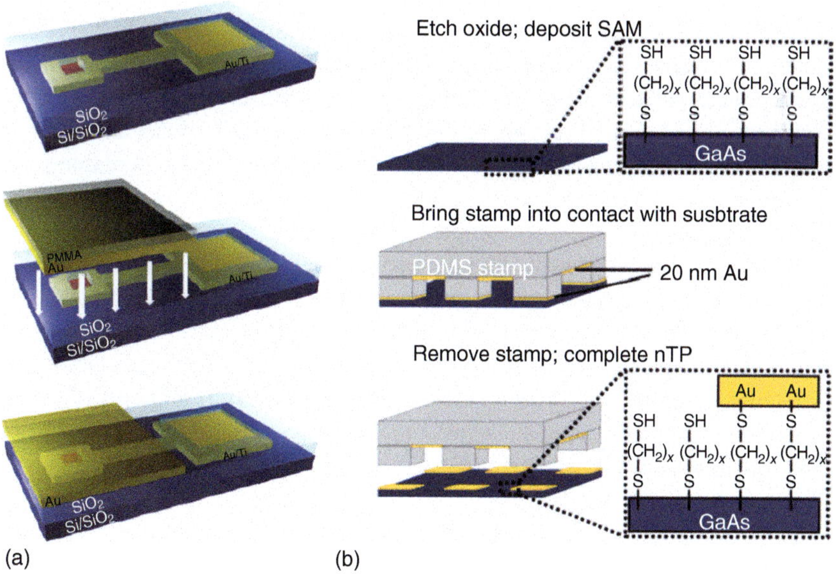

Figure 7.6 (a) Schematic illustration of direct metal transfer method. Source: Jeong et al. [61]. 2014, IOP Publishing. (b) Schematic of the nanotransfer printing (nTP) procedure. Source: Loo et al. [60]. 2003, American Chemical Society.

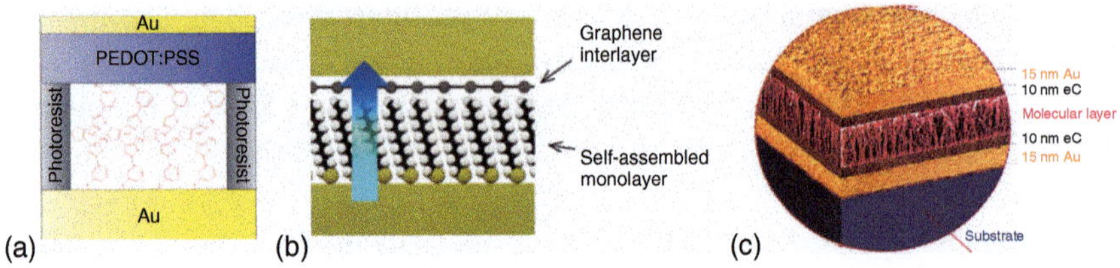

Figure 7.7 (a) Molecular junction using conductive polymer PEDOT:PSS as buffer layer. Source: Kronemeijer et al. [63]. 2008, John Wiley & Sons. (b) Molecular junction using graphene as buffer layer. Source: Li et al. [65]. 2012, John Wiley & Sons. (c) Molecular junction using amorphous carbon as buffer layer. Source: Morteza Najarian et al. [42]. 2016, American Chemical Society.

monolayers that a sticky end group with chemical affinity to the top electrodes are needed, as showed in Figure 7.6b [60]. Moreover, for such transferring technologies, the cleanness of the interface is always in doubt.

Non-metallic Electrodes With the above-mentioned approaches, metal-monolayer-metal molecular junction can be obtained with high yield and excellent reproducibility, yet the stability of them, especially for monolayer with sufficient motional freedom to allow structural deformations, could be influenced by the migration of metal atoms. This migration of metal atoms driven by electromagnetic field induced by applying a voltage or light is one of the trickiest problems, especially for devices based on metallic nanogaps, which could easily cause short circuit. To solve this issue, the fourth strategy was developed that non-metal materials are introduced onto the monolayer through a non-destructively way before the deposition of the top electrodes. Such methods are demonstrated to be very effective. First, under the protection of this conducting layer, the impinging metal atoms can no longer damage the structure of the monolayer during metal deposition process. Second, the conducting layer also act as a barrier layer to stop the migration of the metal atoms under electrical field. In this way, not only the production yield but also the stability of the device is greatly improved. If the conductivity of the non-metallic materials is high enough, no further metal deposition is needed. Very impressive improvement on the production (yield and reproducibility) of the molecular junctions and its performance was demonstrated by this strategy using poly(3,4-ethylenedioxythiophene):poly(4-styrenesulphonic acid) (PEDOT:PSS) [62, 63], graphene [64, 65], and amorphous carbon [41, 42], as showed in Figure 7.7, yet challenges remains, for example, it is still a challenge to control the exact atomic configuration of the interface. Reliable procedures to get rid of the defects or hot spots or a way to eliminate their contribution to the net current are needed.

7.3 Potential Integration Solutions

In the past decades, many efforts have been devoted to molecular electronics to pursue the ultimate aim to build ultra-high density integration circuits through organic molecules in a bottom-up approach. Although we are still far away from the ultimate aim, significant progress has been made. In the molecular level, molecules with various advanced functionalities have been designed, synthesized, and studied electronically [66–68]. In the device level, molecular devices with functions of memories [1, 2], rectifiers [3, 4], transistors [5–7], switches [8, 9], etc. were demonstrated with improved yield, reproducibility and reliability. At the production level, various methods were explored considering the manufacturing compatibility with existing technologies and their integration [62, 69]. In the integration circuits level, memory and logic functions has been demonstrated with performance, functionality and integration level improving year by year [1, 2].

Although remarkable progresses have been made, challenges remain. First, the poor stability of organic molecules makes the process window very narrow that special fabrication processes are always needed, not only for mass production of the elemental device in a reliable and reproducible way but also for wiring and packaging them into integrated circuits. Second, excellent operational stability and satisfactory functional performance are required to assure the functions of the entire integrated circuits. Up to now, remarkable progresses have been made in mass production of molecular devices with robust performance, one of the major goal in molecular electronics next is to wiring up such functional molecular devices into integrated circuits. For this aim, many efforts have been devoted and progresses have been made. In Section 7.3, three most promising integration strategy in using molecular device for high-density integration circuits will be discussed.

7.3.1 Carbon Nanotube or Graphene Interconnects

High-density integration circuits consist of lots of devices, so the first step toward integration is mass production of molecular devices in a reliable and reproducible way. Among various single molecular devices, the ones constructed by graphene or SWCNTs nanogap electrodes are the most promising candidates in high-density integration for several reasons. First, apart from mechanical/chemical/thermal stability, extraordinary electronic properties, nature compatibility with organic/biological molecules, graphene and SWCNTs show higher structure stability than metallic nanogap electrodes. Second, the technology for progressing graphene and SWCNTs for high-density integration is developing at a very fast rate, which can be readily used for molecular device construction and integration [21, 70]. Third, the connection between molecules and graphene and SWCNTs can be realized through a reliable way by covalent bond [21, 23]. Such covalent bond was demonstrated to be strong enough to withstand chemical treatments and external stimuli, thus a high device stability is ensured [22, 71].

Apart from the above-mentioned advantages, SWCNTs are one-dimensional (1D) structures, showing perfect size match with single molecules, offering a very promising test bed for single molecules. Reliable process has been successfully developed by Professor Guo et al. to produce SWCNT nanogap electrodes by oxygen plasma ion etching through an e-beam lithography defined polymethylmethacrylate (PMMA) mask as showed in the left image of Figure 7.8a [23]. By the oxygen plasma ion

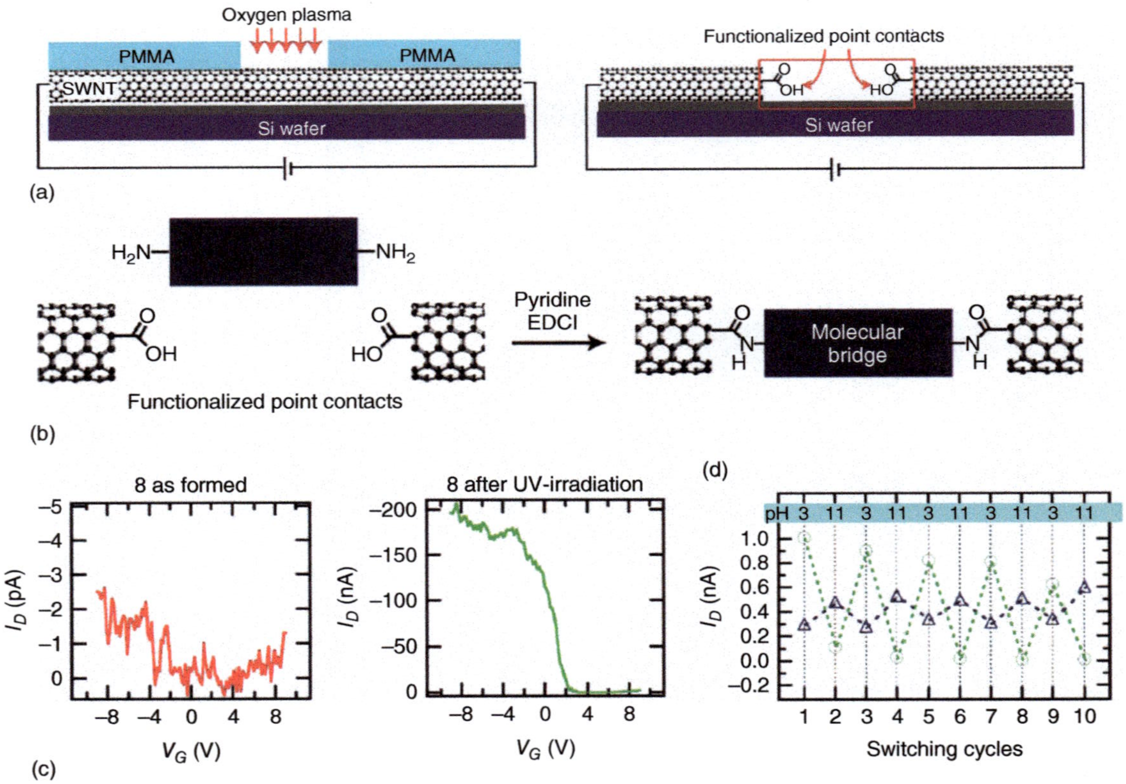

Figure 7.8 SWCNT based molecular devices: construction and application. (a) The fabrication of carboxylic acid-functionalized nanogap on SWCNT by oxygen plasma under a PMMA mask. Source: Guo et al. [23]. 2006, American Association for the Advancement of Science. (b) The formation of the contact between the molecule and the SWCNT by amine bridge. (c) SWCNT based molecular junction as light switch. (d) SWCNT based molecular junction used for pH sensing showing a change in conductance under difference pH environment. Source: Feldman et al. [18]. 2008, American Chemical Society.

etching process, the edge of the formed SWCNTs nanogap electrodes will be dominated by carboxylic acid groups (right image of Figure 7.8a). Molecules with amine anchoring groups can be covalently bonded to the electrodes by robust amide bonds (Figure 7.8b) [18], which are very stable to large environmental changes. With those electrodes, device with various functionalities are demonstrated [18], such as photogated switching (Figure 7.8c), pH sensing (Figure 7.8d), metal-ion recognition, DNA hybridization sensor, etc.

In the society of molecular electronics, great processes have been demonstrated using SWCNT to construct single molecule device. However, challenges remains, the biggest one of which is the low connection yield (~3–5%) and the high device-to-device property variation from the variability of SWCNTs' structures [22]. However, in the society of nanoelectronics, great progress is taking place. On one hand, various synthesis and tube sorting methods are developed to produce SWCNTs with narrow structure distributions. For example, Avetik R. Harutyunyan et al. have developed the method to produce metallic SWCNTs with a proportion as high as 91% [72]. Feng Yang et al. have developed the method to produce SWCNTs with specific structures with a percentage as high as 92% (Figure 7.9a) [73]. On the other hand, various new processes have been developed for integration of SWCNTs into circuits. For example, SWCNTs can be precisely wired between pre-designed electrodes (Figure 7.9b) [74], and large area well aligned SWCNTs with controllable density can be successfully grown [75]. The technologies developed for SWCNTs processing can be readily used for molecular device construction. Combining all the processes mentioned above [18], we could be one step further near the ultimate aim to build ultra-high density integration circuits out of single molecular device, as showed in Figure 7.9c, d [19].

Compared to SWCNTs, graphene is two-dimensional monolayer, which is advantageous in facile fabrication of nanogap electrodes arrays and thus molecular devices. In the society of molecular electronics, various methods for graphene nanogap electrodes are developed. Especially, Professor Guo Xuefeng' group have developed an efficient method (dash-line lithography) to produce graphene nanogap electrodes with carboxylic acid end groups (Figure 7.10a) [22]. Through amide bond, robust single molecular device can be constructed with yield as high as 50%. Using this approach single-molecule electrical switch with outstanding performance were successfully demonstrated showing unprecedented levels of accuracy (on/off ratio of ~100), stability (over a year), and reproducibility (Figure 7.10b) [24]. In the society of nanoelectronics, remarkable progresses have been made in producing high quality graphene and processing them for device and integration. For example, the production of single crystal graphene with size as large as $5 \times 50\ cm^2$ has become a reality now (Figure 7.10c) [70]. Moreover, bottom-up approach has also made progress that functional hybrid molecular system composed of a single molecule and graphene nanoribbons can be synthesized with atomically precise contacts [66]. Combined those progresses together, single molecular device constructed in this way are very promising in realizing high-density integration circuits in the near future, as showed in Figure 7.10d [76].

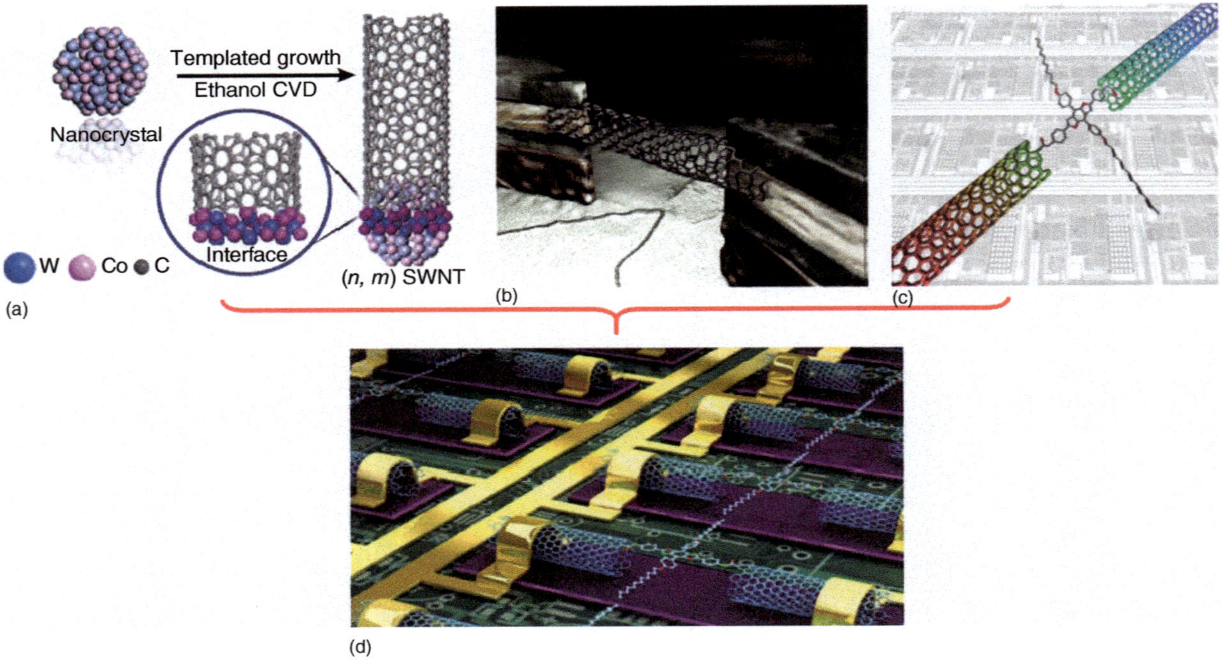

Figure 7.9 Various advances in SWCNT technologies and molecular electronics which can be combined for the development of molecular device based integration circuits. (a) Templated growth of a SWCNT with specified structure. Source: Yang et al. [73]. 2014, Springer Nature. (b) Controlled integration of SWCNTs into pre-designed nano-architectures. Source: Zhou et al. [74]. 2010, IOP Publishing Inc. (c) Molecular junctions by bond molecule to SWCNTs. Source: Feldman et al. [18]. 2008, American Chemical Society. (d) Integration circuits composed of SWCNTs based molecular device. Source: Shen et al. [19]. 2010, Wiley-VCH.

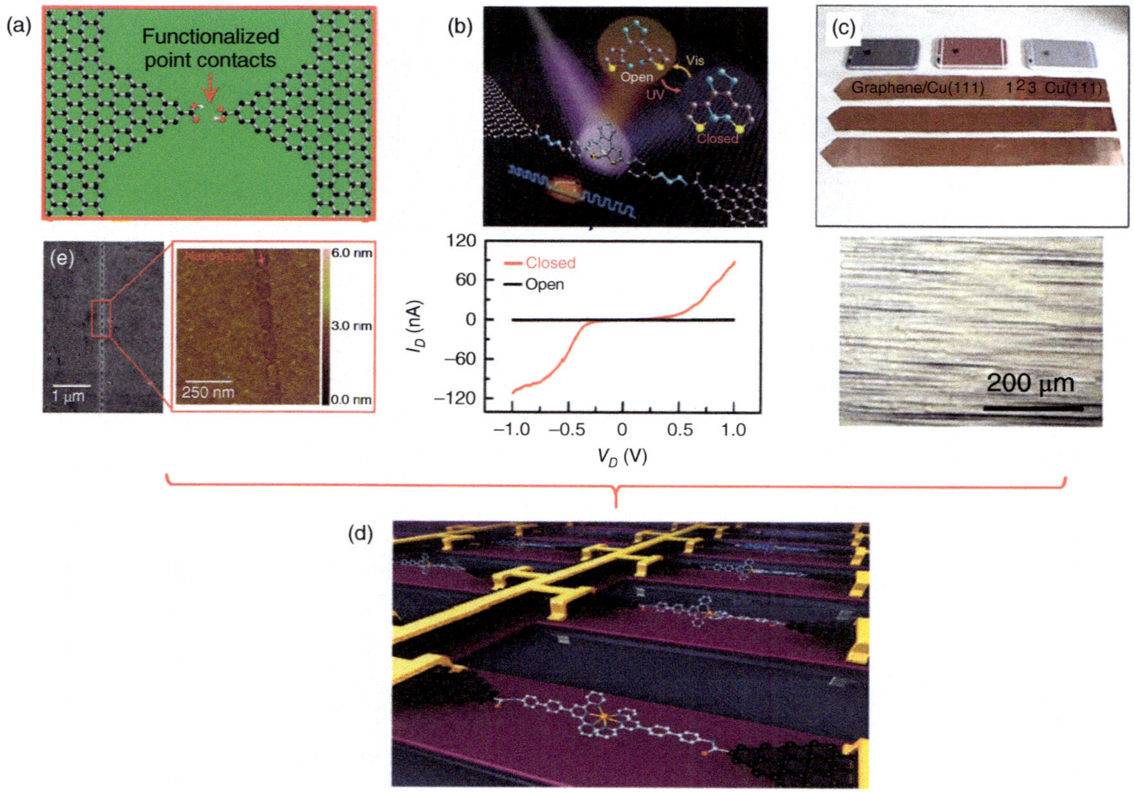

Figure 7.10 Various advances in molecular electronics and graphene technologies, which can be combined for the development of molecular device based integration circuits. (a) Controllable fabrication of indented graphene point contact arrays. Source: Cao et al. [22]. 2012, John Wiley & Sons. (b) Molecular device as stable and reversible photoswitch. Source: Jia et al. [24]. 2016, American Association for the Advancement of Science. (c) Ultrafast epitaxial growth of meter-sized single-crystal graphene. Source: Reproduced with permission from Ref. [70]. Copyright 2017, Elsevier. (d) Integration circuits composed of graphene based molecular device. Source: Reproduced with permission from Ref. [76]. Copyright 2013, The Royal Society of Chemistry.

7.3.2 Self-Assembled Monolayers for Integrated Molecular Junctions

From the prospective of practical application, molecular device based on SAM is quite promising due to several reasons. First, the fabrication process is much less challenging compared to single molecular device because nanoscale size is required only in one direction. Second, the properties of SAM based molecular junction are far more reproducible since transport is a sum of many electron transfer events. Third, the stability of the device is much better. Fourth, the SAM based molecular devices can be in vertical structures, which is advantageous in realizing high-density integration. Up to now, several effective methods to construct molecular junctions have been demonstrated with mass-production capability, high yield and good reproducibility. With those methods, the stability of the device is also greatly improved. Moreover, the first step in using those molecular junctions for integration circuits has been demonstrated.

One of the most effective way to construct SAM based molecular junctions is reported firstly by Hylke B. Akkerman et al. [62]. The key process of their method is to introduce a conductive polymer layer (PEDOT:PSS) onto the SAM before the deposition of the metallic top electrodes. With such method, a high device yield (>95%) as well as excellent reproducibility was achieved. Moreover, molecular device constructed by using PEDOT:PSS has shown excellent stability that stability of large-area molecular junctions under ambient conditions for more than two years is presented [77]. In addition, the good transparent properties of the PEDOT:PSS film enable its application on optoelectronic molecular devices, for example, a reversible conductance switching triggered by optical signals was demonstrated [63]. When combined with flexible substrate, flexible molecular devices can be obtained, showing stable transport properties under severe bending conditions (radius ≤1 mm) and a large number of repetitive bending cycles (≥1000) [78].

On one hand, the molecular device constructed this way shows excellent performance and specific functionalities can be achieved. On the other hand, the fabrication method is reliable showing high yield and reproducibility. Moreover, the fabrication process is a parallel process (Figure 7.11a) and mass production is possible, which is one of the most requirement for integration. Paul A. Van Hal et al. has demonstrated the first result in using PEDOT:PSS based molecular junctions for integration circuits, as showed in Figure 7.11 [79]. In their demonstration, over 20 000 molecular junctions on a single 150-mm wafer were simultaneously fabricated. Integration circuit where up to 200 junctions were connected in series were successfully demonstrated (Figure 7.11c), showing good reproducibility and a yield of unity.

Another effective way for the construction of SAM based molecular junction is using graphene as top electrodes to contact the SAM directly. With splendid thermal/chemical/mechanical stability, atomic thickness, high transparency and good conductivity, graphene is a perfect electrode material. In 2011, Takhee Lee and coworkers reported the application of graphene in molecular junctions [64]. In their method as showed in Figure 7.12A,B, a multilayer graphene

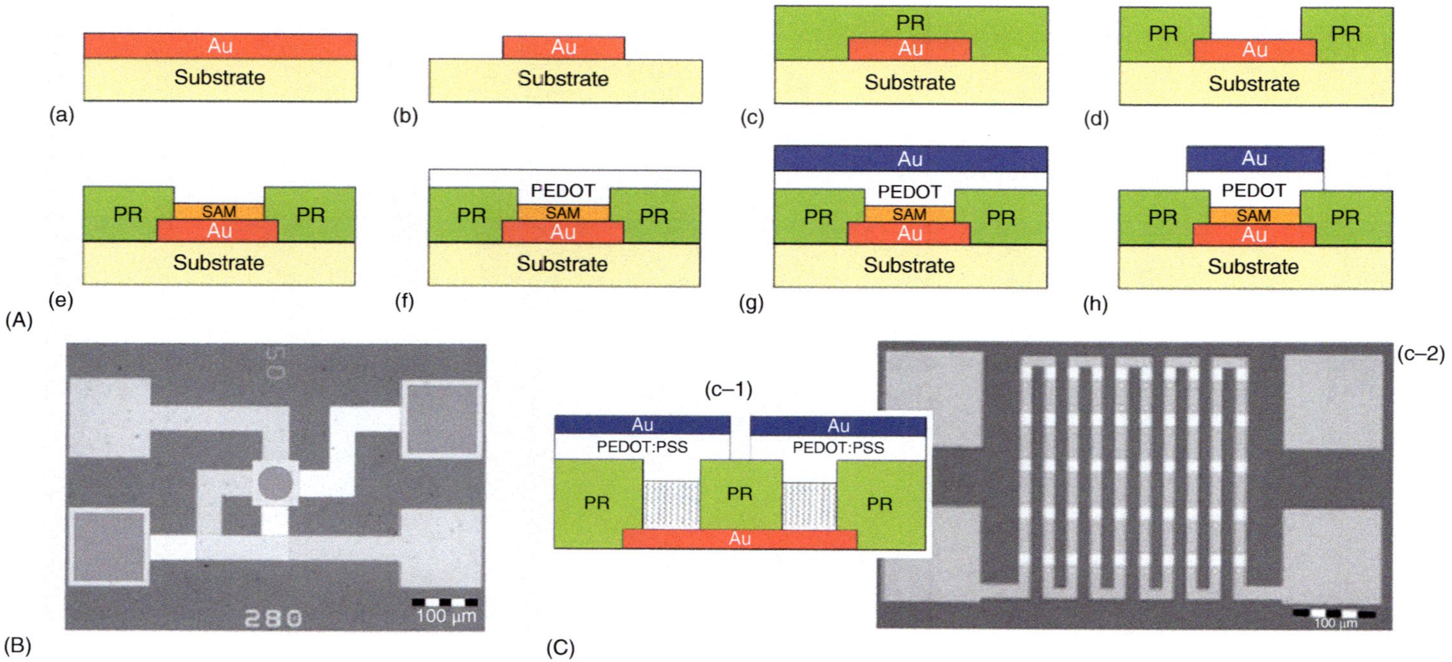

Figure 7.11 Molecular device using PEDOT:PSS as a barrier layer: construction and integration. (A) Schematic presentation of the process flow chart to construct molecular junction, showing mass production capability. (B) Optical micrograph of a circular molecular junction with the leads of a four-point Kelvin probe. (C) Integration of junctions in series. (c-1) Schematic cross-section of a part of a string showing two connected junctions. (c-2) Layout of strings in which a number of molecular junctions are connected in series. Source: Van Hal et al. [79].

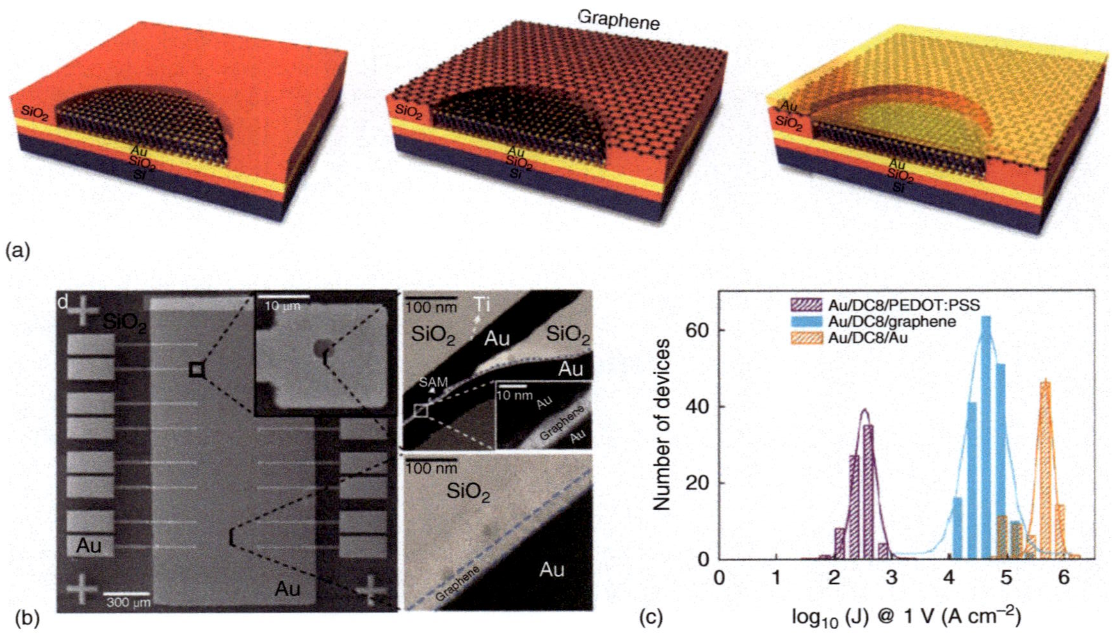

Figure 7.12 Molecular device using multilayer graphene as electrodes. (a) Schematic presentation of the fabrication process. (b) Scanning electron microscopy (SEM) and TEM images of the devices. (c) Histogram of logarithmic current densities at 1 V for working molecular devices with top electrode of PEDOT:PSS, graphene, or Au. Source: Wang et al. [64]. 2010, John Wiley & Sons.

(MLG) film was transferred to the substrate to act as top electrode, after which Au layer was deposited on top of the graphene to enhance the conductivity. A device yield as high as 90% was reported, as well as excellent durability, thermal and operational stabilities, and device lifetimes, all of which are crucial for using molecular devices for integration. Moreover, graphene-based molecular junction shows several advantages over PEDOT:PSS-based molecular devices. First, the contact between graphene and molecule is better as showed in Figure 7.12C, showing larger current density. Second, graphene-based molecular junction has greater thermal stability. Third, graphene-based molecular junction can be constructed with a wide range of molecules regardless of the properties of the isolating layer and the contact groups (hydrophobic vs. hydrophilic).

In the method of Wang et al., MLG films by chemical vapor deposition (CVD) was used. Up to now, the production of high quality graphene with large area has become a reality, and single crystal graphene with size as large as $5 \times 50\ \text{cm}^2$ has been demonstrated [70], so there is no material limitation in using such molecular junction for integrated circuits. However, the manipulation of ultrathin graphene films is still a challenge for molecular device. Graphene transferring process are normally taking place under the support of a polymer layer to maintain the integrity of the graphene. However, it is very hard to get rid of all the polymer residuals on graphene and the process for polymer removal could result in deterioration of SAM. Regarding to this issues, reduced graphene oxide (rGO) film is adopted instead of graphene in molecular devices [65, 80, 81].

Compared to graphene, rGO shows a serial advantage. First, graphene oxide (GO) or rGO are dispersible in various solvents, thus can be processed into films using spin coating, and the size of the film is in principle unlimited. Using a reduction process (either thermal or chemical), GO film can be easily reduced to rGO films with a high conductivity. Second, compared to the production of graphene, rGO can be mass-produced with low cost. Third, rGO film is strong enough that a transferring process can be achieved without the usage of a support polymer layer. Moreover, rGO has very good chemical stability and a work function comparable to that of gold. Therefore, rGO is a very promising material to be used as top electrodes in molecular devices. Two ways are reported to form contact between rGO and SAM. One is by spin coating rGO solution onto SAM [82], the other is by transferring rGO films onto SAM to form the contact [65]. In both methods, device yield as high as more than 90% and good stability are successfully achieved. Using such rGO top contact, devices with functions of light-switch and nonvolatile memory were demonstrated [69, 82]. More importantly, the method used for the fabrication of molecular junctions with rGO are compatible with traditional microfabrication technologies. Parallel fabrication was demonstrated with thousands of devices obtained at same time on a single small chip as showed in Figure 7.13 [65]. The mass production capability and the vertical configuration make such molecular junction very suitable for next generation integration circuits.

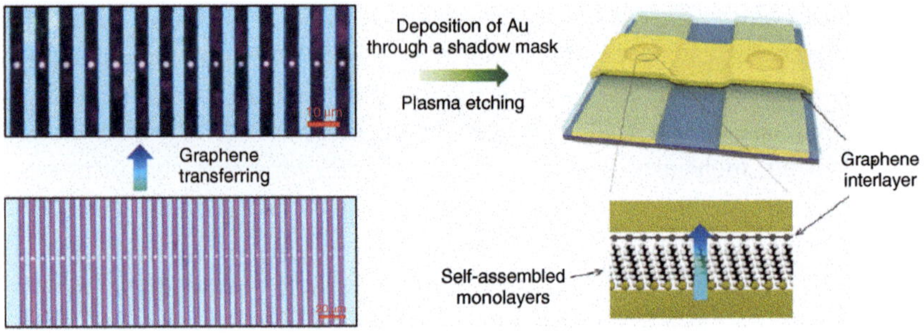

Figure 7.13 Molecular device using rGO film to form the top contact. Processing steps of molecular junctions with solutions-processed graphene films as soft top contacts, showing the capability of mass production and integration. Source: Li et al. [65]. 2012, John Wiley & Sons.

7.3.3 Cross Bar Architecture

Cross bar architecture is one the most famous integration architecture especially for storage/memory, in which multiple input and output lines are arranged in a cross way to form a matrix configuration. The cross point between each input line and output line can be seen as the component device of the matrix, such as a switch. In this way, the matrix is a switch matrix. Cross bar architecture is one of the most frequently used integration strategies in molecular electronics due to several advantages. First, the crossbar geometry makes the fabrication much easier that non-traditional method can be employed, which is quite advantageous for molecular devices. Second, the integration density it could afford is ultra-high because the wire dimensions can be reduced continuously while the number of wires can be increased arbitrarily. Third, with proper design such as integration of devices with diode characteristics with each component, each device in the matrix can be independently addressed. This will not only allow efficient communication with external circuits and systems, but also enable the entire matrix to be fully tested and reconfigured into a working circuit, which makes the cross bar architecture defect tolerant.

In 2003, Yong Chen et al. demonstrated the construction and testing of a crossbar circuits comprising molecular junctions [2]. The fabrication process is quite straightforward and could be used for mass production. Briefly, first, imprint lithography and metal evaporation were used for the fabrication of the bottom electrodes arrays. Second, Langmuir–Blodgett (LB) method was used to deposit a molecular monolayer over the entire substrate. Third, metal evaporation was used to cover the entire substrate with a protective layer of Ti. Fourth, imprint lithography and metal evaporation were used to fabrication the top electrodes arrays, in this process, alignment is required so that the top electrodes arrays are oriented perpendicularly to the bottom electrode to form a cross bar architecture. Finally, a reactive etching process was used to remove the exposed protective layer Ti, leaving the top electrodes arrays intact. With this process, 8×8 crossbar circuit within a 1 μm^2 area with the molecular monolayer sandwiched between the bottom electrodes arrays and the top

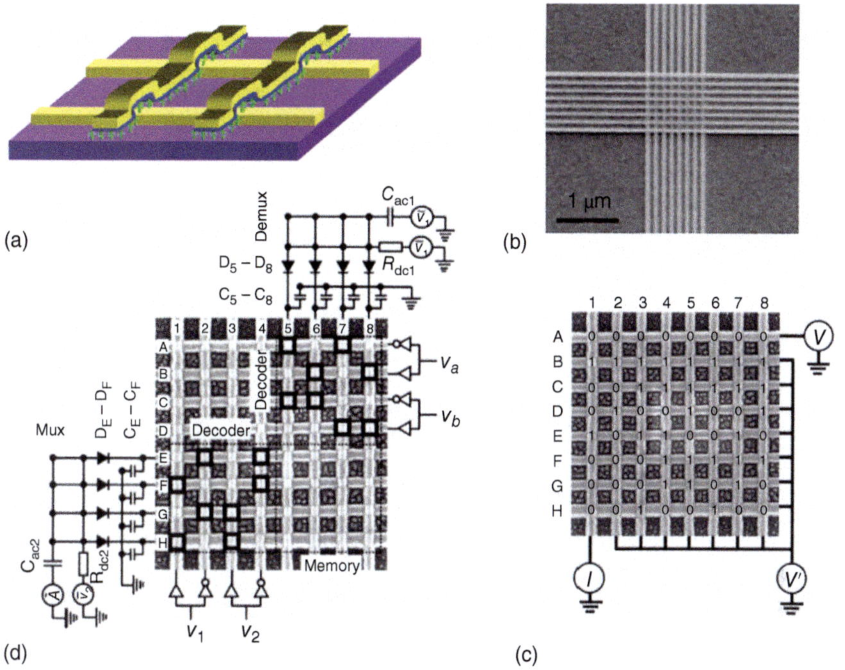

Figure 7.14 Nanoscale molecular-switch crossbar circuits. Schematic representation (a) and SEM image (b) of the crossbar circuit structure. (c) The crossbar as a 64-bit random access memory. (d) The crossbar as a combination of a 4 × 4 demultiplexer (top right), a 4 × 4 memory (bottom right) and a 4 × 4 multiplexer (bottom left). Source: Chen et al. [2]. 2003, IOP Publishing.

electrodes arrays were obtained, as showed in Figure 7.14a. Due to the properties of the molecules ([2]rotaxanes) they used, each device formed at the cross point can be used as a reversible and nonvolatile switch. By using each device as an active memory cell, the 8 × 8 crossbar circuits (Figure 7.14b,c) with a density of 6.4 Gbits cm^{-2} were tested as 64-bit random access memories. The result showed that among the 24 tested 8 × 8 crossbar circuits, ~85% of the devices showed switching properties. Besides high-density memory circuits, a demultiplexer/multiplexer logic integrated with memory (Figure 7.14d) was also demonstrated by such crossbar structures.

In 2007, cross bar circuit composed of 160 000-bit molecular electronic memory circuit were reported and a device integration density as high as 10^{11} bits cm^{-2} were demonstrated (Figure 7.15a–c) [1], the device size of which could be comparable to the dimension of a dynamic random access memory (DRAM) circuit projected to be available by 2020. In their cross bar circuits, molecules ([2]rotaxane) is also used as the data storage elements. Among the 128 devices under test, about 50% of the bits yielded some sorts of switching response (Figure 7.15d). Although the cross bar circuit has large proportion of defects as showed in Figure 7.15d, those defects could be easily identified through electronic testing. By isolating those defects through software coding, the working bits can be reconfigured into a fully functional random access memory circuit.

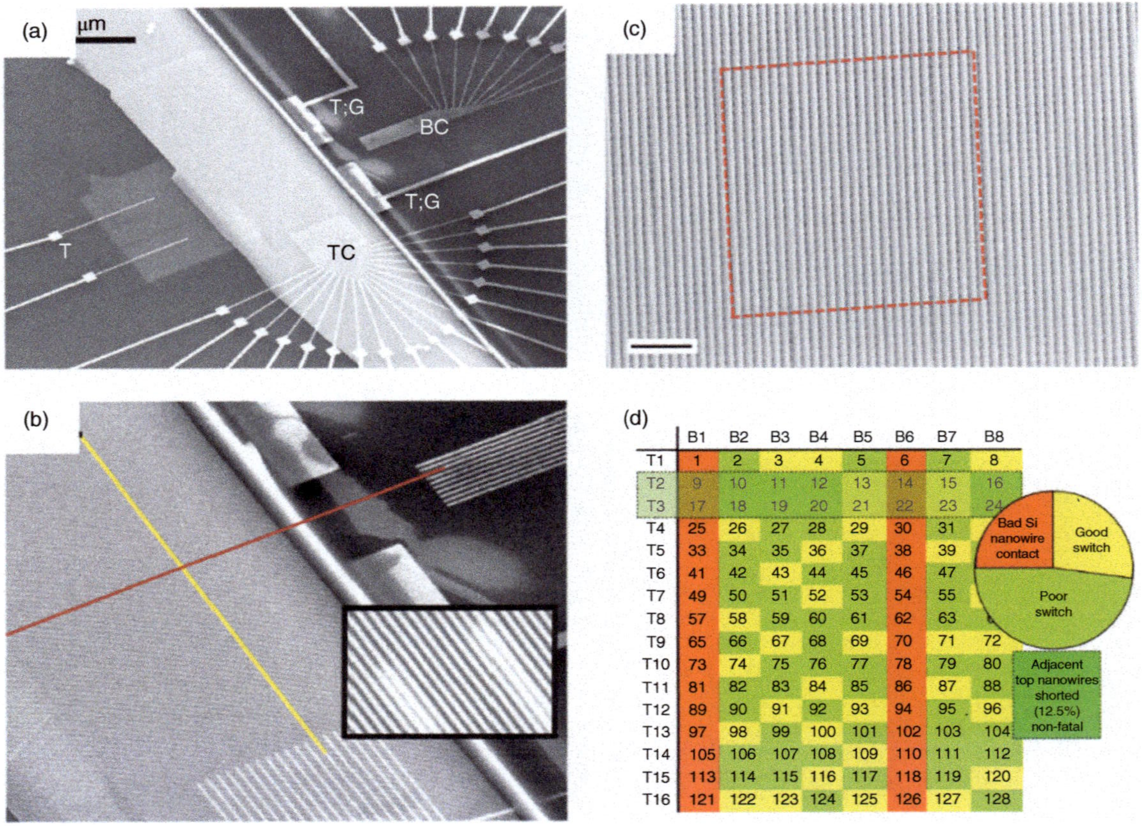

Figure 7.15 Molecular junction based nanowire crossbar memory. (a) SEM image of the entire circuit. (b) An SEM image showing the cross-point of top-(red) and bottom-(yellow) nanowire electrodes. Each cross-point corresponds to an ebit in memory testing. (c) High-resolution SEM of approximately 2500 junctions out of a 160 000-junction nanowire crossbar circuit. (d) Data from evaluating the performance of the 128 ebits within the crossbar memory circuit. A map of the defective and useable ebits, along with a pie-chart giving the testing statistics. Source: Green et al. [1]. 2007, Springer Nature.

Although high-density integration of molecular junction in a crossbar architecture and a defects tolerant scheme are demonstrated, challenges remain to be addressed before the real application of these crossbar circuits. An inherent drawback of crossbar architectures is the crosstalk problem that when a particular bit was selected for writing, other bits will also be written with a relatively high probability. Various useful advices were proposed to overcome this issue, one of the most promising solution is to incorporate diodes at each cross point, so that each device can be addresses independently. Such diode characteristics could arise from the molecules or from the contact. However, their real demonstration in experiment is still shorted. Another significant problem is that the performance of each device is far from satisfying, that the on/off resistance ratios for different devices decay in a quick speed and most device will lose the functionality after a few times of cycles. Thus, intensive researches are still needed in this field before the real application of molecular device for practical integration circuits.

7.4 Beyond Simple Charge Transport

7.4.1 Mechanics

The electrical properties of molecular junctions are determined by a series of parameters, including the highest occupied molecular orbital (HOMO)/lowest unoccupied molecular orbital (LUMO) of molecules, the fermi level of the electrodes, the atomic structure of the electrode–molecule interface, the conformation of the molecules in the gap, etc. Thus, in principle, any operation that can modulate these parameters will introduce changes in the electrical properties of molecular junctions. Using mechanical methods such as applying force on molecular junctions or changing the distance between electrodes (displacements), the position and conformation of molecules and metal–molecule interface could be changed which enable us to study the relationship between the electrical properties and structure of the molecular junctions.

Up to now, advanced measurement techniques (such as STM, C-AFM and a combined system of STM/atomic force microscopy [AFM]) have been applied to study the electrical properties of molecular junctions when subjected to forces or displacements, which can be divided into two categories. One is through simultaneous electrode displacement and electrical measurement (including conductance, current–voltage [I–V], inelastic electron tunneling spectroscopy [IETS], etc.), which is mainly carried out by STM. The other is similar but with force measurement capability provided by a C-AFM or a combined transmission electron microscope (TEM)/AFM system, with AFM provides the capability for force measurement.

In the first approach, an STM probe acts as one electrode and a conductive substrate such as gold substrate acts as the other electrode. After introducing molecules onto the conductive substrate, repeated break-junction cycles are performed to collect the data. In each cycle, the STM probe will be brought into contact with the substrate and pulled back while all the measurements are taken during the pulling process. The contact between the STM probe and the substrate can be indicated by a conductance larger than a few quantum conductance (G_0). During the pulling process, the distance between the two electrodes increases while the conductance decreases. If there are no molecules between the two electrodes (STM probe and the substrate), there will be no step on the curve of current vs. distance trace with conductance smaller than G_0 after the two electrodes are separated. If there are molecules in the contact areas, during the pulling process, there is a chance that one molecule connects between the two electrodes and form a molecular junction. In this case, steps will appear in the current vs. distance trace with conductance smaller than G_0. In addition to current vs. distance trace, electrical measurement such as I–V and IETS at different electrode displacements can also be carried out. By correlating the electrode displacement and the conformation of the molecular junction, under the help of theoretical calculation, the relation between the conformation of the molecular junction and its electrical properties can be explored.

Using this approach, significant progress has been made. For example, Su Ying Quek et al. demonstrated a mechanically controlled reversible binary conductance switch in a single molecule junction [83], as showed in Figure 7.16a. In their experiment, through repeated junction elongation and compression, the conductance of the gold–4,4′-bipyridine–gold single-molecule junctions can be reversibly switched between two states. Using first-principle calculations, the change in the conductance was attributed to the change in the contact geometries of the nitrogen–gold bond. Joshua Hihath et al. demonstrated the application of this method to study the electron–phonon interactions of molecular junction under different strain states through the IETS measurement [85]. Christopher Bruot et al. demonstrated the control over molecular orbital alignment in single molecule junctions through mechanical methods [84], as showed in Figure 7.16b. In their experiment, the conductance of the 1,4′-benzenedithiol (BDT) molecular junction increases by more than 1 order of magnitude during stretching, and decreases during compressing. Through a series of measurements including I–V, conductance–voltage characteristics and IETS, strain-induced shifting of HOMO of the molecules was identified as the reason of the conductance change during stretching and compressing.

In the second approach, either a C-AFM or a combined system of STM/AFM is used for the measurement [86]. Repeated break-junction cycles are also performed for the measurement of the molecular junctions with a probe acts as one electrode and a conducting substrate acts as the other. However, in the second approach, the force during the process can be measured simultaneously, thus enable us to get more information on the molecular junctions. On one hand, force measurement can be carried out directly by AFM system. On the other hand, AFM system also has the capability for electrical measurement. Thus, it can be used for simultaneous measurement of electrical and mechanical properties of the molecular junction [86], as showed in Figure 7.17a, b [87]. Since the size of AFM probe is usually bigger than molecules, more than one molecule could be connected between the two electrodes, the feature of the experiment data could be different from that carried out in an STM system. Specifically, with the increase of electrode displacement, the conductance

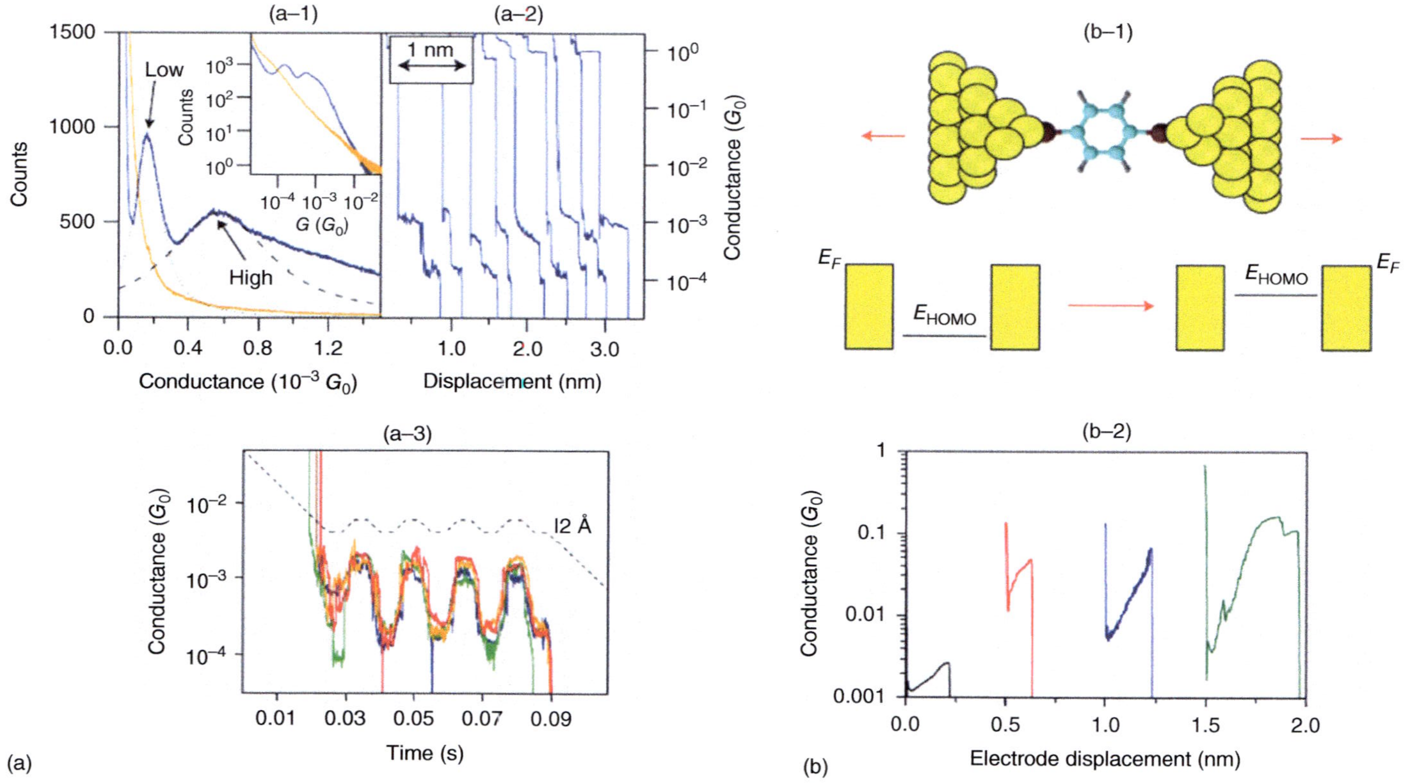

Figure 7.16 Mechanically controlled conductance in molecular junction: (a) Mechanically controlled binary conductance switching of a single-molecule junction: (a-1) Normalized conductance histogram for 4,4′-bipyridine molecule junction (blue trace), showing two conductance peaks. (a-2) Sample conductance traces as function of electrodes displacement, showing two conductance steps in succession. (a-3) Controlled conductance switching by mechanical manipulation of electrodes displacement. Source: Quek et al. [83]. 2009, Springer Nature. (b) Stretching induced conductance change of gold–1,4′-BDT–gold junctions. (b-1) Schematic of the molecular junction (top) and energy diagram (bottom) showing how the energy of the HOMO changes relative to EF of the electrodes as electrode separation increases. (b-2) shows the conductance increases with displacement. Source: Bruot et al. [84]. 2011, Springer Nature.

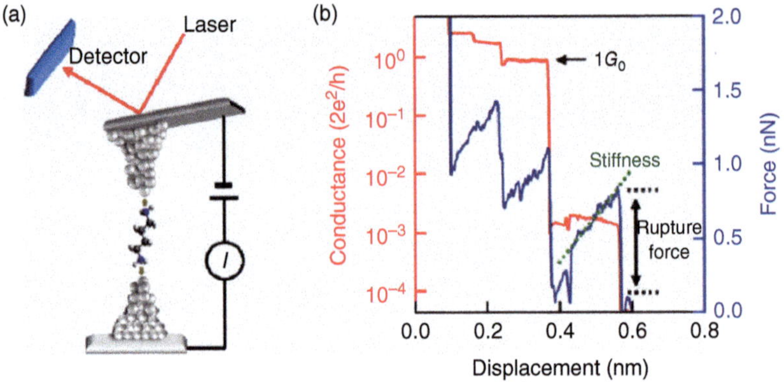

Figure 7.17 (a) Schematic of the C-AFM setup for simultaneously measurement of conductance and force of molecular junction. A molecular junction is formed between the AFM tip and substrate. (b) Sample conductance (red) and force (blue) trace during the pulling process. The both side black arrow indicates rupture force. The slope of the force ramp (as indicated by the dashed green line) indicates the stiffness. Source: Kim et al. [87]. 2015, Korean Chemical Society.

decreases in discrete steps while the force decreases like sawtooth-waves, with most abrupt force drops coincide with conductance changes. Each linear force ramp indicates elastic stretching of the junction and the abrupt drop on the force corresponds to bond rupture or junction rearrangements. Correspondingly, each decrease step in conductance indicates the breakdown of one molecule from the junction or junction rearrangements. As the pulling process continues, the last molecule will break from the electrodes, and there will be no more abrupt decrease in conductance or force. By analyzing the last several conductance steps and force sawtooth-waves, electromechanical properties of single molecular junction can be obtained, including the rupture force (the force needed to breakdown the molecular junction to an open one) and the stiffness of the molecular junction (slope of the force ramps), dependence of the conductance on the applied force, etc.

Using this approach, a lot of information on the properties of the molecular junctions was successfully obtained. For example, Bingqian Xu et al. demonstrated the application of AFM to study of the electromechanical properties of molecular junctions [86]. They found that the Au—Au bond is responsible for the breakdown of the 1,8′-octanedithiol junction, while N–Au is responsible for the breakdown of 4,4′-bipyridine junction. Michael Frei et al. studied the breaking force for amine and pyridine terminated molecules, and found that the strength of the N—Au bond depends on the electronic structure of the molecular backbone [88]. Also in another study of them, junctions formed by alkanes terminated with four different chemical link groups (amine, methylsulfide, diphenylphosphine, and thiol) were studied and the results demonstrated that molecular junctions with thiol-terminated alkane showed different conductance and force features because strong Au—S bond can cause structural rearrangements in the electrodes [89]. Taekyeong Kim compared molecular junctions with same molecule (1,4-butanediamine) but different electrode material, and found that rupture force and stiffness of the molecular junctions for Ag are smaller than those for Au electrodes [87]. Molecular junctions with same electrode material but different molecules are also studied in his report that the rupture force of the amine–Ag bond in conjugated molecule (1,4-benzenediamine) is smaller than in saturated molecule (1,4-butanediamine). In addition to the electromechanical properties of the molecular junctions, more information can be obtained by combined experiment data with theory simulations. For example, Sriharsha V. Aradhya et al. found that, in addition to chemically specific N—Au donor–acceptor bond, pyridine derivatives can bind to Au electrodes through VdW interactions. Such VdW interactions between the pyridine ring and Au electrodes play a key role in the properties of the junctions [90].

In addition to AFM system, a combined system of STM/AFM was also reported for the measurement of both electrical properties and the mechanical properties at the same time. In this experimental configuration, mechanical properties of the molecular junctions are measured through a qPlus tuning fork design (a kind of AFM cantilever, with high spring constant k_0, oscillating with resonance frequency f_0 and a small amplitude of A_0). A similar process with break-junction cycles is also performed to study the molecular junctions. In the pulling process of the junction, a frequency shift Δf will be generated under force, which can be measured at high precision as a function of junction elongation. Such frequency shift can be related to the stiffness k (or force gradient) of the molecular junction in the limit of small oscillation amplitudes through the equation: $k = dF/dz = -(2k_0/f_0)\Delta f$, where F is the force acting on the apex of the tip. Through the stiffness data, the force F and potential energy U of the molecular junction could be obtained through calculations.

Using this approach, significant progress has been made in understanding molecular junctions. For example, N. Fournier et al. successfully demonstrated the change of stiffness, force and potential energy with the evolution of molecule configuration and contacts during the process of lifting a single molecule off the surface [91], as showed in Figure 7.18. Through the experiment data, not only can the relationship between the forces, conductance and junction geometries be obtained, but also critical points can be identified, such as the points where individual bonds to the surface are broken or the molecules are prone to change their contact configuration. C. Wagner et al. demonstrated the quantitative identification of different bonding channels of molecules on metal surface [92]. In their report, not only can the precise shape of the total binding potential be determined but also the contributions from, e.g. local chemical bonds, metal-molecule hybridization, and VdW interactions can be determined through the mechanical data of the molecular junctions.

7.4.2 Thermoelectronics

Thermoelectric effect describes the conversion between temperature differences and electricity. The significance of studying thermoelectricity is self-evident in energy conversion, yet the study of such effect in molecular junction have extra significances. On one hand, studies of thermoelectric effect of molecular junctions not only give more fundamental insight into structure–function relationship, but also pave the way to the development of thermally robust molecular circuits, organic

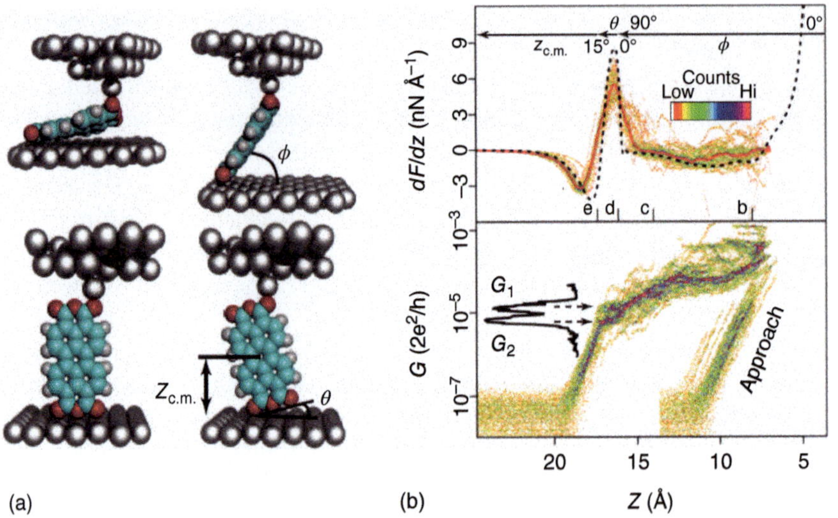

Figure 7.18 Correlation of forces, conductance and junction geometries. (a) Simulated junction geometries during tip retraction process. (b) The force and conductance of the molecular junction during the tip retraction process. Source: Fournier et al. [91]. 2011, American Physical Society.

thermoelectric devices, on-chip cooling scheme, etc. On the other hand, it is predicted that the best thermoelectric efficiency can be achieved when charge transport occurs through a single energy level [93]. In this sense, molecular junction is ideal in maximize the thermoelectric efficiency. Thus, widespread attention has been attracted in studying the thermoelectric properties of the molecular junctions.

The thermoelectric effect of molecular junctions is similar to that in bulk material that when the two sides of a molecular junction are at different temperatures, a voltage difference ΔV, also known as thermovoltage, proportional to the temperature difference ΔT appears when there is no current flow between them. These two parameters could be related by $\Delta V = -S\Delta T|_{I=0}$, where S is the thermopower or Seebeck coefficient. If the two sides of a molecular junction are connected to form a closed loop without applied bias, a thermocurrent will flow between the two electrodes $I_{\text{th}} = GS\Delta T$. Thus by measuring the conductance of the molecular junction G, the temperature difference ΔT, and the thermovoltage or thermocurrent, the Seebeck coefficient of the molecular junction can be obtained experimentally. On the other hand, transport through a single molecular junction can be described by Landauer's formula that the current across them is determined by the transmission probability of electrons and their Fermi distributions on the left and right electrodes. Based on this, the Seebeck coefficient of the single molecular junction can be deducted as: [94]

$$S = -\frac{\pi^2 k_B^2 T}{3|e|} \frac{\partial \ln(\tau(E))}{\partial E}\bigg|_{E=E_F}$$

where T is the mean temperature of the two contacts $T = (T1 + T2)/2$, $\tau(E)$ is the transmission probability of an electron at energy E, e is the electron's charge, E_F

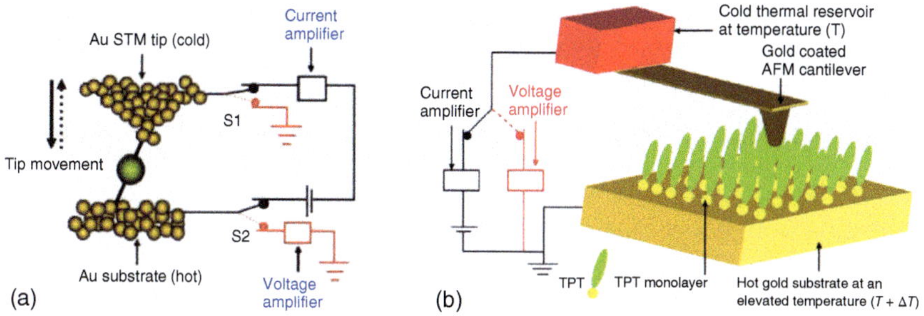

Figure 7.19 Schematic of the experimental set up based on probe-based techniques. (a) Schematic of the experimental set up based on an STM. Molecules are trapped between the Au STM tip kept at ambient temperature and a heated Au substrate kept at temperature ΔT above the ambient. Source: Reddy et al. [95]. 2007, American Association for the Advancement of Science. (b) Schematic of the experimental set up based on an C-AFM. SAM are trapped between the Au AFM tip kept at temperature T and a heated Au substrate kept at temperature $T + \Delta T$. Source: Tan et al. [96]. 2010, American Institute of Physics.

is the fermi level of the electrodes, k_B is the Boltzmann constant. Thus, the most important parameter in determining S is the transmission coefficient around the Fermi level.

The setup for measuring the Seebeck coefficient bears a lot similarity with those developed to measure their conductance except that a temperature difference is needed across the molecular junction. STM based techniques are usually used for the measurement of single molecular junction and C-AFM are usually used for SAM based molecular junctions, as showed in Figure 7.19 [95, 96]. For those probe-based techniques, the most common setup to introduce a tip-substrate temperature difference is keeping the probe at room temperature by connect the tip to a large thermal reservoir at room temperature while heat the substrate with an electric heater to a desired temperature. Apart from those probe-based techniques, mechanically break junctions and electromigration break junctions were also reported to explore the thermoelectric properties of molecular junctions. Compared to probe-based techniques, the technique based on mechanically break junctions and electromigration break junctions can apply a gate voltage to explore the effect of electrostatic on the thermoelectric properties of molecular junction. Through those techniques, after a well-defined temperature difference are established on the two electrodes and a molecular junction are formed which can be judged from the electrical properties, the thermovoltage or thermocurrent can be measured with a voltage or current amplifier. Then the Seebeck coefficient of the molecular junction can be calculated with the value of the conductance of the molecular junction and the temperature difference.

Since Seebeck coefficient is mainly determined from the slope of the transmission at the Fermi level, any parameters that could change the transmission around the Fermi level could modulate the value of Seebeck coefficient. On one hand, these parameters can be used to modulate the Seebeck coefficient of molecular junction. On the other hand, the Seebeck coefficient of the molecular junction can be used to explore the effect of those parameters on the electronic structure and charge transport of molecular junctions. A series of parameters can influence the transmission of the molecular junction such as molecule related parameters (structures, length, conformation, etc.), electrodes related parameters (fermi level of the electrodes), the coupling between molecule and electrodes, electrostatic effect through gate voltage, etc. In this respect, molecular junction is advantageous in thermoelectric device application because great tunability of their properties is possible.

Among those parameters, one of the most important one in determining the transmission of the molecular junction is the relative position between the Fermi energy and molecular levels (mainly HOMO and LUMO levels of the molecules). One pioneer work by Magnus Paulsson and Supriyo Datta demonstrated that for molecular junction, the thermoelectric voltage is large enough to be measured and the Seebeck coefficient could give information on the sign of the charge carrier and the location of the Fermi energy relative to the molecular levels [94]. Such conclusion is of great importance because such electronic information is crucial in understanding the electrical properties of molecular junction yet not readily accessible through other experiments. In 2007, Pramod Reddy et al. reported the determination of the relative position between E_F and molecular levels [95]. On one hand, they experimentally measured the Seebeck coefficients of molecular junction. On the other hand, they calculated the transmission and Seebeck coefficients as a function of the relative position of the Fermi level with respect to the HOMO and LUMO levels through nonequilibrium Green's function formalism in conjunction with extended Hückel theory, as showed in Figure 7.20. As can be seen from Figure 7.20 that S is positive if E_F is closer to the HOMO level (p-type) and negative if E_F closer to the LUMO level (n-type). By comparing the experimental data with the calculation results in Figure 7.20, the relative position between E_F and HOMO was determined. After those pioneer works, Seebeck coefficient measurement has become a popular approach to explore the transport properties of molecular junctions. For example, Jonathan R. Widawsky et al. reported the determination of the dominant transport orbital for amine–Au and pyridine–Au linked molecular junctions [97]. They found that the Seebeck coefficient is negative for pyridine–Au linked molecular junctions and positive for amine–Au linked molecular junctions, suggesting that for pyridine-terminated molecules, the dominant transport in the molecular junction is through LUMO while for amine-terminated molecules, the dominant transport is through HOMO.

A lot of work has been done in studying the thermoelectric properties of molecular junction to explore the relationship between the Seebeck coefficient and the structure of molecular junctions, so as to find a way to modulate the thermoelectric properties of molecular junction on one hand and get in touch with the fundamental electronic information of them on the other hand. Up to now, remarkable progress has been made that the influence of molecular length, molecular structures, electrode materials, gate voltage, etc. on the thermoelectric properties of molecular junctions has been explored [98–103], which could be used further to extract electronic information of molecular junctions. Comprehensive reviews have

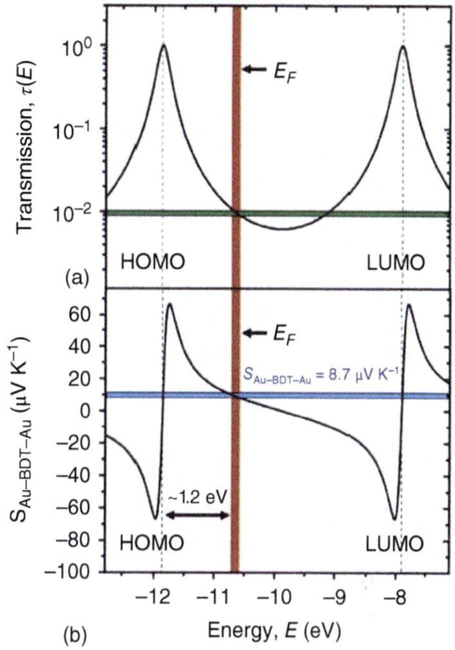

Figure 7.20 Relating the Seebeck coefficient of Au–BDT–Au junction to the position of Fermi level. Theoretical prediction of the transmission function (a) and Seebeck coefficient (b) of a Au–BDT–Au junction plotted as a function of the relative position of the Fermi level of the Au electrodes with respect to the HOMO and LUMO levels. When the measured value of $S_{Au-BDT-Au} = +8.7 \pm 2.1\,\mu V\,K^{-1}$ (blue band) is shown, it is clear that the Fermi level is ~1.2 eV above the HOMO level. Source: Reddy et al. [95]. 2007, American Association for the Advancement of Science.

been published in this regard [104, 105]. Here we will just give a few examples. Kanhayalal Baheti et al. studied the effect of chemical structure of molecules on the properties of molecular junction [102]. As can be seen in Figure 7.21, they found that substituents on 1,4-BDT molecule generate small changes in transport no matter whether they are electron-withdrawing or -donating groups. However, when the thiol end groups have been replaced by cyanide end groups, the transport is radically changed that the dominate orbit is LUMO instead of HOMO in BDT molecular junction. Aaron Tan et al. studied the coupling between molecule and electrodes on the properties of the molecular junction and found that reduction in coupling could reduce the conductance greatly while have a much smaller influence on the Seebeck coefficient, as can be seen from Figure 7.21A [106]. In their study, the relationship between the length of molecules and the Seebeck coefficient of the molecular junction was also explored experimentally that Seebeck coefficient of thiol-terminated aromatic molecular junctions increased linearly with length as can be seen from Figure 7.21B,C.

Although we are still far away from using molecular junctions as an energy converter in practical applications, the studies on the thermoelectric properties of molecular junctions are of great significance in molecular electronics because of the extra information on the transport of molecular junctions it can afford is irreplaceable and hard to obtain through other approaches. In this case, the structure-transport properties of molecular junction can be revealed in a larger degree, which could give new sight in molecular electronic device designing.

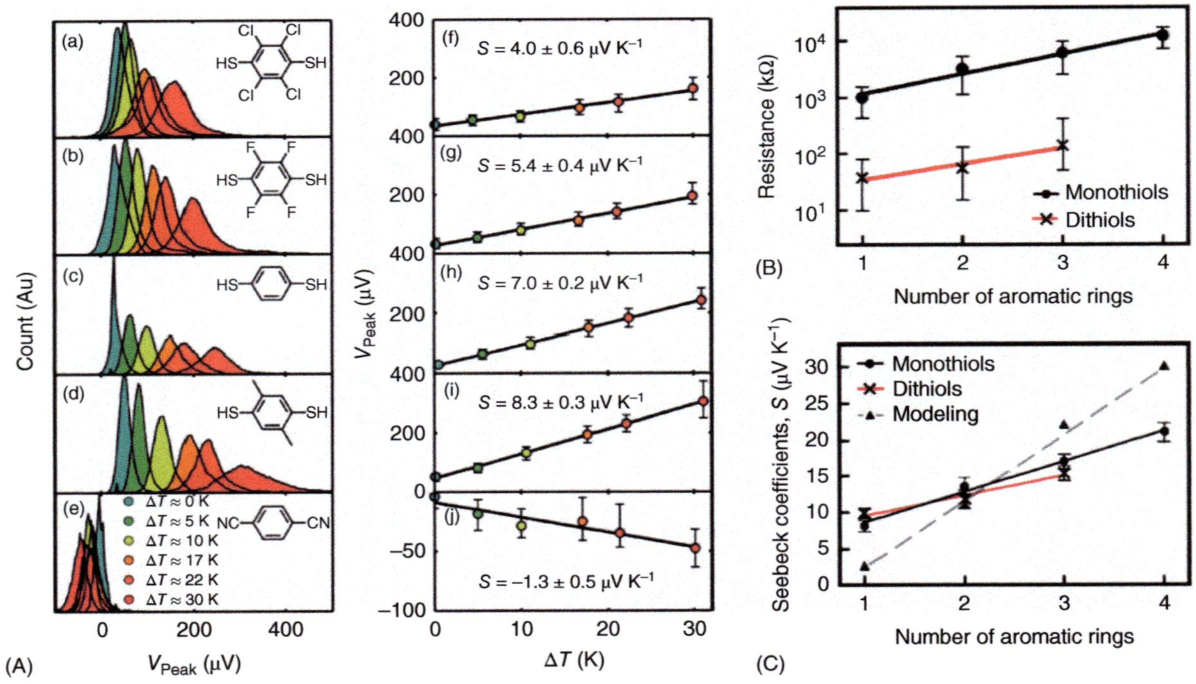

Figure 7.21 (a) The influence of chemical structure of molecules on the Seebeck coefficient of molecular junction. Left: normalized histograms of thermoelectric voltage under each ΔT; Right: histogram peaks of Seebeck coefficient as a function of temperature differential. (b, c) The influence of coupling between molecule and electrodes on the conductance and thermoelectric of the molecular junctions. Source: Tan et al. [106]. 2011, American Chemical Society.

7.4.3 Quantum Interference

Due to the wave nature of electrons and the small dimensions of molecular junctions respect to the electronic phase coherence length, quantum interference would arise when there are more than one pathways for the electron waves to propagate. When such effect takes place in a molecular junction, the overall conductance of the molecular junction is not anymore the conductance sum of each pathways, instead, a higher conductance would be resulted if constructive interference happens and a lower conductance if destructive interference happens. Different from the interference of light, which happens between contributions from difference paths in space, the interference of electrons in molecular junction happens between contributions from difference paths in energy (transmission through different molecular orbitals) [107, 108].

Molecular junctions with various molecules has been shown to exhibit quantum interference [109]. One of the most classic system studied composed of the para, meta, and ortho substituted benzene rings, for which the general conclusion by both theoretical calculation and experiments is that molecular junction of meta substituted benzene shows lower conductance than that of ortho and para [110, 111]. The lower conductance in meta substituted benzene rings is very counter intuitive because the distance between the two electrodes is shorter and should lead to exponentially increased tunneling current. Such big difference of conductance was attributed to the occurring of destructive interference in meta substituted benzene rings. Based on the studies of this molecular system, the signature of destructive interference in molecular junction was identified.

For a molecular junction without quantum interference, the transmission $T(E)$ peaks normally exhibit only around the orbital energies of the molecules because of resonant tunneling, and in the energy gap, $T(E)$ is featureless without showing any peaks. However, when quantum interference takes place in a molecular junction, the transmission will show a big difference. From theoretical calculation by Hückel-model calculations, which only consider the π electrons in the system, two striking signatures of destructive interference in molecular junctions were identified in the energy gap: a much lower transmission probability and a sharp dip in $T(E)$ in the logarithmic scale (also known as transmission anti-resonance) with the resonance transport undisturbed as showed in Figure 7.22a. Compared to conductance, the transmission anti-resonance is a more striking signature for destructive interference since the conductance can be influenced by various factors. Hückel-model calculations is accurate enough for molecular system where the dominate transport are the π electrons. However, for a molecular system where σ electrons also contribute significantly to charge transport, according a more realistic calculation [111], the overall transmission of the molecular system would be the result of both π electrons and σ electrons. In this situation, the sharp dip in the transmission of the π electrons would be offset to some degree by the σ electrons, as showed in Figure 7.22b.

Experimentally, the occurrence of destructive interference in molecular junction is mainly identified through electrical measurement. For example, cross-conjugated

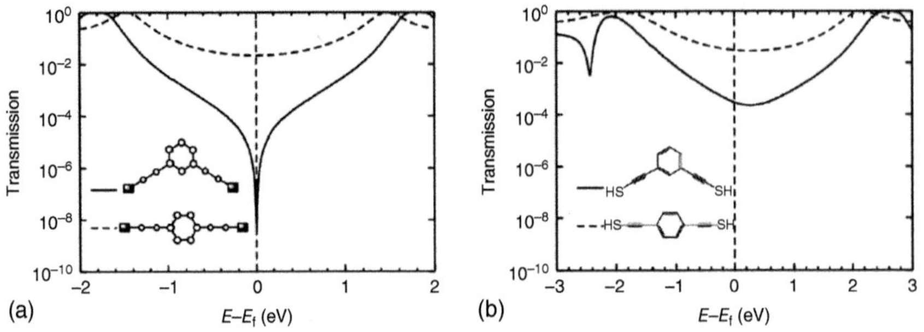

Figure 7.22 The transmission through para and meta substituted benzene rings as calculated by the Hückel method (a) and by gDFTB (b). Source: Solomon et al. [111]. 2009, John Wiley & Sons.

molecules are the most famous molecules showing destructive interference while linear-conjugated molecules with similar structure and length are not. By comparing the electrical properties (mainly conductance and differential conductance (dI/dV) vs. bias voltage (V) curves) of molecular junctions of cross-conjugated molecules to that of linear-conjugated molecules with similar molecule structure and length, the signature of destructive interference can be identified. For example, Constant M. Guedon et al. reported the direct observation of destructive quantum interference at room temperature [112]. First, five kinds of rigid π-conjugated molecules were synthesized, the structure of which were shown in Figure 7.23a. The key comparison was made among AQ-DT, AQ-MT and AC-DT, in which AQ-DT and AQ-MT are both cross-conjugated, whereas AC-DT are linearly conjugated. The three molecules have a similar length and similar HOMO–LUMO gap. Then SAM on gold substrate were obtained out of these molecules. Through conducting AFM, the electrical properties of these SAM based molecular junctions formed between the AFM tip and gold substrate (as showed in Figure 7.23b) were obtained. From the experimental data, two main phenomenal were observed. First, the magnitude of conductance of cross-conjugated molecules (2,6-bis[(4-acetylthiophenyl)ethynyl]-9,10-anthraquinone [AQ-DT] and 2-[(4-acetylthiophenyl)ethynyl]-6-(phenylethynyl)-9,10-anthraquinone [AQ-MT]) are almost two orders lower than that of linearly conjugated molecule 2,6-bis[(4-acetylthiophenyl)ethynyl]anthracene (AC-DT), as showed in Figure 7.23c,d. Second, a strong dip on the plots of log(dI/dV) vs. bias voltage was observed in AQ-MT, as showed in Figure 7.23e, which is a direct evidence for the dip on the transmission, thus a direct and strong signature of destructive interference. Such dip is a very nice signature for destructive interference, however, it is hard to get because several condition must be met. First, the contribution from other transport channels such as σ electrons can offset the contribution from π electrons, making the anti-resonance feature not obvious. Second, the dip on the plot of log(dI/dV) vs. bias voltage can only be observed when the anti-resonance feature lies very close to the fermi level. Because of this, in most experiment, the only signature that can be observed for destructive quantum interference is a lower conductance.

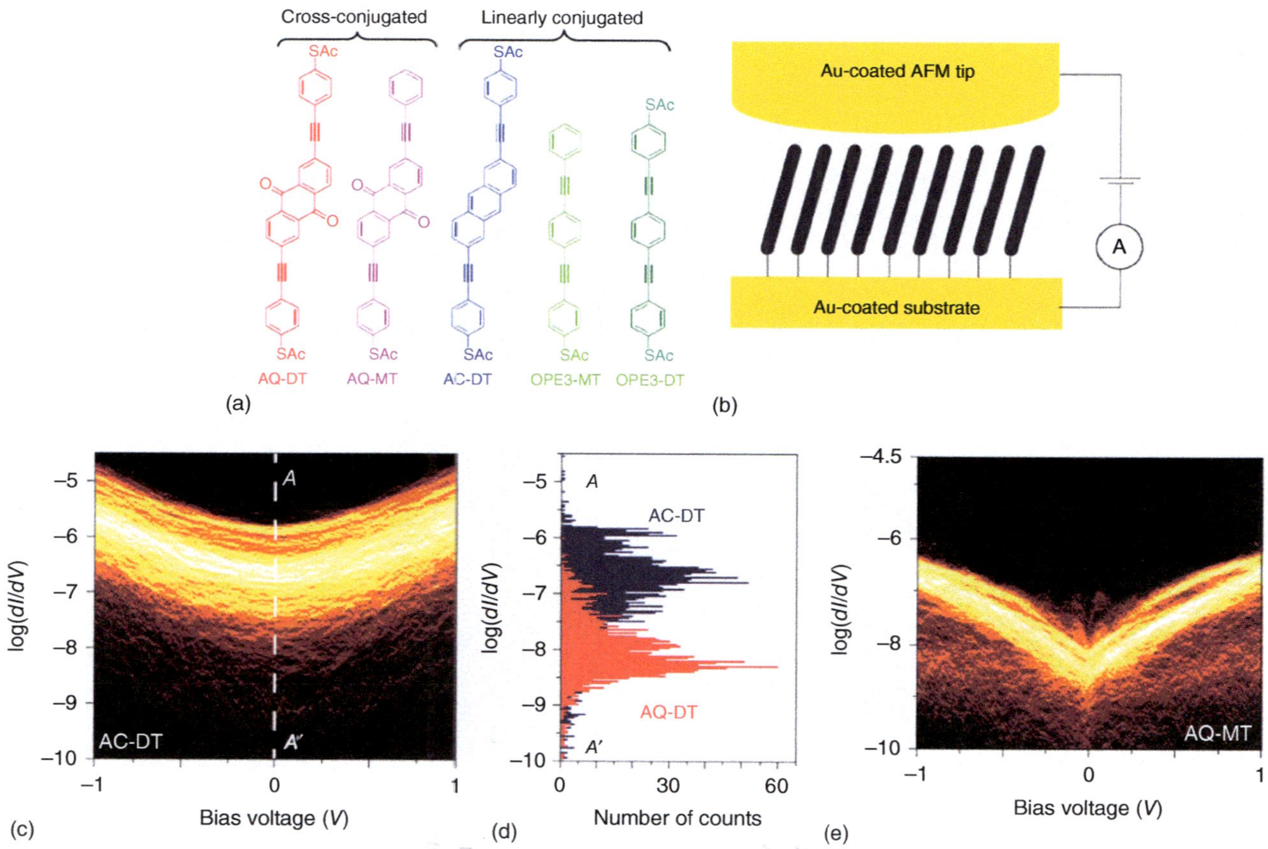

Figure 7.23 The identification of destructive interference in molecular junction. (a) The chemical structure of molecules used in which AQ-DT and AQ-MT are cross-conjugated, whereas AC-DT is linearly conjugated. (b) Schematic view of the molecular junction for electrical measurement. (c) Logarithmically binned two-dimensional histogram for dI/dV values vs. bias voltage V for AC-DT; the color scale indicates the number of counts (black, no counts; white, more than 40 counts). (d) A one-dimensional histogram of zero-bias conductance for AC-DT and AQ-DT. (e) Logarithmically binned two-dimensional histograms of dI/dV vs. bias voltage V for AQ-MT. Source: Guédon et al. [112]. 2012, Springer Nature.

Up to now, destructive quantum interference effect has been identified in various molecules both experimentally and theoretically [107, 110–113]. The fundamental factor that determines the occurrence of such effects is the chemical structure of the molecule. Using theoretical calculation, the occurrence of such effect in a specific molecule could be identified by checking the transmission. However, if a direct connection between the structure of the molecules and the destructive interference effect could be made, the occurrence of such effect can be predicted easily by just looking at the chemical structure of the molecules. Efforts have been devoted in this consideration and great advances have been made. A comprehensive summary for predicting such effect through the molecule structure was made in Ref. [108]. Here we will introduce one of those works to show how to relate the structure to the occurrence of destructive interference.

A simple graphical prediction scheme to predict the occurrence of destructive interference from the chemical structure of molecule of conjugated hydrocarbon systems was developed by Troels Markussen et al. [114]. In this method, first, a continuous path line from the sites connected to the left electrode to the sites connected to the right electrode through the backbone of the molecule was drawn. Second, if there are sites not covered by the path line, the remaining sites can be paired up with its neighbors when possible. After these two operations, if there are remaining isolated sites, then destructive interference effect is expected to show in this molecule, otherwise, there will be no destructive interference. The first verification of this scheme was made on the famous ortho, meta, and para connected benzene rings, as showed in Figure 7.24. Using this scheme, destructive interference is expected to take place only in meta connected benzene rings as there is remaining site, which is in accordance with previous conclusion experimentally or theoretically. The second verification of this scheme was made on 10 different configurations of anthraquinone (labeled as L1–L5 and N1–N5), which differ only in the position of the two oxygen atoms, as showed in Figure 7.25a. Through the graphical prediction scheme, destructive interference is expected to happen only in L1–L5 among the 10 configurations, as showed in Figure 7.25a, b. Through a series of calculations, a much smaller conductance and an anti-resonance transmission were indeed observed in L1–L5 instead of N1–N5, as showed in Figure 7.25c, which confirmed the prediction.

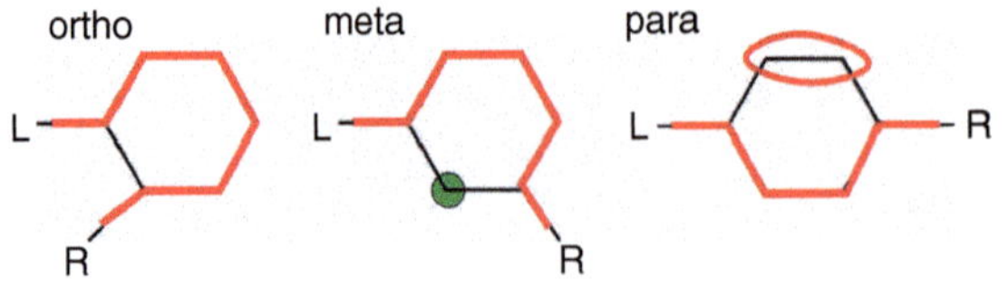

Figure 7.24 Demonstration of using the graphical prediction scheme in ortho, meta, and para connected benzene rings. After drawing a continuous path and pairing up the remaining site, there is no remaining site left for ortho and para connected benzene rings, while there is one for meta connected benzene rings, as marked by the green spot. Source: Markussen et al. [114]. 2010, American Chemical Society.

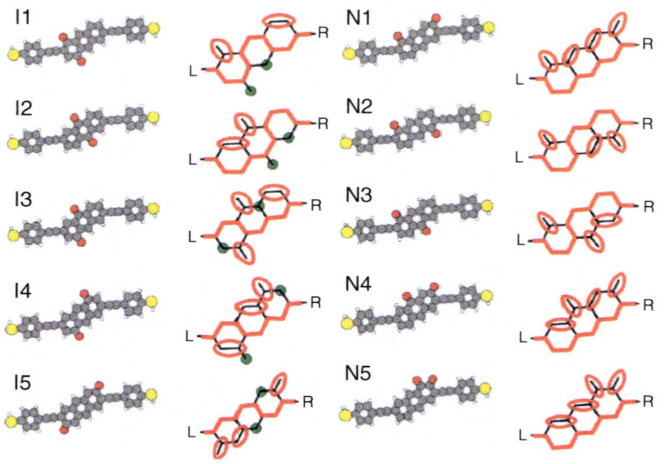

(a)

Molecule	G (G_0)	Molecule	G (G_0)
I1	7.71×10^{-8}	N1	1.38×10^{-2}
I2	1.82×10^{-8}	N2	8.57×10^{-2}
I3	3.03×10^{-4}	N3	7.09×10^{-3}
I4	3.55×10^{-4}	N4	8.98×10^{-3}
I5	8.84×10^{-4}	N5	6.19×10^{-3}

(b)

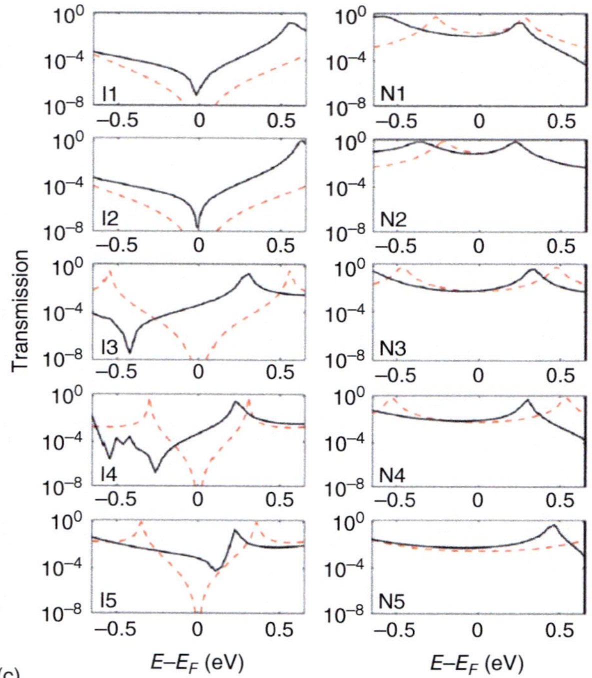

(c)

Figure 7.25 Confirmation of the graphical prediction scheme through a serial anthraquinone-based molecule. (a) Configurations of the 10 molecule labeled as I1–I5 and N1–N5. For I1–I5 it is not possible to connect the contact sites without creating isolated sites (green spots). For N1–N5 one can draw a continuous path along the sites and pair up all the remaining sites. (b) Density functional theory (DFT) conductance values of the 10 molecules. (c) Transmission functions calculated from DFT (black solid lines) and tight binding (red dashed lines) for the structures I1–I5 and N1–N5, respectively. Source: Markussen et al. [114]. 2010, American Chemical Society.

From the previous researches, it can be seen that the fundamental factor to determine whether destructive interference will happen or not in molecules is the chemical structure of them. Thus, quantum interference is of great importance in molecular electronics because such effect could provide new insight in controlling the electrical transport through chemical designing of the molecules and lead to high performance molecular devices with various functionalities such as switches [67, 115], transistors [116, 117], logic gates, etc.

7.4.4 Spintronics

The constituent elements of organic molecules are mostly low-weight atoms, and spin-orbit couplings scales with Z^4, in which Z is the atomic number, thus the spin-orbit couplings in organic molecules are weak. Moreover, hyperfine (HF) interactions are also weak in organic materials. Due to the weak spin–orbit coupling and hyperfine interaction, the spin of a carrier only interacts with the organic environment weakly, thus long coherence time was expected in organic molecules based device. Moreover, through structure engineering and advanced synthesis methods, molecules with electronic properties and functionalities spanning an extremely large range could be obtained, offering opportunities unachievable with inorganic materials. Therefore, organic molecules have drawn more and more attention in the field of spintronic and subfields named as organic spintronic and molecular spintronic have been developed to study the injecting, manipulating and detecting spins in organic molecules based devices.

One of the most basic spintronic devices in inorganic world is a three-layer structure with a non-magnetic layer embedded between two ferromagnetic (FM) electrodes. In such a device configuration, one FM electrode acts as spin injector, the other FM electrode acts as spin detector, and the non-magnetic layer acts as a transport channel for the spin polarized current. The resistance of the whole structure depends on the relative orientation of the magnetization direction of the two electrodes, thus enables two resistance states for detection. If the non-magnetic layer is insulating and thin enough, the transport between the two electrodes will be dominated by tunneling. In this case, the device is named as magnetic tunnel junction (MTJ). Otherwise, if the non-magnetic layer can transport electrons not through tunneling, then the device is named as SV. Similar structure configuration was also adopted to study organic molecules to exploit the freedom of spin, in which the non-magnetic layer was organic materials. Apart from that, molecular spintronic devices with single or few molecules were also studied, which have no equivalent in inorganic spintronic device. Here in this part, the spintronic devices based on organic thin films will not be discussed, instead we will mainly focus on spintronic devices based on SAM or single or few molecules.

7.4.4.1 SAM-Based Magnetic Tunnel Junctions

Although long coherence time makes SAM-based MTJs very promising, the construction of them used to be a great challenge. Apart from the challenges mentioned in Section 7.2 in making contact to SAM in a reliable way, the oxidation issue associated with FM electrodes makes the construction of them even harder. Nevertheless, persistent efforts have been made and progresses have been demonstrated. In 2004, J.R. Petta et al. reported the study of MTJs of Ni–octanethiol–Ni, in which SAM of molecule octanethiol acts as the barrier between two FM electrodes, as showed in Figure 7.26a [118]. Through low temperature measurement, magnetoresistance up to 16% was demonstrated (Figure 7.26b), showing that spin-conserving transport in molecular devices is possible. Wenyong Wang and Curt A. Richter also reported the measurement of Ni–octanethiol–Ni junction [119]. In their report, IETS was used to provide unambiguous experimental evidence that molecules does play a role in the junction. However, a quite different behavior regarding the relationship between the magnetoresistance and magnetic field was reported, as shown in Figure 7.26c. In 2012, great progress has been made by Marta Galbiati et al. that stable and efficient spin transport properties were identified in SAM-based MTJs, in which $La_{2/3}Sr_{1/3}MnO_3$ and Co act as the two FM electrodes and dodecylphosphonic acid SAM is used to define the tunnel junction, as showed in Figure 7.26d [120], and a magnetoresistance as high as 35% were demonstrated, as showed in Figure 7.26e.

7.4.4.2 Molecule Based Spin-Valves or Magnetic Tunnel Junctions

Different from inorganic spintronic devices, where the states detection depends on the relative orientation of the magnetization direction of the two electrodes, in molecule based spin-valves or MTJs, the two magnetization directions come from either electrodes or molecules. Depends on the origins of the magnetization, those device can be divided into three classes: (i) two FM electrodes with non-magnetic molecule, (ii) one FM electrode with magnetic molecule, and (iii) non-magnetic electrodes with at least two magnetic molecules or one molecule possessing two magnetic centers. With analogous to its inorganic counterpart, the detection in molecular spin-valves also relies on the mutual alignments of the magnetization directions of the two magnetic elements.

For the first class of devices that composed of two FM electrodes with non-magnetic molecules, one of the mostly studied system is carbon nanotube. By connecting carbon nanotube between two FM electrodes, a molecular spin-valve can be obtained. In such system, an obvious magnetoresistance was observed in different reports using single-wall carbon nanotube or multiwall carbon nanotube with different kinds of FM electrodes [121–123]. Especially, a large magnetoresistance as high as 61% were reported by Luis E. Hueso et al. in a device made up of a multiwall carbon nanotube that spans between two $La_{0.7}Sr_{0.3}MnO_3$ FM electrodes, as showed in Figure 7.27a,b [124]. The good performance was enabled by the long spin lifetime and high Fermi velocity in nanotubes, the high spin polarization in the manganite electrodes, and the resistance of the interfacial barrier for spin injection. Apart from carbon nanotube, C_{60} based molecular devices were also explored showing large magnetoresistance [126]. Other kinds of organic molecules were also explored, but mostly through an STM system consists of two FM electrodes bridged by an organic molecule. This is because it is still a challenge to contact single molecule in a device configuration even with nonmagnetic electrodes. Nevertheless, advances have been

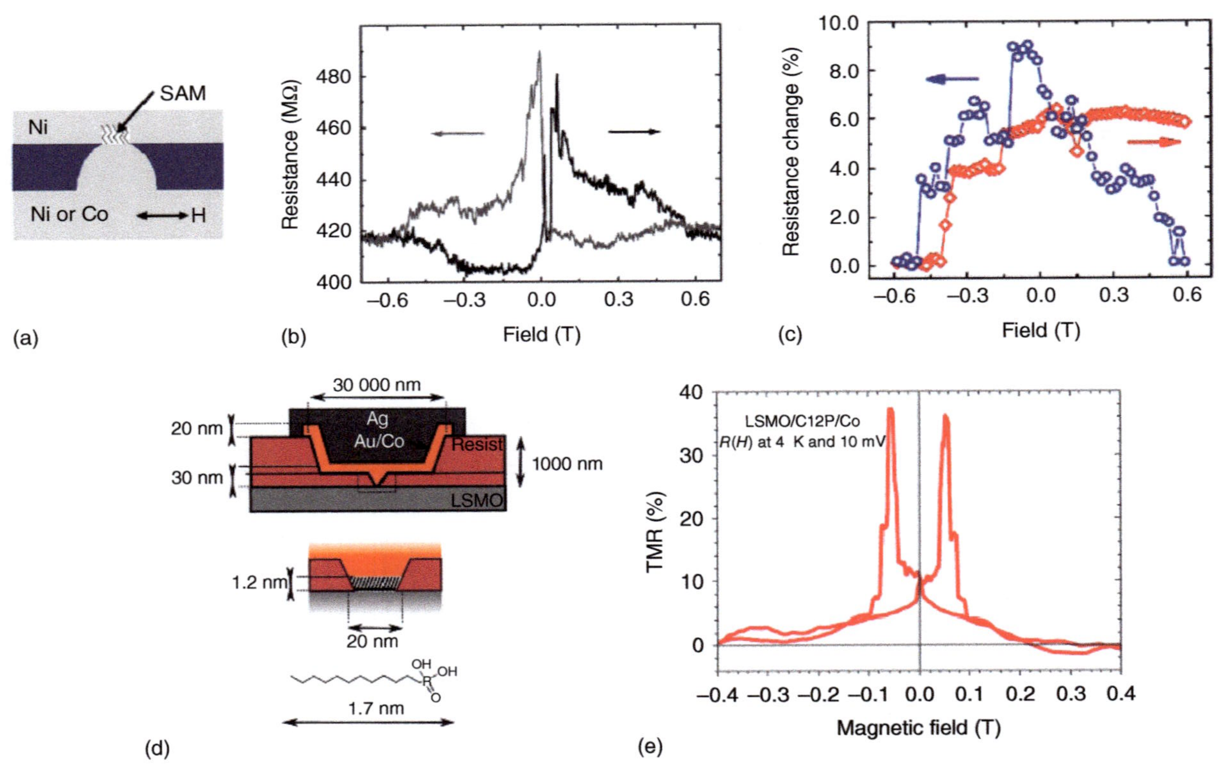

Figure 7.26 SAM-based magnetic tunnel junctions. (a) Device structure through nanopore technique, with SAM sandwiched between two electrodes. (b) Resistance vs. magnetic field for one Ni–octanethiol–Ti–Ni junction, showing an obvious magnetoresistance. Source: Petta et al. [118]. 2004, American Physical Society. (c) Resistance vs. magnetic field for one Ni–octanethiol–Co junction. Source: Wang et al. [119]. 2006, American Physical Society. (d, e) Magnetic tunnel junctions made up of dodecylphosphonic acid SAM sandwiched between $La_{2/3}Sr_{1/3}MnO_3$ and Co electrodes, and its resistance vs. magnetic field curve. Source: Galbiati et al. [120]. 2012, John Wiley & Sons.

made. For example, through mechanically controllable break junction, Ryo Yamada et al. succeeded in the construction of Ni/single benzene–1,4-dithiol/Ni molecular junctions (Figure 7.27c), and a negative magnetoresistance as large as 30% was observed (Figure 7.27d) [125].

For the second class, the two magnetization directions are provided by one FM electrode and one magnetic molecule. The construction of single molecular device is still a big challenge, let alone single molecular device with electrodes of two different materials (one is magnetic and the other is non-magnetic). Thus, up to now, most of the studies on this kind of devices are carried out though spin polarized scanning tunneling microscopy (SP-STM), where the magnetic molecule is probed by an FM STM probe, and great progress has been made [127, 128]. However, due the large STM system, such device cannot be used for practical applications, thus will not be discussed here.

For the third class, the devices are composed of non-magnetic electrodes and the magnetoresistance comes from the spin state of the organic molecules instead of electrodes. The most classical model system is made up of at least two magnetic molecules or one molecule possessing two magnetic centers, which act as both magnetic polarizers and analyzers [129]. With such system, a magnetoresistance as large as 300% was demonstrated by M. Urdampilleta et al. [26]. In their device, a SWCNT was bridged between two non-magnetic electrodes, and two SMM $TbPc_2$ are grafted to the carbon nanotube through π–π interactions. By controlling the magnetic momenta directions of the two grafted molecules through magnetic field, the conductance of the carbon nanotube can be greatly changed through supramolecular interactions, as showed in Figure 7.28a. Such a big magnetoresistance was attributed to different magnetic momenta alignment of the two absorbed SMM molecules that parallel alignment of the magnetic momenta of the two SMM molecules allows electron with one spin to flow easily while antiparallel alignment hinders electron flow of both spins, as showed in Figure 7.28b.

7.4.4.3 Single Molecular Spin Transistor

With no equivalence in inorganic spintronic devices, in single molecular spin transistor, the conductance is determined only by the spin state of the single molecule. In this kind of device, non-magnetic electrodes such as Au are used to connect a magnetic molecule in a transistor configuration. Depending on the coupling strength between the molecule and the electrodes, different interesting phenomena were observed, such as Kondo effect [6, 130], complete current suppression and negative differential conductance [131], nonlinear variations of energy levels with magnetic field [132], etc. More importantly, fundamental progresses have been made in the past decade on the addressing and manipulation of spin state. In 2010, Alexander S. Zyazin et al. demonstrated electric-field control over the magnetic anisotropy of a single magnetic molecule in a single molecular transistor of Fe_4 SMM, as showed in Figure 7.29a [133]. Edgar A. Osorio et al. demonstrated the capability to electrically control the spin state of a single-molecule incorporating a single Mn^{2+} ion coordinated by two terpyridine ligands, as showed in Figure 7.29b [134]. In 2017, Clément Godfrin et al. reported a method to read out the state of

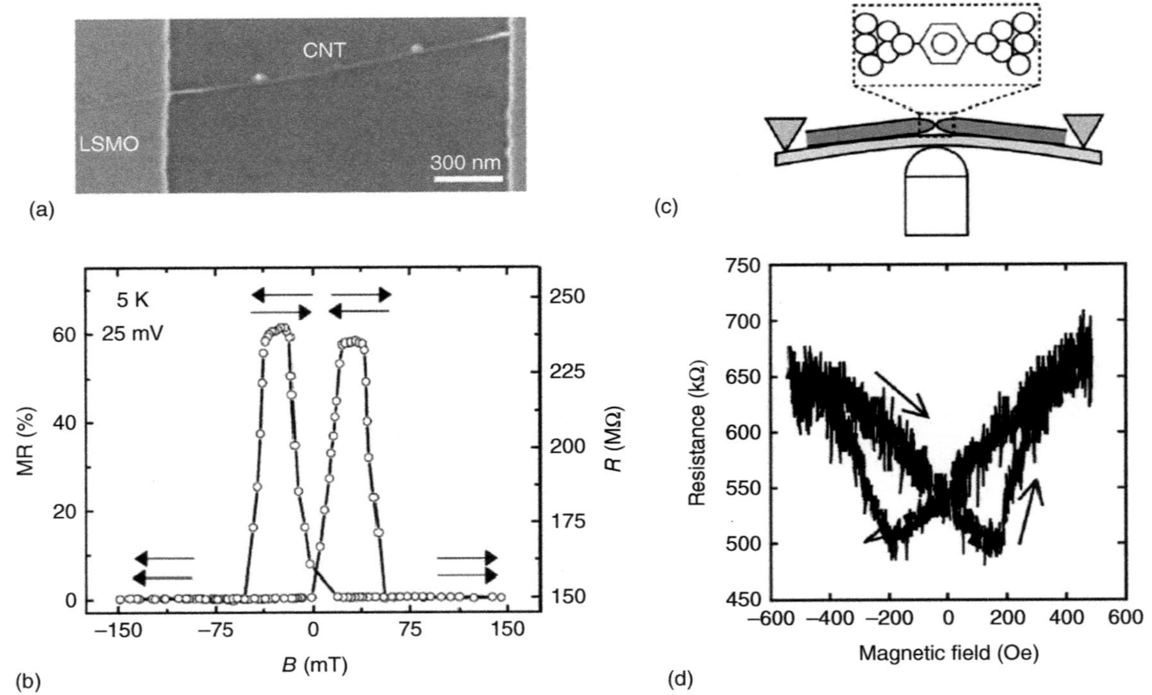

Figure 7.27 Molecule based spin-valve or magnetic tunnel junction using two FM electrodes. (a) SEM image of a carbon nanotube running between LSMO electrodes. (b) Magnetoresistance of the device in (a), showing two distinct states of resistance. The arrows indicate the relative magnetic orientation of the two FM electrodes. Source: Hueso et al. [124]. 2007, Springer Nature. (c) Schematic of the molecular device constructed by mechanically controllable break junction. (d) The magnetoresistance loops of the Ni/single benzene–1,4-dithiol/Ni junctions. Source: Yamada et al. [125]. 2011, American Institute of Physics.

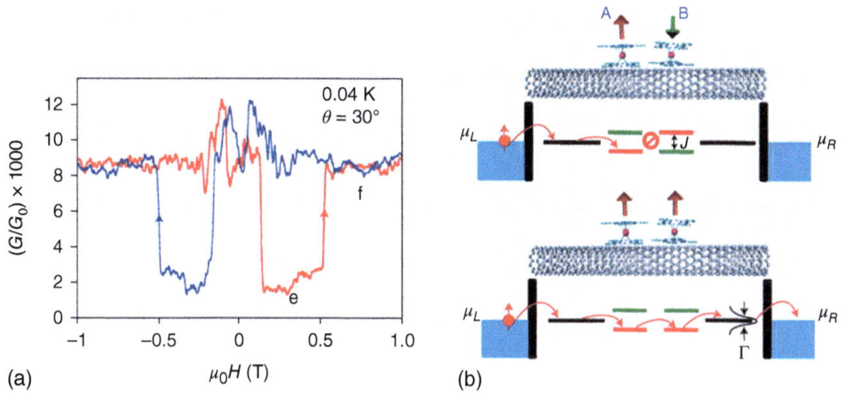

Figure 7.28 Supramolecular spin valves without using FM electrodes. (a) Zero-bias conductance of the device as a function of magnetic field. (b) Schematic representation of the mechanism involving two $TbPc_2$ SMM molecules (A and B) grafted on a carbon nanotube. With increasing magnetic field, molecule A switches first, leading to an antiparallel alignment of the magnetic momenta of the two SMM molecules with low conductance. When molecule B switches, a parallel configuration was recovered, leads to high conductance. Source: Urdampilleta et al. [26]. 2011, Springer Nature.

Figure 7.29 Single molecular spin transistor with various molecules. (a) Fe_4Ph. Source: Zyazin et al. [133]. 2010, American Chemical Society. (b) ($[Mn(terpy\text{-}O\text{-}(CH_2)_6\text{-}SAc)_2)]^{2+}$). Source: Osorio et al. [134]. 2010, American Chemical Society. (c) Bis(phthalocyaninato)terbium(III) ($TbPc_2$). Source: Godfrin et al. [135]. 2017, American Chemical Society.

a single electronic spin in a single molecular transistor of $TbPc_2$, as showed in Figure 7.29c [135]. Such system can be simplified into a two elements coupled quantum system. One is an electronic spin carried by the terbium Tb^{3+} ion. The other is a read-out quantum dot created by the Pc ligands. The readout quantum dot is tunnel-coupled to source and drain to perform transport measurements. The existence of a strong exchange interaction between the two systems resulted in a spin-dependent conductance of the read-out quantum dot, enabled readout of a single electronic spin electrically without affecting its magnetic properties. A read-out fidelity as high as 99.5% was demonstrated.

7.4.4.4 Single Molecular Nuclear Spin Transistor

Compared to electrons, which usually have strong coupling with environment, nuclear spins are well isolated from environment, thus have longer coherence time. In this respect, the nuclear spins are better since longer coherence time is

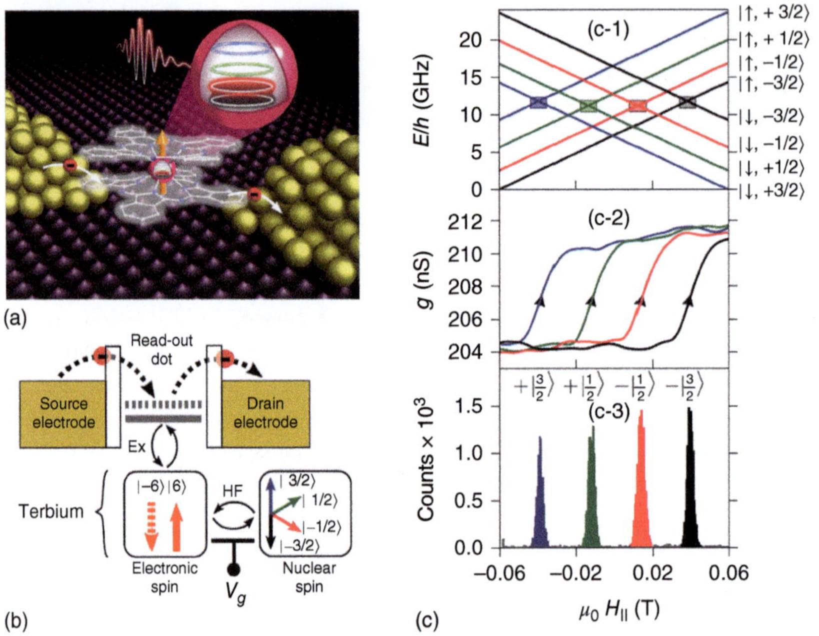

Figure 7.30 Single molecule nuclear spin transistor and its detection. (a) A nuclear spin qubit transistor based on a single $TbPc_2$ molecular magnet. (b) The device can be seen as a three coupled quantum system made up of (i) The four-level nuclear spin qubit is HF coupled to (ii) an electronic spin, which in turn is exchange (Ex) coupled to (iii) a readout quantum dot. (c) (c-1) Zeeman diagram of the $TbPc_2$ molecular magnet, showing the hyperfine split electronic spin ground state doublet as a function of the external magnetic field. The ligand-field-induced avoided level crossings (colored rectangles) allow for tunneling of the electron spin. (c-2) The jumps of the conductance of the readout quantum dot during magnetic-field sweeps are nuclear-spin dependent. (c-3) Histograms of the positions of about 75 000 conductance jumps, each conductance jump can be assigned to a nuclear spin state. Source: Thiele et al. [136]. 2014, American Association for the Advancement of Science.

beneficial in realizing quantum coherence and stable entanglement. However, weak coupling also means it is challenging to address and manipulate the state of nuclear spins by external stimuli. With regard to this, significant progress has been made in the past few years using molecular spin transistor with an SMM named as bis(phthalocyaninato)terbium(III) ($TbPc_2$), as showed in Figure 7.30a. Not only the addressing of nuclear spins has been demonstrated with a long lifetime (tens of seconds) [137], but also electrical manipulation of single nuclear spin were successfully realized in a very exquisite system [136].

Since significant progresses are achieved with $TbPc_2$ or its derivatives in this respect, the detail of this molecular system will be briefly introduced here [136, 137]. Briefly, $TbPc_2$ based molecular transistor can be seen as a three elements coupled quantum system composed of a nuclear spin qubit, an electronic spin, and a read-out quantum dot, as showed in Figure 7.30b. (i) The atomic core of Tb^{3+} ion offers a nuclear spin qubit of $I = 3/2$, which involves into four unequally spaced levels when the degeneracy is lifted. (ii) Terbium's 4f electrons offers the electronic spin. The spin-orbit coupling and the Pc ligand field results in an electron spin ground state doublet that is well isolated from the excited states and a uniaxial anisotropy axis perpendicular to the Pc plane. Through hyperfine coupling, the degeneracy of the doublet is lifted and each splits into four levels. (iii) The Pc ligands act as the read-out quantum dot, which tunnel-coupled to the source and drain electrodes through their delocalized π electron system. The overlap between the π electron system and the terbium's 4f wave functions resulted in an exchange coupling between the read-out dot and the electronic spin.

When an external magnetic field is applied to the device, avoided level crossings will be generated as showed in Figure 7.30c-1. When sweeping magnetic field slowly over such an anticrossing, quantum tunneling of magnetization (QTM) can be activated with the reverse of the electronic spin. This reverse can influence the read-out dot's conductance through the exchange interaction. Thus, a conductance jump can be detected on the readout dot, as showed in Figure 7.30c-2, c-3. Since the magnetic field to generate the conductance jump is nuclear spin dependent, the conductance jump can be unambiguously assigned to a nuclear qubit state, enable us to read out the single-nuclear spin state. Moreover, by using the hyperfine Stark effect as a magnetic field transducer, electrical manipulation of single nuclear spin were also successfully demonstrated which allows for fast switching and spatially confined spin control [136].

7.4.4.5 Molecule Based Hybrid Spintronic Devices

Apart from the device configuration of single molecular transistor, hybrid device configuration was also adopted to explore the spin state of molecules. In this kind of device, magnetic molecule is not connected to electrodes directly as in single molecular transistor, instead, it is coupled to a non-magnetic quantum dot such as carbon nanotubes or graphene structures, which is connected to the source and drain electrodes directly. In this hybrid system, the current passes through the non-magnetic quantum dot and its value can be influenced by the spin state of the magnetic molecule. Thus, through the current of the non-magnetic quantum conductor, the spin state of the magnetic molecule can be read out. In this sense, such hybrid system is similar with single molecular spin transistor of $TbPc_2$. The difference is that $TbPc_2$ molecule itself can be seen as a multidot system while in this hybrid system, the multidots come from different materials.

One classical hybrid spintronic device is demonstrated by Wolfgang Wernsdorfer and coworkers [138]. In their device, an individual pyren-substituted $TbPc_2$ SMM was grafted to a suspended carbon nanotube through π–π interaction, and the carbon nanotube was connected to the source and drain electrodes, as showed in Figure 7.31a. For $TbPc_2$ based molecules, as we mentioned earlier, the electronic spin can be reversed by quantum tunneling of the magnetization at the four avoided level crossings, as showed in Figure 7.31b. However, when the magnetic field is high enough, a direct quantum tunneling process could take place with the excitation of a phonon (Figure 7.31b). Since the molecule is physically decoupled with the substrate and the electrodes, it can only couple to the phonons of the suspended carbon

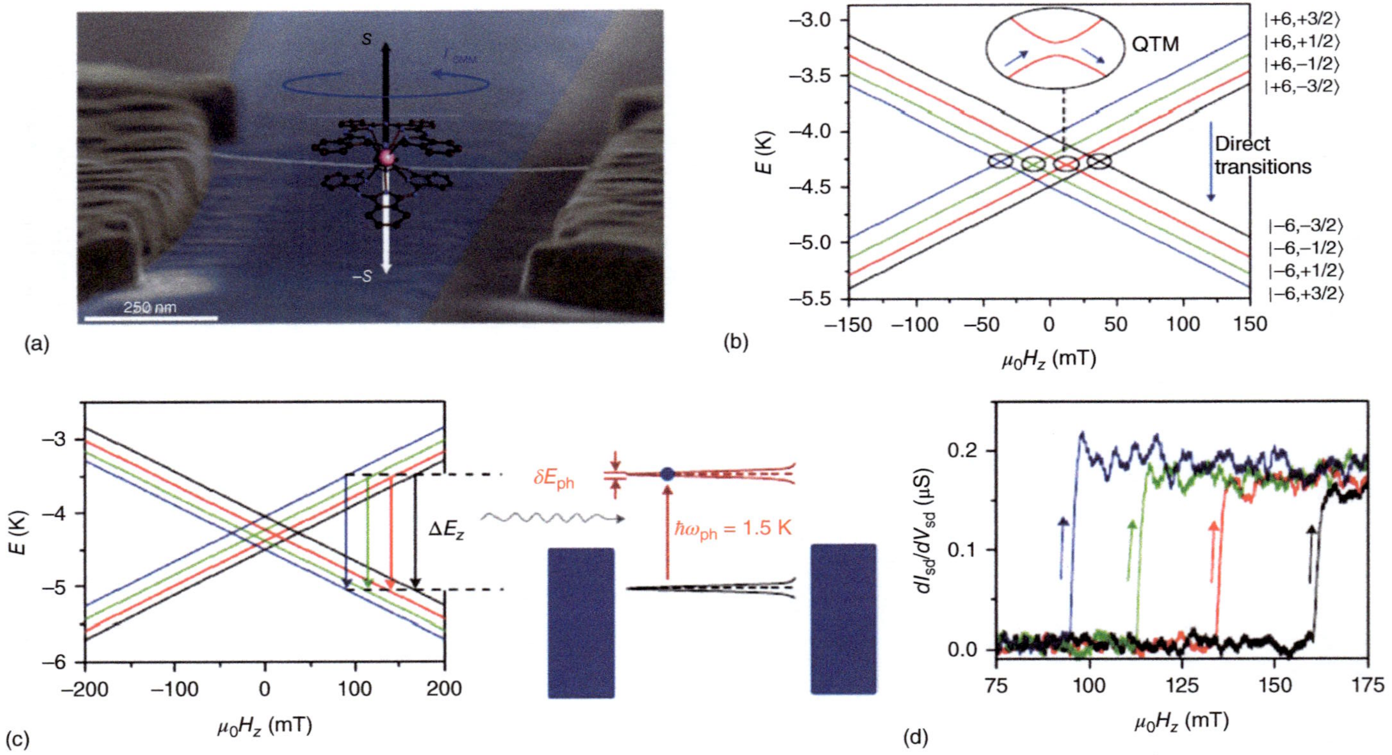

Figure 7.31 Molecule based hybrid spintronic devices composed of SMM and carbon nanotube. (a) Schematic view of the device with SMM shown as a chemical structure that is overlaid on a false-colored SEM image of a carbon nanotube between two contacts. (b) Zeeman diagram presenting the energy *E* of the two ground states as a function of magnetic field. The strong hyperfine coupling with the nuclear spin in the Tb^{3+} ion splits the ground-state doublet into four states. Magnetization reversal could take place at the four avoided level crossings (black circles) through quantum tunneling of magnetization or through direct transitions, involving the excitation of a phonon. (c) The magnetization reversal of the $TbPc_2$ via direct transition results in excitation of the electron into a vibrational state. (d) Sweeping the magnetic field will induce magnetization reversal of the Tb^{3+} ion, resulting in an abrupt increase in the differential conductance through the dot. The switching field depends on the nuclear spin state occupied before magnetization reversal. Source: Ganzhorn et al. [138]. 2013, Springer Nature.

nanotube. In the stability diagram of the carbon nanotube, equidistant excited states in the region of single-electron tunnelling (SET) were observed, which were attributed to quantized longitudinal phonons excitation. Thus in the device, direct quantum tunneling process could take place with the emission of a longitudinal phonon with a specific energy (Figure 7.31c). Since there are four nuclear spin states, four different transitions could take place with a transition energy equal to a longitudinal phonon. As showed in Figure 7.31c, each transition occurs at a specific magnetic field, corresponding to the nuclear spin states. Because the magnetization reversal of the molecule grafted to a carbon nanotube could induced an abrupt change in differential conductance, thus, by sweeping the magnetic field and monitor the differential conductance of the nanotube (Figure 7.31d), the switching magnetic field can be identified, by which the nuclear spin state can be identified.

7.5 Electrochemistry with Nanogap Electrodes

Apart from application in molecular devices and surface enhanced Raman spectroscopy (SERS) for the detection and characterization of single molecules, nanogap electrode can also be used in electrochemistry for the detection and characterization of single molecule. From the practical application point of view, the method of electrochemistry does not need molecules to bridge between electrodes as in molecular devices or the distance between the nanogap electrodes has to be sub-3 nm as in SERS detection, thus is much easier for implementation. Moreover, the capability to detect single or a few molecules in solution is of great importance both in fundamental research and in practical applications.

The detection of molecule through electrochemistry usually carries out through the detection of the charge transferred in electrochemical processes. However, for single molecule, only a few electrons can be transferred in one electrochemical reaction, and detection of the charges involved in the process is beyond the capability of conventional electrochemical instrumentation. In order to detect signal from single molecule, the signal must be amplified, which can be done through a process called redox cycling. For redox species that are chemically reversible with their reduced and oxidized forms stable over the time scale of measurement, the process of redox cycling can be used to enhance the electrochemical current. In a redox cycling experiment, two electrodes are soaked in the solution of electroactive molecule, which can diffuse randomly between the two electrodes. The two electrodes are biased independently with one below the reduction potential and the other above the oxidation potential of the electroactive molecule. In this way, the product of a redox reaction at one electrode serves as reactant at the other electrode, and vice versa, thus the electroactive molecule can be reduced and oxidized repeatedly. Each oxidation and reduction allow electrons to transport between the two electrodes, thus the repeated oxidation and reduction will enhance the detected electrochemical current.

When the mass transport in a solution is dominated by diffusion, the motion of a molecule in solution can be described by Brownian motion theory. In Brownian motion theory, the average time t for a molecule to diffuse from one electrode to the other can be described by $t = L^2/2D$, where D is the diffusion coefficient and L is the distance between the two electrodes. Thus the current transport in this process can be written as $I = enD/L^2$, where e is the electron charge and n is the number of electrons transferred in one redox reaction. Therefore, in order to achieve measurable single molecule signals, one effective way is to decrease the distance between the two electrodes, which makes nanogap electrodes a perfect candidate in single molecule electrochemistry.

Significant advances have been made in single molecule detection with various forms of nanogap electrodes based electrochemistry setups. Roughly, there are two kinds of nanogap electrodes based electrochemistry setups. One is scanning electrochemical microscope system, in which nanogap is formed between a scanning probe and a conductive substrate. The other is on-chip nanogap electrodes fabricated through microfabrication technology. Through scanning electrochemical microscope system, great progresses has been made in single molecule detection [139–141]. For example, the pioneering work by Fu-Ren F. Fan and Allen J. Bard showed that single molecule can be trapped and detected electrochemically by using nanogap electrodes formed between a tip and a conductive substrate [139, 140]. The tip used in their experiment was wrapped in an insulating sheath, as illustrated in Figure 7.32a, which offers the confinement in the plane parallel to the conductive substrate. Through a scanning electrochemical microscope system, the tip can approach a conductive substrate and form a confined volume with nanoscale size in three dimension. Due to the small distance between the two electrodes, an amplification as high as 10^7 can be realized with redox cycling, enabling the detection of single molecule electrochemically. Yet challenges remain. For example, it is still a challenge to produce the tip with controllable size and geometry in a reproducible way, which makes systematic measurements hard to carry out. Moreover, independent characterization of the tip geometry is still shorted. In addition, for practical applications, the existing of the huge scanning electrochemical microscope system makes them hard to be integrated with other devices on chip.

Compared to scanning electrochemical microscope system, on-chip nanogap electrodes can be mass-produced in a more controllable way. Moreover, they can be characterized independently, making the comparison with theoretical results possible. In addition, on-chip nanogap electrodes are more convenient for practical application. Thus, more and more interests arises in using on-chip nanogap electrodes in electrochemistry and various form of nanogap electrodes, as showed in Figure 7.32b–e, are fabricated and tested [142–150]. With those on-chip nanogap electrodes, great progress has been made in amplifying the electrical current through redox cycling.

With on-chip nanogap electrodes, single molecule detection has been successfully demonstrated [142, 151]. In 2011, Marcel A. G. Zevenbergen et al. demonstrated single molecule sensitivity in an electrochemical device made up of two parallel electrodes with distance of 70 nm, as showed in Figure 7.33a [142]. The distance between the two electrodes are predefined by a sacrificial layer, and by controlling the thickness of the sacrificial layer, nanoscale distance between the two electrodes

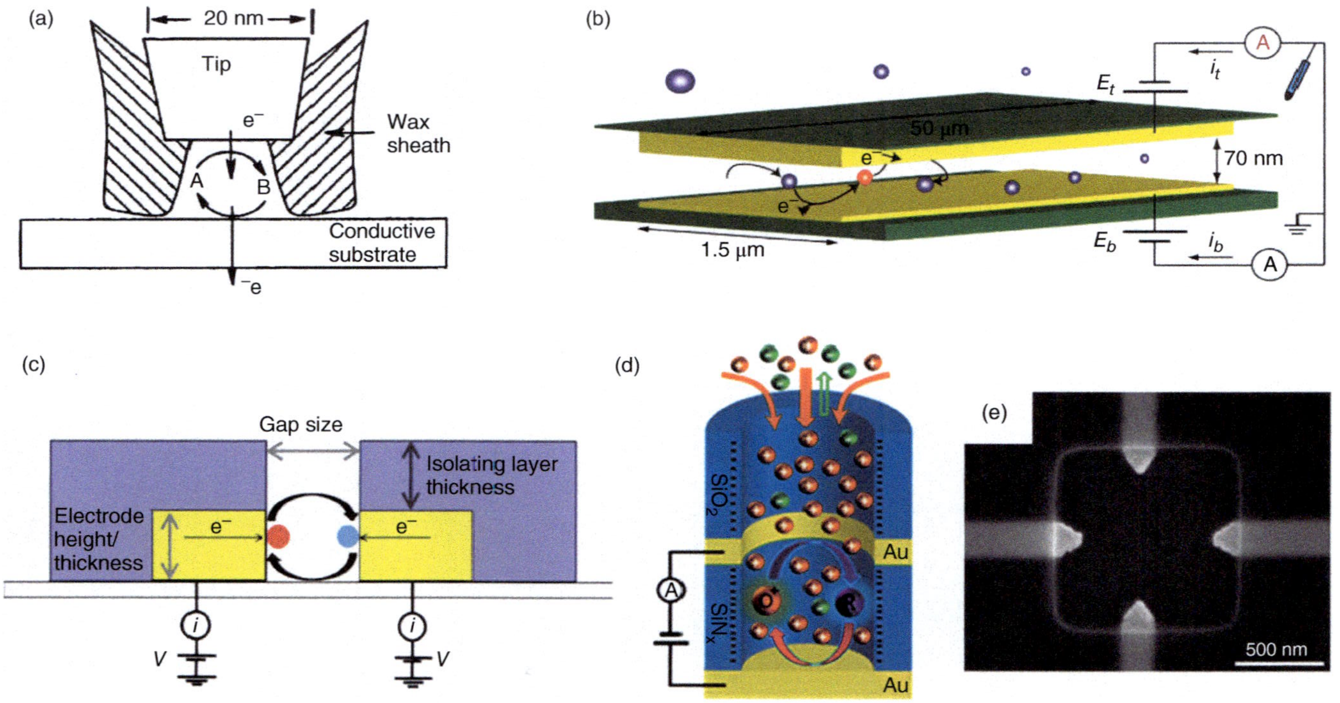

Figure 7.32 Various forms of nanogap electrodes used in electrochemistry. (a) Nanogap formed between a tip and a conductive substrate. The tip is wrapped in an insulating sheath. Source: Bard et al. [140]. 1996, American Chemical Society. (b–e) On-chip nanogap electrodes with various configurations. Source: (b) Zevenbergen et al. [142]. 2011, American Chemical Society. (c) Zafarani et al. [143]. 2016, Elsevier. (d) Ma et al. [144]. 2016, American Chemical Society. (e) Li et al. [145]. 2010, Springer.

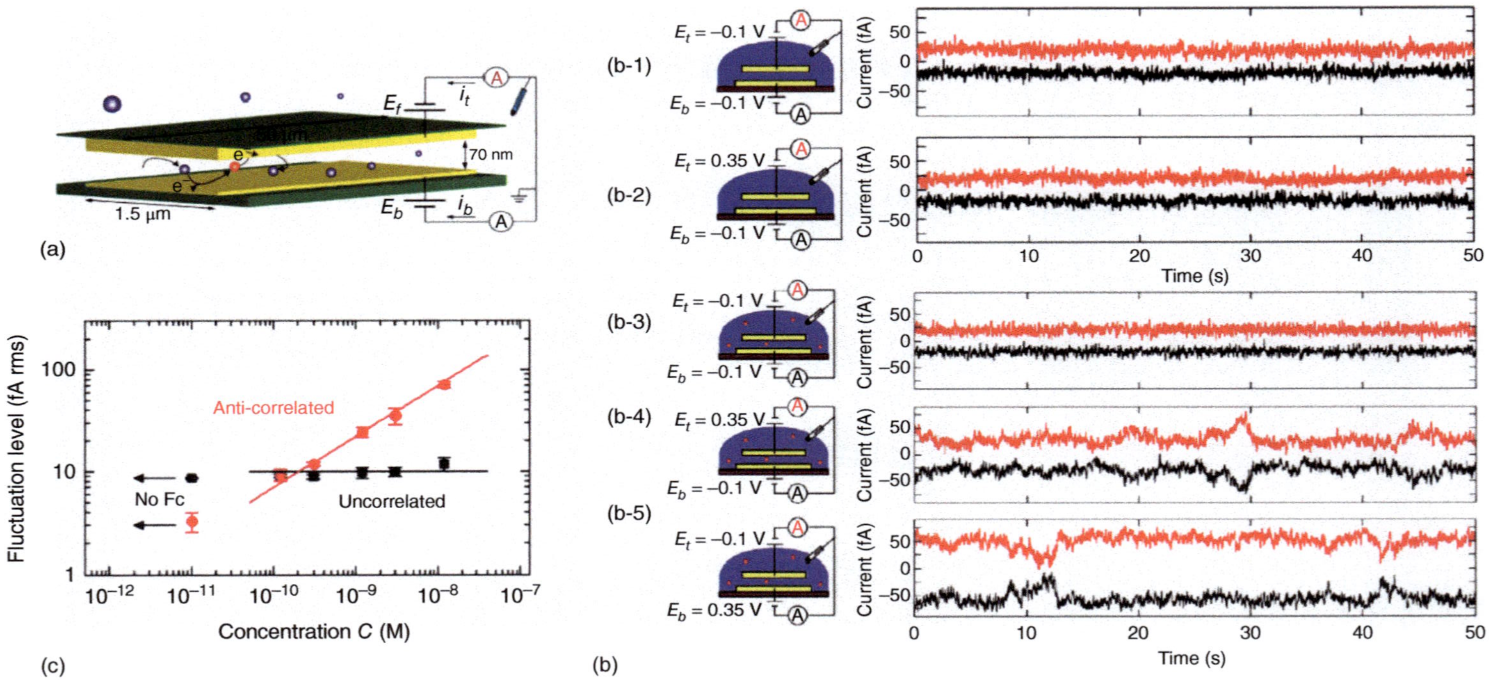

Figure 7.33 Detection of single molecule through electrochemistry in nanogap electrodes. (a) Illustration of the detection principle: electroactive molecules undergoing Brownian motion are oxidized and reduced repeatedly at the two electrodes, leading to a measurable current. (b) Current vs. time of the device in different condition: (b-1) in the absence of redox species and with both electrodes set to the same bias; (b-2) in the absence of redox species and under bias conditions that allow redox cycling; (b-3) in the present of redox species and with both electrodes set to the same bias; (b-4, b-5) in the present of redox species and under bias conditions that allow redox cycling. (c) Uncorrelated and anticorrelated parts of the measured fluctuations in the current vs. Fc concentration. The anticorrelated signal is proportional to $C(1/2)$ (red line has slope = 1/2). Source: Zevenbergen et al. [142]. 2011, American Chemical Society.

can be obtained after removing the sacrificial layer through wet etching. When filling the gap with low concertation ferrocene (Fc) solution to a degree that the average number of molecule existing between the electrodes is only 0.38, biasing the two electrodes under redox cycling condition of Fc, and measuring the current generated at the bottom and top electrode separately, excursions in current were observed compared to device without Fc or not biased in the redox cycling condition, as showed in Figure 7.33b. Such excursions in current happens simultaneously for the bottom and top electrodes, showing equal values but opposite signs, thus originates from oxidation and reduction of the Fc molecule. Each excursion indicates the existence of Fc molecule between the electrodes, starting when molecule enters the region and stopping when molecule leaves. For independent molecules in diffusive equilibrium with a reservoir, the root mean square (rms) fluctuation in the number of molecules is expected to scale with $C^{1/2}$, where C is the Fc concentration. Thus, the corresponding fluctuations in current are also expected to scale with $C^{1/2}$. Through a cross-correlation analysis of the signals at the two electrodes, such a relation between the fluctuation level and Fc concentration was indeed identified (Figure 7.33c), further proving that the excursions indeed originate from single molecule fluctuations. This report is of great importance by showing ultra-high sensitivity of electrochemical detection.

With single molecule level electrochemical sensitivity, such nanogap electrodes based electrochemical devices can be used to measure liquid flow rates. For a Poisson-distributed dilute solute, the number N of molecules in a specific volume fluctuates in time due to Brownian motion. For a macro system, such number fluctuations are usually too small to be detected. However, for a system in nanoscale dimension, such number fluctuation can be big enough for detection. Based on this, Klaus Mathwig et al. demonstrated the detection of an ultralow liquid flow rate below 10 pl min^{-1} [152]. The device configuration is similar to the one used in single molecule electrochemical detection, as showed in Figure 7.34a. In this device, two top electrodes were used to define two detection volumes. Through bias control, redox cycling could take place in both of the two volumes, enabling the detection of the electrochemical current of molecules in both of them. When

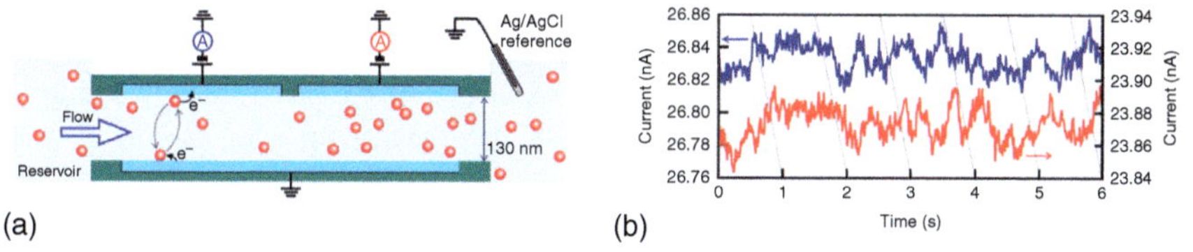

Figure 7.34 Detection of liquid flow rates through electrochemistry in nanogaps electrodes. (a) Schematic of the measurement concept: number fluctuations of the electroactive molecules are measured at two electrodes through redox cycling. Liquid flow velocity is determined by time-of-flight detection of these fluctuations as the molecules are transported along the channel from one top electrodes to the other top electrodes. (b) Raw current-time traces recorded at both electrodes at a flow rate corresponding to a time shift of approximately 0.2 seconds. Source: Mathwig et al. [152]. 2012, American Physical Society.

liquid flow through the channel fast enough, the number of fluctuations can be preserved while they travel along the channel. Thus, the fluctuations measured at the first electrode will be measured again at the second electrodes after a time delay, which is the time required for molecules to flow from one top electrode to the other top electrodes (time-of-flight). Thus by evaluating the correlation between current measured at the two top electrodes, the time-of-flight can be obtained as showed in Figure 7.34b, as well as the average liquid velocity, considering that the time-of-flight is approximation inversely proportional to the average liquid velocity.

References

1 Green, J.E., Choi, J.W., and Boukai, A. (2007). A 160-kilobit molecular electronic memory patterned at 10^{11} bits per square centimetre. *Nature* 445 (7126): 414–417.

2 Chen, Y., Jung, G.Y., Ohlberg, D.A.A. et al. (2003). Nanoscale molecular-switch crossbar circuits. *Nanotechnology* 14 (4): 462–468.

3 Diez-Perez, I., Hihath, J., Lee, Y. et al. (2009). Rectification and stability of a single molecular diode with controlled orientation. *Nat. Chem.* 1 (8): 635–641.

4 Metzger, R.M., Chen, B. Hopfner, U. et al. (1997). Unimolecular electrical rectification in hexadecylquinolinium tricyanoquinodimethanide. *J. Am. Chem. Soc.* 119 (43): 10455–10466.

5 Park, H., Park, J., Lim, A.K.L. et al. (2000). Nanomechanical oscillations in a single-C_{60} transistor. *Nature* 407 (6800): 57–60.

6 Park, J., Pasupathy, A.N., Goldsmith, J.I. et al. (2002). Coulomb blockade and the Kondo effect in single-atom transistors. *Nature* 417 (6890): 722–725.

7 Liang, W.J., Shores, M.P., Bockrath, M. et al. (2002). Kondo resonance in a single-molecule transistor. *Nature* 417 (6890): 725–729.

8 Dulic, D., van der Molen, S.J., Kudernac, T. et al. (2003). One-way optoelectronic switching of photochromic molecules on gold. *Phys. Rev. Lett.* 91 (20): 207402-1–207402-4.

9 Aragones, A.C., Aravena, D., Cerda, J.I. et al. (2016). Large conductance switching in a single-molecule device through room temperature spin-dependent transport. *Nano Lett.* 16 (1): 218–226.

10 Collier, C.P., Wong, E.W., Belohradsky, M. et al. (1999). Electronically configurable molecular-based logic gates. *Science* 285 (5426): 391–394.

11 Xiang, D., Jeong, H., Lee, T., and Mayer, D. (2013). Mechanically controllable break junctions for molecular electronics. *Adv. Mater.* 25 (35): 4845–4867.

12 Perrin, M.L., Verzijl, C.J.O., Martin, C.A. et al. (2013). Large tunable image-charge effects in single-molecule junctions. *Nat. Nanotechnol.* 8 (4): 282–287.

13 O'Neill, K., Osorio, E.A., and van der Zant, H.S.J. (2007). Self-breaking in planar few-atom Au constrictions for nanometer-spaced electrodes. *Appl. Phys. Lett.* 90 (13): 133109-1–133109-3.

14 Heersche, H.B., de Groot, Z., Folk, J.A. et al. (2006). Kondo effect in the presence of magnetic impurities. *Phys. Rev. Lett.* 96 (1): 017205-1–017205-4.

15 Prins, F., Hayashi, T., van Steenwijk, B.J.A.D. et al. (2009). Room-temperature stability of Pt nanogaps formed by self-breaking. *Appl. Phys. Lett.* 94 (12): 123108-1–123108-3.

16 Johnston, D.E., Strachan, D.R., and Johnson, A.T.C. (2007). Parallel fabrication of nanogap electrodes. *Nano Lett.* 7 (9): 2774–2777.

17 Guo, X.F., Whalley, A., Klare, J.E. et al. (2007). Single-molecule devices as scaffolding for multicomponent nanostructure assembly. *Nano Lett.* 7 (5): 1119–1122.

18 Feldman, A.K., Steigerwald, M.L., Guo, X.F., and Nuckolls, C. (2008). Molecular electronic devices based on single-walled carbon nanotube electrodes. *Acc. Chem. Res.* 41 (12): 1731–1741.

19 Shen, Q., Guo, X.F., Steigerwald, M.L., and Nuckolls, C. (2010). Integrating reaction chemistry into molecular electronic devices. *Chem. Asian J.* 5 (5): 1040–1057.

20 Candini, A., Klyatskaya, S., Ruben, M. et al. (2011). Graphene spintronic devices with molecular nanomagnets. *Nano Lett.* 11 (7): 2634–2639.

21 Jia, C.C., Ma, B.J., Xin, N., and Guo, X.F. (2015). Carbon electrode-molecule junctions: a reliable platform for molecular electronics. *Acc. Chem. Res.* 48 (9): 2565–2575.

22 Cao, Y., Dong, S.H., Liu, S. et al. (2012). Building high-throughput molecular junctions using indented graphene point contacts. *Angew. Chem. Int. Ed.* 51 (49): 12228–12232.

23 Guo, X.F., Small, J.P., Klare, J.E. et al. (2006). Covalently bridging gaps in single-walled carbon nanotubes with conducting molecules. *Science* 311 (5759): 356–359.

24 Jia, C.C., Migliore, A., Xin, N. et al. (2016). Covalently bonded single-molecule junctions with stable and reversible photoswitched conductivity. *Science* 352 (6292): 1443–1445.

25 Urdampilleta, M., Nguyen, N.V., Cleuziou, J.P. et al. (2011). Molecular quantum spintronics: supramolecular spin valves based on single-molecule magnets and carbon nanotubes. *Int. J. Mol. Sci.* 12 (10): 6656–6667.

26 Urdampilleta, M., Klyatskaya, S., Cleuziou, J.P. et al. (2011). Supramolecular spin valves. *Nat. Mater.* 10 (7): 502–506.

27 Lefebvre, J., Radosavljevic, M., and Johnson, A.T. (2000). Fabrication of nanometer size gaps in a metallic wire. *Appl. Phys. Lett.* 76 (25): 3828–3830.

28 Nagase, T., Kubota, T., and Mashiko, S. (2003). Fabrication of nano-gap electrodes for measuring electrical properties of organic molecules using a focused ion beam. *Thin Solid Films* 438: 374–377.

29 Hatzor, A. and Weiss, P.S. (2001). Molecular rulers for scaling down nanostructures. *Science* 291 (5506): 1019–1020.

30 Tang, J., Wang, Y., Nuckolls, C., and Wind, S.J. (2006). Chemically responsive molecular transistors fabricated by self-aligned lithography and chemical self-assembly. *J. Vac. Sci. Technol., B* 24 (6): 3227–3229.

31 Cui, A., Liu, Z., Dong, H.L. et al. (2016). Mass production of nanogap electrodes toward Robust resistive random access memory. *Adv. Mater.* 28 (37): 8227–8233.

32 Tang, J., De Poortere, E.P., Klare, J.E. et al. (2006). Single-molecule transistor fabrication by self-aligned lithography and in situ molecular assembly. *Microelectron. Eng.* 83 (4–9): 1706–1709.

33 Fursina, A., Lee, S., Sofin, R.G.S. et al. (2008). Nanogaps with very large aspect ratios for electrical measurements. *Appl. Phys. Lett.* 92 (11): 113102-1–113102-3.

34 McCarty, G.S. (2004). Molecular lithography for wafer-scale fabrication of molecular junctions. *Nano Lett.* 4 (8): 1391–1394.

35 Liu, S.H., Tok, J.B.H., and Bao, Z.N. (2005). Nanowire lithography: fabricating controllable electrode gaps using Au–Ag–Au nanowires. *Nano Lett.* 5 (6): 1071–1076.

36 Qin, L.D., Park, S., Huang, L., and Mirkin, C.A. (2005). On-wire lithography. *Science* 309 (5731): 113–115.

37 Luber, S.M., Strobel, S., Tranitz, H.P. et al. (2005). Nanometre spaced electrodes on a cleaved AlGaAs surface. *Nanotechnology* 16 (8): 1182–1185.

38 Tyagi, P., Li, D.F., Holmes, S.M., and Hinds, B.J. (2007). Molecular electrodes at the exposed edge of metal/insulator/metal trilayer structures. *J. Am. Chem. Soc.* 129 (16): 4929–4938.

39 Luber, S.M., Zhang, F., Lingitz, S. et al. (2007). High-aspect-ratio nanogap electrodes for averaging molecular conductance measurements. *Small* 3 (2): 285–289.

40 Chen, X.D., Braunschweig, A.B., Wiester, M.J. et al. (2009). Spectroscopic tracking of molecular transport junctions generated by using click chemistry. *Angew. Chem. Int. Ed.* 48 (28): 5178–5181.

41 Yan, H.J., Bergren, A.J., and McCreery, R.L. (2011). All-carbon molecular tunnel junctions. *J. Am. Chem. Soc.* 133 (47): 19168–19177.

42 Najarian, A.M., Szeto, B., Tefashe, U.M., and McCreery, R.L. (2016). Robust all-carbon molecular junctions on flexible or semi-transparent substrates using "process-friendly" fabrication. *ACS Nano* 10 (9): 8918–8928.

43 Todeschini, M., Fanta, A.B.D., Jensen, F. et al. (2017). Influence of Ti and Cr adhesion layers on ultrathin Au films. *ACS Appl. Mater. Interfaces* 9 (42): 37374–37385.

44 de Boer, B., Frank, M.M., Chabal, Y.J. et al. (2004). Metallic contact formation for molecular electronics: interactions between vapor-deposited metals and self-assembled monolayers of conjugated mono- and dithiols. *Langmuir* 20 (5): 1539–1542.

45 Kim, T.W., Wang, G.N., Lee, H., and Lee, T. (2007). Statistical analysis of electronic properties of alkanethiols in metal-molecule-metal junctions. *Nanotechnology* 18 (31): 315204-1–315204-8.

46 Walker, A.V., Tighe, T.B., Cabarcos, O.M. et al. (2004). The dynamics of noble metal atom penetration through methoxy-terminated alkanethiolate monolayers. *J. Am. Chem. Soc.* 126 (12): 3954–3963.

47 Mann, B. and Kuhn, H. (1971). Tunneling through fatty acid salt monolayers. *J. Appl. Phys.* 42 (11): 4398–4405.

48 Chiechi, R.C., Weiss, E.A., Dickey, M.D., and Whitesides, G.M. (2008). Eutectic gallium-indium (EGaIn): a moldable liquid metal for electrical characterization of self-assembled monolayers. *Angew. Chem. Int. Ed.* 47 (1): 142–144.

49 Nerngchamnong, N., Yuan, L., Qi, D.C. et al. (2013). The role of van der Waals forces in the performance of molecular diodes. *Nat. Nanotechnol.* 8 (2): 113–118.

50 Du, W., Wang, T., Chu, H.S. et al. (2016). On-chip molecular electronic plasmon sources based on self-assembled monolayer tunnel junctions. *Nat. Photon.* 10 (4): 274–280.

51 Haick, H., Ambrico, M., Ghabboun, J. et al. (2004). Contacting organic molecules by metal evaporation. *Phys. Chem. Chem. Phys.* 6 (19): 4538–4541.

52 Metzger, R.M., Xu, T., and Peterson, I.R. (2001). Electrical rectification by a monolayer of hexadecylquinolinium tricyanoquinodimethanide measured between macroscopic gold electrodes. *J. Phys. Chem. B* 105 (30): 7280–7290.

53 Bonifas, A.P. and McCreery, R.L. (2010). 'Soft' Au, Pt and Cu contacts for molecular junctions through surface-diffusion-mediated deposition. *Nat. Nanotechnol.* 5 (8): 612–617.

54 Maitani, M.M., Allara, D.L., Ohlberg, D.A.A. et al. (2010). High integrity metal/organic device interfaces via low temperature buffer layer assisted metal atom nucleation. *Appl. Phys. Lett.* 96 (17): 173109-1–173109-3.

55 Shimizu, K.T., Fabbri, J.D., Jelincic, J.J., and Melosh, N.A. (2006). Soft deposition of large-area metal contacts for molecular electronics. *Adv. Mater.* 18 (12): 1499–1504.

56 Stein, N., Korobko, R., Yaffe, O. et al. (2010). Nondestructive contact deposition for molecular electronics: Si-Alkyl//Au junctions. *J. Phys. Chem. C* 114 (29): 12769–12776.

57 Krabbenborg, S.O., Wilbers, J.G.E., Huskens, J., and vander Wiel, W.G. (2013). Symmetric large-area metal-molecular monolayer-metal junctions by wedging transfer. *Adv. Funct. Mater.* 23 (6): 770–776.

58 Guerin, D., Merckling, C., Lenfant, S. et al. (2007). Silicon-molecules-metal junctions by transfer printing: chemical synthesis and electrical properties. *J. Phys. Chem. C* 111 (22): 7947–7956.

59 Ojima, K., Otsuka, Y., Matsumoto, T. et al. (2005). Printing electrode for top-contact molecular junction. *Appl. Phys. Lett.* 87 (23): 234110-1–234110-3.

60 Loo, Y.L., Lang, D.V., Rogers, J.A., and Hsu, J.W.P. (2003). Electrical contacts to molecular layers by nanotransfer printing. *Nano Lett.* 3 (7): 913–917.

61 Jeong, H., Kim, D., Kim, P. et al. (2015). A new approach for high-yield metal-molecule-metal junctions by direct metal transfer method. *Nanotechnology* 26 (2): 025601-1–025601-10.

62 Akkerman, H.B., Blom, P.W.M., de Leeuw, D.M., and de Boer, B. (2006). Towards molecular electronics with large-area molecular junctions. *Nature* 441 (7089): 69–72.

63 Kronemeijer, A.J., Akkerman, H.B., Kudernac, T. et al. (2008). Reversible conductance switching in molecular devices. *Adv. Mater.* 20 (8): 1467–1473.

64 Wang, G., Kim, Y., Choe, M. et al. (2011). A new approach for molecular electronic junctions with a multilayer graphene electrode. *Adv. Mater.* 23 (6): 755–760.

65 Li, T., Hauptmann, J.R., Wei, Z.M. et al. (2012). Solution-processed ultrathin chemically derived graphene films as soft top contacts for solid-state molecular electronic junctions. *Adv. Mater.* 24 (10): 1333–1339.

66 Li, J.C., Merino-Diez, N., Carbonell-Sanroma, E. et al. (2018). Survival of spin state in magnetic porphyrins contacted by graphene nanoribbons. *Sci. Adv.* 4 (2): 1–6.

67 van Dijk, E.H., Myles, D.J.T., van der Veen, M.H., and Hummelen, J.C. (2006). Synthesis and properties of an anthraquinone-based redox switch for molecular electronics. *Org. Lett.* 8 (11): 2333–2336.

68 Vazquez, H., Skouta, R., Schneebeli, S. et al. (2012). Probing the conductance superposition law in single-molecule circuits with parallel paths. *Nat. Nanotechnol.* 7 (10): 663–667.

69 Li, T., Jevric, M., Hauptmann, J.R. et al. (2013). Ultrathin reduced graphene oxide films as transparent top-contacts for light switchable solid-state molecular junctions. *Adv. Mater.* 25 (30): 4164–4170.

70 Xu, X.Z., Zhang, Z.H., Dong, J.C. et al. (2017). Ultrafast epitaxial growth of metre-sized single-crystal graphene on industrial Cu foil. *Sci. Bull.* 62 (15): 1074–1080.

71 Cao, Y., Dong, S.H., Liu, S. et al. (2013). Toward functional molecular devices based on graphene molecule junctions. *Angew. Chem. Int. Ed.* 52 (14): 3906–3910.

72 Harutyunyan, A.R., Chen, G.G., Paronyan, T.M. et al. (2009). Preferential growth of single-walled carbon nanotubes with metallic conductivity. *Science* 326 (5949): 116–120.

73 Yang, F., Wang, X., Zhang, D.Q. et al. (2014). Chirality-specific growth of single-walled carbon nanotubes on solid alloy catalysts. *Nature* 510 (7506): 522–524.

74 Zhou, Y.S., Xiong, W., Gao, Y. et al. (2010). Towards carbon-nanotube integrated devices: optically controlled parallel integration of single-walled carbon nanotubes. *Nanotechnology* 21 (31): 3011–3017.

75 Hong, S.W., Banks, T., and Rogers, J.A. (2010). Improved density in aligned arrays of single-walled carbon nanotubes by sequential chemical vapor deposition on quartz. *Adv. Mater.* 22 (16): 1826–1830.

76 Jia, C.C. and Guo, X.F. (2013). Molecule-electrode interfaces in molecular electronic devices. *Chem. Soc. Rev.* 42 (13): 5642–5660.

77 Akkerman, H.B., Kronemeijer, A.J., Harkema, J. et al. (2010). Stability of large-area molecular junctions. *Org. Electron.* 11 (1): 146–149.

78 Park, S., Wang, G., Cho, B. et al. (2012). Flexible molecular-scale electronic devices. *Nat. Nanotechnol.* 7 (7): 438–442.

79 Van Hal, P.A., Smits, E.C.P., Geuns, T.C.T. et al. (2008). Upscaling, integration and electrical characterization of molecular junctions. *Nat. Nanotechnol.* 3 (12): 749–754.

80 Kim, D., Jeong, H., Hwang, W.T. et al. (2015). Reversible switching phenomenon in diarylethene molecular devices with reduced graphene oxide electrodes on flexible substrates. *Adv. Funct. Mater.* 25 (37): 5918–5923.

81 Min, M., Seo, S., Lee, S.M., and Lee, H. (2013). Voltage-controlled nonvolatile molecular memory of an azobenzene monolayer through solution-processed reduced graphene oxide contacts. *Adv. Mater.* 25 (48): 7045–7050.

82 Seo, S., Min, M., Lee, J. et al. (2012). Solution-processed reduced graphene oxide films as electronic contacts for molecular monolayer junctions. *Angew. Chem. Int. Ed.* 51 (1): 108–112.

83 Qek, S.Y., Kamenetska, M., Steigerwald, M.L. et al. (2009). Mechanically controlled binary conductance switching of a single-molecule junction. *Nat. Nanotechnol.* 4 (4): 230–234.

84 Bruot, C., Hihath, J., and Tao, N.J. (2012). Mechanically controlled molecular orbital alignment in single molecule junctions. *Nat. Nanotechnol.* 7 (1): 35–40.

85 Hihath, J., Arroyo, C.R., Rubio-Bollinger, G. et al. (2008). Study of electron-phonon interactions in a single molecule covalently connected to two electrodes. *Nano Lett.* 8 (6): 1673–1678.

86 Xu, B.Q., Xiao, X.Y., and Tao, N.J. (2003). Measurements of single-molecule electromechanical properties. *J. Am. Chem. Soc.* 125 (52): 16164–16165.

87 Kim, T. (2015). Molecular bonding force and stiffness in amine-linked single-molecule junctions formed with silver electrodes. *J. Korean Chem. Soc.-Daehan Hwahak Hoe Jee* 59 (2): 132–135.

88 Frei, M., Aradhya, S.V., Koentopp, M. et al. (2011). Mechanics and chemistry: single molecule bond rupture forces correlate with molecular backbone structure. *Nano Lett.* 11 (4): 1518–1523.

89 Frei, M., Aradhya, S.V., Hybertsen, M.S., and Venkataraman, L. (2012). Linker dependent bond rupture force measurements in single-molecule junctions. *J. Am. Chem. Soc.* 134 (9): 4003–4006.

90 Aradhya, S.V., Frei, M., Hybertsen, M.S., and Venkataraman, L. (2012). Van der Waals interactions at metal/organic interfaces at the single-molecule level. *Nat. Mater.* 11 (10): 872–876.

91 Fournier, N., Wagner, C., Weiss, C. et al. (2011). Force-controlled lifting of molecular wires. *Phys. Rev. B* 84 (3): 035435-1–035435-5.

92 Wagner, C., Fournier, N., Tautz, F.S., and Temirov, R. (2012). Measurement of the binding energies of the organic-metal perylene-teracarboxylic-dianhydride/Au(111) bonds by molecular manipulation using an atomic force microscope. *Phys. Rev. Lett.* 109 (7): 076102-1–076102-5.

93 Mahan, G.D. and Sofo, J.O. (1996). The best thermoelectric. *Proc. Natl. Acad. Sci. U.S.A.* 93 (15): 7436–7439.

94 Paulsson, M. and Datta, S. (2003). Thermoelectric effect in molecular electronics. *Phys. Rev. B* 67 (24): 241403-1–241403-4.

95 Reddy, P., Jang, S.Y., Segalman, R.A., and Majumdar, A. (2007). Thermoelectricity in molecular junctions. *Science* 315 (5818): 1568–1571.

96 Tan, A., Sadat, S., and Reddy, P. (2010). Measurement of thermopower and current-voltage characteristics of molecular junctions to identify orbital alignment. *Appl. Phys. Lett.* 96 (1): 013110-1–013110-3.

97 Widawsky, J.R., Darancet, P., Neaton, J.B., and Venkataraman, L. (2012). Simultaneous determination of conductance and thermopower of single molecule junctions. *Nano Lett.* 12 (1): 354–358.

98 Evangeli, C., Gillemot, K., Leary, E. et al. (2013). Engineering the thermopower of C_{60} molecular junctions. *Nano Lett.* 13 (5): 2141–2145.

99 Widawsky, J.R., Chen, W., Vazquez, H. et al. (2013). Length-dependent thermopower of highly conducting Au–C bonded single molecule junctions. *Nano Lett.* 13 (6): 2889–2894.

100 Li, Y.Q., Xiang, L.M., Palma, J. et al. (2016). Thermoelectric effect and its dependence on molecular length and sequence in single DNA molecules. *Nat. Commun.* 7 (11294): 1–8.

101 Pauly, F., Viljas, J.K., Burkle, M. et al. (2011). Molecular dynamics study of the thermopower of Ag, Au, and Pt nanocontacts. *Phys. Rev. B* 84 (19): 195420-1–195420-13.

102 Baheti, K., Malen, J.A., Doak, P. et al. (2008). Probing the chemistry of molecular heterojunctions using thermoelectricity. *Nano Lett.* 8 (2): 715–719.

103 Kim, Y., Jeong, W., Kim, K. et al. (2014). Electrostatic control of thermoelectricity in molecular junctions. *Nat. Nanotechnol.* 9 (11): 881–885.

104 Cui, L.J., Miao, R.J., Jiang, C. et al. (2017). Perspective: thermal and thermoelectric transport in molecular junctions. *J. Chem. Phys.* 146 (9): 092201-1–092201-26.

105 Rincon-Garcia, L., Evangeli, C., Rubio-Bollinger, G., and Agrait, N. (2016). Thermopower measurements in molecular junctions. *Chem. Soc. Rev.* 45 (15): 4285–4306.

106 Tan, A., Balachandran, J., Sadat, S. et al. (2011). Effect of length and contact chemistry on the electronic structure and thermoelectric properties of molecular junctions. *J. Am. Chem. Soc.* 133 (23): 8838–8841.

107 Andrews, D.Q., Solomon, G.C., Goldsmith, R.H. et al. (2008). Quantum interference: the structural dependence of electron transmission through model systems and cross-conjugated molecules. *J. Phys. Chem. C* 112 (43): 16991–16998.

108 Moth-Poulsen, K. (2016). Handbook of single-molecule electronics introduction. In: *Handbook of Single-Molecule Electronics*, 1–4. New York: Jenny Stanford Publishing.

109 Mayor, M., Weber, H.B., Reichert, J. et al. (2003). Electric current through a molecular rod – relevance of the position of the anchor groups. *Angew. Chem. Int. Ed.* 42 (47): 5834–5838.

110 Solomon, G.C., Andrews, D.Q., Hansen, T. et al. (2008). Understanding quantum interference in coherent molecular conduction. *J. Chem. Phys.* 129 (5): 054701-1–054701-8.

111 Solomon, G.C., Andrews, D.Q., Van Duyne, R.R., and Ratner, M.A. (2009). Electron transport through conjugated molecules: when the pi system only tells part of the story. *ChemPhysChem* 10 (1): 257–264.

112 Guedon, C.M., Valkenier, H., Markussen, T. et al. (2012). Observation of quantum interference in molecular charge transport. *Nat. Nanotechnol.* 7 (5): 304–308.

113 Arroyo, C.R., Frisenda, R., Moth-Poulsen, K. et al. (2013). Quantum interference effects at room temperature in OPV-based single-molecule junctions. *Nanoscale Res. Lett.* 8: 1–6.

114 Markussen, T., Stadler, R., and Thygesen, K.S. (2010). The relation between structure and quantum interference in single molecule junctions. *Nano Lett.* 10 (10): 4260–4265.

115 Markussen, T., Schiotz, J., and Thygesen, K.S. (2010). Electrochemical control of quantum interference in anthraquinone-based molecular switches. *J. Chem. Phys.* 132 (22): 224104-1–224104-6.

116 Stafford, C.A., Cardamone, D.M., and Mazumdar, S. (2007). The quantum interference effect transistor. *Nanotechnology* 18 (42): 424014-1–424014-6.

117 Kocherzhenko, A.A., Siebbeles, L.D.A., and Grozema, F.C. (2011). Chemically gated quantum-interference-based molecular transistor. *J. Phys. Chem. Lett.* 2 (14): 1753–1756.

118 Petta, J.R., Slater, S.K., and Ralph, D.C. (2004). Spin-dependent transport in molecular tunnel junctions. *Phys. Rev. Lett.* 93 (13): 136601-1–136601-4.

119 Wang, W.Y. and Richter, C.A. (2006). Spin-polarized inelastic electron tunneling spectroscopy of a molecular magnetic tunnel junction. *Appl. Phys. Lett.* 89 (15): 153105-1–153105-3.

120 Galbiati, M., Barraud, C., Tatay, S. et al. (2012). Unveiling self-assembled monolayers' potential for molecular spintronics: spin transport at high voltage. *Adv. Mater.* 24 (48): 6429–6432.

121 Tsukagoshi, K., Alphenaar, B.W., and Ago, H. (1999). Coherent transport of electron spin in a ferromagnetically contacted carbon nanotube. *Nature* 401 (6753): 572–574.

122 Jensen, A., Hauptmann, J.R., Nygard, J., and Lindelof, P.E. (2005). Magnetoresistance in ferromagnetically contacted single-wall carbon nanotubes. *Phys. Rev. B* 72 (3): 035419-1–035419-5.

123 Tombros, N., van der Molen, S.J., and van Wees, B.J. (2006). Separating spin and charge transport in single-wall carbon nanotubes. *Phys. Rev. B* 73 (23): 233403-1–233403-4.

124 Hueso, L.E., Pruneda, J.M., Ferrari, V. et al. (2007). Transformation of spin information into large electrical signals using carbon nanotubes. *Nature* 445 (7126): 410–413.

125 Yamada, R., Noguchi, M., and Tada, H. (2011). Magnetoresistance of single molecular junctions measured by a mechanically controllable break junction method. *Appl. Phys. Lett.* 98 (5): 053110-1–053110-3.

126 Kawahara, S.L., Lagoute, J., Repain, V. et al. (2012). Large magnetoresistance through a single molecule due to a spin-split hybridized orbital. *Nano Lett.* 12 (9): 4558–4563.

127 Iacovita, C., Rastei, M.V., Heinrich, B.W. et al. (2008). Visualizing the spin of individual cobalt-phthalocyanine molecules. *Phys. Rev. Lett.* 101 (11): 116602-1–116602-4.

128 Brede, J., Atodiresei, N., Kuck, S. et al. (2010). Spin- and energy-dependent tunneling through a single molecule with intramolecular spatial resolution. *Phys. Rev. Lett.* 105 (4): 047204-1–047204-4.

129 Bogani, L. and Wernsdorfer, W. (2008). Molecular spintronics using single-molecule magnets. *Nat. Mater.* 7 (3): 179–186.

130 Scott, G.D. and Natelson, D. (2010). Kondo resonances in molecular devices. *ACS Nano* 4 (7): 3560–3579.

131 Heersche, H.B., de Groot, Z., Folk, J.A. et al. (2006). Electron transport through single Mn_{12} molecular magnets. *Phys. Rev. Lett.* 96 (20): 206801-1–206801-4.

132 Jo, M.H., Grose, J.E., Baheti, K. et al. (2006). Signatures of molecular magnetism in single-molecule transport spectroscopy. *Nano Lett.* 6 (9): 2014–2020.

133 Zyazin, A.S., van den Berg, J.W.G., Osorio, E.A. et al. (2010). Electric field controlled magnetic anisotropy in a single molecule. *Nano Lett.* 10 (9): 3307–3311.

134 Osorio, E.A., Moth-Poulsen, K., van der Zant, H.S.J. et al. (2010). Electrical manipulation of spin states in a single electrostatically gated transition-metal complex. *Nano Lett.* 10 (1): 105–110.

135 Godfrin, C., Thiele, S., Ferhat, A. et al. (2017). Electrical read-out of a single spin using an exchange-coupled quantum dot. *ACS Nano* 11 (4): 3984–3989.

136 Thiele, S., Balestro, F., Ballou, R. et al. (2014). Electrically driven nuclear spin resonance in single-molecule magnets. *Science* 344 (6188): 1135–1138.

137 Vincent, R., Klyatskaya, S., Ruben, M. et al. (2012). Electronic read-out of a single nuclear spin using a molecular spin transistor. *Nature* 488 (7411): 357–360.

138 Ganzhorn, M., Klyatskaya, S., Ruben, M., and Wernsdorfer, W. (2013). Strong spin-phonon coupling between a single-molecule magnet and a carbon nanotube nanoelectromechanical system. *Nat. Nanotechnol.* 8 (3): 165–169.

139 Fan, F.R.F. and Bard, A.J. (1995). Electrochemical detection of single molecules. *Science* 267 (5199): 871–874.

140 Bard, A.J. and Fan, F.R.F. (1996). Electrochemical detection of single molecules. *Acc. Chem. Res.* 29 (12): 572–578.

141 Sun, P. and Mirkin, M.V. (2008). Electrochemistry of individual molecules in zeptoliter volumes. *J. Am. Chem. Soc.* 130 (26): 8241–8250.

142 Zevenbergen, M.A.G., Singh, P.S., Goluch, E.D. et al. (2011). Stochastic sensing of single molecules in a nanofluidic electrochemical device. *Nano Lett.* 11 (7): 2881–2886.

143 Zafarani, H.R., Mathwig, K., Sudholter, E.J.R., and Rassaei, L. (2016). Electrochemical redox cycling in a new nanogap sensor: design and simulation. *J. Electroanal. Chem.* 760: 42–47.

144 Ma, C.X., Xu, W., Wichert, W.R.A., and Bohn, P.W. (2016). Ion accumulation and migration effects on redox cycling in nanopore electrode arrays at low ionic strength. *ACS Nano* 10 (3): 3658–3664.

145 Li, T., Su, L., Hu, W.P. et al. (2010). Femtoliter and attoliter electrochemical cells on chips. *Anal. Chem.* 82 (4): 1521–1526.

146 Zevenbergen, M.A.G., Krapf, D., Zuiddam, M.R., and Lemay, S.G. (2007). Mesoscopic concentration fluctuations in a fluidic nanocavity detected by redox cycling. *Nano Lett.* 7 (2): 384–388.

147 Wolfrum, B., Zevenbergen, M., and Lemay, S. (2008). Nanofluidic redox cycling amplification for the selective detection of catechol. *Anal. Chem.* 80 (4): 972–977.

148 Rassaei, L., Singh, P.S., and Lemay, S.G. (2011). Lithography-based nanoelectrochemistry. *Anal. Chem.* 83 (11): 3974–3980.

149 Mathwig, K. and Lemay, S.G. (2013). Mass transport in electrochemical nanogap sensors. *Electrochim. Acta* 112: 943–949.

150 McCarty, G.S., Moody, B., and Zachek, M.K. (2010). Enhancing electrochemical detection by scaling solid state nanogaps. *J. Electroanal. Chem.* 643 (1–2): 9–14.

151 Kang, S., Nieuwenhuis, A., Mathwig, K. et al. (2013). Electrochemical single-molecule detection in aqueous solution using self-aligned nanogap transducers. *ACS Nano* 7 (12): 10931–10937.

152 Mathwig, K., Mampallil, D., Kang, S., and Lemay, S.G. (2012). Electrical cross-correlation spectroscopy: measuring picoliter-per-minute flows in nanochannels. *Phys. Rev. Lett.* 109 (11): 118302-1–118302-5.

Index

Nanogap Electrodes, First Edition. Edited by Tao Li.

t